Human Factors and Ergonomics in Latin American Countries

While the discipline of Human Factors and Ergonomics (HFE) has expanded globally, the specific sociotechnical realities of developing nations have long remained underrepresented in scientific literature. *Human Factors and Ergonomics in Latin American Countries: Developments, Applications and Approaches* bridges this gap, offering the first authoritative synthesis of research, applications, and legal frameworks dedicated entirely to the region.

Featuring a foreword by the Presidents of the International Ergonomics Association (IEA), this volume presents Latin America not merely as a site for applying existing theories but as a "living laboratory" for generating context-sensitive solutions. This book curates 24 chapters that traverse the full spectrum of the discipline—from the high-tech challenges of Industry 4.0 and passive exoskeletons to the critical social issues of the informal economy and inclusive design.

Readers will explore how HFE principles are adapted to diverse sectors, including healthcare, where system dynamics are used to manage ICU nursing workloads; manufacturing; addressing "quiet quitting" and burnout in export industries; public safety; and analyzing coordination in forest fire control centers. Uniquely, the text also examines ergonomics for vulnerable populations, presenting studies on indigenous women involved in street vending, pregnant workers, and the elderly in digital banking.

This book is an essential resource for researchers, practitioners, and policymakers. It moves beyond diagnosis to offer actionable strategies for improving productivity, safety, and well-being in one of the world's most dynamic emerging markets.

Human Factors and Ergonomics in Latin American Countries

Developments, Applications and Approaches

Edited by
Aide Aracely Maldonado-Macías, Yordán Rodríguez
and Juan Luis Hernández Arellano

CRC Press
Taylor & Francis Group
Boca Raton London New York

CRC Press is an imprint of the
Taylor & Francis Group, an **informa** business

Designed cover image: Aide Aracely Maldonado-Macías

First edition published 2027
by CRC Press
2385 NW Executive Center Drive, Suite 320, Boca Raton FL 33431

and by CRC Press
4 Park Square, Milton Park, Abingdon, Oxon, OX14 4RN

CRC Press is an imprint of Taylor & Francis Group, LLC

ISBN: 9781032947723 (hbk)
ISBN: 9781032949420 (pbk)
ISBN: 9781003582014 (ebk)

DOI: 10.1201/9781003582014

Typeset in Times
by codeMantra

This book is dedicated to God, my family, my friends, and my students.

Dear God, I am fortunate to be a breast cancer survivor and to have accomplished the dream of editing this excellent book.

Thanks to my parents for being an example of effort and love. Thanks to my husband and sons, who are the best motivation for being better each day. I am thankful to my colleagues Yordán and Juan, who supported me in my career and contributed to the accomplishment of this ambitious project. My best wishes to you ever.

To all the workers, educators, healthcare professionals, and innovators across Latin America who strive every day to create safer, healthier, and more humane environments—this book is for you.

To the communities that remind us that progress must always honor dignity and inclusion.

And to the future generations who will inherit the spaces we design today—may these pages inspire solutions that put people first.

—Dr Aide Aracely Maldonado-Macías

To Carmen and Alan, for always being there for me and supporting me.

—Dr Juan Luis Hernández Arellano

Dedicated to the professionals who use Human Factors and Ergonomics to help others through science, and to all the workers who benefit from the application of this discipline's principles. This work is inspired by our shared commitment to human well-being.

—Dr Yordán Rodríguez

Contents

Foreword

It is an honor and a privilege, as Presidents of the International Ergonomics and Human Factors Association (IEA), to introduce this foundational text: *Human Factors and Ergonomics in Latin American Countries: Developments, Applications, and Approaches.*

The discipline of Human Factors and Ergonomics (HFE) has seen remarkable global expansion in recent decades, evolving from its industrial origins into a holistic field that permeates nearly every aspect of human interaction with systems, technology, and environments. However, much of the published academic discourse has been dominated by perspectives from the Global North. This book is a timely contribution, given that the Latin American region is currently witnessing enormous growth in the HFE discipline.

This volume, expertly edited by Dr. Aide Aracely Maldonado-Macías, Dr. Yordán Rodríguez, and Dr. Juan Luis Hernández-Arellano, represents a vital and long-overdue contribution to the HFE field. It fills a significant gap by consolidating research, applications, and developments carried out by and for the Latin American region.

The editors have structured this book into 24 chapters organized into five thematic sections that together offer an extensive panorama of HFE in the region.

SECTION 1: ERGONOMICS IN HEALTHCARE AND MEDICAL CONTEXTS

This section focuses on the application of HFE in healthcare settings, highlighting patient safety, clinical performance, and medical training environments.

SECTION 2: ERGONOMICS AND ACCESSIBILITY IN TECHNOLOGY AND EDUCATION

This section explores human-centered design in technology and education, emphasizing accessibility and inclusive solutions for diverse populations.

SECTION 3: PHYSICAL ERGONOMICS AND WORKPLACE HEALTH

This section addresses physical demands, musculoskeletal risks, and ergonomic interventions to improve occupational health and safety.

SECTION 4: PSYCHOSOCIAL AND ORGANIZATIONAL ERGONOMICS

This section examines psychosocial risks, mental workload, leadership, and organizational climate within modern work environments.

SECTION 5: MACROERGONOMICS AND POLICY

This section expands to system-level perspectives, including policy frameworks, organizational systems, and the role of ergonomics in sustainable development.

This book is more than an academic contribution; it demonstrates the relevance and impact of HFE in Latin America.

This publication will inspire professionals, practitioners, and policymakers, contributing to a more human-centered, sustainable, and inclusive future.

Prof. Andrew Thatcher, PhD,
President of the International Ergonomics Association (2024–2027)

Prof. José Orlando Gomes, PhD,
President of the International Ergonomics Association (2021–2024)

About the Editors

Dr. Aide Aracely Maldonado-Macías is currently a full-time Professor-Investigator at the Autonomous University of Ciudad Juárez (UACJ) and was the President of the Mexican Ergonomics Society (2022–2024), as well as a member of the Executive Committee of the Latin American Ergonomics Union. Her research interests include evaluation and ergonomic design, cognitive ergonomics, anthropometric studies, and studies on work-related stress. Additionally, she focuses on Ergonomic Evaluation for the Supply Chain, Critical Success Factors for Ergonomics implementation, Multicriteria and Multi-attribute Decision-Making, utilizing fuzzy logic and axiomatic design, as well as ergonomic evaluation for the selection of advanced manufacturing technology (AMT), and Structural Equation Modeling. She has also collaborated on other books and book chapters. She received the 2018 State Science and Technology Award and the "Chihuahuense Destacada" distinction in the field of science in México. She is recognized as an investigator SNI 2 by the National Science and Technology Institute.

Yordán Rodríguez, PhD, CPE, is a Full Professor and Senior Researcher at the National School of Public Health, University of Antioquia, Colombia. Additionally, he is the CEO of Ergoyes, a startup dedicated to advancing ergonomics through digital and artificial intelligence (AI) technologies.

Dr. Rodríguez holds a PhD in Ergonomics and Master's degrees in Human Resource Management and Occupational Risk Prevention. He has authored over 60 scientific publications and delivered more than 150 invited lectures at international congresses. He holds the Certified Professional Ergonomist (CPE) designation from the Board of Certification in Professional Ergonomics (BCPE) and serves on the Board of Directors of the Foundation for Professional Ergonomics (FPE) in the USA. Furthermore, he is a Full Member of the Human Factors and Ergonomics Society (HFES) and an Honorary Member of the Chilean Foundation for Patient Safety.

He is a recipient of the Cuban Academy of Sciences National Award and the HFES International Ergonomics Month Competition Award.

Dr. Juan Luis Hernández Arellano is a Distinguished Professor and Researcher at the Autonomous University of Ciudad Juárez (UACJ), where he specializes in the critical fields of ergonomics and occupational health. With a robust academic background culminating in a PhD, he is recognized as a Level 1 member of the National System of Researchers (SNI) and possesses the PRODEP profile, highlighting his commitment to high-quality education and scientific production.

Dr. Hernández has contributed extensively to the discipline through impactful research focused on advanced ergonomic design, industrial risk assessment, and comprehensive occupational safety strategies. His work primarily aims to optimize human well-being and system performance in complex environments. Bridging the gap between theoretical knowledge and practical application, he also serves as the Founder and CEO of ErgoTech Mexico. In this executive role, he leads initiatives to implement ergonomic solutions in the industry, demonstrating a unique balance of academic rigor and entrepreneurial leadership.

Contributors

Andrés Arce Guadalupe
Peruvian Ergonomics Society
Perú

Cesar Omar Balderrama-Armendariz
Architecture, Art, and Design Institute
Universidad Autónoma de Ciudad Juárez,
Ciudad Juárez, México

Luis Eduardo Bautista
Industrial Designer School
Universidad Industrial de Santander
Santander, Colombia

Igor Bello
Venezuelan Society on Occupational Health
Venezuela

Oscar Bernal
Industrial Engineering Department
Pontificia Universidad Javeriana
Colombia

Shyrle Berrio
Industrial Engineering Department
Pontificia Universidad Javeriana
Colombia

Judith Cavazos-Arroyo
Business School Department
Universidad Popular Autónoma del Estado de Puebla
Puebla, México

Eduardo Cerda-Díaz
Ergonomics Laboratory, Department of Kinesiology
Faculty of Medicine
University of Chile
Santiago, Chile

Denis I. Chavez
OdeCom
Pontificia Universidad Católica del Ecuador
Quito, Ecuador

Francisca Chávez M.
Department of Ergonomics
Universidad de Concepción
Concepción, Chile

David Cortés Sáenz
Architecture, Art, and Design Institute
Universidad Autónoma de Ciudad Juárez
Ciudad Juárez, México

Anabel De la Rosa Gómez
Coordinación de Universidad Abierta y Educación Digital
Universidad Nacional Autónoma de México
Mexico City, México

Enrique Javier De la Vega Bustillos
Industrial Engineering Department
Tecnológico Nacional de México / I.T. Hermosillo
Hermosillo, México

José Ángel Díaz-Alcaraz
Industrial Engineering Department
Tecnológico Nacional de México / I.T. Tijuana
Tijuana, México

Ana Paula Díaz Pinal
Business School
Universidad Panamericana, Guadalajara
México

Jorge Espinoza B.
Department of Ergonomics
Universidad de Concepción
Concepción, Chile

Esteban Eugenio Esquivel Santoveña
Department of Social Sciences
Universidad Autónoma de Ciudad Juárez
Ciudad Juárez, Chihuahua, México

Marcelo Farías A.
Department of Ergonomics
Universidad de Concepción
Concepción, Chile

Lorena Alejandra Flores Plata
Facultad de Estudios Superiores Iztacala
Universidad Nacional Autónoma de México
State of Mexico, México

Jorge Luis García-Alcaraz
Industrial Engineering and Manufacturing Department
Universidad Autónoma de Ciudad Juárez
Ciudad Juárez, México

Berenice García-Vázquez
Industrial Engineering and Manufacturing Department
Universidad Autónoma de Ciudad Juárez
Ciudad Juárez, México

Caridad González-Maldonado
Faculty of Habitat, Infrastructure and Creativity
Pontifical Catholic University of Ecuador
Quito, Ecuador

Elvia Luz González-Muñoz
Centro Universitario de Arte, Arquitectura y Diseño
Universidad de Guadalajara, Guadalajara
México

Mónica Gabriela Gutiérrez-Hernández
Industrial Engineering Department
Tecnológico Nacional de México / IT, Ciudad Juárez, México
Engineering and Technology Institute
Universidad Autónoma de Ciudad Juárez
Ciudad Juárez, Chihuahua, México

Juan Luis Hernández Arellano
Architecture, Art and Design Institute
Universidad Autónoma de Ciudad Juárez
Ciudad Juárez, México

Guadalupe Hernández-Escobedo
Industrial Engineering Department
Tecnológico Nacional de México / I.T. Tijuana
Tijuana, México

Gabriela Jacobo-Galicia
Engineering Faculty
Universidad Autónoma de Baja California
Baja California, México

Juan Sebastián Jaén Posada
Analytics and Research for Decision Making-ALIADO, Department of Industrial Engineering
Universidad de Antioquia
Medellín, Colombia

Daniel Lizama Álvarez
Kinesiologist and Ergonomics Specialist
INACAP
Santiago, Chile

Virginia Guadalupe López-Torres
Faculty of Administrative and Social Sciences
Universidad Autónoma de Baja California
Baja California, México

Aide Aracely Maldonado-Macías
Industrial Engineering and Manufacturing Department
Universidad Autónoma de Ciudad Juárez
Ciudad Juárez, México

David Antonio Manjarrez-Zazueta
Industrial Engineering Department
Tecnológico Nacional de México / I.T. Tijuana
Tijuana, México

Fernanda Maradei
Industrial Designer School
Universidad Industrial de Santander
Colombia

Erwin Adán Martínez-Gómez
Engineering and Technology Institute Director
Universidad Autónoma de Ciudad Juárez
Ciudad Juárez, México

Fabiola Maureira C.
Department of Ergonomics
Universidad de Concepción
Concepción, Chile

Aurora Irma Máynez-Guaderrama
Industrial Engineering and Manufacturing Department
Universidad Autónoma de Ciudad Juárez, Ciudad Juárez, México

Ismael Mendoza Muñoz
Engineering Faculty
Universidad Autónoma de Baja California
México

Felipe Meyer C.
Department of Ergonomics
Universidad de Concepción
Concepción, Chile

Mildrend Ivett Montoya Reyes
Engineering Faculty
Universidad Autónoma de Baja California
México

Pablo Antonio Muñoz Gutiérrez
Architecture, Art, and Design Institute
Universidad Autónoma de Ciudad Juárez
Ciudad Juárez, México

Jonathan Osorio-Vasco
Ergonomics Research and Development Group
Ergoyes: Comunidad de Ergonomía
Medellín, Colombia

Mayra Pacheco-Cardín
Industrial Engineering Department
Tecnológico Nacional de México
Merida, México

Mayra Ivette Peña Ontiveros
Architecture, Art, and Design Institute
Universidad Autónoma de Ciudad Juárez
Ciudad Juárez, México

Elizabeth Pérez
Ergonomics Research and Development Group
Ergoyes: Comunidad de Ergonomía
Medellin, Colombia

Leonardo Quintana
Industrial Engineering Department
Pontificia Universidad Javeriana
Colombia

Julio César Ramos Rodríguez
Engineering and Technology Institute
Universidad Autónoma de Ciudad Juárez
Ciudad Juárez, México

Carlos Real
Sandalwood Engineering and Ergonomics
Livonia, MI, USA

Arturo Realyvásquez-Vargas
Department of Industrial Engineering
Tecnológico Nacional de México / I.T. Tijuana
Tijuana, México

Martín Rodríguez
Latin American Ergonomics Organization
Unión Latinoamericana de Ergonomía
Buenos Aires, Argentina

Yordán Rodríguez
National School of Public Health
Universidad de Antioquia
Medellín, Colombia

Carolina Angélica Rodríguez Herrera
Ergonomics Laboratory, Department of Kinesiology
Faculty of Medicine
University of Chile
Santiago, Chile

David Saénz Zamarron
Industrial Engineering Department
Tecnológico Nacional de México
México

Bertha Ivonne Sánchez Luján
Department of Basic Sciences
Tecnológico Nacional de México / I.T. Cd. Jiménez
Jiménez, Chihuahua, México

Mauricio Santos Morales
Engineering Faculty
Universidad de las Américas
Santiago, Chile

Moisés Adolfo Solorio Benítez
Engineering Faculty
Universidad Autónoma de Baja California
México

Patricia Eugenia Sortillón González
Engineering Faculty
UACJ / Universidad de Sonora
México

Yaniel Torres
National School of Public Health
Universidad de Antioquia
Medellín, Colombia
Department of Mechanical Engineering
École de technologie supérieure
Montréal, Canada

Pablo D. Valencia
Coordinación de Universidad Abierta y Educación Digital
Universidad Nacional Autónoma de México
Mexico City, México

Juan Ángel Vargas
Industrial Designer School
Universidad Industrial de Santander
Colombia

Olivia Yessenia Vargas Bernal
Engineering Faculty
Universidad Autónoma de Baja California
México

María Marisela Vargas-Salgado
Administrative Sciences Department
Universidad Autónoma de Ciudad Juárez
Ciudad Juárez, México

Preface

Human Factors and Ergonomics (HFE) are more than technical disciplines—they are catalysts for change, shaping environments where people can thrive. Around the world, these fields have transformed workplaces, improved safety, and elevated the quality of life. Yet, in Latin America, where cultural richness meets unique economic and social challenges, the stories of innovation and resilience in HFE have often remained untold. This book seeks to change that.

Human Factors and Ergonomics in Latin American Countries: Developments, Applications and Approaches is a celebration of knowledge, creativity, and commitment. Across 24 chapters, it brings together voices from diverse countries and disciplines, united by a common purpose: to design systems that respect human capabilities and limitations while fostering well-being and performance. From addressing mental workload in education and healthcare to embracing Industry 4.0 technologies and inclusive design, these pages reveal how ergonomics can be a bridge between progress and humanity.

This work is not just for specialists but for anyone who believes that technology and work should serve people, not the other way around. It invites readers to imagine workplaces where innovation and empathy coexist, where regulations protect dignity, and where design opens doors for vulnerable populations. It challenges us to adapt global solutions to local realities and to create new ones that reflect our region's identity.

We hope this book inspires collaboration, sparks ideas, and strengthens the role of HFE in building a sustainable future for Latin America. We thank the authors for their dedication. We thank the readers for joining us on this journey toward safer, healthier, and more inclusive environments.

The Editors

Acknowledgments

I want to acknowledge our contributors for their valuable contributions to this book. I also recognize the prestigious reviewers, who are a key element in providing improvements to each chapter; the academic institutions and forums that provide us with valuable scientists, academics, and practitioners to review this book; and the applications that disseminate the call for chapters and make the reviews and book meetings possible. I acknowledge our assistants, Julio César Ramos Rodríguez and particularly Ivan Carvajal Carlos, for their time, dedication, and effort that have made this book possible.

Special Acknowledgments

We extend our deepest gratitude to the following:

- **The contributing authors**, whose expertise and passion made this work possible.
- **Academic institutions and research groups** across Latin America that supported the development of these studies, including the Autonomous University of Ciudad Juárez, Del Valle University, and the Ergoyes research group.
- **Professional associations and networks**, including the International Ergonomics Association, Latin American Union of Ergonomics and Human Factors (ULAERGO), for fostering collaboration and knowledge exchange.
- **Our families and colleagues**, for their unwavering support and encouragement throughout this journey.
- Finally, **every reader**, for believing in the power of ergonomics to transform lives and organizations.

—Dr Aide Aracely Maldonado-Macías

I am deeply grateful to the authors for entrusting us (the editors) with the completion of this project. Likewise, I thank Iván Carbajal for his invaluable work and support during the editing of this book.

—Dr Juan Luis Hernández Arellano

I would like to thank everyone who has helped me grow, both as an ergonomist and as a human being. I also thank my family, friends, colleagues, professors, and students for giving me the drive, support, and strength to dedicate over 20 years of my life to this beautiful and passionate profession. "This book is a blend of passion, effort, and deep conviction that a better world is possible through ergonomics."

—Dr Yordán Rodríguez

Section 1

Ergonomics in Healthcare and Medical Contexts

1 Analyzing ICU Nursing Workload Using System Dynamics Modeling

Elizabeth Pérez, Juan Sebastián Jaén Posada, and Yordán Rodríguez

1.1 INTRODUCTION

Intensive care units (ICUs) in hospital centers are highly complex sociotechnical systems where the timely and accurate use of monitoring and life support are crucial in reversing life-threatening conditions (Moghadam, Chehrzad, Masouleh, Mardani, et al. 2021; Quintairos, Pilcher, and Salluh 2022). The high workload experienced by ICU nurses has been identified as a significant concern in the healthcare sector (Farid, Purdy, and Neumann 2020; Hoonakker et al. 2011; Ivziku et al. 2022), directly impacting nursing staff decision-making, patient safety, and quality of care (Gurses, Carayon, and Wall 2009; Qureshi, Purdy, Greig, et al. 2022; WHO 2020).

One of the direct effects of the high workload in healthcare is the lack of time to perform tasks that require a high degree of attention, leading to the abandonment or postponement of these activities (Hoonakker et al. 2011). This lack of time can also impact communication between caregivers, patients, and family members (Hoonakker et al. 2011; Ivziku et al. 2022). Fatigue is another direct consequence of a high workload that can increase human error among nursing staff, jeopardizing their health and safety, as well as that of their patients (Ghasemi, Samavat, and Soleimani 2019). Additionally, the high workload of nursing staff has been identified as one of the primary causes of burnout, which is associated with staff shortages, absenteeism, and turnover (Farid, Purdy, and Neumann 2020; Hoonakker et al. 2011). All these aspects negatively impact healthcare quality as they can cause adverse effects on patients (Weinger, Reddy, and Slagle 2004). Additionally, they decrease job satisfaction, well-being, and motivation while increasing fatigue and stress among healthcare personnel and costs associated with absenteeism and turnover (Carayon and Gürses 2005; Farid, Purdy, and Neumann 2020; Hoonakker et al. 2011; Qureshi, Purdy, and Neumann 2021; Qureshi, Purdy, Greig, et al. 2022).

In hospitals, typically, more than 70% of healthcare personnel are women (WHO 2025a; PAHO 2025), and nurses account for approximately 60% of the hospital staff (Moghadam, Chehrzad, Masouleh, Mardani, et al. 2021; PAHO 2025). These figures make the health sector a significant source of employment for the global female population (Hignett et al. 2013). Recent statistics indicate that low-income countries have a nurse density of 9.3 nurses per 10,000 inhabitants, while high-income countries have 100.4 nurses per 10,000 inhabitants (WHO 2025b). This gap is expected to widen in the coming years as a global shortage of 5.7 million nurses is predicted by 2030, particularly in low- and lower-middle-income countries (WHO 2020).

Data from the Pan American Health Organization (PAHO) indicate a significant difference between the subregion of North America, with a density of 117.10 nurses per 10,000 inhabitants, and the subregion of Central America, the Latin Caribbean, and Mexico, which has only 26.32 nurses per 10,000 inhabitants (PAHO 2025; WHO 2025c). Countries such as Paraguay, Uruguay, Argentina, and Brazil have nursing staff densities above or close to the regional average (44.37) (PAHO 2025). In contrast, a significant group of countries in the region have figures well below the

DOI: 10.1201/9781003582014-2

average, such as Bolivia, Peru, Venezuela, and Colombia. The situation is more critical in countries like Haiti and Honduras, with densities of 3.84 and 7.13, respectively (PAHO 2025). In many of these countries, the low density of nursing staff is related to precarious working conditions and high rates of mobility and migration (PAHO 2025).

Given this scenario, it is necessary to approach workload analysis systematically, considering that workload is a multidimensional concept due to the different factors that affect it (Carayon and Gürses 2005; Gurses, Carayon, and Wall 2009; Hignett et al. 2013; Qureshi, Purdy, Greig, et al. 2022; Salmon et al. 2020). Ergonomics/human factors (E/HF) is a scientific discipline that has been proven to contribute effectively to solving this problem (Hignett et al. 2013; WHO 2021) as it employs a systems approach to address, understand, design, and optimize interactions between humans and elements of work systems (IEA 2000).

The problem of high nursing workload in ICUs has been addressed by recognized authors from the systems perspective of E/HF (Farid, Purdy, and Neumann 2020; Ghasemi, Samavat, and Soleimani 2019; Hoonakker et al. 2011; Moghadam, Chehrzad, Masouleh, Maleki, et al. 2021; Qureshi, Purdy, and Neumann 2021; Qureshi, Purdy, Greig, et al. 2022). However, these studies highlight some gaps in the research, such as the need to include other determinants of nursing workload and to recognize, define, and quantitatively model the most important systemic interactions between them (Farid, Purdy, and Neumann 2020; Qureshi, Purdy, and Neumann 2020). It also highlights the need to utilize tools from other disciplines, not currently provided by systems E/HF, to model and reflect the dynamism, understood as the temporal variation of variables, and the uncertainty of sociotechnical healthcare systems (Salmon et al. 2020).

Integrating E/HF with computational modeling methods, such as simulation under system dynamics (SD), can overcome the aforementioned gaps (Salmon et al. 2020). This integration can benefit applications in the field of E/HF as it will provide dynamic simulations of system behavior over time, allowing for prospective evaluation and analysis of ergonomic interventions (Salmon et al. 2020). Additionally, it will enable current shortcomings in ICU nursing staff workload design to be addressed without exposing workers to potentially unsafe and untested work environments (Farid, Purdy, and Neumann 2020; Perez et al. 2014; Qureshi, Purdy, and Neumann 2021; Qureshi, Purdy, Greig, et al. 2022; Sujan, Pickup, Bowie, et al. 2021).

To the extent that E/HF continues to adopt principles of complexity science and systems thinking, integrating new modeling and analysis tools will be necessary to overcome the limitations of current E/HF methods for systems (Salmon et al. 2020). This chapter proposes a novel approach to addressing the high workload in ICU nursing from a multidimensional perspective using SD. We analyze the causal relationships between the factors that affect ICU nursing workload and identify leverage points for policy design.

1.2 WORKLOAD IN ICU NURSING SERVICES

The concept of nursing workload has a dynamic nature as it is associated with multiple factors that constantly vary its condition, such as patient health (Hoonakker et al. 2011; Morris et al. 2007; Qureshi, Purdy, and Neumann 2020). This concept has sometimes been incorrectly interchanged with the concept of nursing work. Nursing work refers to the functions and actions performed in nursing, which may or may not be directly related to the patient, e.g., patient care, disease prevention, health promotion, and administrative tasks (Alghamdi 2016). In contrast, nursing workload refers to the extent of performance required to perform those nursing work activities in a specific period (Alghamdi 2016; Morris et al. 2007).

In acute care, the workload in nursing services is highly relevant as patient care is characterized by highly demanding tasks (Hoonakker et al. 2011; Moghadam, Chehrzad, Masouleh, Mardani, et al. 2021). Swiger et al. (2016), in performing a conceptual analysis of workload in ICU nursing, define: "Nursing workload is the amount of time and physical and/or cognitive effort required to accomplish direct patient care, indirect patient care, and nonpatient care activities." It is understood, then, that workload will be influenced by aspects that go beyond the clinical conditions of the patient and encompass the characteristics of the nursing staff, the patient, the care unit, and the work system in general and that, in turn, these characteristics will have a direct effect on the time and effort required to provide high-quality nursing care (Gurses, Carayon, and Wall 2009; Moghadam, Chehrzad, Masouleh, Mardani, et al. 2021; Swiger, Vance, and Patrician 2016).

On the other hand, Gurses et al. (2009) refer that: "Workload is nurses' subjective experiences reflecting the combined effects of both task demands while performing direct and indirect care activities and demands imposed on them by the obstacles in the work system." Obstacles being defined as: "… the work system design characteristics that inhibit performance and are closely associated with the immediate work setting." This definition contains three key aspects: first, it specifies that the individual's perception may reflect the effects of various work-related demands, such as physical, cognitive, emotional, and temporal (Gurses, Carayon, and Wall 2009). Second, it includes among the demands the characteristics of the work system's design that constitute an obstacle to adequate performance, which, in the context of this chapter, are related to the lack of an ergonomic approach in the design. For example, Gurses et al. (2009) demonstrated that a large part (40%) of the variability of nursing workload in ICUs was explained by the work system's characteristics, even without considering the patient's criticality status. Finally, it extends the boundaries of nursing work beyond direct care.

1.3 E/HF IN ICU NURSING WORKLOAD ANALYSIS

Healthcare systems are complex, dynamic, and uncertain environments that operate under multiple constraints and pressures to ensure the well-being and safety of patients, their families, and healthcare workers (Sujan, Pickup, Vosper, et al. 2021; WHO 2021). These characteristics make healthcare systems resilient and adaptive (Carayon et al. 2015; Rodríguez and Hignett 2021). Complex adaptive systems exhibit structural and dynamic characteristics that adjust adaptively in response to internal and external perturbations. These adaptations occur through a quasi-organic process of redistributing activities and responsibilities throughout the system (Carayon et al. 2015). Healthcare systems must respond effectively and efficiently to the diverse demands and priorities of patients, their families, and healthcare workers (Sujan, Pickup, Vosper, et al. 2021).

The complexity of today's sociotechnical systems and the uncertainties generated in them require a systemic approach to E/HF to address them (Wilson 2014). Renowned researchers and institutions worldwide have emphasized the need to integrate E/HF into healthcare systems to optimize human interaction (Aceves-González et al. 2021; Buckle et al. 2006; Hignett et al. 2013; IEA 2021; Mouloua et al. 2021; Sujan, Pickup, Bowie, et al. 2021; WHO 2021). E/HF has advocated for a systems approach to address these interactions in complex work systems, such as the aviation sector and the nuclear industry (Buckle et al. 2006; Sujan, Pickup, Bowie, et al. 2021; Torres, Rodríguez, and Pérez 2022). These industries have realized the risks of considering system elements in isolation. An ICU is also a complex and dynamic sociotechnical system; therefore, integrating a systems approach in analyzing problems such as high nursing workload is a challenge and a necessity, primarily due to the potential impact on its different users (Buckle et al. 2006; CIEHF 2021; Sujan, Pickup, Vosper, et al. 2021).

The multidisciplinary and integrative approach of E/HF considers the person as integrated in a sociotechnical context, taking into account the workers (nurses) in the work environment (ICU) and the patient within the care pathway (Carayon et al. 2020; WHO 2021). The lack of E/HF, on the other hand, results in inflexible and error-prone healthcare systems (WHO 2021). The principles underlying E/HF recognize that the safety, performance, and well-being of clinical staff are intrinsically linked to patient safety and the performance and efficiency of the healthcare organization (Sujan, Pickup, Bowie, et al. 2021).

The variability in the behavior of healthcare systems, especially in ICUs, is a reality that is difficult to change; therefore, if the aim is to analyze, design, and improve the system, it is essential to understand the context, its elements, and interactions (Buckle et al. 2006; Sujan, Pickup, Vosper, et al. 2021). In the work context, E/HF focuses on the design of work and work systems (Hignett et al. 2013; Sujan, Pickup, Bowie, et al. 2021); however, in ICU nursing services, the design has mainly focused on the individual (nurse and patient) instead of analyzing the interactions in the whole system (Hignett et al. 2013). Many studies have been conducted to analyze and quantify ICU nursing workload, mainly from the patient's perspective and needs, leaving out aspects related to the nursing staff, the characteristics of the work system, and the temporality of nursing activities (Gurses, Carayon, and Wall 2009; Hoonakker et al. 2011; Quintairos, Pilcher, and Salluh 2022). The traditional approach based on the patient's clinical status has been commonly used by healthcare institutions seeking to optimize the nurse–patient ratio (Alghamdi 2016; Griffiths et al. 2020; Gurses, Carayon, and Wall 2009). Additionally, improvement solutions derived from this approach focus on increasing the number of nursing staff, which does not seem feasible in the Latin American context, given the current shortage and predicted reduction of nursing staff (Gurses, Carayon, and Wall 2009; WHO 2025a).

It must be understood that patients' healthcare needs have become more complex, so ICU nurses will face significant challenges in the future, such as responding to the needs of an increasingly aging patient population, global health crises, facing new diseases, staff shortages and burnout, and work system obstacles (Qureshi, Purdy, and Neumann 2021; Sujan, Pickup, Bowie, et al. 2021). Additionally, changing complex systems, such as ICU nursing workload design, can present barriers due to costs, potential risks to workers and patients, and potential system disruptions (Farid, Purdy, and Neumann 2020). The healthcare sector must also consider new ways to proactively address these issues from a systems E/HF approach, without exposing workers to potentially unsafe and untested work environments (Farid, Purdy, and Neumann 2020; Perez et al. 2014; Qureshi, Purdy, and Neumann 2021; Qureshi, Purdy, Greig, et al. 2022; Sujan, Pickup, Bowie, et al. 2021).

1.4 SYSTEMS DYNAMIC MODELING TO ADDRESS WORKLOAD IN NURSING

Current tools offered by systems E/HF present limitations that do not allow reflecting the dynamism and uncertainty of sociotechnical healthcare systems (Salmon et al. 2020). The main limitations of these tools are as follows: (a) the ergonomic analyses performed are detailed but are static and do not allow simulating the behavior of the sociotechnical system over time, (b) it is difficult and sometimes subjective to assess the likely impact of the proposed interventions, and (c) although they provide an in-depth analysis of the system, they do not allow knowing how the variables interact to influence the behavior of the system as a whole (Salmon et al. 2020; Salmon and Read 2019). In this regard, Salmon et al. (2020) and Farid et al. (2020) recommend the integration of SD with current systems' E/HF tools as a suitable, novel, and more appropriate approach than trial and error in real healthcare systems.

SD is an intuitive tool that is ideal for multidisciplinary work, so learning it can be more efficient than learning other methodologies (Zanker, Bureš, and Tučník 2021). SD is based on systems thinking to understand the dynamic behavior of complex systems (Faezipour and Ferreira 2013; Farid, Purdy, and Neumann 2020). SD accepts the complexity of systems, nonlinearity, and feedback loop structures inherent in social and physical systems (Forrester 1994). SD uses causal diagrams as a graphical method of representing system factors and their relationships (Faezipour and Ferreira 2013). Subsequently, it allows the simulation of the causal model to test dynamic hypotheses and determine the system-wide impact of changing factors (Jetha, Kernan, and Kurowski 2017). In this chapter, SD will be employed to analyze the causal relationships between workload variables in ICU nursing services.

Although applications of SD can be found in healthcare, few studies in the literature have used SD to analyze the effects of determinants on workload in ICU nursing services from the E/HF perspective (Farid, Purdy, and Neumann 2020; Choi, Noh, and Rha 2019; Jin et al. 2023). A recent study by Farid et al. (2020) employed the SD technique to address the problem of workload and nurse staffing, its effects on staff health and well-being, and the quality of patient care. The model results showed that long shifts and workweeks for nursing staff doubled fatigue levels and increased burnout by up to six times, absenteeism by up to five times, and medical errors by up to 150%.

The application of SD in healthcare can aid in system understanding and decision-making (Jetha, Kernan, and Kurowski 2017), for example, in understanding how nursing workload affects the health of nursing staff and the quality of patient care (Farid, Purdy, and Neumann 2020) or in identifying the factors that contribute to long waiting times in emergency units (Lane, Monefeldt, and Rosenhead 2000).

1.5 METHODOLOGY FOR MODEL CONSTRUCTION

The model presented in this chapter was developed following an SD approach, carried out in two main phases: (a) the construction of a conceptual framework and (b) the development of a causal loop diagram (CLD).

The first phase consisted of a systematic literature review to identify the key dimensions and factors of nursing workload in the ICU. The search focused on the Scopus, PubMed, and Web of Science databases, using search algorithms that combined keywords such as “nursing workload,” “modeling,” “ergonomics,” and “human factors,” limiting the results to scientific articles in English. From the analysis of the most relevant works, the three central dimensions of the problem—patient characteristics, work system characteristics, and nurse characteristics—and their main effects were identified, thus structuring the conceptual framework that guides this study.

In the second phase, a CLD was constructed based on this framework. The specific variables of the model were extracted from the reviewed literature. To ensure the relevance and validity of these variables and their interconnections, a validation was carried out through a semi-structured interview with a head nurse with more than 15 years of experience in an ICU of a high-complexity hospital. During this consultation, the causal relationships were confirmed, and the polarity (positive ‘+’ or negative ‘–’) of each one was defined, based on both scientific evidence and practical experience. The CLD was developed using the academic version of the Vensim PLE Version 10.2.2 software.

1.6 PROPOSAL OF A CONCEPTUAL FRAMEWORK FOR WORKLOAD ANALYSIS IN ICU NURSING SERVICES

This conceptual framework is based on the work and hypotheses put forward by Gurses et al. (2009) and Carayon et al. (2005) (Gurses, Carayon, and Wall 2009; Carayon and Gürses 2005). Both hypotheses have as their starting point and causes of workload the obstacles and facilitators of performance in the work system. These obstacles and facilitators affect the workload, impacting patients, family members, nurses, economic indicators, and quality of care (Gurses, Carayon, and Wall 2009; Carayon and Gürses 2005). In these models, causes and consequences are presented in a linear approach, facilitating the understanding of the problem of ICU nursing workload. However, this linear approach may overlook important relationships in these complex systems.

Figure 1.1 presents our proposed conceptual framework for analyzing ICU nursing workload, and Table 1.1 describes its conceptual elements. This conceptual framework aims to study the impact of ICU nursing workload on the system, occupational diseases among nurses, and their perception of workload, considering three key dimensions: patient characteristics, work system characteristics, and nursing characteristics. This conceptual framework provides an integrative and general overview of the ICU nursing workload problem; the detailed interactions among the variables and their polarities will be addressed more precisely in the next section using a causal diagram.

Other researchers have analyzed the elements of this conceptual framework, but on an individual basis, considering, in most cases, only those elements that the researcher can measure or intervene in their research (Farid, Purdy, and Neumann 2020; Qureshi, Purdy, and Neumann 2020).

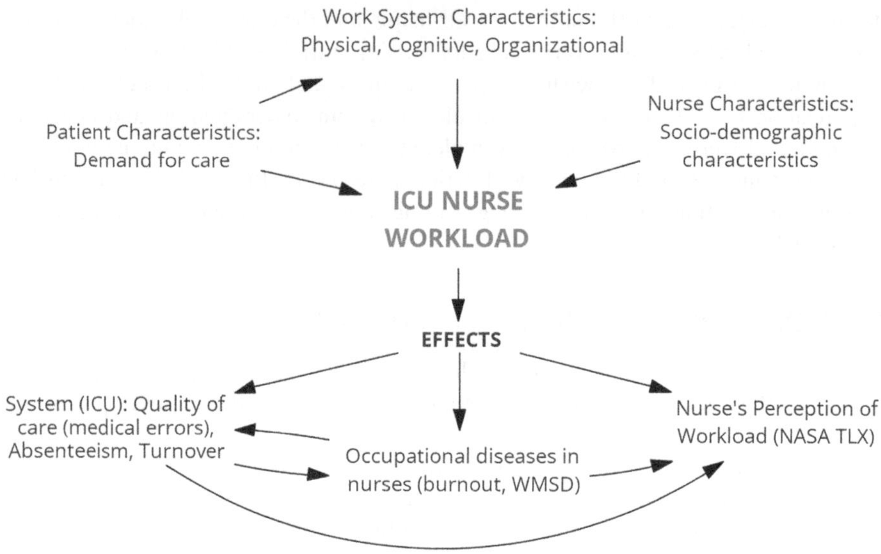

FIGURE 1.1 Conceptual framework for nursing ICU workload analysis.

TABLE 1.1
Elements of the Conceptual Framework for ICU Nursing Workload Analysis

Framework Elements	Descriptions
ICU nurse workload	Amount of physical, cognitive, and temporal effort required by the nurse to perform nursing work (Swiger, Vance, and Patrician 2016).
Effects	Impact of workload.
Patient characteristics	Individual patient conditions that lead to greater or lesser demand for nursing care.
Work system characteristics	Design of physical, cognitive, and organizational interactions between ICU and nursing staff (Gurses, Carayon, and Wall 2009; Carayon and Gürses 2005; IEA 2000).
Nurse characteristics	Individual nurse conditions that may impact workload and performance.
System	Refers to the work system, in this case, the ICU.
Occupational disease in nurses	Work-related disease reported by ICU nurses. Two types of disease will be considered: musculoskeletal disorders (WMSD) and burnout.
Nurses' perception of workload	Nurses' perception of their workload, measured by the subjective NASA Task Loading Index (NASATLX) tool (Hoonakker et al. 2011; Hart and Staveland 1988).

Additionally, some studies focus only on evaluating the severity of the disease, the success of therapeutic interventions, and the performance of intensive care to assign or measure the workload (Griffiths et al. 2020; Hoonakker et al. 2011). All this has limited the approach to workload from a more integrative approach (Hoonakker et al. 2011).

The conceptual framework proposed in this chapter establishes three key dimensions that directly determine ICU nursing workload: the patient's characteristics, the work system's characteristics, and the nurse's sociodemographic characteristics. This approach highlights the importance of intervening in these dimensions to manage and decrease workload. Another contribution of this framework is recognizing that workload directly affects the care system, affecting critical indicators such as quality of care through medical errors, absenteeism, and staff turnover (Gurses, Carayon, and Wall 2009; Qureshi, Purdy, Greig, et al. 2022; WHO 2020). Also, workload affects the nurse physically and mentally, increasing the risk of occupational diseases such as work-related musculoskeletal disorders (WMSDs) and burnout syndrome. It also impacts the nurse's perception of their workload, affecting motivation and job performance (Farid, Purdy, and Neumann 2020; Hoonakker et al. 2011).

Our conceptual framework also recognizes relationships between dimensions; for example, specific patient characteristics, such as severity or clinical acuity, may necessitate organizational adjustments, such as staffing or modifications in care processes (Hoonakker et al. 2011; Ivziku et al. 2022). Similarly, there are dynamic relationships between the effects generated by workload. For example, nurses who are exhausted, have physical ailments, or face the consequences of medical errors may experience a greater perception of the workload. Sick nurses may unintentionally make more medical errors, experience absenteeism, and turnover (Farid, Purdy, and Neumann 2020; Griffiths et al. 2020; Ivziku et al. 2022; Qureshi, Purdy, Greig, et al. 2022).

In summary, this conceptual framework offers a comprehensive approach to analyzing ICU nursing workload, considering the triggers and their interactions and effects. This approach can facilitate the development of effective interventions to reduce workload and, second, the measurement of their impact.

1.7 CAUSAL LOOP DIAGRAM FOR WORKLOAD ANALYSIS IN ICU NURSING SERVICES

Building on the conceptual framework, the main original contribution of this chapter is the development of a CLD. A CLD is a tool used in SD to graphically represent systems' feedback structures (Sterman 2009). CLDs allow modelers to identify and express essential relationships that may be causal to a problem and to hypothesize about the dynamic behavior of these relationships (Sterman 2009).

In CLDs, the connections between system variables are represented using arrows with polarity, which indicates how the dependent variable (effect) changes when the independent variable (cause) changes (Sterman 2009). A positive polarity (+) indicates that if the cause increases (decreases), the effect increases (decreases) above (below) what it would otherwise have been (Sterman 2009). A negative polarity (−) indicates that if the cause increases (decreases), the effect decreases (increases) below (above) what it would otherwise have been (Sterman 2009). Connections between several variables can form positive (reinforcing change) or negative (balancing change) feedback loops (Sterman 2009). A quick way to identify them is to count the number of negative links in the loop: if the number of negative links is even, the loop is reinforcing; if the number is odd, the loop is balancing (Sterman 2009).

The CLD is a very valuable SD tool. However, it must be considered that causality and not correlation should always prevail in the relationships between variables (Sterman 2009). Two variables may be highly correlated, but there may be no causality between them; however, after causality has been established, it is possible that correlations may arise. SD models should be simple and understandable, yet they should simulate the structure of the real system well enough so that the model behaves as the real system would and can be used as a basis for developing actions and implementing policies (Sterman 2009; Cavana and Mares 2004).

Figure 1.2 shows the CLD built from the conceptual framework, and Table 1.2 presents its variables. The variables and causal relationships of this model were defined, considering first the evidence available in the scientific literature on these variables and their relationships, second the opinion of an ICU head nurse from a hospital institution, and third statistics on occupational diseases among ICU nurses from the same institution. As a result of this analysis, we found 12 reinforcing loops and two balancing loops. Figures 1.3–1.6 highlight four loops considered to be of interest in the analysis of ICU nursing workload.

Figure 1.3 shows the R1 reinforcing loop and explains how increased occupational diseases, specifically WMSDs and burnout, increase ICU nursing workload. We can describe it this way:

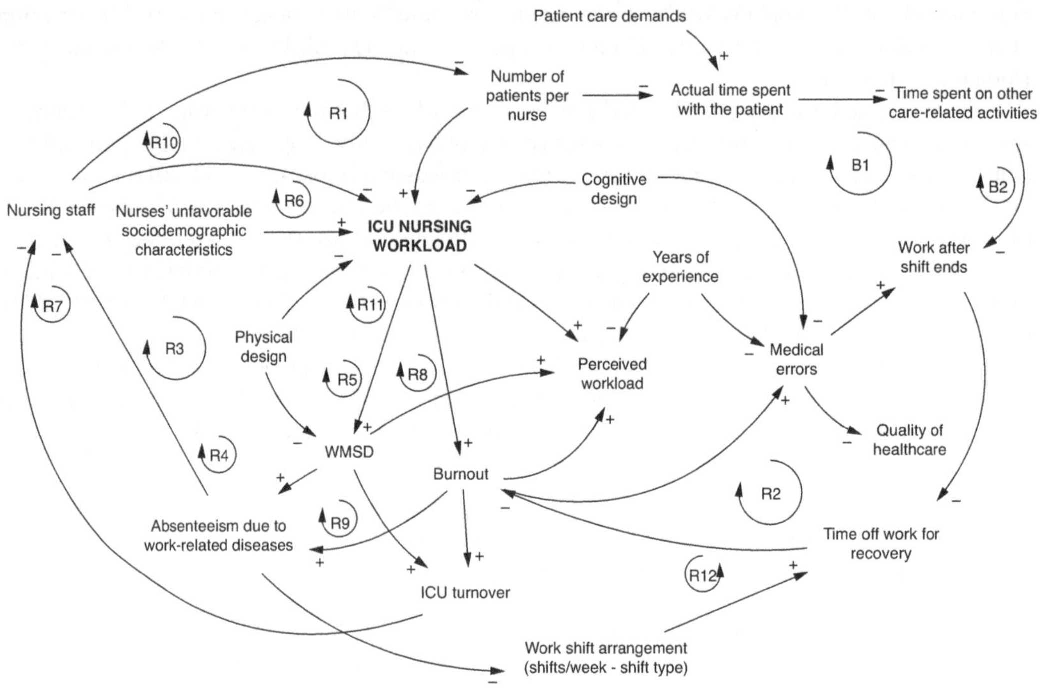

FIGURE 1.2 Causal loop diagram for analyzing the workload in the ICU nursing services.

TABLE 1.2
Variables for the Analysis of ICU Nursing Workload

Framework Elements	Variables
Patient characteristics	**Patient care demands** indicates the patient's need for nursing care given a medical condition (Ivziku et al. 2022). The demand for care is not necessarily related to the severity of the patient's condition; for example, a patient in a noncritical condition but with diabetes may require a constant glucometer during the shift.
Work system characteristics	**Physical design** refers to the design of the nurse's physical interactions with the elements of the ICU (Gurses, Carayon, and Wall 2009; IEA 2000), for example, the workstations' design, the patient's geographical allocation (Qureshi, Purdy, Greig, et al. 2022), and the physical effort required when handling a patient.
	Cognitive design refers to the nurse's cognitive or mental interactions with the elements of the ICU (Gurses, Carayon, and Wall 2009; IEA 2000), for example, decision-making under pressure, cognitive effort when switching between patients with different health conditions, and interaction between the nurse and the health record software.
	Organizational design refers to the design of the organizational and sociotechnical interactions of the nurse with the elements of the ICU (Gurses, Carayon, and Wall 2009; IEA 2000):
	Work shift arrangement: shifts per week and type of shift (day or night) (Farid, Purdy, and Neumann 2020).
	Time off work for recovery: days of rest outside the ICU (Farid, Purdy, and Neumann 2020; Ghasemi, Samavat, and Soleimani 2019).
	Number of patients per nurse (Qureshi, Purdy, Greig, et al. 2022).
	Actual nursing work time: the actual time spent with the patient during the shift, plus the time spent on other care-related activities during the shift, plus the time spent after the end of the shift to complete postponed or incomplete tasks (Farag et al. 2022; Ghasemi, Samavat, and Soleimani 2019; Rogers et al. 2004).

(Continued)

TABLE 1.2 (*Continued*)
Variables for the Analysis of ICU Nursing Workload

Framework Elements	Variables
	Nursing staff: number of nursing staff available to work a given shift in the ICU (Farid, Purdy, and Neumann 2020; Griffiths et al. 2020; Ivziku et al. 2022; Qureshi, Purdy, Greig, et al. 2022).
Nurse characteristics	**Years of experience:** years working as an ICU nurse (Moghadam, Chehrzad, Masouleh, Maleki, et al. 2021; Moghadam, Chehrzad, Masouleh, Mardani, et al. 2021).
	Other sociodemographic characteristics, such as the number of children, time spent with children and family, hours of sleep, and means and times of transportation to and from the hospital, can affect workload (Moghadam, Chehrzad, Masouleh, Maleki, et al. 2021; Moghadam, Chehrzad, Masouleh, Mardani, et al. 2021).
System	**Medical errors** (Weinger, Reddy, and Slagle 2004): statistics of medical errors associated with ICU nursing staff. These statistics are an indicator of the **quality of healthcare.**
	Absenteeism (Farid, Purdy, and Neumann 2020; Hoonakker et al. 2011)**:** human resources reports of occupational disease absenteeism rates (burnout and WMSDs) among ICU nurses.
	ICU turnover (Farid, Purdy, and Neumann 2020; Hoonakker et al. 2011)**:** number of nurses leaving the ICU. They can withdraw from the healthcare institution or only from the ICU.
Occupational disease in nurses	**Burnout** (Farid, Purdy, and Neumann 2020; Hoonakker et al. 2011)**:** human resources reports on medical disabilities resulting from burnout or burnout diagnoses among ICU nurses.
	WMSD (Qureshi 2020)**:** human resources reports on medical disabilities or diagnoses of diseases related to musculoskeletal disorders in ICU nurses.
Nurses' perception of workload	This element can be assessed using the NASA TLX, a tool widely used among the nursing workforce that combines mental demand, physical demand, temporal demand, performance, effort, and frustration level (Hoonakker et al. 2011; Hart and Staveland 1988). This tool assumes that the combination of these dimensions will likely represent workers' perceived workload (Hoonakker et al. 2011; NASA 2020).

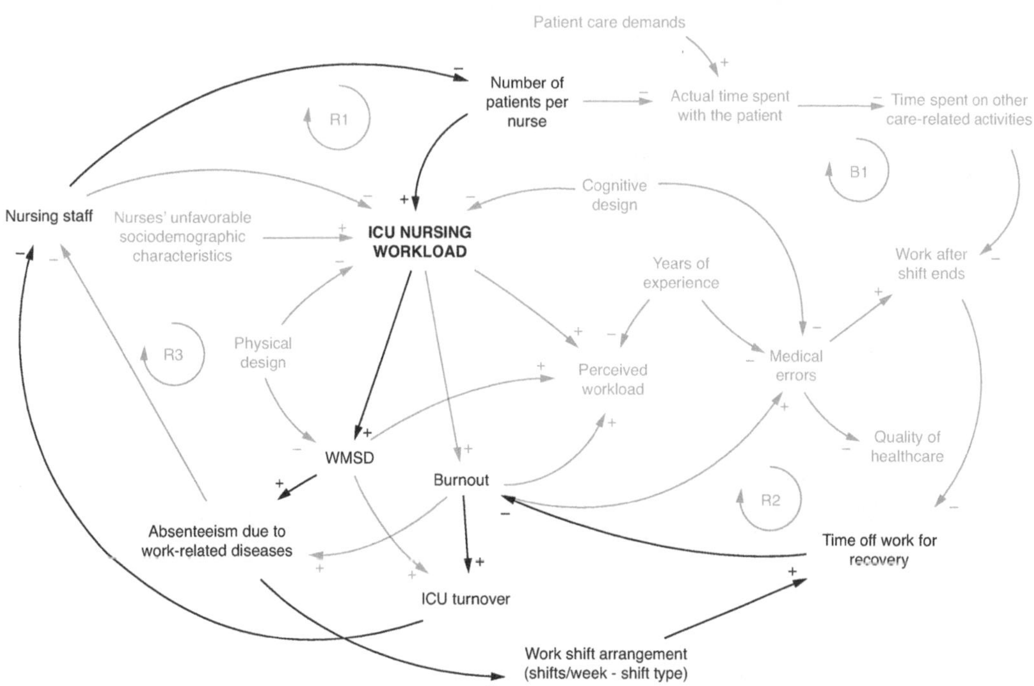

FIGURE 1.3 R1 reinforcing loop.

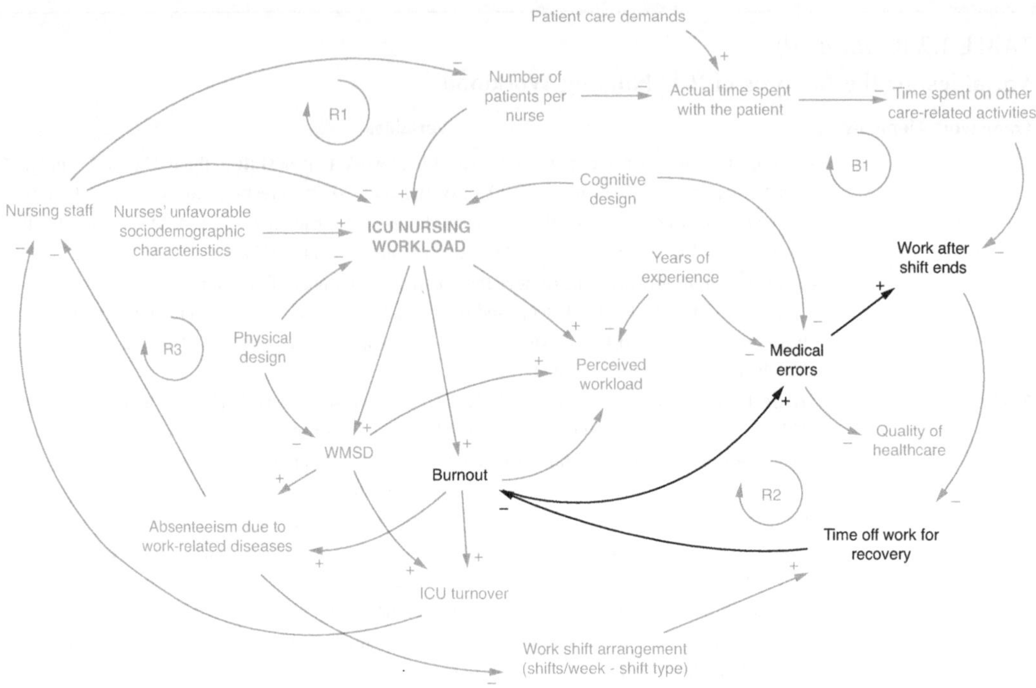

FIGURE 1.4 R2 reinforcing loop.

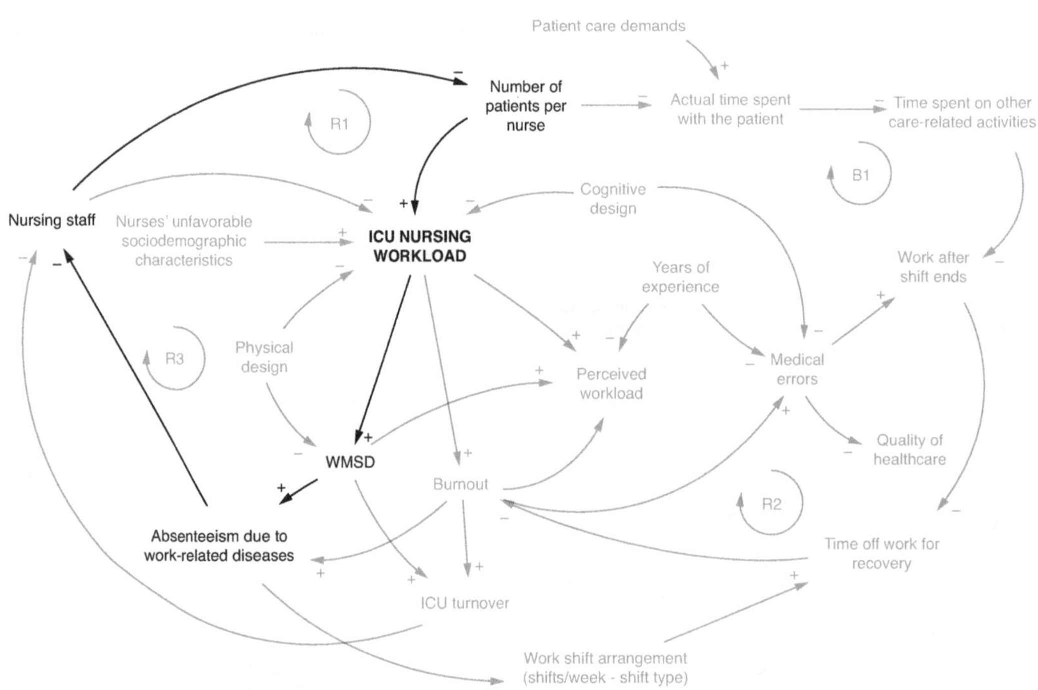

FIGURE 1.5 R3 reinforcing loop.

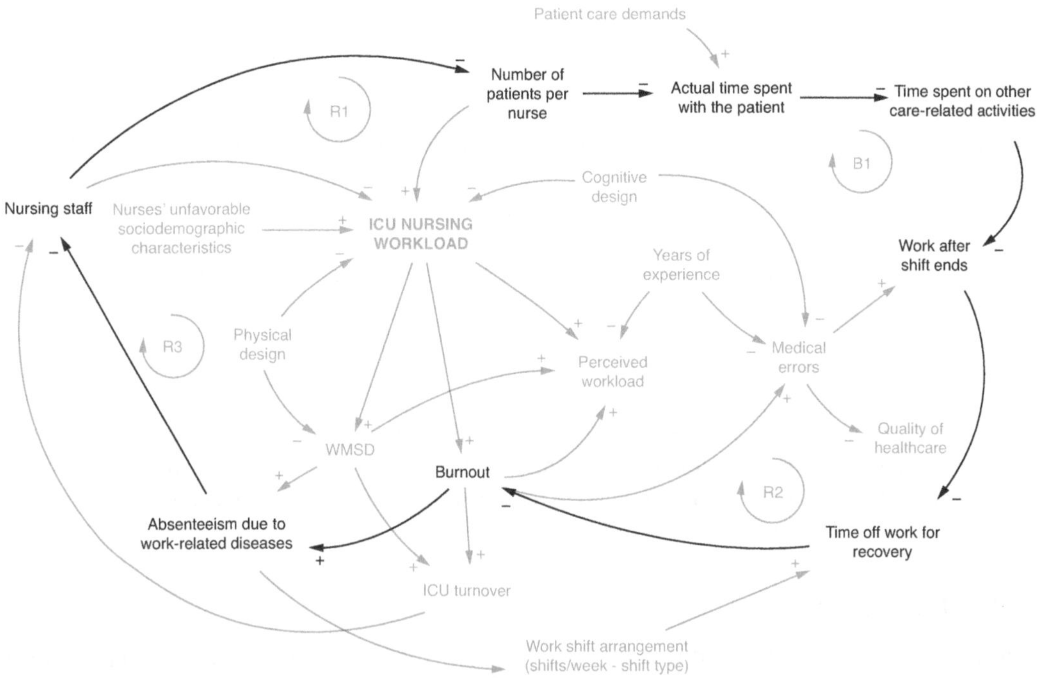

FIGURE 1.6 B1 balancing loop.

the increase in WMSDs among ICU staff leads to increased absenteeism in the ICU. Increased absenteeism generates deconfigurations (unfavorable configurations) in work shifts, e.g., a greater number of shifts assigned per week, which reduces the time off work available for recovery: the greater (less) shift configuration, the more (less) recovery time. Less (greater) time off work for recovery increases (decreases) burnout rates of nursing staff. As burnout increases, staff turnover increases, leading to decreased available nursing staff per work shift (Farid, Purdy, and Neumann 2020; Hoonakker et al. 2011). A decrease (increase) in nurse staffing increases (decreases) the number of patients assigned per nurse, leading to an increase (decrease) in workload (Farid, Purdy, and Neumann 2020; Griffiths et al. 2020; Ivziku et al. 2022; Qureshi, Purdy, Greig, et al. 2022). When analyzing this loop, a common policy in hospital institutions is to increase the nursing staff. However, we have already seen that this may be a short-term solution, but it is not very sustainable in the medium to long term. We must ask ourselves: What strategies can be implemented to increase the time off work available for nursing staff recovery? What other variables related to the characteristics of the work system can mitigate the occurrence of WMSDs and burnout in this context? For example, an ergonomic intervention in the physical design of nursing workstations can decrease exposure to WMSD risk factors.

Figure 1.4 shows the R2 reinforcing loop, which explains how increased burnout among nurses leads to increased medical errors and, consequently, decreased healthcare quality. Nurses with burnout may be more prone to errors or omissions as they may be emotionally exhausted, feel depersonalized, and demotivated by their work (Farid, Purdy, and Neumann 2020; Choi, Noh, and Rha 2019). The increase in medical errors in the ICU results in an increase in the amount of time nurses must stay after the end of their shift to correct errors or make reports (Farid, Purdy, and Neumann 2020). As the amount of time spent after the shift increases, the time nurses have to recuperate and spend personal time decreases, thus closing the reinforcing loop with increased burnout. In addition to the need for strategies to increase time off work for recovery, we must also ask: Can cognitive design decrease medical errors in the ICU? For example, the ergonomic design of monitor interfaces (e.g., font sizes and colors and amount of information) of patients in the ICU may help make the information captured by nurses during their rounds easier to read, and fewer errors are made.

WMSDs among ICU nurses can be generated mainly due to factors such as the handling of patients who, due to their health condition, cannot collaborate; also, they can be generated by long working days in a standing posture or inadequate designs of workstations and work elements such as wheelchairs. When analyzing the reinforcing loop R3 presented in Figure 1.5, the hypothesis arises again: What other variables related to the physical characteristics of the work system can mitigate the occurrence of WMSDs?

In balancing loop B1 shown in Figure 1.6, we can see that increasing nurses' time off work for recovery can be an important variable and strategy in balancing this loop, leading to decreased burnout among ICU nurses.

From this causal analysis, three variables that can mitigate the effects of workload on ICU nurses are observed: physical design, cognitive design, and time off work for recovery, which is part of the organizational design. Interventions related to these variables may have a greater and more sustainable impact in the long term.

1.8 LIMITATIONS

The causal model presented in this chapter has limitations inherent to its conceptual nature. Its primary purpose is to represent the feedback structures and dynamic hypotheses that underlie the nursing workload in the ICU, based on the existing literature and expert consultation. As such, it is a qualitative tool designed to enhance the systemic understanding of the problem; however, it has not been quantitatively validated with empirical data from a specific ICU setting. Therefore, the model does not generate numerical predictions; rather, it identifies structural relationships and potential points of intervention.

Furthermore, this model is a simplification of a highly complex sociotechnical system. Although variables identified as critical in the literature have been included, other factors not considered in this model could influence nursing workload. The model's boundaries were intentionally set to maintain clarity and focus on the most significant reported interactions.

Finally, the variables and causal relationships were defined based on a literature review and the opinion of a single ICU head nurse. A more robust model could be developed with the participation of an expert panel.

1.9 CONCLUSIONS

This chapter proposes a conceptual framework and a causal model to analyze the workload of ICU nurses using SD. The primary contribution of our conceptual framework lies in its explicit integration of the patient's characteristics, the work system's characteristics, and the nursing staff's characteristics, extending beyond the traditional view that focuses mainly on the patient's clinical status. The causal representation of the model through CLD allows us to identify the dynamic relationships between the variables that affect ICU nursing workload and their systemic effects, such as absenteeism, turnover, and the occurrence of occupational diseases. This facilitates the identification of critical points for effectively implementing ergonomic interventions. This approach introduces new methodological perspectives to the study of workload in ICU nursing, providing a tool to analyze these complex systems and avoiding direct experimentation on the real system, which reduces costs and risks for the nursing staff.

REFERENCES

Aceves-González, Carlos, Rodríguez, Yordán, Escobar-Galindo, Carlos Manuel, Pérez, Elizabeth Gutiérrez-Moreno, Beatriz, Hignett, Sue and Lang, Alexandra Rosewall. 2021. "Frontiers in Human Factors: Integrating Human Factors and Ergonomics to Improve Safety and Quality in Latin American Healthcare Systems." *International Journal for Quality in Health Care* 33 (Supplement_1): 45–50. doi:10.1093/intqhc/mzaa135.

Alghamdi, Mohammed G. 2016. "Nursing Workload: A Concept Analysis." *Journal of Nursing Management* 24 (4): 449–57. doi:10.1111/jonm.12354.

Buckle, Peter, Clarkson, John, Coleman, Roger, Ward, James and Anderson, Janet. 2006. "Patient Safety, Systems Design and Ergonomics." *Applied Ergonomics* 37 (4): 491–500. doi:10.1016/j.apergo.2006.04.016.

Carayon, Pascale, Wooldridge Abigail, Hoonakker Peter, Hundt Ann Schoofs and Kelly Michelle M. 2020. "SEIPS 3.0: Human-Centered Design of the Patient Journey for Patient Safety." *Applied Ergonomics* 84 (April): 103033. https://doi.org/10.1016/j.apergo.2019.103033

Carayon, Pascale, and Gürses Ayşe P. 2005. "A Human Factors Engineering Conceptual Framework of Nursing Workload and Patient Safety in Intensive Care Units." *Intensive and Critical Care Nursing* 21 (5): 284–301. doi:10.1016/j.iccn.2004.12.003.

Carayon, Pascale, Hancock, Peter, Leveson, Nancy, Noy, Ian, Sznelwar, Laerte and van Hootegem, Geert. 2015. "Advancing a Sociotechnical Systems Approach to Workplace Safety – Developing the Conceptual Framework." *Ergonomics* 58 (4): 548–64. doi:10.1080/00140139.2015.1015623.

Cavana, Robert Y., and Mares, Edwin D. 2004. "Integrating Critical Thinking and Systems Thinking: From Premises to Causal Loops." *System Dynamics Review* 20 (3): 223–35. doi:10.1002/sdr.294.

Choi, Donghyun, Noh, Yonghwi and Rha, Jin Sung. 2019. "Work Pressure and Burnout Effects on Emergency Room Operations: A System Dynamics Simulation Approach." *Service Business* 13 (3): 433–56. doi:10.1007/s11628-018-00390-1.

CIEHF. 2021. Human Factors and Ergonomics in Healthcare AI. Chartered Institute of Ergonomics & Human Factors. https://www.ergonomics.org.uk/common/Uploaded%20files/Publications/CIEHF-AI-in-Healthcare-White-Paper.pdf.

Faezipour, Misagh, and Ferreira, Susan. 2013. "A System Dynamics Perspective of Patient Satisfaction in Healthcare." *Procedia Computer Science* 16: 148–56. doi:10.1016/j.procs.2013.01.016.

Farag, Amany, Scott, L. D., Perkhounkova, Y., Saeidzadeh, S. and Hein, M. 2022. "A Human Factors Approach to Evaluate Predictors of Acute Care Nurse Occupational Fatigue." *Applied Ergonomics* 100 (April): 103647. doi:10.1016/j.apergo.2021.103647.

Farid, Mashal, Purdy, Nancy, and Neumann, W. Patrick. 2020. "Using System Dynamics Modelling to Show the Effect of Nurse Workload on Nurses' Health and Quality of Care." *Ergonomics* 63 (8): 952–64. doi: 10.1080/00140139.2019.1690674.

Forrester, Jay W. 1994. "System Dynamics, Systems Thinking, and Soft OR." *System Dynamics Review* 10 (2–3): 245–56. doi:10.1002/sdr.4260100211.

Ghasemi, Fakhradin, Samavat, Parnia and Soleimani, Fatemeh. 2019. "The Links among Workload, Sleep Quality, and Fatigue in Nurses: A Structural Equation Modeling Approach." *Fatigue: Biomedicine, Health & Behavior* 7 (3): 141–52. doi:10.1080/21641846.2019.1652422.

Griffiths, Peter, Saville, Christina, Ball, Jane, Jones, Jeremy, Pattison, Natalie and Monks, Thomas. 2020. "Nursing Workload, Nurse Staffing Methodologies and Tools: A Systematic Scoping Review and Discussion." *International Journal of Nursing Studies* 103 (March): 103487. doi:10.1016/j.ijnurstu.2019.103487.

Gurses, Ayse P., Carayon, Pascale, and Wall, Melanie.2009. "Impact of Performance Obstacles on Intensive Care Nurses' Workload, Perceived Quality and Safety of Care, and Quality of Working Life." *Health Services Research* 44 (2p1): 422–43. doi:10.1111/j.1475-6773.2008.00934.x.

Hart, Sandra G., and Staveland, Lowell E. 1988. "Development of NASA-TLX (Task Load Index): Results of Empirical and Theoretical Research." In Advances in Psychology, edited by Peter A. Hancock and Najmedin Meshkati, 52: 139–83. Human Mental Workload. North-Holland. doi:10.1016/S0166-4115(08)62386-9.

Hignett, Sue, Carayon, Pascale, Buckle, Peter, and Catchpole, Ken. 2013. "State of Science: Human Factors and Ergonomics in Healthcare." *Ergonomics* 56 (10): 1491–503. doi:10.1080/00140139.2013.822932.

Hoonakker, Peter, Carayon, Pascale, Gurses, Ayse P., Brown, Roger, Khunlertkit, Adjhaporn, McGuire, Kerry and Walker, James M. 2011. "Measuring Workload of ICU Nurses with a Questionnaire Survey: The NASA Task Load Index (TLX)." *IIE Transactions on Healthcare Systems Engineering* 1 (2): 131–43. doi: 10.1080/19488300.2011.609524.

IEA. 2000. "What Is Ergonomics (HFE)?" *International Ergonomics Association.* https://iea.cc/what-is-ergonomics/.

IEA. 2021. "World Patient Safety Day, September 17." The International Ergonomics Association Is a Global Federation of Human Factors/Ergonomics Societies, Registered as a Nonprofit Organization in Geneva, Switzerland. https://iea.cc/world-patient-safety-day-september-17/.

Ivziku, Dhurata, Ferramosca, Federica Maria Pia, Filomeno, Lucia, Gualandi, Raffaella, De Maria, Maddalena and Tartaglini, Daniela. 2022. "Defining Nursing Workload Predictors: A Pilot Study." *Journal of Nursing Management* 30 (2): 473–81. doi:10.1111/jonm.13523.

Jetha, Arif, Kernan, Laura and Kurowski, Alicia. 2017. "Conceptualizing the Dynamics of Workplace Stress: A Systems-Based Study of Nursing Aides." *BMC Health Services Research* 17 (1). doi:10.1186/s12913-016-1955-8.

Jin, Haizhe, Xiao, Zhibin, Yao, Junhan, Gong, Zibo, Wang, Haiying and Zhao, Yinan. 2023. "Effects of Workload on Medication Administration Errors in Nursing: An Analysis Based on System Dynamics Modeling." *Simulation* 99 (9): 885–902. doi:10.1177/00375497231168631.

Lane, David, Monefeldt, Camila and Rosenhead, Jonathan. 2000. "Looking in the Wrong Place for Healthcare Improvements: A System Dynamics Study of an Accident and Emergency Department." *Journal of the Operational Research Society* 51 (5): 518–31. doi:10.1057/palgrave.jors.2600892.

Moghadam, Kolsoom Nasirizad, Mino, Mitra Chehrzad, Reza Masouleh, Shademan, Abbas Mardani, Maleki, Maryam, Akhlaghi, Elham and Harding, Celia. 2021. "Nursing Workload in Intensive Care Units and the Influence of Patient and Nurse Characteristics." *Nursing in Critical Care* 26 (6): 425–31. doi:10.1111/nicc.12548.

Moghadam, Kolsoom Nasirizad, Mino, Mitra Chehrzad, Reza Masouleh, Shademan, Maleki, Maryam, Abbas Mardani, Atharyan, Shahaboddin and Harding, Celia. 2021. "Nursing Physical Workload and Mental Workload in Intensive Care Units: Are They Related?" *Nursing Open* 8 (4): 1625–33. doi:10.1002/nop2.785.

Morris, Roisin, MacNeela, Padraig, Scott, Anne, Treacy, Pearl and Hyde, Abbey. 2007. "Reconsidering the Conceptualization of Nursing Workload: *Literature Review." Journal of Advanced Nursing* 57 (5): 463–71. doi:10.1111/j.1365-2648.2006.04134.x.

Mouloua, Salim A., Ball, Rachel V., Ferraro, James C. and Mouloua, Mustapha. 2021. "The History of Human Factors in Healthcare: From Its Emergence 50 Years Ago to COVID-19." *Proceedings of the International Symposium on Human Factors and Ergonomics in Health Care* 10 (1): 165–69. doi:10.1177/2327857921101089.

NASA. 2020. NASA TLX Task Load Index. National Aeronautics and Space Administration. https://human-systems.arc.nasa.gov/groups/TLX/index.php.

PAHO. 2025. The Health Workforce in the Americas: Regional Data and Indicators. Washington, DC: Pan American Health Organization. doi:10.37774/9789275129708.

Pérez, Jorge, de Looze, Michiel Pieter, Bosch, Tim and Neumann, Patrick. 2014. "Discrete Event Simulation as an Ergonomic Tool to Predict Workload Exposures during Systems Design." *International Journal of Industrial Ergonomics* 44 (2): 298–306. doi:10.1016/j.ergon.2013.04.007.

Quintairos, Amanda, Pilcher, David and Salluh, Jorge I. F. 2022. "ICU Scoring Systems." *Intensive Care Medicine*, October. doi:10.1007/s00134-022-06914-8.

Qureshi, Sadeem Munawar, Purdy, Nancy, Greig, Michael A., Kelly, Helen, vanDeursen, Anne and Neumann, W. Patrick. 2021. "Developing a Modelling Approach to Quantify Quality of Care and Nurse Workload — Field Validation Study." *Operations Research for Health Care* 29. doi:10.1016/j.orhc.2021.100301.

Qureshi, Sadeem Munawar, Purdy, Nancy and Neumann, W. Patrick. 2020. "Development of a Methodology for Healthcare System Simulations to Quantify Nurse Workload and Quality of Care." *IISE Transactions on Occupational Ergonomics and Human Factors* 8 (1): 27–41. doi:10.1080/24725838.2020.1736692.

Qureshi, Sadeem Munawar, Purdy, Nancy, Greig, Michael A, Kelly, Helen, vanDeursen, Anne and Neumann, W. Patrick. 2022. "Developing a Simulation Tool to Quantify Biomechanical Load and Quality of Care in Nursing." *Ergonomics*: 1–18. doi:10.1080/00140139.2022.2113921.

Qureshi, Sadeem. 2020. "Developing an Approach to Quantify Nurse Workload and Quality of Care Using Discrete Event Simulation." PhD, Toronto, Canada: Ryerson University. https://www.researchgate.net/publication/342051602_Developing_an_Approach_to_Quantify_Nurse_Workload_and_Quality_of_Care_using_Discrete_Event_Simulation.

Rodríguez, Yordán, and Hignett, Sue. 2021. "Integration of Human Factors/Ergonomics in Healthcare Systems: A Giant Leap in Safety as a Key Strategy during Covid-19." *Human Factors and Ergonomics in Manufacturing & Service Industries* 31 (5): 570–76. doi:10.1002/hfm.20907.

Rogers, Ann E., Hwang, Wei-Ting, Scott, Linda D, Aiken, Linda H and Dinges, David F. 2004. "The Working Hours of Hospital Staff Nurses and Patient Safety." *Health Affairs* 23 (4): 202–12. doi:10.1377/hlthaff.23.4.202.

Salmon, Paul M., and Read, Gemma J. M. 2019. "Many Model Thinking in Systems Ergonomics: A Case Study in Road Safety." *Ergonomics* 62 (5): 612–28. doi:10.1080/00140139.2018.1550214.

Salmon, Paul M., Read, Gemma J. M., Thompson, Jason, McLean, Scott and McClure, Rod. 2020. "Computational Modelling and Systems Ergonomics: A System Dynamics Model of Drink Driving-Related Trauma Prevention." *Ergonomics* 63 (8): 965–80. doi:10.1080/00140139.2020.1745268.

Sterman, John D. 2009. Business Dynamics: Systems Thinking and Modeling for a Complex World. Nachdr. Boston: Irwin/McGraw-Hill.

Sujan, Mark, Pickup, Laura, Vosper, Helen and Catchpole, Ken. 2021. "Making Human Factors and Ergonomics Work in Health and Social Care: Chapter 1." In Human Factors in Health and Social Care, 25. UK: CIEHF. https://ergonomics.org.uk/resource/hf-in-health-and-social-care-ebook-chapter-1.html.

Sujan, Mark, Pickup, Laura, Bowie, Paul, Hignett, Sue, Ives, Fran, Vosper, Helen and Rashid, Noorzaman. 2021. "The Contribution of Human Factors and Ergonomics to the Design and Delivery of Safe Future Healthcare." *Future Healthcare Journal* 8 (3): e574–79. doi:10.7861/fhj.2021-0112.

Swiger, Pauline A., Vance, David E. and Patrician, Patricia A. 2016. "Nursing Workload in the Acute-Care Setting: A Concept Analysis of Nursing Workload." *Nursing Outlook* 64 (3): 244–54. doi:10.1016/j.outlook.2016.01.003.

Torres Yaniel, Rodríguez, Yordán and Pérez, Elizabeth "¿Cómo mejorar la calidad de los servicios de salud y la seguridad del paciente adoptando estrategias del sector de la aviación?" *Journal of Healthcare Quality Research* 37 (3): 182–90. doi:10.1016/j.jhqr.2021.10.009.

Weinger, Matthew B., Reddy, Swapna B. and Slagle, Jason M. 2004. "Multiple Measures of Anesthesia Workload During Teaching and Nonteaching Cases." *Anesthesia & Analgesia* 98 (5): 1419–25. doi:10.1213/01.ANE.0000106838.66901.D2.

WHO. 2020. State of the World's Nursing 2020: Investing in Education, Jobs and Leadership. Geneva: World Health Organization. https://apps.who.int/iris/handle/10665/331677.

WHO. 2021. Global Patient Safety Action Plan 2021–2030: Towards Eliminating Avoidable Harm in Health Care. Geneva: World Health Organization. https://apps.who.int/iris/handle/10665/343477.

WHO. 2025a. Health Worforce. World Health Organization. https://www.who.int/data/gho/data/themes/health-workforce.

WHO. 2025b. State of the World's Nursing 2025: Investing in Education, Jobs, Leadership and Service Delivery. Geneva: World Health Organization. https://iris.who.int/bitstream/handle/10665/381329/978924011 0236-eng.pdf?sequence=1.

WHO. 2025c. The National Health Workforce Accounts Database. Geneva: World Health Organization. https://www.who.int/activities/improving-health-workforce-data-and-evidence.

Wilson, John R. 2014. "Fundamentals of Systems Ergonomics/Human Factors." *Applied Ergonomics, Systems Ergonomics/Human Factors*, 45 (1): 5–13. doi:10.1016/j.apergo.2013.03.021.

Zanker, Marek, Bureš, Vladimír and Tučník, Petr. 2021. "Environment, Business, and Health Care Prevail: A Comprehensive, Systematic Review of System Dynamics Application Domains." *Systems* 9 (2): 28. doi:10.3390/systems9020028.

2 A Comprehensive Ergonomic Model to Reduce Risk in Maxillofacial Surgery Training

Moisés Adolfo Solorio Benítez, Mildrend Ivett Montoya Reyes, Gabriela Jacobo-Galicia, Ismael Mendoza Muñoz, and Olivia Yessenia Vargas Bernal

2.1 INTRODUCTION

Musculoskeletal disorders (MSDs) are a group of injuries that represent a highly prevalent occupational hazard due to the magnitude and frequency with which they occur. This type of injury is considered one of the main causes of sick leave, a reduction in service quality and, under certain circumstances, resignation from professional practice or early retirement (Bustos 2018; Harutunian et al. 2011; Ziaeefar et al. 2024). In dentistry, professionals perform various types of surgeries that require great precision, as well as specific environmental conditions, such as adequate lighting, to carry them out. These activities are usually performed over long periods of time and involve static postures and a wide variety of repetitive hand and fingers movements that can result in the development of long-term musculoskeletal injuries (Calderón-Vallejo et al. 2024; Faust et al. 2021). Moreover, dentists are exposed to psychosocial risks associated with mental stress that may promote the appearance of MSDs mainly in the neck, back, shoulders, hands and wrists (Secretaría de Salud Laboral y Desarrollo Territorial 2019).

In Mexico, most studies on dentists focus on the stress generated by their activities. This research has identified stressors (such as noise from the instruments used), symptoms of stress (insomnia, fatigue, anxiety) and their repercussions, which can affect both physical and mental states, as well as work performance (Pozos-Radillo et al. 2016). Regarding MSDs, the research carried out has focused on the recognition of the main areas affected by injuries, where discomfort in the neck, dorsal or lumbar region, shoulders, hands and wrists predominates. These studies report that the tools used to evaluate the worker and his or her work environment use the Rapid Entire Body Assessment (REBA) method for postural load, surveys on the work environment, and the Stress Symptomatology and Burnout Inventories, as well as interviews and questionnaires (Fimbres et al. 2016).

As for oral and maxillofacial surgery practice, studies carried out on dental surgeons found that the appearance of discomfort or even MSDs in areas such as the neck, upper and lower back, shoulders and wrists was predominant (Howarth et al. 2019; Uppada et al. 2020). Another analysis of the working conditions of different dental specialties used the Rapid Upper Limb Assessment (RULA) method, which showed that oral and maxillofacial surgeons present greater risks of exposure to MSDs than other specialties because of the degree of complexity of the surgeries they perform (Holzgreve et al. 2022).

The number of studies in the field of oral and maxillofacial surgery that report ergonomic evaluations to improve the quality of life and work environment of these professionals is relatively scarce, and the results of the analyses have not been used to improve working conditions (Holzgreve et al. 2022; Jahanimoghadam et al. 2018; Terán and Izquierdo 2020). Therefore, this paper reports a more

 DOI: 10.1201/9781003582014-3

comprehensive study, where various evaluation techniques and tools are applied with a holistic approach to the diagnosis of physical and mental risks, such as the presence of chronic pain in the soft tissues, joints, muscles and tendons or those that may result from the cognitive load of the task, to generate a proposal for the improvement of working conditions that can be implemented.

2.2 METHODOLOGY

2.2.1 Materials and Methods

A case study was conducted on a convenience sample of 30 dentistry students to evaluate the risk of occurrence of MSDs throughout the academic year of 2024. Data were collected through observation, interviews, structured questionnaires and measurements of environmental conditions such as temperature, illumination and noise. Data collection and analysis were carried out in the first three phases, and improvement measures were implemented in the fourth phase.

The research was conducted on dentistry students who train in the surgery clinic at the Faculty of Dentistry at a Mexican State University. The students perform tooth extractions and surgeries of a considerable degree of difficulty but appropriate for their skills and knowledge. Procedures involve the use of instruments specific to the type of surgery to be performed. Students perform the interventions under the supervision of a teacher, who verifies that the technique is implemented correctly and that the appropriate procedure is followed.

This analysis includes semi-manual operations, as well as a study of the students' working conditions and their interaction with patients, in order to identify improvements. For this reason, this study required that participants be filmed and photographed during the performance of their work since ergonomic risk assessment methodologies need the use of these means to elaborate an adequate and valid analysis. To this end, permission was requested from each of the participants by signing an informed consent form.

A four-phase comprehensive ergonomic assessment was implemented consisting of the following: pre-diagnosis, assessment, diagnosis of the current situation and improvement actions. The pre-diagnosis phase consisted of different evaluations: first, a preliminary interview was conducted to collect general data such as age, sex, eating habits, and general healthcare and illness. The Nordic Standardized Questionnaire (also known as Kuorinka Questionnaire) was then applied to identify perceived musculoskeletal discomfort. Finally, the analysis of the surgery process was carried out by means of a flow diagram to identify and understand the main activities the students must perform during a surgery.

The assessment phase consisted of three different analyses: environmental, physical ergonomics and cognitive. The environmental analysis considered temperature, noise and lighting. To assess them, measurement points were established and distributed throughout the work area. The specific Mexican Official Regulations that establish the acceptable environmental conditions for the workplace were consulted. The measurements obtained were recorded in an Excel table and then plotted.

Next, the physical ergonomics analysis was performed. Two methods were applied: REBA to evaluate posture and the Revised Strain Index (RSI) to assess the distal upper extremity physical exposure to exertion. First, the REBA method was applied using photographs and videos of the performance of the subject to be evaluated. With the help of these, the duration of the activities was measured, and the postures that presented a greater ergonomic risk were selected. In addition, the necessary angular measurements were taken to assign scores to each part of the body on the field sheet and thus establish the level of action necessary in each case. The RSI method was then applied, which requires to identify the number of movements and the total observation time for each upper extremity (left and right) separately. The value of the multiplier for each risk factor was determined, and the data were captured into the field sheet, which generated the risk factor. It is important to note that pinch strength was measured with a pinch gauge, also known as pinch meter, in three different positions: tip-to-tip pinch (press the pinch meter between the tips of the index finger and the thumb), lateral pinch (place the pinch meter between the radial side of the index finger and the thumb and press) and three-point pinch (place the pinch meter between the pulp of the thumb and pulp of the index and middle fingers and press). These measurements were made at the beginning and end of the activity to identify if there were changes in the grip strength due to fatigue.

The last analysis of this phase evaluated the factors that influence mental load using the NASA Task Load Index (TLX) method. This method yields scores for six different dimensions that correspond to the level of mental load at work: mental demand, physical demand, time demand, effort, performance and level of frustration. This method was carried out in two stages: the weighting stage in which the subjects determined the level of the mental load (from 0 to 5) that each of the factors contributed and the rating stage, where the subjects evaluated the magnitude (on a scale of 1 to 20) of the load for the activities carried out in each of the six dimensions.

In the third phase, i.e., diagnosis of the current conditions, the results obtained in the evaluation phase were used to identify the factors that affected the students, such as inappropriate postures that are maintained for prolonged periods of time and/or with high repetitiveness, the environmental conditions in which they work and those dimensions considered sources of mental load.

Finally, for the fourth phase, after diagnosing the students' working conditions, a physiotherapist and a psychologist were consulted on the type of exercises and other improvement measures recommended to be implemented, with the aim of reducing the factors identified by the study, which have the potential to affect the health of the students.

2.3 RESULTS

2.3.1 Pre-Diagnosis

Preliminary interviews were conducted with 30 students (19 women and 11 men). Their answers indicated that 63.3% usually eat breakfast before the start of their school day. Regarding hours of sleep, 36.7% reported sleeping 6 hours, 26.7% 7 hours, 23.3% 8 hours and 13.3% 5 hours. Regarding physical activity, 53.3% of students performed some physical activity regularly. In the area of hydration, 76.7% of students indicated that they are constantly hydrated. In addition, 23.3% of the students indicated that they had suffered some type of injury before, among which they reported fractures in arms, wrists and feet. Finally, in relation to the clinical experience of the students, 50% had 4 to 6 months of experience, 23.3% had 1 to 2 years, 20% had 7 to 12 months and 6.7% had 1 to 3 months of experience.

The results of the Nordic Standardized Kuorinka Questionnaire indicated that the ages of the students range from 20 to 26 years, with 50% of students born between 2000 and 2001, followed by 30% who were born between 2002 and 2003. In the case of weight, 23.3% of the students reported weighing between 70 and 80 kg and 20% between 50 and 60 kg. In relation to height, most students were 1.71 to 1.80 m tall (36.7%), followed by those 1.51 to 1.60 m tall (23.3%). The majority (43.3%) of the students reported working in the clinic for 1 to 3 hours per week, but it is important to note that 10% of students work there 12 hours a week or more. Regarding the parts of the body in which they had felt some pain or discomfort, students reported problems in the neck, shoulders, elbows, wrists, upper and lower back, legs and knees (see Figure 2.1). Of these, neck, shoulder, upper and lower back and leg problems occurred in days close to the date of the interview but had no impact on the development of their work activities at the time.

Of the parts of the body that presented some type of discomfort, the one that was reported the most times was the lumbar area with 13 cases. Students reported that the discomfort manifested itself for a few hours (53.8%), from 1 to 7 days (23.1%) or every day in some cases (23.1%). Of the cases that reported a duration of more than 1 day, only one student received care from a physiotherapist. Another body part with reported discomfort is the neck, with nine students affected, of which 55.6% have lived with discomfort for a period of 1 to 7 days. Finally, six cases with shoulder discomfort were detected, and in two of these subjects, the problems persisted for approximately 1 to 7 days.

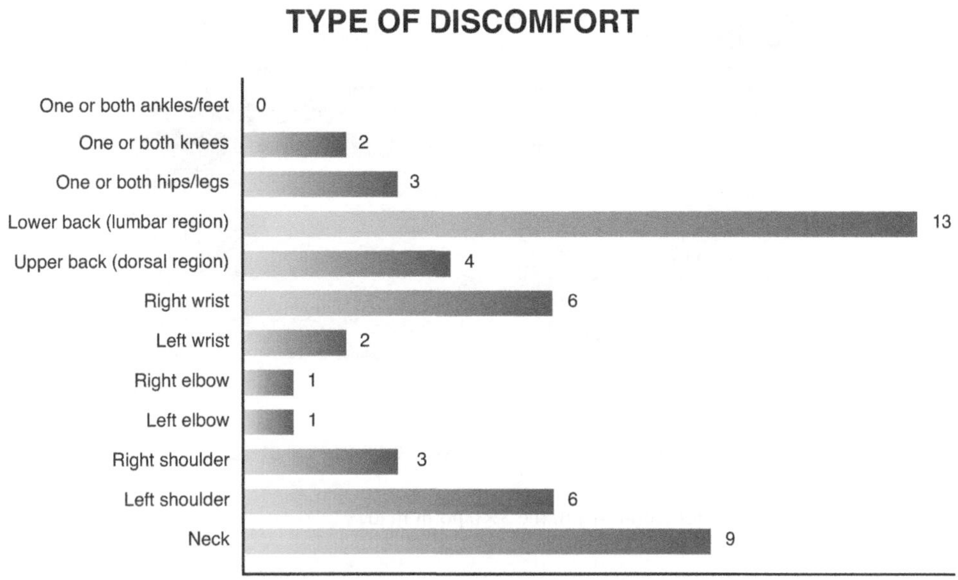

FIGURE 2.1 Reported discomfort by students through the Nordic Standardized Questionnaire.

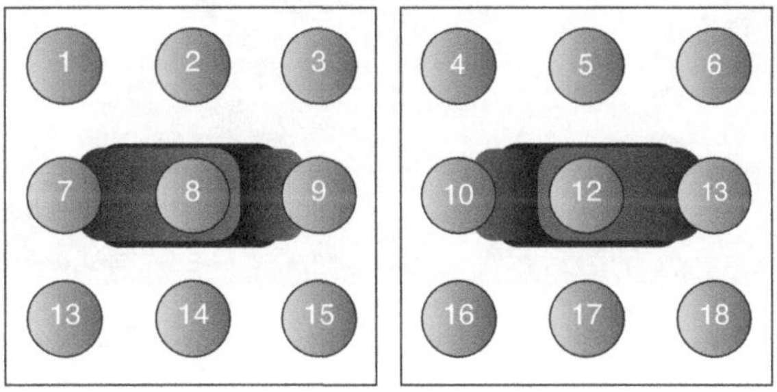

FIGURE 2.2 Sample 3×6 matrix.

2.3.2 Evaluation

To perform the temperature, noise and lighting measurements more accurately, the layout area was divided into two, and a matrix with 3 x 6 test points was built, as shown in Figure 2.2.

An infrared thermometer was used to measure temperature in the work area. Figure 2.3 shows the distribution of temperatures captured in the 3×6 point matrix. In this representation, black indicates temperatures of 18°C to 19°C, dark grey represents temperatures of 19°C to 20°C, light grey corresponds to temperatures of 20°C to 21°C, and white is used for temperatures of 21°C to 22°C.

The illumination was measured with a luxmeter that provides readings in luxes, without casting a shadow on the instrument and at each point of the matrix. The maximum recorded level of illumination was 11,820 lx when measured using the facility's light in conjunction with the work unit's light. Figure 2.4 shows the distribution of the readings collected in intervals of 0 to 5,000 lx (gray), 5,000 to 10,000 lx (light gray) and 10,000 to 15,000 lx (gray).

The distribution of noise levels is shown in Figure 2.5. Maximum noises were detected in a range of 80 to 85 decibels (dBA) in dark gray, although readings of 70 to 75 dBA (gray) were predominant. The measurements were taken with a sound level meter on a day when all 10 work units were in use.

It was found, regarding to the environmental ergonomic analysis, that the temperature, lighting and noise factors evaluated in the matrices were within the limits established by NOM-015-STPS-2001,

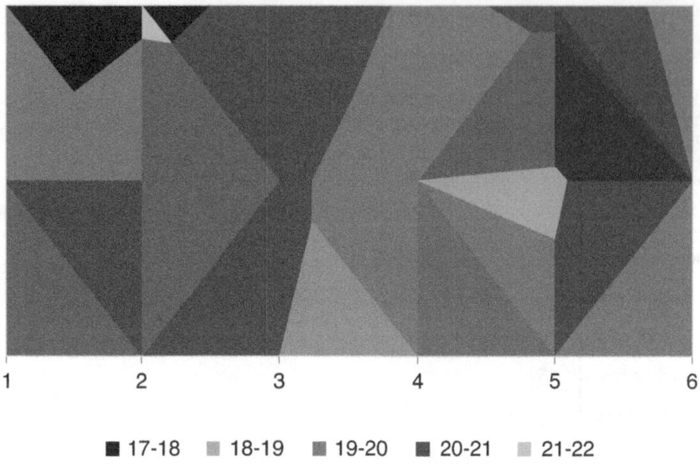

FIGURE 2.3 Distribution of temperatures in the 3×6 point matrix.

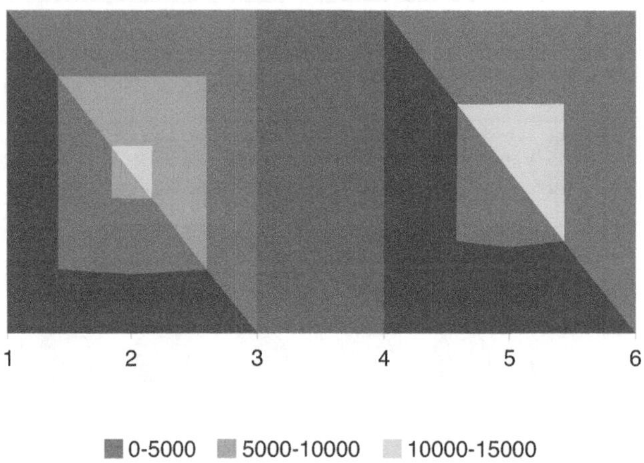

FIGURE 2.4 Distribution of lighting readings.

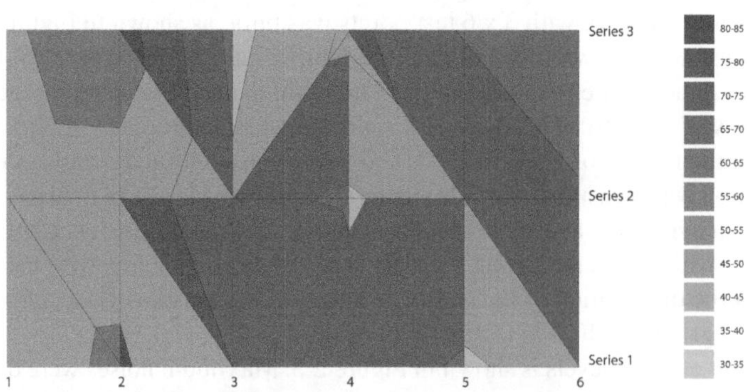

FIGURE 2.5 Distribution of noise measurements.

NOM-025-STPS-2008 and NOM-011-STPS-2001, respectively, as shown in Table 2.1, according to the type of work and the exposure time.

The analysis of the pinch strength was performed with a pinch meter at two different time points: before the students conducted the intervention and at the end of it. Of the total number of students evaluated, 37% presented a reduction in the strength of the tip-to-tip pinch, 33% recorded less strength in lateral pinch and 27% are cases with reduced three-point pinch strength. Variations in the average strength by sex are shown in Table 2.2. As can be seen, men had higher levels of reduced strength in all three types of pinch.

For the postural assessment, the REBA and RSI methods were used to evaluate students´ postures. Videos of the students performing surgeries were recorded to identify the highest-risk posture for each body part to be assessed. These postures were then captured as photographs to determine the joint angles required by the REBA method, using the RULER tool available on the "Ergonautas" website (Ergonautas, n.d.). Figure 2.6 shows an example of these images.

Subsequently, a REBA score was assigned to each body part, as well as to load and grip, to obtain a final score indicating the level of risk the students were working with. The results are shown in Table 2.3 and indicate that 3% of the students presented a very high risk level, 50% a high risk level and 47% a medium risk level. These results make it necessary to implement preventive measures.

TABLE 2.1
Analysis of Environmental Factors

Factor	Applicable Regulation		Results	Comply?	At Risk?
Noise levels	NOM-011-STPS-2001, safety and hygiene conditions in workplaces where noise is generated, establishes the maximum permissible exposure limits to noise in the workplaces for workers.	For a maximum permissible exposure time of 2 hours, a noise exposure level of 96 dB is allowed.	Minimum level: 68.7 dB. Maximum level: 83.6 dB.	☑	No
Temperature	NOM-015-STPS-2001, high or abatement thermal conditions and safety and hygiene conditions, determines the maximum levels of exposure to high thermal conditions.	Maximum permissible temperature in °C for a light working regime: 30°C.	Temperature: 18.9°C and 23.9°C.	☑	No
Lighting	NOM-025-STPS-2008, lighting conditions in workplaces, indicates the levels of lighting necessary for each type of visual activity carried out by workers.	Minimum level Inside: 50 luxes High accuracy in distinguishing details: 1,000 luxes.	Inside: 240 luxes High accuracy in distinguishing details: 1,084 luxes.	☑	No

TABLE 2.2
Pinch Strength Measurements

		Tip-to-Tip Pinch Strength (lb)	Lateral Pinch Strength (lb)	Three-Point Pinch Strength (lb)
Female	Before	11.2	17.1	15.3
	After	10.5	15.8	14.1
	Reduction	0.7	1.3	1.2
Male	Before	15.2	24.2	21.4
	After	13.3	22.7	19.4
	Reduction	1.9	1.5	2.0

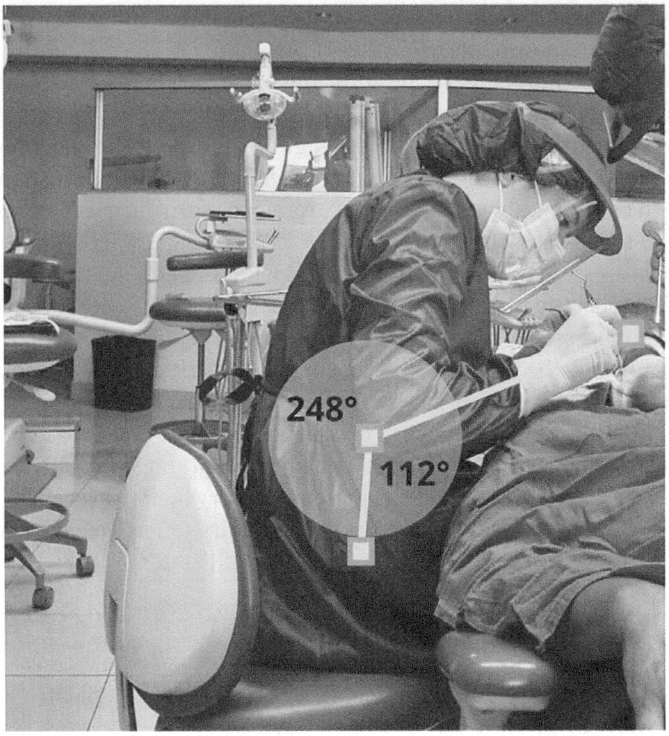

FIGURE 2.6 Right forearm posture assessment.

TABLE 2.3
Results from Physical Evaluations

Physical Analysis

Risk Level (REBA)	Results	Actions	A Risk?
Negligible Low	0% of students.	No immediate action is necessary.	No
Medium	47% of students.	Corrections to student postures are required.	Yes
High	50% of students.	It is necessary to intervene to correct the position of the students and raise awareness about the possible damage to health as soon as possible.	Yes
Very high	3% of students.	Immediate intervention is required to preserve the health of students.	Yes
Risk level (RSI)	Results	Actions	At risk?
Low	Right hand-wrist: 27% of students. Left hand-wrist: 93% of students.	No ergonomic intervention is necessary.	No
Medium	Right hand-wrist: 60% of students. Left hand-wrist: 3% of students.	Ergonomic intervention is necessary to prevent long-term health damage.	Yes
High	Right hand-wrist: 13% of students. Left hand-wrist: 3% of students.	Immediate intervention is necessary; it is recommended to carry out an analysis of the critical factors and implement exercises to release tension in the limbs.	Yes

TABLE 2.4
Results from Cognitive Analysis

Cognitive Analysis

Mental Load	Results	Actions	At Risk?
Low	13% of students	Immediate action is not necessary, but dissemination of mental load at work is recommended.	No
Medium	37% of students	Intervention is necessary; an analysis of the conditions in which workers operate must be carried out to determine the main causes that raise their level of mental load at work.	Yes
High	50% of students	Immediate intervention in the workplace is necessary; it is necessary to implement a program that promotes mental health, which includes exercises for stress control and mental fatigue.	Yes

To perform the calculations required by the RSI method, a video segment was used, in which the greatest effort required from the tooth extraction/surgery activity was detected. A value was assigned to the intensity of the effort, its duration was measured and the movements per minute were counted for each hand. The angles of the postures were determined, as well as the speed with which they work. Also, the duration of the activities per day was calculated. The results shown in Table 2.3 indicate that in 60% of the cases, the students show a moderate risk of developing a condition in the right hand-wrist area, 13% show a high risk in the same area and a considerably lower risk is observed in the left hand-wrist, with 3% of students at moderate risk and 3% at high risk.

The NASA-TLX method was used to analyze the mental load. The evaluation was divided into two stages. First, before starting the interaction with the patient, the students were asked to order the dimensions according to their load level while they performed their activities. In the second stage, at the end of the activities, they were asked to assign a grade from 1 to 20 to each dimension. The student's experience was considered. Table 2.4 shows that 50% of the students have high levels of mental load, 37% have medium levels of mental load and 13% have low levels of mental load. The factor with which they felt most identified is performance (degree of satisfaction with one's own performance) with 40%, followed by frustration (degree of stress, irritation, etc., felt during the performance of the task) with 23%, effort (degree of mental and physical effort that must be made to obtain their level of performance) with 17% and mental demand with 10%.

2.3.3 Diagnosis

After performing all evaluations and reviewing the data, the following aspects were identified:

- The noise, temperature and lighting conditions are within those required by the applicable regulations.
- In the pre-diagnosis, the lower back and neck were identified as the main areas of discomfort.
- REBA, RSI and strength measurements with the pinch meter indicated that most students are at medium to high and up to very high level of risk (REBA); therefore, they may develop MSDs, and it is necessary to take preventive and corrective measures to reduce the level of risk.
- The evaluation of mental load allowed us to identify that 37% of the interviewed students were at a medium risk level and 50% at a high risk level in relation to the perceived mental load. According to the data, immediate measures must be implemented so that students learn to resolve their dissatisfaction with their own performance and their frustration.

2.3.4 Improvements

With the help and advice of experts in the areas of physiotherapy and psychology, solutions to the identified discomforts were sought. The experts recommended exercises to release tension, and for frustration and problems with performance, frustration management techniques were advised. For the issue of physical risks, a plan of stretching and mobility exercises was developed for students of the Faculty of Dentistry. The plan considers factors such as space, time and the ease of performing those exercises within the clinical facilities. To work with the risks associated with mental load, a group of short- and long-term exercises was incorporated with an emphasis on the needs of university students, with the intention to reduce the feeling of overwhelming and stress caused by their daily activities in the clinic.

To correct inappropriate postures, it is recommended to emphasize in class and put into practice the importance of adopting good ergonomic practices when working in a standing and/or sitting position, even with new students, to facilitate the development of proper work habits. Some of the habits to be corrected include properly positioning the patient using an adjustable dental chair, according to the type of operation, adjusting the auxiliary lighting of the dental chair to ensure a clear view of the oral cavity and avoiding awkward or forced head posture.

2.4 DISCUSSION

The results obtained indicate that, although there are no ergonomic environmental risks in accordance with the Normas Oficiales Mexicanas used as a guide for this research, ergonomic risks of a physical and cognitive nature are found in the sample evaluated. Through the REBA and RSI methods, as well as the strength test performed with the pinch meter, it was possible to identify that the extraction and oral surgery students working in the clinic are at ergonomic risk levels ranging from medium to very high. These factors indicate the likelihood that students will develop disorders and other physical or mental health problems. Therefore, by not implementing preventive measures and changes in the development of their activities, their development as professionals may be affected in the future, in addition to limiting their well-being.

Through the cognitive analysis carried out with the NASA-TLX method, it was found that, of the 30 students evaluated, only 13% have low levels of mental load, while half have high levels of mental load. The dimensions to which students attribute a greater weight are performance and frustration, which is why there is a probability that stress, burnout, depression and other mental health disorders may be present in the student population and interfere with the development of their daily and work activities. These conditions pose a risk to all parties involved: the students, the professors who supervise them and patients who are undergoing oral surgery or extraction treatments.

Due to the potential risk to the health and well-being of those involved, it is advisable to carry out routine evaluations not only of the physical but also cognitive and environmental conditions of the workplace, as well as going to specialists in the areas of physiotherapy and psychology to determine and implement the best practices to reduce the appearance of possible physical disorders and help with the different cognitive problems that may arise among students. If students are able to internalize these practices, the possibility that they will continue them during their professional practice increases, and their working life may be longer and healthier.

REFERENCES

Bustos, Z. J. 2018. La Importancia de la ergonomía en la práctica odontológica. Bach diss., Universidad Nacional Autónoma de México.

Calderón-Vallejo, J., A. A. Noguera-Cundar, L. Castro-Cepeda, and M. Moreno-Barriga. 2024. Evaluation of ergonomic risks factors in a dental center located in the historical center of the City of Riobamba. *EPOCH Congress: The Ecuadorian Journal of STEAM* 3, no. 3:60–85. https://doi.org/10.18502/espoch.v3i3.16613

Ergonautas. n.d. RULER - Medición de ángulos en fotografías. https://www.ergonautas.upv.es/herramientas/ruler/ruler.php

Faust, A. M., S. N. Ahmed, L. B. Johnston, and J. B. Harmon. 2021. Teaching methodologies for improving dental students' implementation of ergonomic operator and patient positioning. *Journal of Dental Education* 85, no. 3:370–378. https://doi.org/10.1002/jdd.12438

Fimbres, K. L., J. A. García, R. M. Tinajero, R. E. Salazar, and M. O. Quintana. 2016. Trastornos musculoesqueléticos en odontólogos. *Benessere Revista de Enfermería* 1, no. 1:35–46. https://revistas.uv.cl/index.php/Benessere/article/view/1337/1374 2016.1337

Harutunian, K., J. Gargallo-Albiol, R. P. Barbosa de Figueiredo, and C. Gay Escoda. 2011. Ergonomics and musculoskeletal pain among postgraduate students and faculty members of the School of Dentistry of the University of Barcelona (Spain). A cross-sectional study. *Medicina Oral, Patología Oral y Cirugia Bucal* 16, no. 3:425–429. https://doi.org/10.4317/medoral.16.e425

Holzgreve, F., L. Fraeulin, W. Betz, C. Erbe, E. M. Wanke, D. Brüggmann and D. Ohlendorf. 2022. A RULA-based comparison of the ergonomic risk of typical working procedures for dentists and dental assistants of general dentistry, endodontology, oral and maxillofacial surgery, and orthodontics. *Sensors* 22, no. 805:1–17. https://doi.org/10.3390/s22030805

Howarth, A. L., M. S. Hallbeck, V. Lemaine, D. J. Singh and S. S. Noland, S. S. 2019. Work-related musculoskeletal discomfort and injury in craniofacial and maxillofacial surgeons. *Journal of Craniofacial Surgery* 30, no.7:1982–1985. https://doi.org/10.1097/SCS.0000000000005631

Jahanimoghadam, F., A. Horri, N. Hasheminejad, N. H. Nejad, and M. R. Baneshi. 2018. Ergonomic evaluation of dental professionals as determined by rapid entire body assessment method in 2014. *Journal of Dentistry* 19, no. 2:155–158. https://pmc.ncbi.nlm.nih.gov/articles/PMC5960736/

Pozos-Radillo, B. E., M. Preciado-Serrano, M. Acosta-Fernández, M. Aguilera -V. and A. R. Plascencia-Campos. 2016. Síntomas psicofisiológicos predictores del estrés en odontólogos. *Revista Médica del Instituto Mexicano del Seguro Social* 54, no.2:151–158. https://www.medigraphic.com/cgi-bin/new/resumen.cgi?IDARTICULO=65198

Secretaría de Salud Laboral y Desarrollo Territorial. 2019. *Manual Informativo de PRL: Ergonomía*. Riesgos Ergonómicos, UGT-Madrid.

Terán, A. A. and A. E. Izquierdo. 2020. Valoración del riesgo ergonómico de estudiantes de odontología mediante el método OWAS. *Odontología* 22, no. 2:60–71. https://revistadigital.uce.edu.ec/index.php/odontologia/article/view/2386

Uppada, U. K., R. Sinha, S. Madishetti, S. G. Pampana, S. Kumar, S. and P. R. Chidagam. 2020. Ergonomics among oral and maxillofacial surgeons in the Indian States of Telangana and Andhra Pradesh-An evaluative study. *Annals of Maxillofacial Surgery* 10, no. 2:325–329. https://doi.org/10.4103/ams.ams_39_20

Ziaeefar, P., H. Hatami, D. Panahi, M. Poursadeghiyan and A. Salehi-Sahlabadi. 2024. Neck pain severity and work ability index: A study of Tehran dental students. *Iranian Rehabilitation Journal* 22, no. 2:265–276. https://doi.org/10.32598/irj.22.2.2004.2

3 TAPTE Model
Ergonomics in Pregnant Workers

Carolina Rodríguez-Herrera and Eduardo Cerda-Díaz

3.1 OVERALL WORKLOAD ON PREGNANCY WORKERS

3.1.1 Women in Work

Work dynamics have been changing with the inclusion of women in the labor market, with an increasing number of women of reproductive age participating in the paid labor market. In some cases, this participation has been greater in certain productive sectors known as "feminized". Despite this, the intention of women and companies has been to increase the participation of women in sectors that were previously exclusive to men. This leads to challenges for companies in generating more inclusive and gender-equitable jobs.

On the other hand, there are differences in the guidelines of the companies and productive sectors, with multinational companies sometimes being the ones that place in their guidelines the incorporation of women in their jobs in the industry sector, thus looking for gender equity. Therefore, over the next few years, the increased participation of women in the workplace should be considered, decreasing gender gaps, which could increase the number of women in a transitory situation, such as pregnancy, with the creation of work environments with an ergonomic approach favorable for the compatibility between pregnancy and the environment.

3.1.2 Anatomical and Physiological Changes

"Human pregnancy is characterized by profound anatomic and physiologic changes that affect virtually all systems and organs in the body". This condition is normal, transient, and personalized, and the changes are variable according to the trimester of gestation (Abdelwahab et al., 2022).

3.1.3 Overall Workload: Physical and Mental Aspects

The overall workload perceived by the worker is the set of physical and mental requirements to which she is exposed during her working day. It is composed of six dimensions as described in NASA TLX (Hart & Staveland, 1988): mental demand, physical demand, temporal demand (time demand), performance, effort, and frustration dimensions. The physical and physiological changes in the pregnant woman may cause difficulties in performing activities and/or work tasks, which will hinder her ability to respond to the same demands, even according to her week of gestation, with the relation of the pregnant woman with her work environment and work system being even more variable.

Pregnant women in their workplace will be exposed to the demands of the environment (work) and to the demands of pregnancy itself, with its respective physiological and physical changes. Therefore, the work capacity will be influenced by workplace, individual, and pregnancy factors (Figure 3.1).

DOI: 10.1201/9781003582014-4

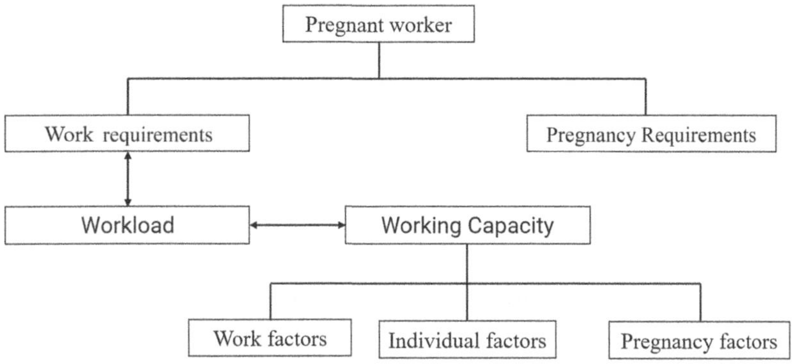

FIGURE 3.1 Pregnant worker model.

3.2 LEGISLATION IN CHILE ON PREGNANT WOMEN

In the labor area, the guidelines are focused on the Labor Code (Article 202), Supreme Decree 594, and guidelines of the International Labor Organization (ILO). Article 194 of the Labor Code (2001) cannot condition the hiring of women in the absence or existence of pregnancy or request certificates to confirm their pregnancy status.

Article 202 of the Chilean Labor Code mentions aspects that are considered detrimental to pregnant women; these are as follows: night work (22:00 PM to 7:00 AM), lifting, dragging or pushing, existence of physical effort, standing for a long time, overtime work, or any activity that the competent authority declares inconvenient for the state of pregnancy.

According to Law 20.001, Law 20.949, and Supreme Decree N°63, the maximum weight of manual load handling is limited for pregnant women to 0 kg; that is, they cannot perform manual load handling, i.e., handling loads (animate or inanimate objects) greater than 3 kg.

Other rights of pregnant women are related to the prenatal and postnatal periods; the prenatal period is 6 weeks before delivery, and the postnatal period corresponds to 12 weeks after delivery.

3.3 MODEL TAPTE

The Conceptual Model of Job Adaptation Time for Pregnant Women was developed by Rodriguez, C in 2016 to determine the variability of labor-active pregnant women from a physical and mental scope, considering the workload. In the physical realm, as Paul mentions, "the special spatial requirements for pregnant women will be especially relevant between the third and ninth month of pregnancy" (Paul et al., 1995). Conversely, there is a lack of ergonomic evaluation instruments specifically oriented to the variables associated with pregnant workers and their work system according to the trimester of gestation.

The conceptual model called TAPTE is developed by generating a sequence of analyses associated with the woman's gestation period, considering the study of several variables such as workload, description and analysis of the difficulties that pregnant women have when developing their work activity and/or work task, measurement of anthropometric body dimensions, perception of discomfort, and perception of musculoskeletal discomfort.

In order to create the TAPTE Model, in the first stage, specific aspects of each pregnant woman are considered since it is essential to determine their limits and capabilities according to the trimester of gestation. On the other hand, physical and mental ergonomic factors are fundamental to consider in addition to organizational aspects (see Figure 3.2).

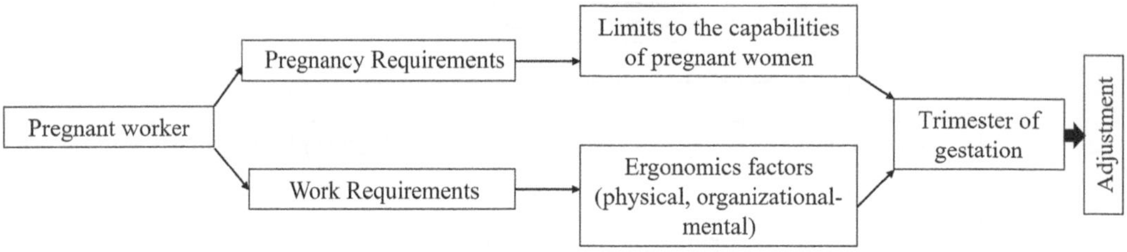

FIGURE 3.2 Pregnant women and human factors and ergonomics.

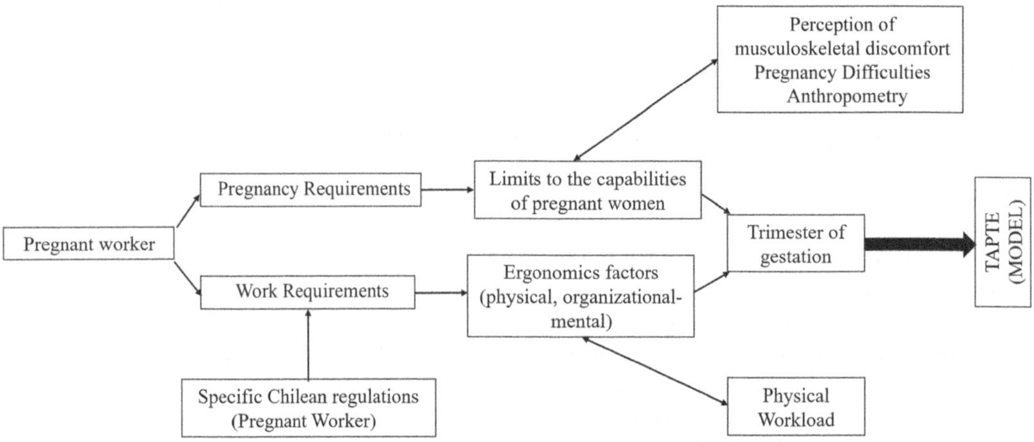

FIGURE 3.3 The caption is Conceptual model of human factors and ergonomics for pregnant workers.

In addition to the above, the relevance of subjective perceptions of discomfort, difficulties, and anthropometric changes is highlighted. Based on the current regulations and requirements, it should be noted that other factors such as noise, vibrations, and chemical agents, to which pregnant women should not be exposed, should be considered (see Figure 3.3).

Gestational conceptual model TAPTE with perception and demands (by the author): The TAPTE Model is based on a six-stage diagram to decide on the key aspects to define the best ergonomic conditions for the pregnant woman and her workplace, considering that ergonomics seeks to adapt the workplace to the person (see Figure 3.4).

3.4 EXPERIENCE ON THE ERGONOMICS PROGRAM TO ADAPTATION PLAN TO PREGNANT WORKERS IN THE CHILEAN MANUFACTURING INDUSTRY

The beginning of action in the Ergonomics Program related to pregnant women occurs with the recognition of the worker in the Preventive Health Unit. After that, the Preventive Health Unit develops the following stages. The Adaptation to Work procedure associated with the Ergonomics Program considers the execution of seven stages:

1. Diagnostic Cases and Functional Capacities Analysis
2. Competency and Skills Requirement (Sociotechnical) Analysis
3. Systemic and Ergonomics Analysis of Jobs, Tasks, and Workplaces

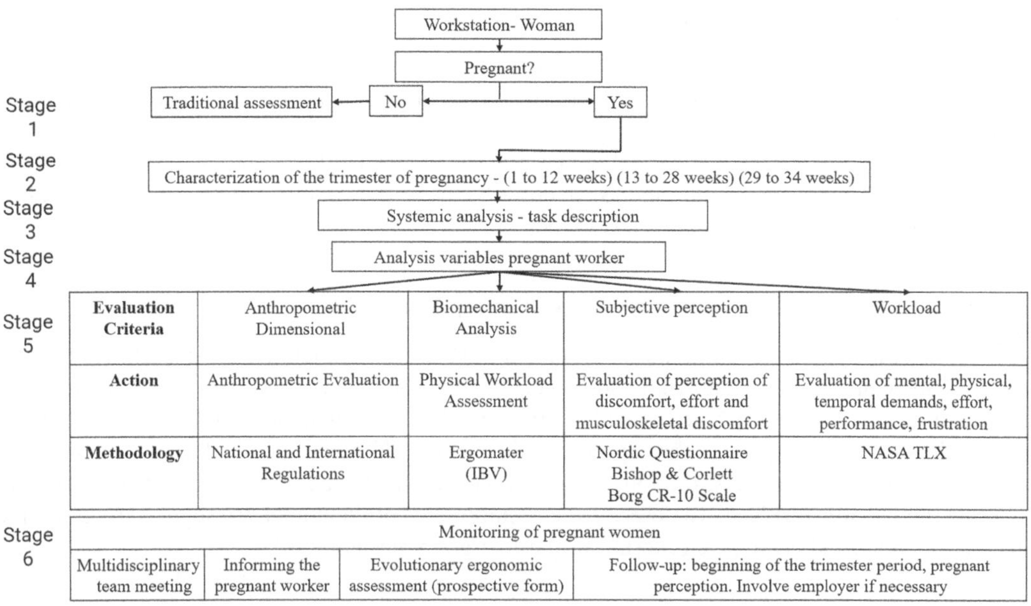

FIGURE 3.4 Workstation for pregnant workers.

4. Security and Prevention Analysis of Jobs, Tasks, and Workplaces
5. Workplace Adaptation Plan
   - Define specific characteristics of the workplace related to individual health conditions and normative requirements
   - Evolutive adaptation criteria
6. Execution and Follow-up Related to the Pregnancy's Specific Characteristics and Work Conditions.
7. Conclusion of the Evolutive Adaptive Process.

This flow is established as a procedure associated with the Ergonomics Program with the objective of achieving a better evolutive adaptation of pregnant workers in the workplace context. In this case, it is addressed and considered an everyday health condition with a strong preventive focus.

3.4.1 Diagnostic Cases and Functional Capacities Analysis

At this stage, the physicians execute a diagnosis related to the health condition and specific functional capacities. This information is notified to the ergonomist, who will define the workplace in the following stages, considering sociotechnological requirements. It is important to note that information communication complies with process confidentiality requirements, including local regulatory requirements.

3.4.2 Competency and Skills Requirement (Sociotechnical) Analysis

At this stage, the ergonomist, together with the organization and main actors, performs an analysis of the competencies and skills required by the job, tasks, and the workplace. This is done to know which tasks the pregnant worker can perform and at which workplaces. The sociotechnical analysis is relevant to understand, with a systemic strategy, the possibilities to adapt the work to the evolution of pregnancy.

3.4.3 Systemic and Ergonomics Analysis of Jobs, Tasks, and Workplaces

At this stage, the ergonomist describes the job, tasks, and workplace conditions. It is relevant to know the characteristics of strength, precision, or mixed task/s; exposure to ergonomic risk factors; and additional specific levels of criticality.

The joint analysis establishes the following criteria as a general framework, considering functional capacities and system requirements. To describe the criteria of the model, we suggest dividing them into general, specific, and preventive types.

3.4.4 Security and Prevention Analysis of Jobs, Tasks, and the Workplace

At this stage, the ergonomist and the Environmental, Health and Safety (EHS) department execute the analysis related to identified risk factors to the health of pregnant workers. It is relevant to know the characteristics of the exposure related to physical, chemical, biological, and other security risk factors. It is pertinent to prevent the exposure and consider additional specific levels of criticality.

Based on the joint analysis, the following criteria are established as a general framework considering the overall workload. To describe the requirements of the model, we suggest dividing into general, specific, and preventive criteria, similar to ergonomics issues.

3.4.5 Workplace Adaptation Plan

At this stage, the ergonomist defines the specific workplace characteristics related to individual health conditions and normative requirements. In this step, the Model TAPTE is applied, which was responsible for communicating changes or maintaining jobs. The ergonomist was directed to the Preventive Health Unit and EHS Headquarters, and once approved, communication was sent to leadership.

The ergonomist must consider the aspects of functional capacities, ergonomics analysis of the probable area to adapt (consider a new area), safety analysis, and an analysis of sociotechnical requirements. Participatory Ergonomics is applied considering the main actors.

3.4.6 The Criteria of the TAPTE Model Are Applied in this Stage

Execution and follow-up related to the pregnancy-specific characteristics and work conditions.

At this stage, the Preventive Health Unit carries out health and risk factor surveillance work. Necessary adjustments are made as adaptive, and evolutionary intervention criteria are implemented.

3.4.7 Conclusion of the Evolutive Adaptive Process

Typically, pregnant workers leave for prenatal care in the 34th week prior to childbirth. At this moment, the Preventive Health Unit team has concluded this process.

REFERENCES

Abdelwahab, Mahmoud, Costantine, Maged, and Pacheco, Luis. 2022. "Physiologic Changes during Pregnancy." In *Clinical Pharmacology during Pregnancy*, 5–17. San Diego: Academic Press. https://doi.org/10.1016/B978-0-12-818902-3.00008-7.

Hart, Sandra, and Staveland, Lowell E. 1988. "Development of NASA-TLX (Task Load Index): Empirical and Theoretical Research Results." In *Human Mental Workload*, edited by P. A. Hancock and N. Meshkati, 139–183. Amsterdam: Elsevier Science Publishers. https://doi.org/10.1016/S0166-4115(08)62386-9

Jeannette A Paul, Frings-Dresen, Monique H.W., Sallé, Herman J. A., and Rozendal, Rients. H. 1995. "Pregnant Women, Working Surface Height, and Working Surface Areas for Standing Manual Work." *Applied Ergonomics* 26, no. 2: 129–133. https://doi.org/10.1016/0003-6870(95)00008-Z

Morgan Donot et Darío Rodríguez, « Carisma mediático y populismo. El caso de Carlos Menem en Argentina (1989-1999) », *Atlante* [En ligne], 21 | 2024, mis en ligne le 01 octobre 2024, consulté le 21 avril 2026. URL: https://journals.openedition.org/atlante/32412; DOI: https://doi.org/10.4000/130p7

Rodríguez Herrera, Carolina Angélica. 2016. Analysis of Variables for the Construction of a Conceptual Model of Job Adaptation Time According to the Evolution of Gestational Development. Doctoral thesis, Universitat Politècnica de Catalunya, Departament d'Organització d'Empreses. https://doi.org/10.5821/dissertation-2117-96192.

4 Usability/Accessibility of a Free Online Telepsychology Intervention by Patients and Trained Psychotherapists in Mexico

Esteban Eugenio Esquivel Santoveña, Lorena Alejandra Flores Plata, Anabel de la Rosa Gómez, and Pablo D. Valencia

4.1 INTRODUCTION

The development of mental health technological applications and public health emergencies (e.g., COVID-19-pandemic) increased telepsychology intervention use. Part of the challenges of technology-assisted mental health interventions are related to costs and accessibility (Eunkyung et al. 2023). Indeed, usability in telepsychology interventions ensures operating mental health efforts adhering to universal principles such equitable, flexible, simple, and intuitive use of online mental health applications (Martínez-García et al. 2024; Pramuka and van Roosmalen 2009). A client-oriented focus bearing in mind usability/accessibility to online psychological interventions can affect mental health practitioners' performance and clients' participation, adherence, and intervention outcomes (Cillessen et al. 2020; Wozney, Baxter and Newton 2015). The present chapter documents the results on the usability/accessibility reported by trained psychotherapists assisting clients enrolled in a free online telepsychology transdiagnostic intervention aimed at emotional, stress, and trauma-related disorders in Mexico. A three-arm randomized controlled trial (RCT) (control, assisted participants, and self-guided groups) was conducted to recruit participants. End-users reported in this study belong to the self-guided intervention group. Usability was evaluated with 36 participants using a 16-item measure (Jaimez-González 2019). The interface design was positively evaluated. Results show that 66.6% of participants consider that it is easy to advance, 90.9% considered that sufficient information was provided to understand the content, and 87.8% indicate that the order of activities and content is optimal. The evaluation indicates adequate usability, but more research is needed in telepsychology applications.

4.2 TECHNOLOGIES IN MENTAL HEALTH: TELEPSYCHOLOGY

Since the 1990s, technologies in the field of mental health started being implemented; however, this trend clearly became highly relevant since and after the COVID-19 pandemic through the application of telepsychology or cyberpsychology. This ongoing trend has since become very important for patients and psychotherapists reinforcing its use (McClellan et al. 2022; Pierce et al. 2021; Sampaio et al. 2021). Such a sanitary crisis favors the organization of procedures, strategies, ethical use, and best practice recommendations in this area. Despite the lack of a consensus over its regulation,

 DOI: 10.1201/9781003582014-5

it is important to move its research and development forward to improve its applications (De Angelis et al. 2025; Haddouk et al. 2025).

Telepsychology, a widely accepted term defined by the American Psychological Association (APA) as the delivery of psychological services through text, video, voice, cellphones, video calls, text messages, virtual reality, webpages, etc.; it can include prevention, psychological education, diagnostics, consultancy, and treatments through synchronous, asynchronous, or hybrid means. Since 2013, the APA has devised standards to ensure ethical practices, security, and efficacy in the application of telepsychology (American Psychological Association 2024).

Mental health in Mexico requires structures and strategies to address training, attention, and follow-up. The National Geography and Statistics Institute (INEGI as it is known in Spanish) reported in 2021 that around 35 million adult Mexicans have experienced depressive symptoms, while 19.3% have experienced acute anxiety symptoms, and 31.3% report such symptoms to some degree. There is a clear need to address mental health problems, as well as training and supervision, not only in psychology-related topics but also in bridging telepsychology and psychologists/psychotherapists (Sampaio et al. 2021). Such an endeavor not only assists with the goal of developing clinical skills through the use of technologies but also identifying more suitable tools, developments, and systems for patients. Such identification can inform designers, developers, and computer programmers for innovations in mental health that consider usability and acceptability (De Angelis et al. 2025).

To understand it better, research in this area has documented a greater need for user-centered design practices to tackle low adherence/high attrition rates. Some of these recommendations on such design can include adding personalization, improving user interfaces, taking adequate steps to ensure anonymity/privacy/security, including peer engagement, and access to mental healthcare professionals (Oti and Pitt 2021). Specifically, some research has shown that mental healthcare professionals' concerns on telepsychology interventions focus on legal aspects and risk management mechanisms, while younger users' usability concerns focus on transparency and personal autonomy (Wozney, Baxter, and Newton 2015).

4.3 USABILITY IN TELEPSYCHOLOGY DESIGNS

From an Ergonomics or Human Factors (HF) approach, mental health interventions can be improved, be redesigned, and implement strategies that allow for more secure systems for patients and in general for people involved in healthcare systems be implemented (International Ergonomics Association (IEA) 2000). However, despite efforts to incorporate ergonomics into evidence-based healthcare systems, in Latin America, such progress in terms of application and dissemination is rather slow and slightly promising (Aceves-González et al. 2021).

Within ergonomics and HF, five main domains integrated in mental health services have been identified: (a) technology usability, (b) patient security and incidence analysis, (c) security and performance of healthcare professionals, (d) system design, and (e) organizational resilience of the system. Regardless of whether these domains have a medical vision, they can also be applied in the field of psychology (Carayon, Xie, and Kianfar 2014).

With such considerations in accordance with telepsychology, usability has been defined as the extent to which a product can be used by specific users to achieve specific goals with effectiveness, efficiency, and satisfaction within a specified context of use (International Organization for Standardization 2018; Martins et al. 2023). Usability within the field of telepsychology has been used to assess the quality of the user's experience in terms of universal design and user-friendliness, when interacting with a product (i.e., Web app) or system (i.e., telepsychology platform).

Several studies have researched usability applied to tools and mental health-related elements. These studies have documented positive experiences in the implementation of such tools in general system assessments, information, or responses from users, showing benefits for end-users and psychotherapists involved in treatments. There is also a growing debate about the importance of addressing methodologies in more depth to improve assessments and telepsychology treatments (Abuyadek et al. 2024; Casu et al. 2024; De Angelis et al. 2025; Newton et al. 2020).

Scarce research has proposed guidelines/principles to assess the usability and universal designs of online-delivered psychological interventions (Yogarajah et al. 2020), so its incorporation for the case of Latin America will be valid. Despite the importance of usability assessment in the delivery of mental health services through telepsychology, there is a lack of consensus on its assessment (Goundar et al. 2022). Currently, questionnaires remain the most common means to assess the usability of electronic health apps. Such means provide a general measure of usability but do not detect specific issues to be addressed (Maramba et al. 2019).

In this regard, the rapid growth of digital mental healthcare solutions presents an opportunity to bridge gaps in mental healthcare access, particularly in regions with limited infrastructure. However, despite the benefits of increased service utilization, shorter waiting times, and improved accessibility, especially in remote areas, many digital mental health interventions still face critical usability challenges (Wibowo et al. 2024). Issues such as confusing navigation, high cognitive load, and lack of personalization contribute to user frustration and disengagement, leading to high attrition rates. Addressing these design flaws is essential to ensuring that digital mental health tools not only reach users but also sustain their engagement over time. By integrating user-centered design (UCD) principles and optimizing user interface (UI) and user experience (UX), these tools can enhance adherence and overall effectiveness.

Hence, this chapter has endeavored to investigate the usability experiences of psychotherapists and end-users of a telepsychology platform developed and disseminated in Mexico. These efforts are in line with the application of knowledge, in this case, of human emotional/cognitive capabilities through the development and application of a telepsychology platform, as part of human factors of interest in mental health and ergonomics (Harte et al. 2017).

4.3.1 Study Aims

To investigate the usability of a telepsychology intervention for emotion-, trauma-, and stress-related disorders by trained Mexican psychologists.

To describe the patients' levels of usability of an online psychological intervention by patients by delivery mode (self-administered versus supported by a trained psychologist).

4.4 METHODS SECTION

Data presented here are part of an RCT registered at ClinicalTrials.gov NCT055225701, which aimed at evaluating the efficacy and acceptability of an online assisted transdiagnostic intervention versus a self-administered online intervention aimed at stress- and trauma-related disorders and a waiting list of Mexican participants. Usability was initially measured by therapists within a training platform and later measured within the transdiagnostic intervention platform by end-users. Ethical approval was gained from the Ethics Committee from the Faculty of Higher Studies Iztacala at the National Autonomous University of Mexico (CE/FESI/082020/1363) with strict adherence to the Psychologist Ethical Code (Sociedad Mexicana de Psicología, 2010) and to the APA's guidelines for telepsychology practice (American Psychological Association 2024). All participants provided informed consent, while all confidentiality and privacy of their personal information were secured.

4.5 STUDY DESIGN AND PARTICIPANTS

Data in the present study were obtained from two samples. One of these samples (Sample 1) is composed of 36 psychotherapists successfully trained in an intervention based on a transdiagnostic model and in an online-delivered psychological support platform. Most of them are female

TABLE 4.1
Sociodemographics of Participants Seeking Psychological Support from a Telepsychology Intervention

Sociodemographic	Men, *n* (%)	Women, *n* (%)
Relationship Status		
Single		
Single	7 (63.6)	14 (27.5)
Cohabiting	1 (9.1)	9 (17.6)
Married	3 (27.3)	17 (33.3)
Divorced	0	5 (9.8)
Separated	0	4 (7.8)
Other	0	2 (3.9)
Educational attainment		
Middle school	1 (9.1)	1 (2)
High school	2 (18.2)	10 (19.6)
Undergraduate	7 (63.6)	32 (62.7)
Master's degree	1 (9.1)	8 (15.7)
Income (monthly, Mexican pesos)		
$7500 or less	4 (36.4)	14 (27.5)
$7500 to $10,000	0	14 (27.5)
10,000 to $15,000	3 (27.3)	6 (11.8)
$15,000 to $20,000	1 (9.1)	4 (7.8)
More than $20,000	2 (18.2)	3 (5.8)
Would not want to disclose	1 (9.1)	10 (19.6)

(83.3%), and 16.7% are male. Most participants (88.9%) are affiliated with the National Autonomous University of Mexico in the State of Mexico. Participants come from other regions in Mexico, including Mexico City (25%), Puebla (11.1%), Oaxaca (8.3%), and Veracruz (2.8%), as well as one Mexican participant from San Diego, California (2.8%).

In terms of age, participants range from 23 to 71 years, with the average age around 41 years. Most females (55.6%) are in the 30- to 50-year age range, while half of the males are in their late 20s (27–29 years old). In terms of educational level, 88.9% of participants have a bachelor's degree, while 11.1% have a postgraduate degree.

A second sample (Sample 2) is composed of 62 individuals (82.3% were female participants; 17.7% were males) who sought psychological support from a free online transdiagnostic intervention aimed at trauma- and stress-related disorders. Most of them were female participants and held a university degree. Around a third of all participants (29%) earned below the minimum wage back in 2023, most male participants were either single, and most women were either married or single (see Table 4.1).

4.6 RECRUITMENT PROCESS

For sample 1, Mexican undergraduate psychology students and mental healthcare professionals declaring knowledge in cognitive behavioral therapy (CBT) were invited/recruited through links in different higher education institutions such as the Faculty of Higher Studies Iztacala and Faculty of Psychology, UNAM, Technological Institute of Sonora, Autonomous University of Ciudad Juárez, Faculty of Health Sciences of the Autonomous University of Baja California, and the Tlaxcala Ministry of Health. Selected participants met a screening process that included pre- and post-training assessments on the intervention online platform.

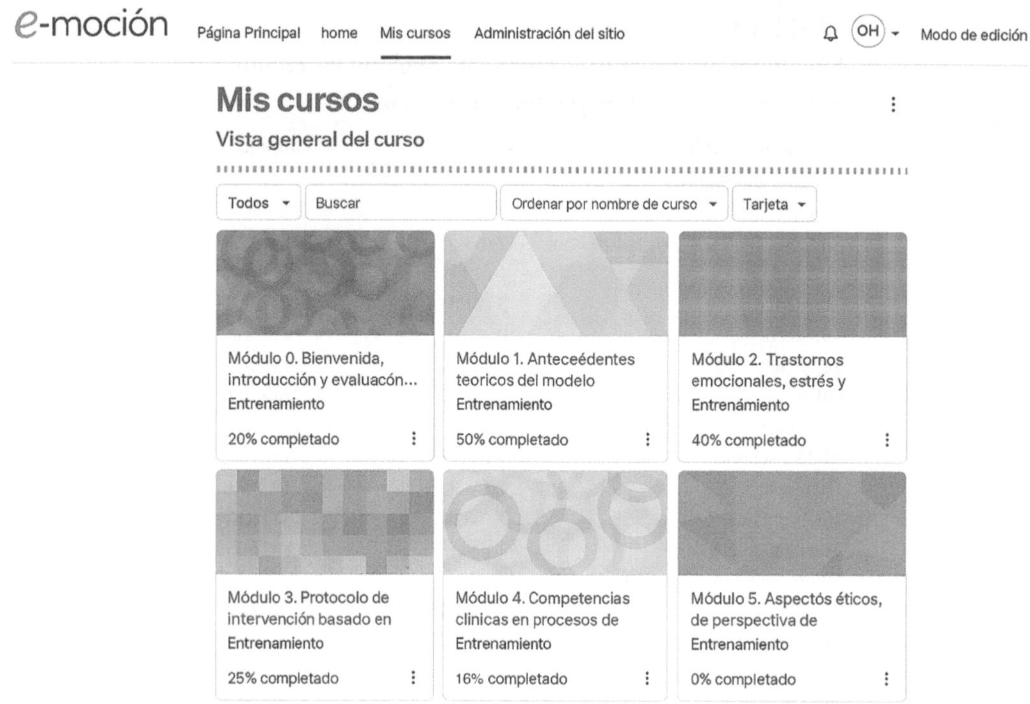

FIGURE 4.1 Recruitment process.

For sample 2, Mexican participants seeking psychological support were recruited via advertisements in digital media and through dissemination on social networks and contacted through e-mail of telephone calls (De la Rosa-Gómez et al. 2022). These participants provided their information at intake.

4.6.1 Psychotherapists Training Platform Design

Training of sample 1 participants was conducted through a self-administered Moodle platform aimed at healthcare professionals. The 40-hour training focused on treatment of trauma- and stress-related disorders based on a transdiagnostic model. The training program was composed of five modules: (0) Introduction and assessment, (1) Transdiagnostic model background, (2) Definition of emotional disorders, (3) Transdiagnostic intervention protocol aimed at stress- and trauma-related disorders, (4) Clinical skills in online interventions processes, and (5) Ethical aspects (gender and human rights focus) in mental health telepsychology services (see Figure 4.1). All modules were (a) laid out presenting evidence-based topics to be covered in every module, (b) with learning reinforcement/follow-up activities to reflect and review necessary topics, (c) with an e-book–delivered intervention protocol of a patient to learn about his/her intervention aims, and (d) a self-assessment. A minimum passing grade of 8 (on a scale of 0–10) was required to successfully complete the training.

4.7 END-USER INTERVENTION PLATFORM DESIGN

The telepsychology intervention system of trauma- and stress-related disorders is an online platform accessible in a variety of electronic devices such as personal computers, laptop computers, tablets, and cellular phones. The system analyzes and matches the profiles of psychotherapists and potential patients, assigning them to the intervention modality according to the research/intervention randomization process. The intervention system is designed to be completed between 8 and

12 weeks, with participants accessing and making progress at their own pace. The modules cover the following areas: “Module 0: Pre-assessment”, “module 1: Goals and motivations for change”, “module 2: Emotional coping skills”, “module 3: Acceptance and sensibilization skills focus in the present”, “modules 4 and 5: Cognitive coping skills”, “modules 6 and 7: Behavioral coping skills”, “module 8: Achievements, maintenance, and relapse prevention”, and “module 9: Post-assessment”. Each module included psychoeducational exercises and practical assignments, while the platform provided alerts to participants to keep up to date with their progress.

4.8 MEASURES

Usability of the training platform (for sample 1) was assessed through a 16-item instrument that evaluates the following aspects: esthetic design, functionality, and utility. The response options range from (a) very suitable, (b) suitable, (c) not very suitable, and (d) not suitable at all (Jaimez-González 2019).

Perception of the course content was assessed with a previously designed questionnaire that consisted of eight questions that enquired about the degree of difficulty in navigating through the five modules, the principles and activities of the transdiagnostic model, order and content of course activities, attention received to resolve doubts, self-perception to serve users under the unified protocol in the online telepsychology intervention, and limitations and general comments.

4.8.1 System Usability Scale: SUS

Brooke (1996) perceived the usability of the intervention platform from end-users (sample 2), which was assessed through a 10-item measure, within a 5-point scale (1=totally disagree to 5=totally agree). SUS scores can range from 0 to 100, with a possible single composite index measuring the overall usability of the system or platform. The Alpha reliability coefficient of the SUS in the present study is $\alpha=0.78$.

4.9 USABILITY

The esthetic design of the interface was generally rated positively. For example, participants considered that the organization of the information, buttons, and windows is adequate, as well as the colors, size, and typography used (see Table 4.2). Images used in animations are also deemed appropriate (see Figure 4.2). The most cited adjectives to describe the images were appropriate, clear, dynamic, illustrative, and eye-catching. Functionality of the didactic material was considered easy to use, with instructions and sequences of actions in appropriate animations. Most participants (97%) considered that the material is adequate in general, including the graphic metaphors used, indicating that the material reinforced their previous knowledge.

4.10 PERCEPTION OF THE COURSE

Two-thirds of participants considered that it was easy to go through the five modules, and 90.9% considered that enough information was provided to understand the principles, activities, and tasks of the transdiagnostic model. In addition, 87.8% estimated that the order of the activities and contents helped achieve the objectives of the course, 72.7% expressed that the attention received to clarify doubts was adequate, while 90.8% of participants considered that they felt sufficiently prepared to assist a patient under the unified protocol. Additional practice turned out to be the most common area of opportunity of the course (50%), followed by supervision and content enrichment (both 15.4%) and session modeling (3% and 8%).

TABLE 4.2
Usability of the Clinical Skills Training Platform

Esthetic Design	Very Suitable, N (%)	Suitable, N (%)	Not Very Suitable, N (%)	Not Suitable At All, N (%)
1. Do you consider that the organization of the information in the interface is suitable?	13 (39.4)	19 (57.6)	1 (3)	–
2. Do you consider that the displayed buttons and windows that are displayed are suitable?	16 (48.5)	17 (51.5)	–	–
3. Do you consider that the colors used are suitable?	20 (60.6)	13 (39.4)	–	–
4. Do you consider that the size and font used are suitable?	18 (54.5)	12 (36.4)	3 (9.1)	–
5. Do you consider the images used in the animations suitable?	23 (69.7)	9 (27.3)	1 (3)	–
6. Do you consider the overall interface suitable?	16 (48.5)	17 (51.5)	–	–
Functionality				
8. Do you consider that the instructions given for the use of the didactic material are suitable?	11 (33.3)	21 (63.6)	1 (3)	–
9. Do you consider that the sequence of actions in each of the animations is suitable?	21 (63.6)	11 (33.3)	1 (3)	–
10. Do you consider the error messages displayed suitable?	12 (36.4)	17 (51.5)	3 (9.1)	1 (3)
Utility				
12. Do you consider that the didactic material is suitable for understanding the subject of Emotional Disorders and Derivatives of Stress and Trauma?	20 (60.6)	11 (33.3)	2 (6.1)	–
Do you consider that the didactic material is suitable for learning the topic of Emotional Disorders and Derivatives of Stress and Trauma?	19 (57.6)	14 (42.4)	–	–
15. Do you think that the use of the graphic metaphors used in the didactic material is suitable?	21 (63.6)	11 (33.3)	1 (3)	–

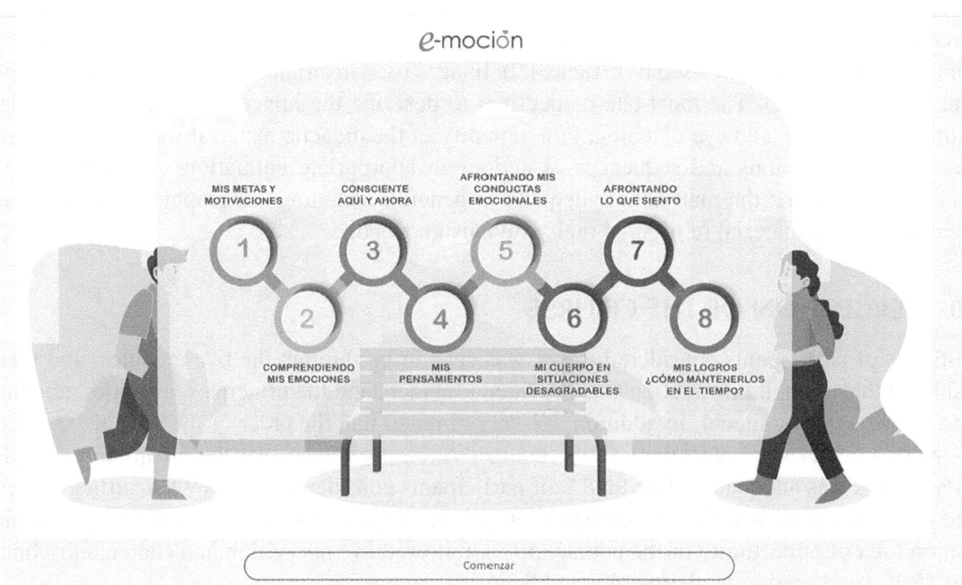

FIGURE 4.2 Emotional self-management program interface for workers.

TABLE 4.3
End-User Usability Levels

SUS Items	Totally Disagree (%)	Disagree (%)	Do Not Agree/ Disagree (%)	Agree (%)	Totally Agree (%)
I think that I would like to use this system frequently	0	1.6	4.8	36.5	55.6
I found the system unnecessarily complex	42.9	33.3	15.9	3.2	3.2
I thought the system was easy to use	0	0	3.2	44.4	50.8
I think that I would need the support of a technical person to be able to use this system	27	22.2	14.3	20.6	14.3
I found the various functions in this system were well integrated	0	3.2	9.5	42.9	42.9
I thought there was too much inconsistency in this system	49.2	36.5	9.5	3.2	0
I would imagine that most people would learn to use this system very quickly	0	0	11.1	49.2	38.1
I found the system very cumbersome to use	38.1	41.3	12.7	4.8	1.6
I felt very confident using the system	0	0	9.5	34.8	54
I needed to learn a lot of things before I could get going with this system	50.8	31.7	9.5	4.8	1.6

4.10.1 End-Users' Usability Reports of the Online Transdiagnostic Psychological Platform

A chi-square analysis (not shown) from 62 participants on items from the SUS by sex of the participant, age, marital status, educational attainment level, and type of occupation revealed no significant differences among the tested groups. Such consistency in participants' scores regardless of the sociodemographic category attests to the positive evaluation that participants express in the SUS. Most participants reported a positive perception of the SUS, particularly in those items regarding user-friendliness and willingness to use it frequently. In general, such a positive outlook of the usability screening tool is accessible for a wide array of end-users.

Most end-users of the online transdiagnostic intervention for stress- and trauma-related disorders, positively rated the intervention platform. Many users agreed that the system was easy to use, its functions were well integrated, that they felt very confident using it, and in general, it was perceived in a user-friendly fashion. Interestingly, some users (around a third of them) believed they would need a person for technical support with the system (see Table 4.3).

In general terms, high scores in items such as "I would like to use this tool frequently" (m=4.5) and "I consider that the tool is easy to use" (m =4.3) show a favorable perception of users. Conversely, low scores on negative items such as "I consider that the tool is unnecessarily complex" (m=1.8) and "using the tool is tedious" (m=2.1) mirror that participants do not perceive the tool to be difficult nor complex to use. A low dispersion of participant's responses on positive items underscores a positive perception, consistent with participants' responses.

Results about usability levels show that 80% of participants scored within an "acceptable" range, while only a marginal 15% scored within a marginal level, and only one individual obtained a "non-acceptable" score level. These results suggest that the tool is in general welcomed by users, although specific areas can be improved to better streamline its use.

4.11 DISCUSSION

The present study conducted in Mexico evaluated the usability of a free online transdiagnostic telepsychology intervention aimed at addressing emotion-, stress-, and trauma-related disorders. The usability assessment, conducted with 36 participating psychotherapists using a 16-item measure

(Jaimez-González 2019), revealed positive results: 66.6% of users found it easy to navigate, 90.9% reported that the information provided was sufficient to understand the content, and 87.8% indicated that the order of activities and content was optimal. Information from end-users echoes this trend, being that most of them found the transdiagnostic telepsychology platform easy to use, they would use it frequently, and felt confident while navigating in it, while some end-users expressed that they might need technological support. While these findings suggest adequate usability, they also highlight the need for continued research and optimization of telepsychology applications to enhance engagement and effectiveness. These findings highlight the importance of usability and accessibility in digital mental health interventions, particularly in the context of telepsychology, and reinforce the critical role of UCD in the development of digital mental health solutions, ensuring that interventions are accessible, user-friendly, and capable of fostering sustained engagement.

On the one hand, the design of effective digital solutions should follow usability standardization principles (Diehl et al. 2022; Goundar et al. 2022) to be inclusive toward a broad and diverse audience. Nonetheless, the ever-increasing heterogeneous audience of digital solutions calls for a context-oriented approach in which careful user requirement analysis is needed. Importantly, the complexity of usability evaluation is underlined by the need to assess the users and their relationship with information technology and the environment, carefully incorporating contextual factors impacting the adoption in natural settings (Yen and Bakken 2012). In this sense, experts have stressed the need for designing highly effective, easy-to-use, engaging, and trustworthy tools to fully benefit from the results of digital mental health solutions (De Angelis et al. 2025; Roland et al. 2020).

This context-driven approach to usability and acceptability testing is even more relevant to the work context, primarily when the technology investigated covers potentially stigmatized topics such as mental health. The importance of UCD becomes particularly evident in digital mental health interventions, as demonstrated in low- and middle-income countries, where usability and accessibility play a crucial role in adoption and engagement. In Latin America, where structural barriers such as limited internet access, digital literacy gaps, and cultural perceptions about mental health can hinder the effectiveness of digital interventions, designing with usability in mind is essential to ensure sustained engagement and reduce high dropout rates. Studies suggest that interventions designed with UCD principles, such as the Double Diamond model, improve adherence by prioritizing user needs, iterating based on real-world feedback, and developing intuitive, culturally adapted interfaces that resonate with local populations (Domínguez-Rodríguez and De la Rosa-Gómez 2022).

Moreover, research on digital behavior change interventions (DBCIs) highlights the need to integrate both intervention design (INTD), based on psychological theories of behavior change, and interaction design (IxD), which focuses on user experience and usability, to ensure effective digital interventions. While DBCI frameworks frequently incorporate psychological principles for intervention design, they often lack structured IxD methodologies such as user research, prototyping, and usability testing. This knowledge gap is particularly relevant in Latin America, where usability issues can significantly impact adoption and engagement. The absence of clear guidance for IxD designers in health-related DBCIs underscores the need for more structured frameworks that integrate usability testing and goal-directed design principles to create more accessible and engaging digital interventions.

Furthermore, high attrition rates, often reaching 50% in self-applied digital mental health interventions (Domínguez-Rodríguez and De la Rosa-Gómez 2022), highlight the importance of a seamless and motivating user experience. A lack of personalization, low perceived relevance, and technological friction contribute to user disengagement. To address this, successful eHealth tools must not only be functionally efficient but also incorporate engagement strategies such as gamification, adaptive content, and feedback mechanisms that foster long-term user commitment. The use of goal-directed design (GDD) principles, which align digital intervention features with user goals and motivations, has been proposed as a promising approach to enhance adherence and impact in health-related DBCIs. Additionally, in a region where mobile internet is often more prevalent than broadband connections, optimizing digital interventions for mobile-first experiences with offline capabilities can further improve accessibility and sustained participation.

Optimizing UI and UX is particularly relevant for mental health apps as these factors directly impact user engagement and adherence. A systematic review of mental health apps revealed common design issues such as confusing navigation, high cognitive load, and lack of personalization, which can contribute to high dropout rates (Wibowo et al. 2024). Addressing these challenges requires design improvements, including intuitive navigation, customization options, age-appropriate gamification, and clear interactive instructions. Importantly, UI and UX improvements in the present study allowed for some of these concerns to be addressed (for example, ease in navigation, clear instructions, and customization of response/interaction buttons). This work, combined with the integration of a transdiagnostic protocol in this intervention, represents a step toward improving access to mental health services. Additionally, essential features such as user connectivity, anonymity, support for logging mental health progress, and robust data privacy protections are necessary to build trust and encourage sustained use of mental health platforms. In Latin America, where access to traditional mental health services is often limited, digital solutions must prioritize these usability factors to enhance accessibility and engagement. By implementing structured UI/UX methodologies and integrating user feedback throughout the development process, mental health apps can become more effective in supporting individuals in their psychological well-being.

As a result, the UCD of eHealth tools, which requires users' involvement in the developmental process, has been increasingly deployed through various methods, including, mining user reviews of mental health apps to understand engagement patterns and usability pain points (Alqahtani and Orji 2019). Ensuring that usability principles are deeply embedded in digital mental health solutions is not just a technical consideration but a fundamental requirement for making these interventions truly effective and impactful in Latin America, where mental health support remains critically underfunded and underutilized.

REFERENCES

Abuyadek, Rowan M., et al. 2024. "Acceptability of Tele-Mental Health Services among Users: A Systematic Review and Meta-Analysis." *BMC Public Health*, 24: 1–18. https://doi.org/10.1186/s12889-024-18436-7

Aceves-González, Carlos, Irma Cecilia Landa-Ávila, Fernando Carvalho, Bertha Alicia Ortega-Ruiz, and Gyuchan Thomas Jun. 2021. "Ergonomía en los sistemas de salud de América Latina: Revisión sistemática de la situación actual, necesidades y desafíos futuros." *Ergonomía, Investigación y Desarrollo*, 3: 10–27.

Alqahtani, Felwah H, and Rita Orji. 2019. "Usability Issues in Mental Health Applications." In *Usability Issues in Mental Health Applications*. Larnaca Cyprus: Association for Computing Machinery, 343–348.

American Psychological Association. 2024. *APA Guidelines for the Practice of Telepsychology*. https://www.apa.org/about/policy/telepsychology-revisions (accessed 01 2025).

Brooke, John. 1996. "SUS—A Quick and Dirty Usability Scale." In *Usability Evaluation in Industry*, by Patrick W Jordan, Bruce Thomas, Bernard A Weerdmeester and Ian L McClelland, 189–194. London: Taylor & Francis.

Carayon, Pascale, Anping Xie, and Sarah Kianfar. 2014. "Human Factors and Ergonomics as a Patient Safety Practice." *BMJ Quality & Safety*, 23: 196–205. https://doi.org/10.1136/bmjqs-2013-001812

Casu, Mirko, Sergio Triscari, Sebastiano Battiato, Luca Guarnera, and Pasquale Capponetto. 2024. "AI Chatbots for Mental Health: A Scoping Review of Effectiveness, Feasibility, and Applications." *Applied Sciences*, 14: 1–23. https://doi.org/10.3390/app14135889

Cillessen, Linda, Monique Om van de Ven, Félix R. Compen, Else M Bisseling, Marije L. van der Lee, and Anne M. Speckens. 2020. "Predictors and Effects of Usage of an Online Mindfulness Intervention for Distressed Cancer Patients: Usability Study." *Journal of Medical Internet Research*, 22: 1–11. https://doi.org/10.2196/17526

De Angelis, Marco, Lucia Volpi, Davide Giusino, Luca Pietrantoni, and Federico Fraboni. 2025 "Acceptability and Usability of a Digital Platform Promoting Mental Health at Work: A Qualitative Evaluation." *International Journal of Human–Computer Interaction*, 41(13): 1246–1259.

De la Rosa-Gómez, Anabel, et al. 2022. "Efficacy of a Transdiagnostic Guided Internet-Delivered Intervention for Emotional, Trauma and Stress-Related Disorders in Mexican Population: Study Protocol for a Randomized Controlled Trial." *BMC Psychiatry*, 22: 1–12. https://doi.org/10.1186/s12888-022-04132-6

Diehl, Ceci, et al. 2022. "Defining Recommendations to Guide User Interface Design: Multimethod Approach." *JMIR Human Factors*, 9(3): 1–12. https://doi.org/10.2196/37894

Domínguez-Rodríguez, Alejandro, and Anabel De la Rosa-Gómez. 2022. "A perspective on how user-centered design could Improve the Impact of Self-Applied Psychological Interventions in Low- or Middle-Income Countries in Latin America." *Frontiers in Digital Health*, 4: 1–6. https://doi.org/10.3389/fdgth.2022.866155

Eunkyung, Jo, Whitney-Jocelyn Kouaho, Stephen Schueller, and Daniel A Epstein 1. 2023. "Exploring User Perspectives of and Ethical Experiences with Teletherapy Apps: Qualitative Analysis of User Reviews." *JMIR Mental Health*, 10: 1–13. doi:10.2196/49684

Goundar, Munil Shiva, Bimal Aklesh Kumar, and A B M Shawkat Ali. 2022. "Development of Usability Guidelines: A Systematic Literature Review." *International Journal of Human–Computer Interaction*, 40: 1298–1316. https://doi.org/10.1080/10447318.2022.2141009

Haddouk, Lise, Benoit Schneider, Tom van Daele, and Fredi Lang. 2025. "Telepsychology in France Following the COVID-19 Crisis." *European Journal of Psychology Open*, 84(1): 13–19. https://doi.org/10.1024/2673-8627/a000063

Harte, Richard, et al. 2017. "A Human-Centered Design Methodology to Enhance the Usability, Human Factors, and User Experience of Connected Health Systems: A Three-Phase Methodology." *JMIR Human Factors*, 16(4): 1–21. doi:10.2196/humanfactors.5443

Instituto Nacional de Estadística y Geografía (INEGI). 2021. *Presenta INEGI resultados de la primera Encuesta Nacional de Bienestar Autorreportado (ENBIARE) 2021* (Press conference number 772/21). https://www.inegi.org.mx/contenidos/saladeprensa/boletines/2021/EstSociodemo/ENBIARE_2021.pdf

International Ergonomics Association (IEA). 2000. *What's Ergonomics?* https://iea.cc/about/what-is-ergonomics/ (accessed 01 2025).

International Organization for Standardization. 2018. ISO 9241-11:2018(en). https://www.iso.org/obp/ui/en/#iso:std:63500:en (accessed 01 2025).

Jaimez-González, Roberto. 2019. "Evaluación de recursos didácticos en línea para apoyar el proceso enseñanza-aprendizaje de la programación web con JavaScript y Java Server Pages". *Revista Dilemas Contemporáneos*, 3(54): 1–23. https://dilemascontemporaneoseducacionpoliticayvalores.com/index.php/dilemas/article/view/1784/1956

Maramba, Inocencio, Arunangsu Chatterjee, and Craig Newman. 2019. "Methods of Usability Testing in the Development of eHealth Applications: A Scoping Review." *International Journal of Medical Informatics*, 126: 95–104. https://doi.org/10.1016/j.ijmedinf.2019.03.018

Martínez-García, Laura, Alba Fadrique-Jimenez, Vanesa Ferreres-Galan, Cristina Robert-Flors and Jorge Osma. 2024. "RegulEm, An Unified Protocol Based-App for the Treatment of Emotional Disorders: A Parallel Mixed Methods Usability and Quality Study." *BMC Medical Informatics and Decision Making*, 24 (267): 1–14. https://doi.org/10.1186/s12911-024-02679-w

Martins, Ana Isabel, et al. 2023. "Consensus on the Terms and Procedures for Planning and Reporting a Usability Evaluation of Health-Related Digital Solutions: Delphi Study and a Resulting Checklist." *Journal of Medical Internet Research*, 25: 1–13. https://doi.org/10.2196/44326

McClellan, Michael J., et al. 2022. "The Effectiveness of Telepsychology with Veterans: A Meta-Analysis of Services Delivered by Videoconference and Phone." *Psychological Services*, 19(2): 294–304. https://doi.org/10.1037/ser0000522

Newton, Amanda, et al. 2020. "A Mobile Phone-Based App for Use during Cognitive Behavioral Therapy for Adolescents with Anxiety (MindClimb): User-Centered Design and Usability study." *JMIR Mhealth and Uhealth*, 8(12): 1–18. https://doi.org/10.2196/18439

Oti, Olugbenga, and Ian Pitt. 2021. "Online Mental Health Interventions Designed for Students in Higher Education: A User-Centered Perspective." *Internet Interventions*, 26: 1–14. https://doi.org/10.1016/j.invent.2021.100468

Pierce, Bradford S., Paul B. Perrin, Carmen M. Tyler, Grace B. McKee, and Jack D. Watson. 2021. "The COVID-19 Telepsychology Revolution: A National Study of Pandemic-Based Changes in U.S. *Mental Health Care Delivery." American Psychologist*, 76(1): 14–25. https://doi.org/10.1037/amp0000722

Pramuka, Michael, and Linda van Roosmalen. 2009. "Telerehabilitation Technologies: Accessibility and Usability." *International Journal of Telerehabilitation*, 1(1): 85–98. https://doi.org/10.5195/ijt.2009.6016

Roland, Jonty, Emma Lawrance, Tom Insel, and Helen Christensen. 2020. *The Digital Mental Health Revolution. Transforming Care through Innovation and Scale-Up*. Doha, Qatar: WISH 2020 Forum on Mental Health.

Sociedad Mexicana de Psicología. 2010. Código Ético del Psicólogo. CDMX: Trillas.

Sampaio, Mariana, Maria Vicenta Navarro-Haro, Bruno De Sousa, Wilson Vieira-Melo, and Hunter G. Hoffman. 2021. "Therapists Make the Switch to Telepsychology to Safely Continue Treating Their Patients During the COVID-19 Pandemic. *Virtual Reality Telepsychology May Be Next." Frontiers in Virtual Reality*, 1: 1–17. https://doi.org/10.3389/frvir.2020.576421

Wibowo, Bagas, Paulus Insap Santosa, and Syukron Abu Ishaq Alfarozi. 2024. "A Survey Study of Strategies for Improving User Interface in Mental Health." 2024 International Conference on Information Technology and Computing (ICITCOM). Yogyakarta, Indonesia: IEEE. 48–53.

Wozney, Lori, Pamela Baxter, and Amanda S. Newton. 2015. "Usability Evaluation with Mental Health Professionals and Young People to Develop an Internet-Based Cognitive-Behaviour Therapy Program for Adolescents with Anxiety Disorders." *BMC Pediatrics*, 15: 1–11. https://doi.org/10.1186/s12887-015-0534-1

Yen, Po-Yin, and Suzanne Bakken. 2012. "Review of Health Information Technology Usability Study Methodologies." *Journal of the American Medical Informatics Association*, 19(3): 413–422. https://doi.org/10.1136/amiajnl-2010-000020

Yogarajah, Aravinthan, Robin Kenter, Yngve Lamo, Viktor Kaldo, and Tine Nordgreen. 2020. "Internet-Delivered Mental Health Treatment Systems in Scandinavia - A Usability Evaluation." *Internet Interventions*, 20: 1–10. https://doi.org/10.1016/j.invent.2020.100314

Section 2

Ergonomics and Accessibility in Technology and Education

5 Ergonomic Approach for Describing the User Experience (UX) in Elder Adults While Interacting with the Graphical Use Interface (GUI) of a Fintech Application

Juan Ángel Vargas, Fernanda Maradei, and Luis Eduardo Bautista

5.1 INTRODUCTION

The modernization of daily activities, along with the expansion of access to technological devices, has allowed people to carry out activities remotely and asynchronously. Some companies have moved their operations to digital environments to optimize their processes, reach previously hard-to-access populations, and streamline procedures that used to require physical presence (Campos, 2014). One sector that has joined this trend is the financial sector, through digital wallets and platforms for the virtual payment of products, goods, and services. The magazine *Forbes* indicates that nearly 53% of the U.S. population makes payments through digital wallets more frequently than with traditional physical payment methods, where younger participants showed nearly double the likelihood of using a digital wallet app compared to their older counterparts (Claypool, 2023).

On the other hand, the adult population over 60 years of age, which constitutes approximately 13.8% of the Colombian population, and by the year 2030, around 16% (Ministerio de Salud y Protección Social, 2021), is characterized by a series of conditions associated with the aging process, reflected in the decline of physical and cognitive abilities. According to Delgado Silveira et al. (2015), individuals over 70 years of age commonly experience muscle weakness, low energy endurance, constant fatigue, slow mobility, and low levels of physical activity, whose physiological changes in the body affect cognitive processes associated with information reception and the deterioration of episodic memory (Curilem et al., 2022). These deficiencies tend to grow over time as aging is a continuous and inevitable process that, in turn, affects the person's emotional state and their relationship with the environment (Vélez et al., 2019). While the digitalization of daily processes represents an advantage for companies, the literature reveals that people over 65 years of age encounter difficulties when using digital platforms. Approximately 43% of this age group perceives websites as complicated to navigate, resulting in feelings of frustration and an unsatisfactory user experience (UX) (Gothelf & Seiden, 2016). Additionally, part of this population perceives these technologies as barriers rather than facilitators (Dubois, 2018), leading to approximately 14% of this population completely avoiding procedures available only through digital means (Gonzalo, 2021). Moreover, they experience feelings of insecurity, concern, and distrust regarding the privacy of their personal data (Ware et al., 2017).

DOI: 10.1201/9781003582014-7

Although the use of these digital platforms does not require mastery of specific skills, it is important that the system provides basic knowledge about the use of the software and the understanding of the tasks to be performed. Therefore, if not carried out correctly, they could partially or entirely limit access to information and communication technologies. In this sense, the ISO 9241-210 standard of 2019 defines the Graphical User Interface (GUI) as the set of interactive components of a system, both in software and hardware, that provides information and control to the user to perform a specific task with the system. In turn, it defines UX within the perception and responses of the person, resulting from their interaction with the product, system, service, design, or manual (ISO Standard, 2019). In this way, UX-based designs require an understanding the expectations, needs, and motivations of users to ensure their ease of use (Gothelf & Seiden, 2016). For this, evaluation methodologies such as product ergonomics, cognitive ergonomics, usability, and web accessibility are employed (W3C, 2024). The literature offers essential guidelines for developers and designers with the aim of creating quality websites and web tools. These guidelines, expressed in the form of templates or design patterns (Pacholczyk, 2014), are formulated around reducing the difficulty in the sequence of actions that a person performs to complete a task within the GUI (Calderón, 2022). However, if not used correctly, they can generate confusion, lower performance, and low user satisfaction (Véliz et al., 2010).

Taking this into account, the aim was to describe and analyze the behaviors of older adults when using a mobile application, based on their desires, needs, abilities, and limitations, using the product ergonomics methodology for special populations. A descriptive study focused on the elderly was conducted, where the individuals' responses were captured while interacting with a mobile application for personal financial management. In this way, through the collection of quantitative and qualitative data, behaviors and emotions of the elderly were identified, related to design patterns, GUI elements, and workflow steps present in the study subject. In this sense, this research provides information that can be used in the development of future projects or digital products with a UX focus as the findings provide a solid foundation for programmers and interface designers, allowing them to better understand the needs of this population sector and implement efficient solutions that aligned with the capabilities and skills of this aging group.

5.2 METHODOLOGY

The objective of this study was to describe the behavior of older adults when using the GUI of a mobile application for financial transactions, in terms of the UX and its relationship with the visual elements, when performing a specific task. The methodology used was based on product ergonomics for special populations, divided into two aspects or parts, as shown in Figure 5.1.

5.2.1 Quantitative Data Measurement

The quantitative data were obtained with the SMI Glasses mobile EyeTracker (IMotions, 2025) and using BeGaze 3.7 software to describe and understand the participant's gaze behavior, obtaining an objective measurement (Rojas López, 2016), expressed in the following metrics:

- *Number of fixations*: Density referred to the total number of gaze points divided by the total minimum area where they converge (Cowen et al., 2002).
- *Duration of fixations*: Fixed gaze on a specific area of a stimulus. For the study, the duration of fixation was considered because if it is high, it indicates a high workload and cognitive effort (Rayner, 1998).
- *Pupillary dilation*: Changes in pupil size. If they are very continuous, they imply a high cognitive load (Beatty, 1982). For the study, the values of pupil diameter were considered within the time range of task execution to determine the value of the difference in its range.

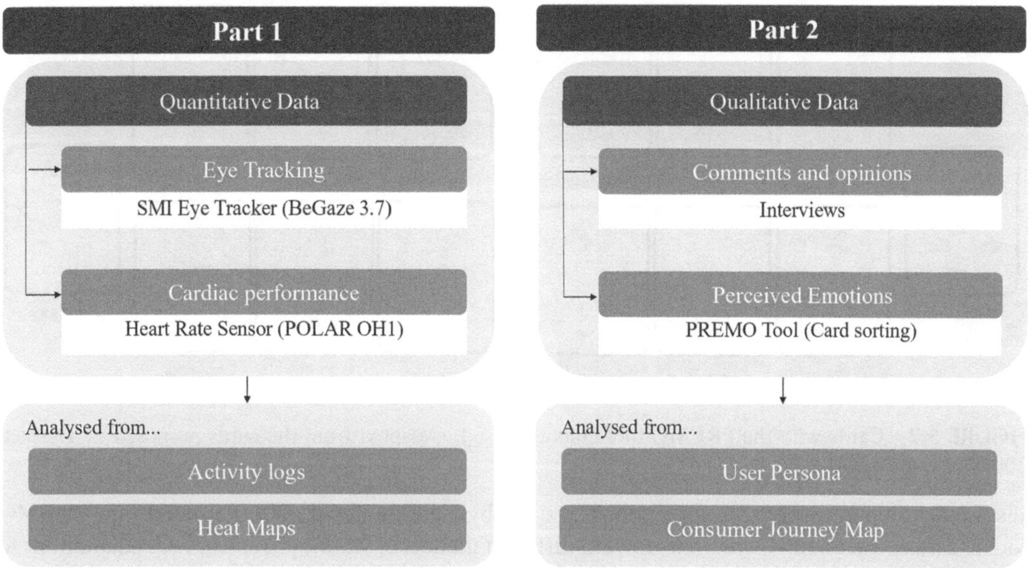

FIGURE 5.1 Description of the data obtained in both sections of the study. (Authors' own creation.)

- *Number of saccades*: Rapid movements of the eye when moving from one fixation to another. They do not process visual information (Shojaeizadeh et al., 2016) but allow tracing a path of eye movement over the stimulus.
- *Number of blinks*: Semi-involuntary action of rapid closure and reopening of the eyelid, produced due to the co-inhibition of the protractor and retractor muscles of the eyelid. This metric can provide useful information about the tendency of the viewer to pay more attention to a specific location in a stimulus (Hollander & Huette, 2022).

Each of these metrics was accounted for and graphically represented in a set of areas of interest (AOI) established in the study subject, grouping the interaction elements of each screen in the workflow of the assigned task.

Likewise, to know the physiological reactions of the participants when they execute the task with the application, cardiac performance data were obtained, with beats per minute (bpm). An optical heart rate sensor in a POLAR OH1 reference bracelet was used as evidence shows a relationship between the variation in the heart rate and the level of stress of the person (Moya-Albiol et al., 2005).

5.2.2 Qualitative Data Measurement

The qualitative data were obtained through semi-structured individual interviews, using physical cards with 14 emotions defined in the PREMO (Product Emotion Measurement Instrument) tool (Desmet, 2004). Although this tool has been implemented for different assessments commonly in Europe and some North American contexts, its use in the population of adults over 60 years of age has been documented in multiple studies, mainly to measure their emotional responses to certain stimuli. (Kremer & den Uijl, 2016). One study used the second version of the PREMO tool to assess the emotions evoked by food in normosmic and hyposmic older adults whose average ages were approximately 67 to 68 years, effectively capturing the differences in emotional profiles between young and older adults (den Uijl et al., 2016).

However, studies such as Gudmundsson et al. (2011) have detailed the need to implement adjustments when applying the PREMO instrument in samples involving older adults, such as reducing the number of emotional ratings, simplifying questionnaire formats, and providing individual

FIGURE 5.2 Cards with the PREMO tool emotions used. (Adapted from the cards proposed by Desmet, 2004.)

sessions to support response completion. The study by Andersen et al. (2011) also used the PREMO tool to assess older adults' emotions toward different models of walking frames, with adaptations to its use such as limiting the time to complete the task, using images printed on large cards, and using a discrete scale. Both studies confirmed the implementation of the PREMO tool with modifications that ensure accurate measurement of emotions despite age-related challenges. For the purposes of this study, these considerations were taken into account, adapting the visual appearance of the tool to the set of cards with illustrations and labels of the 14 main emotions, shown in Figure 5.2.

Both the procedure followed by users when interacting with the application and the online experience represented by comments and emotions were illustrated in a set of Costumer Journey Maps (Fichter & Wisniewski, 2015). It was possible to identify the moments where the user and the product converge, named as 'Touch Point', with the aim of improving the service (Maddox et al., 2022).

5.2.3 Study Subject

For the purposes of the study, the NEQUI Colombia mobile application was used (version 2.1.74 dated February 2024). This mobile app allows money transfers between unbanked individuals, as well as payment of services, withdrawals, and general financial management. To date, NEQUI is the leader in digital wallets in Colombia (Marulanda, 2022). Based on preliminary interviews with older adults and the task they reported using most frequently in the app, a list of possible tasks was determined, and the main functions were narrowed down. In this way, the function 'Send Money' was chosen as the most important and used task for the elderly aging group. Figure 5.3 shows the workflow and the display of screens on the mobile app that allow the development of this task.

5.2.4 Participants

The sample size was calculated from the G*Power 3.1.9.7 software (Cohen, 1988), considering a statistical power of 0.95, an error of 0.5, and a mean effect index (Fidler, 2002). The result in this case was 13 participants. However, Caine's (2016) study recommends that, for studies on human–computer interaction using eye-tracking, the sample size should be 21 people on average.

Therefore, this study considered the collaboration of 20 participants, of whom 20% were male and 80% female. They ranged in age from 64 to 80 years (with an average age of 66 years) and were recruited by word-of-mouth within their community. All were recurrent users of mobile applications for financial transactions. In this way, each participant was asked how often they used the application under study over the course of a week, as well as the functions and situations in which they required its use. This allowed some categories to be established (Table 5.1).

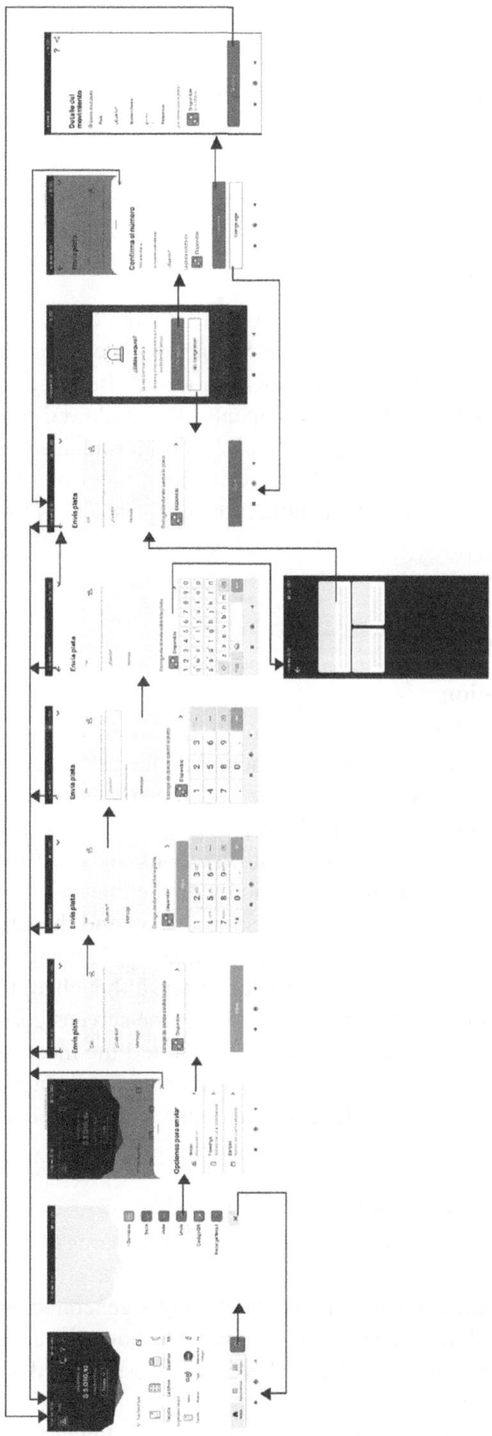

FIGURE 5.3 Workflow to perform the 'Send Money' task in the NEQUI application. (Authors' own creation.)

TABLE 5.1
Categorization of the Participants' Frequency of Use of NEQUI

Frequency of Use	Criteria	Sample Frequency
Low frequency	1 to 3 times per week	5 (25%)
Medium frequency	5 to 7 times per week	8 (40%)
High frequency	More than 7 times per week	7 (35%)
	Total	20

5.2.5 Ethical Consideration

This research is governed by Law 1581 of 2012, Decree 1377 of 2013, and Resolution 1227 of 2013 of the rectorates of the Universidad Industrial de Santander. Therefore, the study guaranteed the proper treatment and protection of the information collected. Before the tests were carried out, informed consent was formalized with the participants. This followed the guidelines for working with human beings, by guaranteeing the privacy of each older adult's data, as well as voluntary participation and the possibility of leaving the experiment at any time. Furthermore, the study is classified as 'No Risk' as no instruments that could threaten the health of the participants were used.

5.3 RESULTS

5.3.1 Quantitative Data

5.3.1.1 Workflow Navigation

With the data collected by the EyeTracker SMI, Activity Logs were constructed for the visual representation of each participant's eye behavior as a function of the task execution time (Rabardel et al., 2002). Figure 5.4 shows an example of an Activity Log constructed for one of the participants. In this way, during each second of the session, it was possible to know the screen the participant was on, the steps he performed, his eye behavior (with fixations and saccadic movements and pupil diameter), and if any error occurred in his interaction with the interface.

With the Activity Logs produced, it was possible to understand the capabilities of the users in terms of the established workflow and its appropriate follow-up. Figure 5.5 shows the total number of entries that each of the screens in the workflow had. Thus, although all the participants managed to complete the task, not all of them followed the order of the screens correctly or omitted at least one of them. It was evident that the screens that were omitted the most were Screens 7, 8, and 9, screens that allow the addition of text messages and additional information, complementing the task to be carried out, as well as confirming the data previously entered, requesting verification of the destination of the money to be sent.

5.3.1.2 Identifying Graphical Elements in the Interface

Table 5.2 presents the descriptive statistical analysis (with measures of central tendency and variability) of the ocular behavior data.

With this information, an attention map representation was generated, with the total fixations of the sample ($N=20$) and their duration per screen. Figure 5.6 shows the attention map of the 12 predefined screens, as well as the signage of the AOI previously established with the graphic elements and information provided by the GUI.

Taking this into account, it is observed that the AOIs with the highest number of fixations belong to the areas containing the elements that interact with users to activate the main functions of each screen. That is to say, the users were able to identify the buttons that correctly execute each step

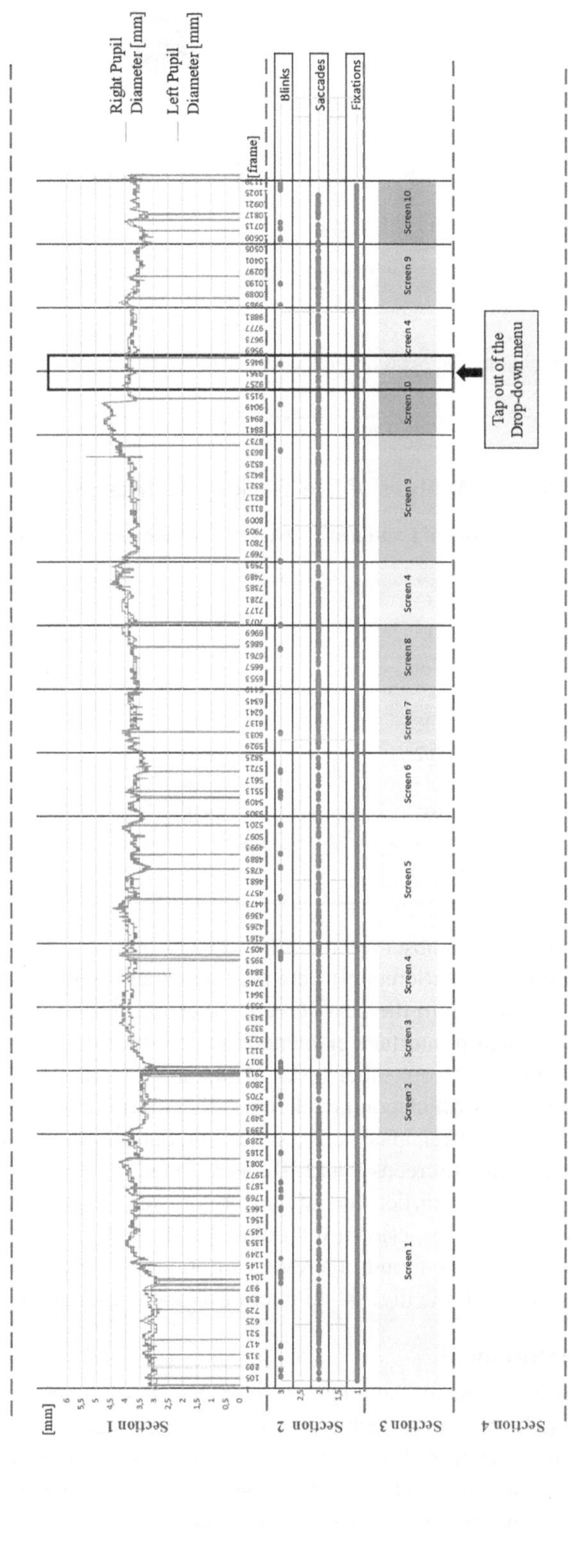

FIGURE 5.4 Example of the Activity Log created for Participant 1. (Authors' own creation.)

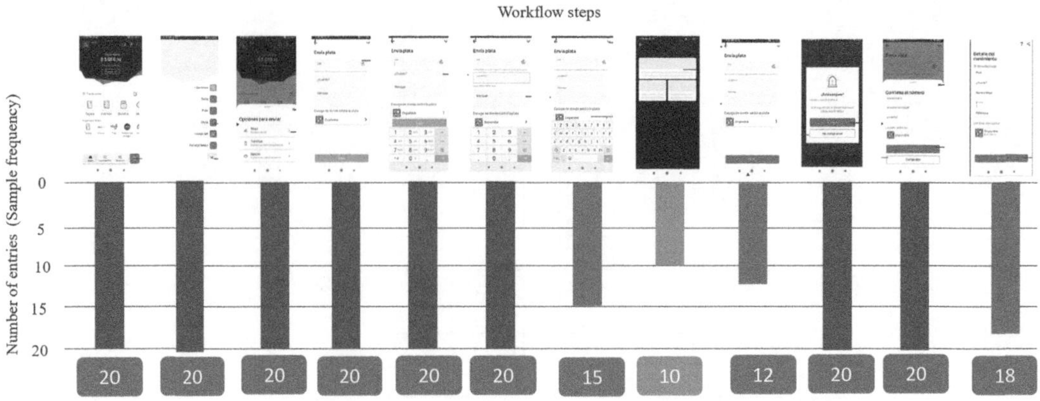

FIGURE 5.5 Number of participants' accesses on each screen in the workflow. $N=20$. (Authors' own creation.)

TABLE 5.2
Descriptive Statistical Analysis of Eye-Tracking Metrics ($N=20$)

	Number of Fixations	Number of Saccades	Number of Blinks
Central Tendency			
Mean	212	173	25
Median	215	155	17
Mode	256	#N/D	11
Variability			
Standard deviation	136	138	21
Range	552	520	71
Minimum	43	18	2
Maximum	595	538	73

Source: Authors' own creation.

in the workflow. For example, it was observed that the Services button (Screen 1), the Send button (Screen 2), and the data confirmation (Screen 9) were the UI elements with the highest fixation, and in turn, they are triggers for moving to the next screen. Likewise, it was observed that areas with a lower number of fixations contain interface elements that do not directly influence the execution of the requested task but rather perform other tasks of the application or provide supplementary information. However, some AOIs which contain elements that do influence the workflow and were not adequately identified by the participants, such as the back button (Screen 8) and the buttons to confirm or correct the entered data (Screens 9 and 10), were found.

On the other hand, as mentioned earlier, the AOIs were created by grouping the graphical elements present on each GUI screen to perform the defined task. In this way, Table 5.3 shows the design patterns focused on the elderly found in a review of previous literature (Vargas et al., 2024) and their presence in the AOIs with the highest number of fixations for each screen.

5.3.1.3 Heart Rate Measurement

The data obtained from the heart rate measurement were processed separately from the Activity Log. It was referred to as the basal heart rate for each participant as the average of the data in bpm that each one presented in the first minute of recording by the monitor. That is, the time between the placement of the bracelet on the forearm and the handover of the mobile device to the participant to start the task was not considered in the study. In this way, having the basal heart rate (bpm) and the mean heart

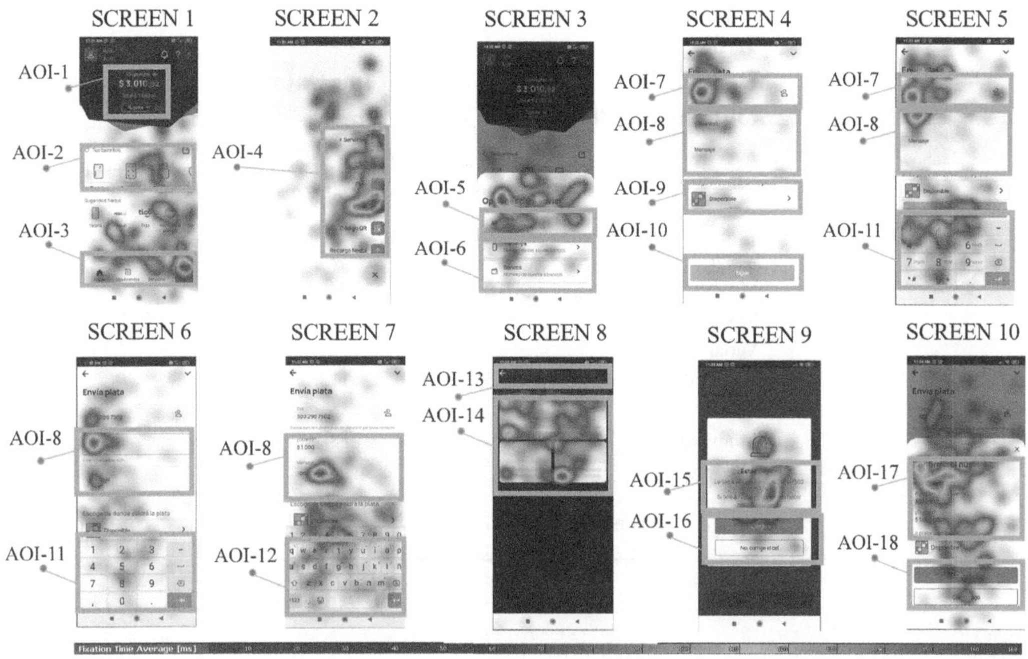

FIGURE 5.6 Attention maps prepared with the workflow screens and each of their areas of interest (AOI) $N=20$. (Authors' own creation.)

TABLE 5.3
Design Patterns Present in Areas of Interest (AOI) with the Highest Number of Fixations

From Systematic Review		From the Obtained Data		
Design Pattern	**Source**	**AOI**	**Screen**	**Graphic Element**
Use buttons that span the horizontal length of the screen.	Del Valle and Gutierrez (2017)	AOI-5	3	Button with 'Send to NEQUI' Label
		AOI-14	8	Button with 'Pockets' Label
Use contrasting text and button colors.	Supanta and Díaz (2020), Zaina et al. (2022), Luna-Garcia et al. (2015)	AOI-3	1	Button with '$' Label
		AOI-4	2	Button with 'Send' Label
		AOI-15	9	Data confirmation text
		AOI-17	10	Data confirmation text
Use standardized icons.	Del Valle and Gutierrez (2017), Cardozo et al. (2020)	AOI-3	1	Button with '$' Label
		AOI-4	2	Button with 'Send' Label
Ensure a balanced interface design, leaving sufficient white space.	Cardozo et al. (2020)	AOI-15	9	Data confirmation text
		AOI-17	10	Data confirmation text
Display the right keyboard at the right time.	Luna-Garcia et al. (2015)	AOI-7	5	Text input field 'Cell phone'
		AOI-8	6	Text input field 'Value'
		AOI-8	7	Text input field 'Message'
Use simple and consistent language.	Supanta and Díaz (2020); Cardozo et al. (2020)	AOI-15	9	Data confirmation text
		AOI-17	10	Data confirmation text

Source: Literature review conducted (2024). Authors' own creation.

TABLE 5.4
Descriptive Statistical Analysis of the Data Obtained by the Heart Rate Monitor

	Basal Heart Rate [bpm]	Average Heart Rate [bpm]
Central Tendency		
Mean	81	77
Median	80	78
Mode	80	79
Variability		
Standard deviation	10.5	11.0
Range	39	45
Minimum	57	58
Maximum	96	103

Source: Authors' own creation.

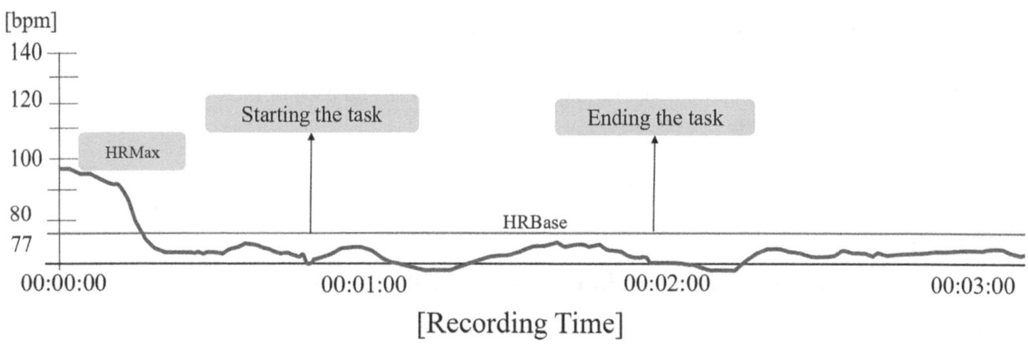

FIGURE 5.7 Heart rate diagram over time. (Authors' own creation.)

rate (bpm) values for each participant, the respective descriptive statistical analyses (with measures of central tendency and variability) for both variables were carried out, as shown in Table 5.4.

Likewise, heart rate diagrams were generated over time for each of the participants. Figure 5.7 shows an example where the Maximum Heart Rate Point (HRMax) recorded by the bracelet is indicated, as well as the start and end times of the task performed. Overall, a white line is observed, which allows for visualizing the variation over time with reference to the baseline previously established for each participant. The collected data show that, in the case of some participants, there is a difference between the initial frequency and the average observed during the execution of the task. It is possible that at the beginning, the frequency is higher, given that a level of stress is generated because of the surprise and alertness that any unexpected situation can provoke. However, during the execution of the task, there was no evidence of an increase in bpm beyond the established basal heart rate over time.

5.3.2 Qualitative Data

5.3.2.1 Perceived Emotions Evaluation

The results obtained with the PREMO tool are shown in Figure 5.8, where the three most frequent emotions were satisfaction (28%), pleasant surprise (17%), and admiration (15%). It is evident that

FIGURE 5.8 Frequency diagram of emotions from the PREMO tool. (Authors' own creation.)

for the elderly evaluated in this project, one of the most valued emotions was pleasant surprise and satisfaction when using technology, which suggests an intrinsic motivation for learning this type of applications.

These results provide an overview of the perception that elderly participants have regarding the use of the application and their experiences with it. It should be considered for the correct analysis that the entire participating population include regular users of financial applications.

5.3.2.2 Comments and Opinions

The comments obtained in the semi-structured interviews, along with the results of the emotional measurement, allowed for the establishment of a set of user personas and their particular use scenarios. In this way, the characteristics collected for each of the previously defined application usage frequency categories were encompassed, as shown in Figure 5.9.

The creation of user persona as profiles allowed for the characterization of the UX, represented through Journey Maps. A Customer Journey Map was created for each profile, relating the phases of the application and the screens (Steps) with the type of interaction and their perceived emotions. Likewise, it is shown based on the scenario presented for each case with the characteristics of each profile in terms of the level of expertise, motivations, and goals. Figure 5.10 shows an example and corresponds to User Persona No. 1.

5.3.3 Integration of Quantitative and Qualitative Findings

The use of multiple tests to obtain data of both types allowed for a multidimensional understanding on how older adults perceive and interact with the mobile study app. From the application of all the tests, some points of notable coincidence were identified between the performance of the task and the user's subjectivity. An example of this is that most of the participants successfully completed the task by following all the steps in the workflow, maintaining a heart rate below the HRMax

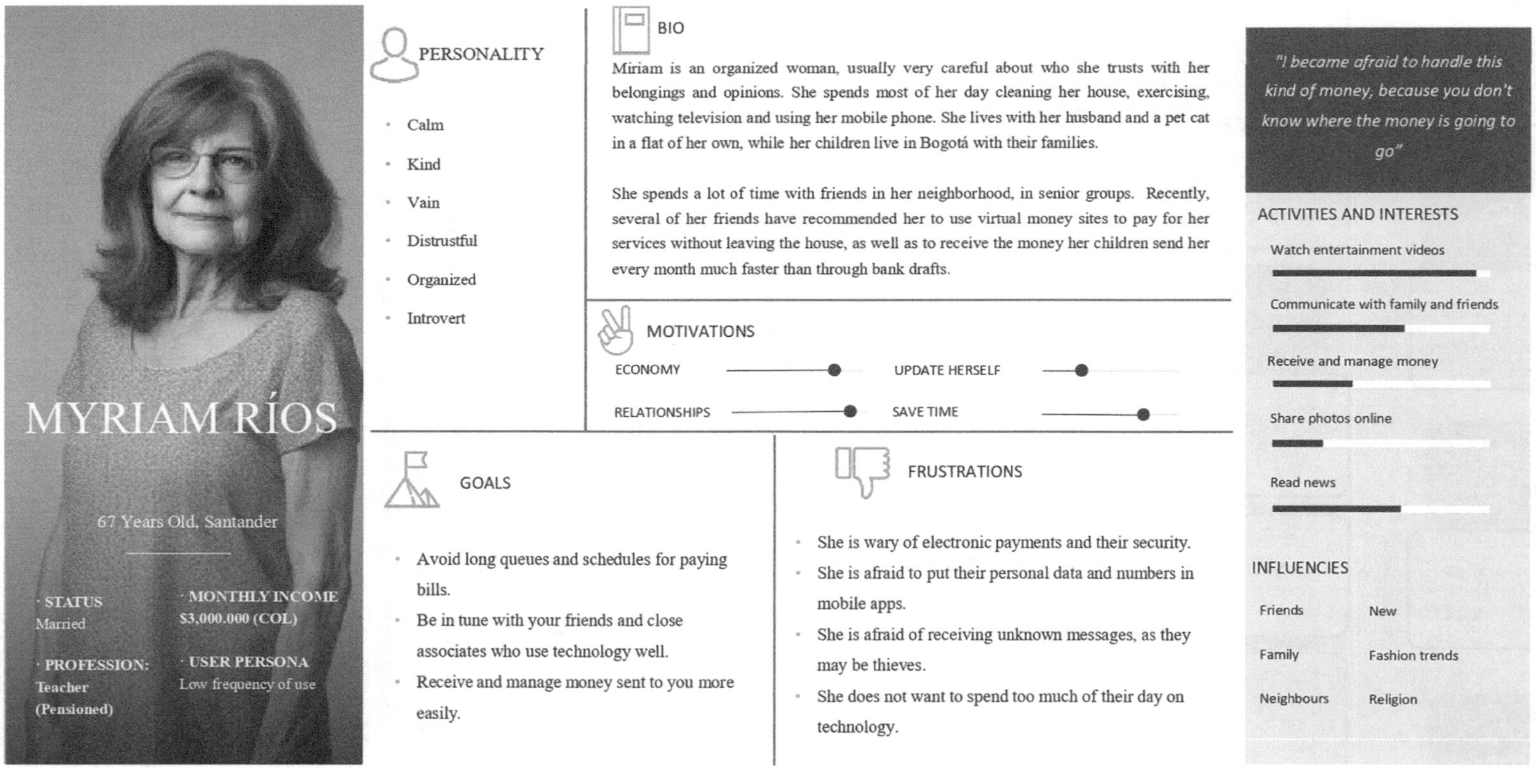

FIGURE 5.9 Example of a developed user persona profile. (Authors' own creation.)

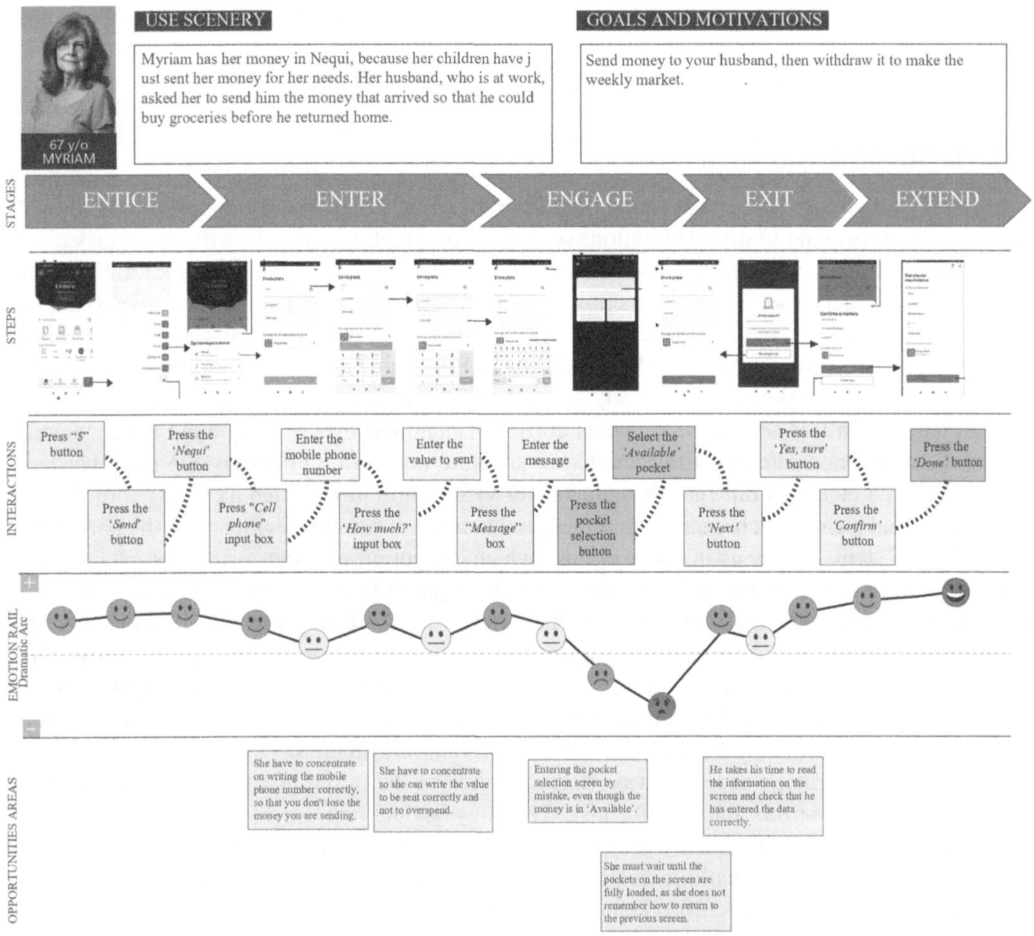

FIGURE 5.10 Example of a Customer Journey Map created. (Authors' own creation.)

and expressing emotions of satisfaction and admiration, with comments about experiences where the use of the app has brought benefits to their daily lives. This integration also revealed critical moments of difficulty, where the omission of optional steps (Screens 7 to 9) in the execution of the task coincided with a low level of recurrence in the use of the app and an increase in the number of fixations expressed by the eye-tracking equipment.

However, some contradictions were also found in the overlay of each result. Some of the participants who skipped screens in the workflow scored positive emotions with the PREMO tool, such as admiration and pleasant surprise, with heart rate values that did not exceed their basal heart rate. Likewise, some participants who expressed emotions such as boredom and distrust with the PREMO tool managed to perform the task correctly, following all the steps and screens of the workflow. These contradictions may be since, although they regularly use the application, they express distrust of the administrators and the capabilities of the system, showing a general dissatisfaction with the technology unrelated to the experience of using the application itself.

The use of both ways for characterizing UX also allowed for the identification of emotionally critical points in the interaction with greater reliability since, according to Sinabell and Ammenwerth (2024), the single use of self-reporting methods to address usability and UX in older adults may not be accurate due to barriers related to aging, from cognition, motivation, perception, and physical abilities. These barriers can make it difficult for participants to verbally reveal their problems during interaction, making established methods such as Think Aloud or cognitive walkthroughs not easily applicable (Wildenbos et al., 2019). Otherwise, this integrative approach contributes to a more comprehensive description considering the requirements of this population.

5.4 DISCUSSION

From the eye-tracking data, it was found that the AOIs of the GUIs that were least observed by the participants correspond to disabled buttons or information that does not directly affect task execution. This is in line with what has been found in the literature, where Shahal et al. (2022) suggest that visual feedback from touch interaction elements promotes greater interaction speed and task execution accuracy, especially in older users as they suffer from inadequate size and color handling more compared to younger adults. Likewise, Elguera Paez and Zapata Del Río (2019) mention that the main usability problems faced by adults can be of three types: visual, psychomotor, or cognitive limitations. In this sense, the elements correctly identified by the sample are not included in any of the mentioned categories as they do not generate interaction problems. On the other hand, the unidentified elements fall into the category of visual limitations, such as issues with the size, clarity of visual elements, characters, graphics, and buttons. It was also evident that the participants omitted some steps or screens in the workflow, which did not prevent the correct completion of the task but did imply that additional information in the requested task was omitted. This aligns with the findings of Luna-Garcia et al. (2015) as they mention that the absence of a GUI element indicating the procedure to follow can cause confusion in the task procedure and significantly extend the navigation time. The inclusion of a progress bar, for example, to visualize the progress in task completion, as a solution or design pattern to this issue, would improve system visibility, reducing negative reactions caused by the lack of clear and relevant information.

Regarding heart rate, a decrease was observed between the initial heart rate and the individual average heart rate over the course of the task. This may be due to the stress generated in the person by the unfamiliarity of the activity they are going to perform. In this sense, the decrease in frequency during the activity could indicate that the participant did not experience stress associated with task execution, as Valdés (2007) mentions that stress can manifest in increased blood pressure and heart rate as a physiological response to external stimuli. This is complemented by the low number of blinks obtained in the average of the sample compared to the average data of the other eye-tracking metrics as various studies suggest that blink frequency is related to emotional and cognitive processing, especially with attention and mental load, reflected in the participant's stress (Maffei & Angrilli, 2019). However, the study did not use nonrecurring participants in the use of financial applications, which could expand the analysis of these results.

On the other hand, regarding the emotional reactions of the elderly population, two different perspectives were evident. First, the results of the evaluation with the PREMO tool indicated that most of the older adults exhibit positive emotions toward the use of these mobile applications, expressing a desire to learn more about managing these technologies and an intention to improve their way of interacting with digital payments. A similar reaction was evidenced in the study by Orlofsky and Wozniak (2022) where, in describing the behavior of older adults in acquiring and using Alexa, it was found that the participants enjoyed the device for selective features and functions they considered novel, despite not fully understanding the spectrum of functionalities and characteristics of the device. Second, a portion of the sample expressed through comments a high concern regarding the entry of their private data, the protection of the money stored in the application, and the legitimacy of using digital money for paying services to external entities. This aligns with the postulates of Bartol et al. (2024), whose study examined how internet skills, along with privacy concerns and perceived control, limit the use and appropriation of mobile technologies by older adults. In this way, these concerns must be considered in the development of digital environments for this audience as they represent a real barrier to online participation and the expansion of their digital engagement.

Both emotional reactions, both positive and negative, align with the findings in the study by Kim & Choudhury (2021) regarding the use of the Google voice assistant, where participants expressed divergent feelings and perspectives when they used it for the first time. Reactions ranged from enthusiasm and curiosity to uncertainty and even rejection because they felt intimidated and insecure with the application. However, the initial results changed over time with use, eventually showing reactions of enjoyment and comfort with the tool.

5.5 CONCLUSION

The present study described the behavior of a sample of 20 adults aged between 64 and 80 years when they used a mobile application from the financial sector and interacting with its GUI. A critical task for the target audience was identified within the application, and a workflow of 12 screens was defined to execute it. Through individual sessions, the actions performed by the participants during their interaction with the application were described, using direct observation and self-report tools, to obtain an overview of the UX for this age group.

In this way, the correct identification of the GUI elements that allow navigation between screens and the effective completion of tasks by the participants was evidenced. Likewise, it was found that a series of design patterns focused on the qualities of this age group are indeed adequately implemented in the GUI, in accordance with what is proposed in the literature. With these patterns and the results obtained, a list of suggestions for the AOIs that had some interaction issues was established to improve its usability from its graphical elements, functions, and navigation, as shown in Table 5.5. In addition to this, among the elderly studied, positive emotions of satisfaction, admiration, and intrigue predominate regarding learning more about the use of these digital platforms.

5.6 LIMITATIONS AND FUTURE WORK

The present research had some limitations. First, there was difficulty in recruiting participants who met the inclusion criteria as knowledge in handling the mobile application without assistance from another person was required. Also, the limited information available regarding the evaluation of UX using EyeTracker equipment made it difficult to compare the data obtained of a quantitative nature. Therefore, it is suggested to continue studying this population sector and their interaction

TABLE 5.5
General Suggestions to Improve the UX and Usability of the Study Subject's GUI

Category	Suggestion
Graphical elements	Use larger fonts for action confirmation buttons.
	Increase the contrast of buttons that return to the previous screen.
	Use standardized icons with less visual complexity.
	Provide visual feedback to the interface elements and successful task completion.
	Complement the designs of some options with descriptive text, mainly those with complex labels (Pockets).
	Keep the actual layout of each screen.
System operation	Delimit clickable areas.
	Provide descriptions that confirm the correct completion of a task as notifications and visual models.
Web navigation	Maintain the same structure in the interface elements.
	Incorporate some screens of the workflow as mandatory.
	Keep the highlight at most active areas on the mobile screen and the buttons.

Source: Authors' own creation.

with mobile applications as they are part of their daily activities. These should employ direct measurement instruments with an emphasis on the elements of the GUI, as well as generate redesign proposals based on the findings obtained.

Future research in this area could expand and contrast these results by including older participants who are unfamiliar with the use of digital financial applications. It is also suggested that significant samples of younger people (under 40) and older adults (over 60) be included in this type of descriptive study. This would allow to distinguish the extent to which factors such as understanding of interface design, product language, and icon complexity affect the interaction, based on the generational gap between each group. Studies from Fernández-Ardèvol et al. (2022) and Taipale et al. (2024) demonstrate these comparisons, implementing mechanisms to collect qualitative data focused on understanding the relationship of both generations with technology, where aspects such as frequency of use, languages, and media ideologies of this generation influenced technological adoption.

The research should aim to reduce accessibility barriers and create a more inclusive design for this population sector. This would allow for an expansion of knowledge on the design of more accessible and reliable digital financial applications.

REFERENCES

Andersen, C. L., Gudmundsson, H. P., Achiche, S., & Boelskifte, P. (2011). Challenges and limitations of applying an emotion-driven design approach on elderly users. In *DS* 68-7*: Proceedings of the 18th International Conference on Engineering Design (ICED 11), Impacting Society through Engineering Design, Vol. 7: Human Behaviour in Design, Lyngby/Copenhagen, Denmark, 15.-19.08. 2011* (pp. 74–84).

Bartol, Jost, Prevodnik, Katja., Vehovar, Vasja., & Petrovčič, Andras. (2024). The roles of perceived privacy control, Internet privacy concerns and Internet skills in the direct and indirect Internet uses of older adults: Conceptual integration and empirical testing of a theoretical model. *New Media & Society,* 26(8), 4490–4510. https://doi.org/10.1177/14614448221122734

Beatty, Jackson. (1982). Task-evoked pupillary responses, processing load, and the structure of processing resources. *Psychological Bulletin,* 91(2), 276–292. https://doi.org/10.1037/0033-2909.91.2.276

Caine, Kelly. (2016). Local standards for sample size at CHI. In *Proceedings of the 2016 CHI Conference on Human Factors in Computing Systems* (pp. 981–992). https://doi.org/10.1145/2858036.2858498

Calderón Ynoñan, Pamela del Carmen, & Prieto Neira, Franck Alberson. (2022). *Integración de un enfoque ágil con técnicas de diseño centrado en usuario (DCU) para la mejora de experiencia de usuario (UX).* https://hdl.handle.net/20.500.12802/10085

Campos, A. S. I. (2014). *El delito de ser pobre: una gestión neoliberal de la marginalidad.* Barcelona. Icaria.

Cardozo, Claudia, Martín, Adriana, Saldaño, Viviana, Gaetán, Gabriela. (2020). Una propuesta para mejorar la experiencia de los adultos mayores con las redes sociales. *Revista de Tecnología, Ciencia y Educación,* (16), 113–142. https://hdl.handle.net/20.500.12226/328

Cohen Jacob. (1988). *Statistical Power Analysis for the Behavioral Sciences.* New York, NY: Routledge Academic.

Cowen, Laura, Ball, Linden Judy, & Delin, J. (2002). An eye movement analysis of web page usability. In *People and Computers XVI-Memorable yet Invisible: Proceedings of HCI 2002* (pp. 317–335). Springer London. https://doi.org/10.1007/978-1-4471-0105-5_19

Curilem Gatica, Cristian, Guandalini Valdete, Regina, Liabeuf Altamirano, Gianella, Bahamondes Ávila, Carlos, Bruneau-Chávez, José, & Berral de la Rosa, Francisco. José. (2022). Efecto del ejercicio en la memoria de adultos mayores con sobrepeso u obesidad: *Revisión sistemática. Revista del Nacional (Itauguá),* 14(1), 58–74. https://doi.org/10.18004/rdn2022.jun.01.058.074

Del Valle Almea, Kahterine Stefanie., & Guano Gutierrez, Johana Karina. (2017). *Análisis de Usabilidad y Prototipo de Mejora en Aplicaciones Móviles de Mensajería Instantánea (Chat) Para Adultos Mayores en Sistemas Operativos Android.* Doctoral dissertation, Universidad de Guayaquil. Facultad de Ciencias Matematicas y Fisicas. Carrera de Ingenieria en Sistemas Computacionales. https://repositorio.ug.edu.ec/handle/redug/19675

Delgado Silveira, Eva, Montero Errasquín, Beatriz, Muñoz García, María, Vélez-Díaz-Pallarés, Manuel, Lozano Montoya, I., Sánchez Castellano, C., & Cruz-Jentoft, Alfonso. (2015). Improving drug prescribing in the elderly: A new edition of STOPP/START criteria. *Revista Española de Geriatría y Gerontología,* 89–96. https://doi.org/10.1016/j.regg.2014.10.005

den Uijl, Louis C., Jager, Gerry, Zandstra, Elizabeth H., de Graaf, Cees, & Kremer, Stefanie. (2016). Self-reported food-evoked emotions of younger adults, older normosmic adults, and older hyposmic adults as measured using the PrEmo2 tool and the Affect Grid. *Food Quality and Preference,* 51, 109–117. https://doi.org/10.1016/j.foodqual.2016.03.002

Desmet, Pieter M.A. (2004). From disgust to desire: How products elicit emotions. In McDonagh, D., Hekkert, P., van Erp, J., & Gyi, D. (Eds.) *Design and Emotion: the Experience of Everyday Things* (pp. 8–12). London: CRC Press.

Dubois, V. (2018). El burócrata y el pobre. Relación administrativa y tratamiento de la miseria. En HAL (Le Centre pour la Communication Scientifique Directe). France: Le Centre pour la Communication Scientifique Directe.

Elguera Paez, Lesly, & Zapata Del Río, Claudia. (2019). Elderly users and their main challenges usability with mobile applications: A systematic review. In *Design, User Experience, and Usability. Design Philosophy and Theory: 8th International Conference, DUXU 2019, Held as Part of the 21st HCI International Conference, HCII 2019,* Orlando, FL, USA, *July 26–31, 2019, Proceedings, Part I* 21 (pp. 423–438). Springer International Publishing. https://doi.org/10.1007/978-3-030-23570-3_31

Fernández-Ardèvol, Mireira, Belotti, Francesca, Ieracitano, Francesca, Mulargia, Simone, Rosales, Andrea., & Comunello, Francesca. (2022). "I do it my way": Idioms of practice and digital media ideologies of adolescents and older adults. *New Media & Society,* 24(1), 31–49. https://doi.org/10.1177/1461444820959298

Fichter, Darlene, & Wisniewski, Jeff. (2015). Customer journey mapping. *Online Searcher,* 39(4), 74–77. https://link.gale.com/apps/doc/A422448471/AONE?u=anon~cb954da&sid=googleScholar&xid=1abd2326

Fidler Fiona. (2002). The fifth edition of the APA Publication Manual: Why its statistics recommendations are so controversial. *Educational and Psychological Measurement,* 62, 749–770. https://doi.org/10.1177/001316402236876

Gonzalo, Marilín. (2021). Brecha digital: El 40% de las personas mayores asegura que nunca ha accedido a internet. *Newtral.* https://www.newtral.es/brechadigitalmayoresinternet/20210720/

Gothelf, J., & Seiden, J. (2016). *Lean UX: Designing Great Products with Agile Teams.* Sebastopol, CA. O'Reilly Media, Inc.

Gudmundsson, Hjalte P., Andersen, Casper L., Achiche, Sofiane, & Boelskifte, Per. (2011). Emotion-driven elicitation of elderly people user needs illustrated by a walking frame case study. In *DS 68-7: Proceedings of the 18th International Conference on Engineering Design (ICED 11), Impacting Society through Engineering Design, Vol. 7: Human Behaviour in Design, Lyngby/Copenhagen, Denmark, 15.-19.08. 2011.*

Hollander, John, & Huette, Stephanie. (2022). Extracting blinks from continuous eye-tracking data in a mind wandering paradigm. *Consciousness and Cognition,* 100, 103303. https://doi.org/10.1016/j.concog.2022.103303

IMotions. (January 24, 2025). SMI Eye Tracking Glasses. SensoMotoric Instruments. Recuperado de https://imotions.com/products/hardware/smi-eye-tracking-glasses/#product-specifications

ISO 9241-210:2019(en), Ergonomics of human-system interaction — Part 210: *Hum*an-centred design for interactive systems. https://www.iso.org/obp/ui/#iso:std:iso:9241:-210:ed-2:v1:e

Kim, Sunyoung, & Choudhury, Abhishek. (2021). Exploring older adults' perception and use of smart speaker-based voice assistants: A longitudinal study. *Computers in Human Behavior,* 124, 106914. https://doi.org/10.1016/j.chb.2021.106914

Kremer, Stefanie, & den Uijl, Louise. (2016). Studying emotions in the elderly. In *Emotion Measurement* (pp. 537–571). Woodhead Publishing. https://doi.org/10.1016/B978-0-08-100508-8.00022-9

Loos, E., & Ivan, L. (2022). Not only people are getting old, the new media are too: Technology generations and the changes in new media use. *New Media & Society,* 26(6), 3588–3613. https://doi.org/10.1177/14614448221101783

Luna-Garcia, Huizilopoztli, Mendoza-Gonzalez, Ricardo, & Alvarez-Rodriguez, Francisco Javier. (2015). Design patterns to enhance accessibility and use of social applications for older adults. *Comunicar,* 23(45), 85–94. https://doi.org/10.3916/C45-2015-09

Maddox, Kyle, Masalonis, Anthony, Motiwala, Aneesa, Adams, Kathleen, Eugene, Nakina, & Speir, Ross. (2022, September). Using Journey Mapping to Visualize Patient Experiences for Quality Improvement Initiatives. In *Proceedings of the International Symposium on Human Factors and Ergonomics in Health Care* (Vol. 11, No. 1, pp. 56–60). Sage, CA: SAGE Publications. https://doi.org/10.1177/2327857922111011

Maffei, Antonio, & Angrilli, Alessandro. (2019). Spontaneous blink rate as an index of attention and emotion during film clips viewing. *Physiology & Behavior,* 204, 256–263. https://doi.org/10.1016/j.physbeh.2019.02.037

Marulanda, Olga Patricia Rendón (2022). *"Nequi tiene 13,5 millones de usuarios y crece al mes en cerca de 400.000".* El colombiano. https://www.elcolombiano.com/negocios/nequi-suma-400000-clientes-cada-mes-y-va-por-mas-como-neobanco-AL18462130

Ministerio de Salud y Protección Social (2021). *Minsalud actualiza política de envejecimiento y vejez.* Consultado el 8 de mayo de 2023. https://www.minsalud.gov.co/Paginas/Minsalud-actualiza-politica-de-envejecimientoyvejez.aspx

Moya-Albiol, Luis, Serrano Rosa, Miguel Ángel, González-Bono, Esperanza, Rodríguez-Alarcón, Gabriel., & Salvador, Alicia. (2005). Respuesta psicofisiológica de estrés en una jornada laboral. *Psicothema,* 17(2), 205–211.

Orlofsky, Sarit, & Wozniak, Kathryn. (2022). Older adults' experiences using Alexa. *Geriatric Nursing,* 48, 247–257. https://doi.org/10.1016/j.gerinurse.2022.09.017

Pacholczyk, Dominik. (2014). *Mobile UI Design Patterns: A Deeper Look at the Hottest Apps Today.* Mountain View, CA: UXPin.

Rabardel, P., Carlin, N., & Chesnais, M. (2002). *Ergonomie: Concepts et méthodes.* Toulouse: Octarès.

Rayner, Keith (1998). Eye movements in reading and information processing: 20 years of research. *Psychological Bulletin,* 124(3), 372.

Rojas López, Juan Carlos. (2016). *Contribución a la evaluación emocional en el diseño de productos mediante la integración de tecnologías de seguimiento de la mirada (Eye tracking), Diferenciales semánticos y potenciales evocados (ERPs).* Doctoral dissertation, Universitat Politècnica de València.

Shahal, Avner, Spang, Robert P, Minge, Michael, Trahms, Carola, & Voigt-Antons, Jan-Niklas. (2022). User-specific touch interfaces: A viable solution for an aging society?. *Behaviour & Information Technology,* 41(9), 1928–1940. https://doi.org/10.1080/0144929X.2021.1906322

Shojaeizadeh, Mina, Djamasbi, Soussan, & Trapp, Andrew C. (2016). Density of gaze points within a fixation and information processing behavior. In *Universal Access in Human-Computer Interaction. Methods, Techniques, and Best Practices: 10th International Conference, UAHCI 2016, Held as Part of HCI International 2016,* Toronto, ON, Canada, *July 17–22, 2016, Proceedings, Part I 10* (pp. 465–471). Springer International Publishing. https://doi.org/10.1007/978-3-319-40250-5_44

Sinabell, Irina, & Ammenwerth, Elske. (2024). Challenges and recommendations for eHealth usability evaluation with elderly users: Systematic review and case study. *Universal Access in the Information Society,* 23(1), 455–474. https://doi.org/10.1007/s10209-022-00949-w

Supanta-Paucay, Danny, y Díaz-Ramírez, Jorge. (2020). Adultos mayores y el uso de Web App para la estimulación cognitiva. *Revista de Ciencias Sociales (Ve),* XXVI(3), 284–296.

Taipale, Sakari, Oinas, Tomi, Ivan, Loredana, & Rosenberg, Dennis. (2024). Mobile phone use before and during the COVID-19 pandemic–A panel study of older adults in seven countries. *Mobile Media & Communication,* 12(1), 23–44. https://doi.org/10.1177/20501579231185479

Valdés Miyar, M. (2007). El estrés En J. Rodés, J. M. Piqué, & A. Trilla (Coords.*), Libro de la salud del Hospital Clínic de Barcelona y la Fundación BBVA* (pp. 427–434). Madrid: Fundación BBVA.

Vargas, Juan Angel, Maradei, Fernanda, & Bautista, Luis. (2024). *Comportamiento del adulto mayor al hacer uso de una Interfaz Gráfica de Usuario en el Área Financiera, a través de la metodología de Ergonomía de Producto para Poblaciones Especiales.* (Tesis de pregrado). Universidad Industrial de Santander. https://noesis.uis.edu.co/handle/20.500.14071/42516

Vélez, Esther Elizabeth Esmeraldas, Centeno, Mariana Rosalia, Zevallos, Mariangel Gabriela, & Vélez, José Adolfo. (2019). El envejecimiento del adulto mayor y sus principales características. *RECIMUNDO: Revista Científica de la Investigación y el Conocimiento,* 3(1), 58–74. https://doi.org/10.26820/recimundo/3.(1).enero.2019.58-74

Véliz, Mónica, Riffo, Bernardo, & Arancibia, Beatriz. (2010). Envejecimiento cognitivo y procesamiento del lenguaje: cuestiones relevantes. *RLA. Revista de lingüística teórica y aplicada,* 48(1), 75–103. https://doi.org/10.4067/S0718-48832010000100005

W3C. (2024). Web Accessibility Initiative (WAI). https://www.w3.org/WAI/about/

Ware, Patrick, Bartlett, Susan J., Paré, Guy., Symeonidis, Iphigenia, Tannenbaum, Cara, Bartlett, Gillian, … & Ahmed, Sara. (2017). Using eHealth technologies: Interests, preferences, and concerns of older adults. *Interactive Journal of Medical Research,* 6(1), e4447. https://doi.org/10.2196/ijmr.4447

Wildenbos, Geke Adriana, Jaspers, Monique, Schijven, Marlies P., & Dusseljee-Peute, Linda W. (2019). Mobile health for older adult patients: Using an aging barriers framework to classify usability problems. *International Journal of Medical Informatics,* 124, 68–77. https://doi.org/10.1016/j.ijmedinf.2019.01.006

Zaina, Luciana Aaparecida, Fortes, Renata Pontin, Casadei, Victor, Nozaki, Leonardo Seiji, & Paiva, Débora Maria Barroso. (2022). Preventing accessibility barriers: Guidelines for using user interface design patterns in mobile applications. *Journal of Systems and Software,* 186, 111213. https://doi.org/10.1016/j.jss.2021.111213

6 Usability of Teaching Resources Used by Students with Visual Disabilities Based on Improvements in Their Accessibility Characteristics and Interface Design

Ana Paula Díaz Pinal and Elvia Luz González-Muñoz

6.1 CONTEXTUALIZATION OF THE PROBLEM

Integrating people with disabilities in education is of utmost importance in any country; inclusive education has been placed on the international agenda as an urgent need for development. The World Report on Disability carried out by UNESCO indicates that more than one billion people in the world have some disability (World Health Organization & World Bank, 2011). Recently, in 2020, 43 million were found to be blind worldwide according to The Lancet Global Health Commission (Burton, et al., 2021). In Mexico, the population that suffers from some type of disability represents 6.8% of the total population, equivalent to 8.9 million people (INEGI, 2024).

The General Law for the Inclusion of Persons with Disabilities (2024) addresses concepts such as special education and inclusive education; the former is aimed at people with disabilities, and the latter should promote their integration into regular schools using materials specific to their characteristics, which facilitate the learning process (Ley General para la Inclusión de las Personas con Discapacidad [LGIPD], 2024). Internationally, technologies available for teaching all students, such as applications, are incorporated into different educational levels, improving efficiency, effectiveness, and accessibility, thereby enhancing personalized teaching, regardless of their physical and intellectual characteristics (Screpnik et al., 2024). It has provided a unique opportunity for people with disabilities to be part of consolidated educational systems, allowing for equal opportunities.

For people with visual impairments, the use of apps represents a potential tool to overcome the difficulties presented by traditional educational tools, related to physical proximity and the rigidity of activities. In the case of teaching mathematics, de Prado et al. (2022) refer to the importance of technology, which allows it to be accessible and attractive to all. They cite the use of tablets as an example, on which mathematical problems are developed and the answers, correct or incorrect, are translated into vibrations or sounds. Although accessibility support has increased recently, the provision of fully accessible applications and learning resources is lacking. The outlook is complicated by the constant updating of the technologies used in education and the need for these to be accessible and easy to use for students.

Here, incorporating ergonomics is paramount in helping achieve technological support by providing guidelines for their design and evaluation, thus ensuring that they comply with usability and accessibility criteria. From the ergonomics perspective, usability is defined as "the extent to which a system, product or service can be used by the specified users to achieve specified goals with

DOI: 10.1201/9781003582014-8

effectiveness, efficiency and satisfaction in a specified context of use" (International Organization for Standardization, 2018). Lewis and Sauro (2021) point out that usability is a complex concept, which can be evaluated through various attributes, such as effectiveness, efficiency, and satisfaction, as well as focusing on other elements of task performance, considering the user's characteristics. As a result, there is a wide variety of evaluation tools, which will be used depending on the focus of the study and the needs of the project.

Validated tools have been used in design and research to check the usability and accessibility of platforms and applications. To define the tools to be used, a scoping study was carried out using the phases proposed by Arksey and O'Malley (2005). The research question was: What are the components considered in the literature for the evaluation of usability in the context of the education of visually impaired students? The search terms were set as follows: "visual impairment" "School platform" OR "learning system" "usability evaluation." The search was carried out in Google Scholar; the inclusion criterion was publications published between 2018 and 2022. Citations, patents, and not researched articles were excluded. Additionally, those that did not allow free access were discarded.

Table 6.1 presents the results to identify the different attributes of the usability concept and the evaluation tools used with the learning systems used by students with visual impairment. Ten articles were obtained, which incorporated the usability components and tools. Reviewing the search results to measure usability in such a specific population as students with visual impairment, we found no established tools that consider the characteristics and needs of this population.

Given this context, a study was carried out to evaluate the usability of a resource for learning mathematics among secondary education students with visual disabilities based on improvements in its accessibility characteristics and interface design.

Mathematics learning was considered due to its cognitive complexity; in this subject, the object of study is conceptual. The student must acquire a semiotic representation of the acquired operations,

TABLE 6.1
Usability Components and Evaluation Tools

Author	Usability Framework	Evaluation Tools
Sahin Kölemen and Akgün (2022)	Efficiency, effectiveness, and satisfaction	Usability test, completed tasks, number of movements, and post-test interviews
Laeeq and Memon (2019)	ISO/IEC 9126-4: Efficiency, effectiveness, and satisfaction	Usability test, percentage of tasks, resources used, and SUS Questionnaire
Kamaghe (2021)	Usefulness and satisfaction	Tareas completadas en tiempo, Cuestionario SUS
Ojeda-Castelo et al. (2021)	Not specified	Cognitive walkthrough combined with thinking aloud. User Experience Questionnaire (UEQ) applied to students
Nahar et al. (2021)	Efficiency, ease of learning, memorability, errors, and satisfaction	Observation during the test; post-test self-completed questionnaire
Alnfiai and Alhakami (2021)	WCAG 2.0 for accessibility	Interview
Bintaleb and Al Saeed (2020)	Efficiency, effectiveness, and satisfaction	Two standardized questionnaires, observation and pretest and post-test interviews, and Morae software
Xie et al. (2020)	Ease of use, satisfaction, and usefulness	Protocol thinking aloud and post-study interviews
Maboe et al. (2019)	Not specified	Time to perform a task
Azeta and Daramola (2018)	ISO 9241-11: Efficiency, effectiveness, and satisfaction	Questionnaire of own elaboration

and this is achieved based on constant practice, which is one of the bases of teaching mathematics (Mateus-Nieves and Devia, 2021).

The work was done in a secondary school in Guadalajara, Jalisco, which incorporated an inclusive education program. It implements an inclusion practice that consists of harmoniously developing students' faculties in conditions of social, economic, physical, intellectual, and emotional vulnerabilities.

Two studies were conducted for this research. The first one aimed to gain a deep understanding of the students, which is essential from an ergonomics perspective due to the characteristics of these users. The second study evaluated the usability and accessibility of the proposals developed based on the first study's results.

6.1.1 Study 1. Understanding Users and Their Needs

A qualitative descriptive study was conducted to understand how students with visual impairments use platforms and other resources. Seven students and seven teachers, two of whom were blind, were interviewed in a semi-structured way. The interviews addressed topics such as general aspects, use of teaching materials, technological resources, learning platforms, accessibility tools, student development, and biotechnology classes. Figure 6.1 presents the sociodemographic description of the students who participated in the study, and Figure 6.2 presents the characterization of the participating teachers.

The following information was obtained from the interviews conducted with teachers and students regarding the various aspects of the teaching–learning process (Table 6.2).

6.1.2 Study 2. Usability Evaluation of a Mathematics Application

A comparative study was conducted to evaluate the characteristics of a math application related to accessibility and usability in the multiplication task to compare a proposal for improvement. The study was divided into two phases. In phase 1, accessibility was evaluated using the Google accessibility suite and the focus order. Phase 2 evaluated usability in three stages: the original application, a prototype with accessibility correction, and a second prototype considering user feedback. The application was selected because it allows voice interaction, which is helpful for blind students.

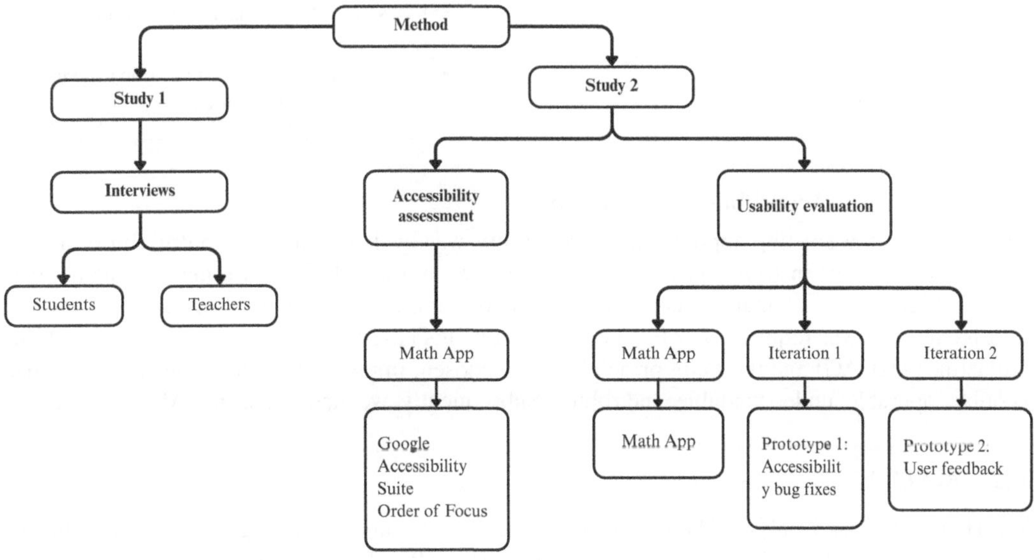

FIGURE 6.1 Method outline.

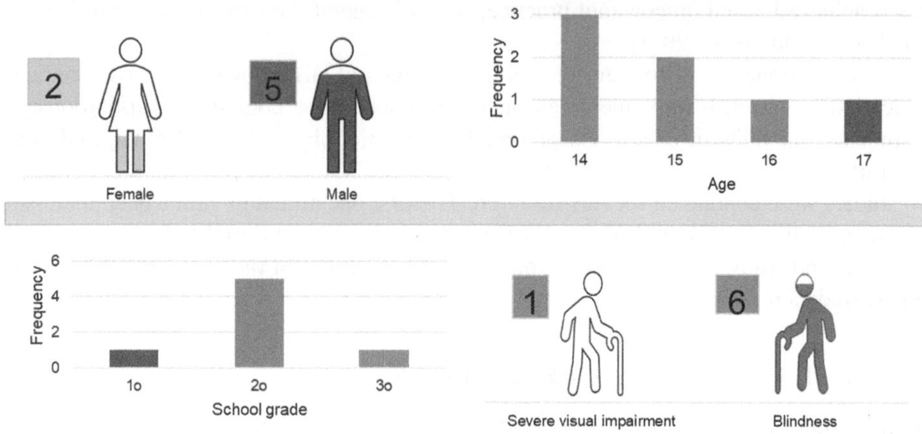

FIGURE 6.2 Demographic characteristics of the students.

TABLE 6.2
Key Themes from In-Depth Interviews

Topic	Teachers	Students
Experience in using technology	• Teach their classes using technology (4)	• Have a smartphone for recreational use (7) • Use Android smartphones (7) • Have a computer for school use (4) • Have a computer for recreational use (3)
Dynamics in class	• Received Braille training (6) • Use typhlotechnology (7)	• Use the Braille system and a ruler to write (7) • Do not like to use the Braille system (4) • Use the Perkins machine (3)
Experience with Google Classroom	• Used classroom (4) • Received training (3)	• Used classroom (7) • Described its use as difficult (6) • Identified problems with screen readers (4)
Use of typhlotechnology		• Use Google Talkback for its ease of use (7) • Use NVDA (Non-Visual Desktop Access) for its free access (6)
Opinion on typhlotechnology classes		• They perceive improvement in writing speed, learning, and computer use (4)

6.1.2.1 Phase 1. Accessibility Assessment

In this phase, the existing application's accessibility was evaluated to carry out usability assessments later. Various instruments were chosen to allow an accessibility assessment of the application, including two automatic accessibility evaluators and a focus order assessment for which a task breakdown was required. For this phase, instruments based on the Web Content Accessibility Guidelines (WCAG) and the main principles were chosen, implying that the designs must be perceptible, operable, understandable, and robust. Subsequently, we made a Journey Map.

6.2 RESULTS

The Hierarchy Task Analysis (HTA) was carried out from the application login to the end of the first training. This was divided into five main tasks: (A) login, (B) select the training, (C) select the topic, (D) select the training mode, and (E) carry out the training. For each of these tasks, the elements that make up the interaction, whether buttons or menus, were reviewed.

TABLE 6.3
Error Count by Subtask and Error Type

Subtask	Number of Error	Type of Error				
		Text Contrast	Element Label	Image Contrast	Item Description	Tactile Element
A	26	24	1	1	–	–
B	23	7	1	3	2	10
C	35	4	6	13	8	4
D	11	1	1	4	1	4
E	20	2	4	6	2	4
Total	115	38	13	27	13	22

By evaluating the focus order, that is, the reading order that the screen reader follows when navigating through the application, this analysis helped identify the user's journey and the possible needs derived from this feature. The biggest problem occurred in Task A since the user must go through all the numbers (153) before reaching the continue button.

In Task B, the reader starts with the main menu, continues with the title, then with the pro-version components, and ends with the practice types. In Task C, Topic Selection, the reader starts with the menu and the title and goes through each level until reaching confirmation.

In analyzing accessibility errors by task, Task C (topic selection) had the most accessibility errors. This can be related to the fact that it has the most considerable number of different screens, which leads to the most number of errors. However, Task D only had one screen, resulting in the lowest number of errors (Table 6.3).

6.2.1 Phase 2. Usability Evaluation

This phase evaluated the application's usability in its original version and two different redesigns. The first stage evaluated the application in its original state; subsequently, modifications were made based on the accessibility evaluation results. In the second stage, the first prototype was evaluated, considering the corrections of the accessibility problems; finally, based on the results of the second stage, a redesign was made, resulting in a second prototype, in which usability tests were also carried out.

Participants: Six visually impaired students from the secondary school participated in the tests.

Materials and equipment: Motorola XT1225 phone with TalkBack activated was used for the first stage. In the second and third stages, an iPad Air (4th generation) was used.

6.3 INSTRUMENTS

SUS Questionnaire. This is a ten-item questionnaire, each with a five-point Likert scale. Odd-numbered items have a positive tone; even-numbered items have a negative tone. According to Brooke (1996), participants should complete the SUS after using the evaluated system but before any briefing or other discussion.

SUM, i.e., single-usability metric, is a standardized, summed, single-usability metric. It was developed to account for most of the variations in four standard usability metrics used in summative usability testing: task completion rate, task duration, satisfaction, and error count.

6.4 PROCEDURE

The test's objective was explained to the participants as well as the general procedure to be performed. The five tasks were explained to the participant, and a questionnaire would be completed at the end of each task, in addition to the SUS questionnaire at the end of the test. During the task, the device screen was used to record the user's path to complete it. HTA (Table 6.4) was used in

TABLE 6.4
Error Count by Subtask and Error Type

Prototype	Subtask A Login	Subtask B Select Practice Type	Subtask C Topic Selection	Subtask D Change Training Mode	Subtask E Training
1	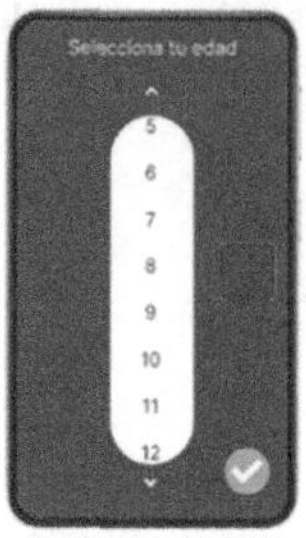 **Buttons:** Labeled "continue" button **Contrast:** Increased the contrast of the background and numbers	 **Contrast:** Contrast between the background and text in the title	 **Contrast:** Contrast between the background and text in the title **Tactile element:** The image and text were unified as a tactile element **Label:** Labeled "continue" button	 **Contrast:** Contrast in buttons and text **Tactile element:** Increased size of buttons **Label:** Labeled missing buttons	 **Contrast:** Contrast in buttons and text **Tactile element:** Increased size of buttons **Label:** Labeled missing buttons
2	The login subtask was eliminated because it does not stand for a useful function for the user since the application does not change its difficulty depending on age; this is because the evaluation is done by topic	 No modifications were made to the task.	 **Button:** The "continue" button was labeled **Tabs:** Labeled as tabs **Descriptions:** Renamed the levels and added an "available" one	 **Buttons:** Added "selected" **Descriptions:** Added descriptions when a # changed and user prompts **Elements:** Removed the question mark	 **Buttons:** Added "active to listen" **Descriptions:** Added descriptions when a # changed and user prompts **Elements:** Removed the question mark

Google Sheets to record the variables of time, errors, task completion, participant comments, and researcher observations. The free calculator in Excel Version 16.75.2 was also used for the SUM. Table 6.4 presents the prototypes used in the evaluation and the proposed modifications in each prototype.

6.5 COMPARATIVE ANALYSIS OF USABILITY IN THE THREE STAGES

In this section, we compare the metrics resulting from the tasks in each stage: the average satisfaction, the number of people who completed the task (completion rate), the average number of errors, and the average time in seconds of the task.

Regarding Task A: Login, the completion rate increased (40%), satisfaction was 0.6, and the time (155 seconds) and percentage of errors (0.4) decreased. In Task B: Select practice type, completion slightly increased (.07), supporting the same number of errors and completion rate during the three stages. This coincides with the decision of not making any changes to this task. In Task C: Topic selection, no positive or negative trends in satisfaction were observed. However, it had an increase in the completion rate (60%), as well as a decrease in errors (0.4) and time (73 seconds). It can be noted that there is still a need to explore new descriptions or designs for the "select level" screen, which generated the most number of problems.

About Task D, changing the training mode, we can notice an increase in satisfaction (1.6) and in the completion rate (80%), as well as a decrease in errors (0.6) and time (70 seconds). This is due to the labeling of buttons in the first instance and the elimination of unnecessary elements.

In Task E: Training, the lowest satisfaction among all the tasks can be noted because the students had difficulty completing the task. Since they had to answer the multiplications in most cases, they mentioned that it took them a long time or seemed difficult. The increase in time in this task (254 seconds) may be because as the stages progressed, more users could complete the task, which took a long time. After all, it involved listening to and answering at least three multiplications. (Figure 6.3).

Regarding the SUM score (Figure 6.4), Stage 1 had a result of 45.8%, placing it in the "worst possible" space. Stage 2 saw an improvement of 26%, placing it in the "average" space. Stage 3 had a percentage of 77.2%, placing it at the starting point of "average" to "good" (Sauro & Kindlund, 2005) (Figure 6.5).

6.6 DISCUSSION AND CONCLUSIONS

The increase in the score obtained in the SUM from the first to the second stage would be related to the accessibility corrections since, comparing them (Figure 6.6), a more significant increase is seen in the satisfaction and completion metrics and a decrease in errors and time. This is due to the

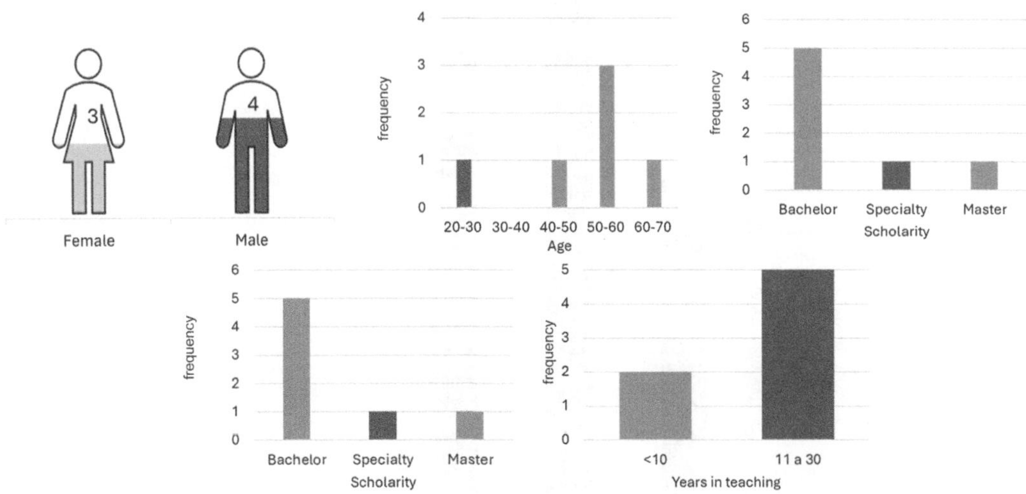

FIGURE 6.3 Demographic characteristics of the teachers.

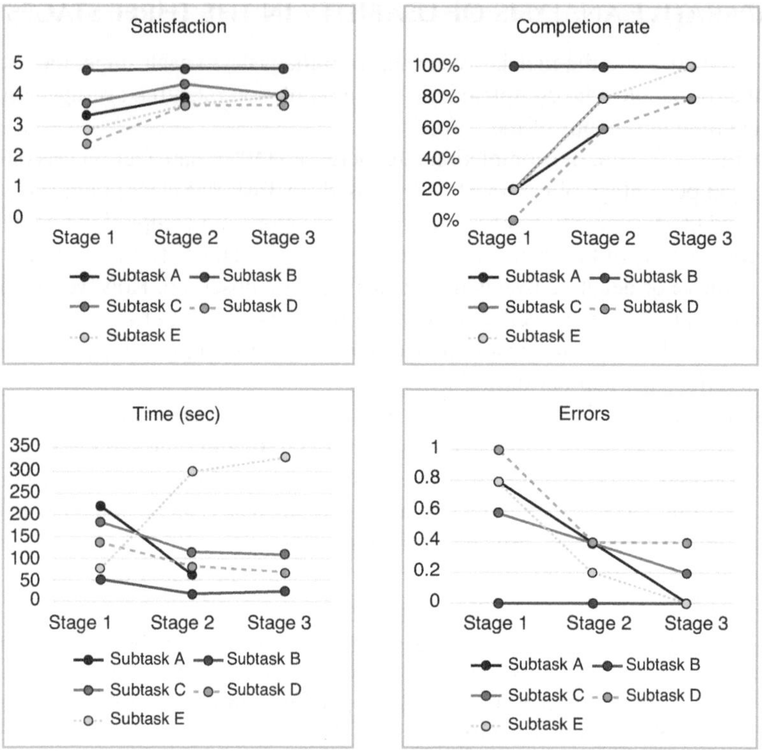

FIGURE 6.4 Comparison of usability metrics.

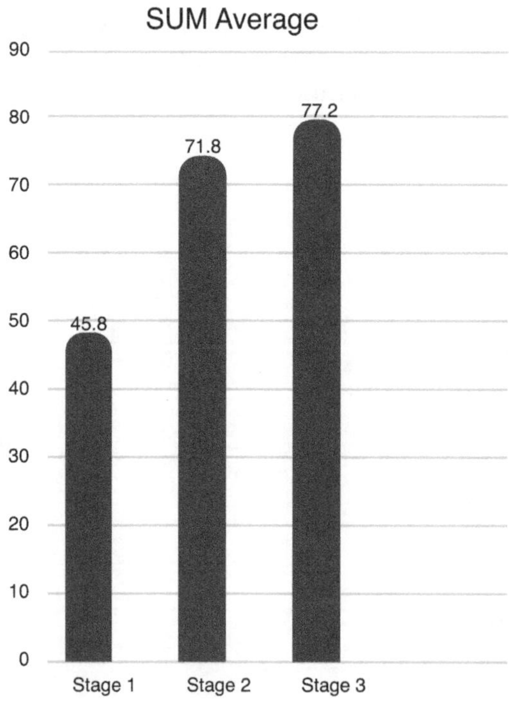

FIGURE 6.5 Comparison of SUM scores from the tests.

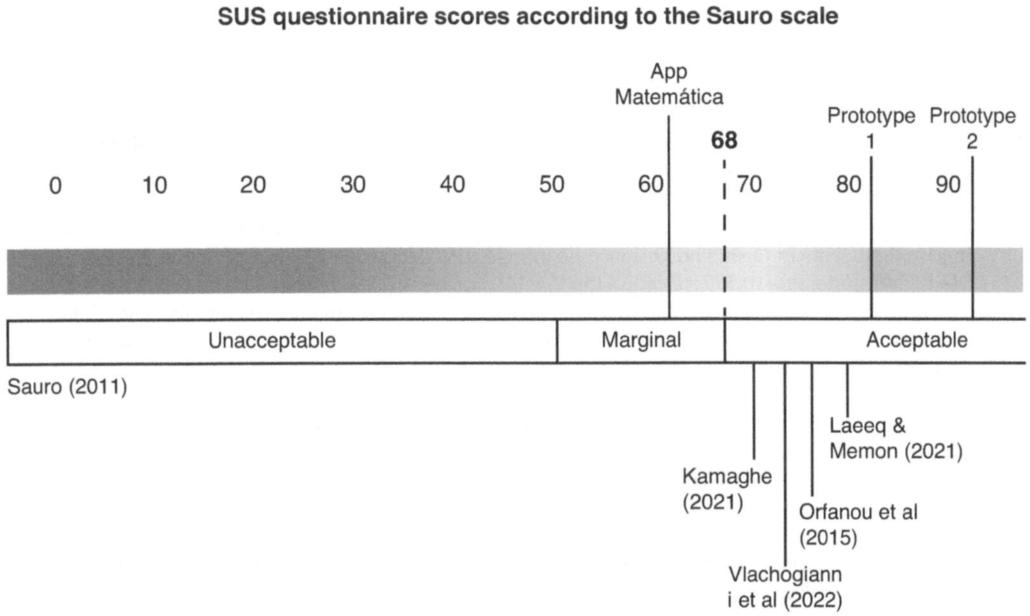

FIGURE 6.6 SUS questionnaire scores to the Sauro scale.

labeling of the buttons, which represented the main problem in Tasks A, C, D, and E. Solving this problem allowed users to complete the tasks, thus resulting in greater satisfaction.

Improvements in the accessibility-related features and interface design of learning resources increase usability in a specific task for secondary school students with visual impairments. Within these characteristics are the information architecture, the design of titles and subtitles, the design of earcons, the design of buttons, elements, and forms of interaction. An example of this is Task A, Login, in which the number selector menu was the element that generated the most significant problem. In Task C, Topic Selection, the names of the levels were modified since it was observed that they were not clear enough for the students. In Task D, Change Training Mode, and E, Training, the design of the different buttons for question-and-answer options and the buttons involved in training were involved.

6.6.1 Limitations and Future Work

During this research, limitations in the method were identified, such as finding more than one working automatic accessibility evaluator for existing applications, recruiting experts to evaluate, and avoiding bias by evaluating it by only one person. Additionally, there were problems with prototyping the application because the most common prototyping tools are not adapted to be tested with a screen reader. This suggests that there is a need for tools that allow us to design and test with different users from an earlier stage in the design process since a design that can work for people with disabilities works for everyone. In future studies or as next steps, more comprehensive evaluations could be carried out in different schools to include visually impaired students from different backgrounds and evaluate the application's functions, including theoretical learning. In addition, they could be evaluated by students with different disabilities to make the application accessible and usable for everyone. This study provides some recommendations to improve usability and accessibility in the application Mathematics: Mental calculation. Although the test was only carried out on one of the topics, these recommendations could be scaled to the other topics of the application in the following operations: addition, subtraction, multiplication, and division. This gives us an idea of the design elements that can specifically affect secondary school students with visual impairments.

REFERENCES

Alnfiai, Mrim, and Alhakami, Wajdi. 2021. "The Accessibility of Taif University Blackboard for Visually Impaired Students." *International Journal of Computer Science and Network Security* 21: 258–68. https://doi.org/10.22937/IJCSNS.2021.21.6.34.

Arksey, Hilary, and O'Malley, Lisa. 2005. "Scoping Studies: Towards a Methodological Framework." *International Journal of Social Research Methodology* 8 (1): 19–32. https://doi.org/10.1080/1364557032000119616.

Azeta, Ambrose A., and Daramola, Olawande. 2018. "A Voice-Based E-Examination Framework for Visually Impaired Students in Open and Distance Learning." *Turkish Online Journal of Distance Education* 19: 34–46. https://doi.org/10.17718/tojde.415635

Bintaleb, Hind Taleb, and Al Saeed, Duaa. 2020. "Extending Tangible Interactive Interfaces for Education: A System for Learning Arabic Braille Using an Interactive Braille Keypad." International Journal of *Advanced Computer Science and Applications* 11: 359–67.

Brooke, John. 1996. "SUS: A Quick and Dirty Usability Scale." In *Usability Evaluation in Industry*, edited by Patrick W. Jordan, B. Thomas, Ian Lyall McClelland, and Bernard Weerdmeester. Boca Raton, FL: CRC Press.

Burton, Matthew J., et al. 2021. "The Lancet Global Health Commission on Global Eye Health: Vision Beyond." *The Lancet Global Health* 9 (4): e489–e551.

de Prado-Llamazares, José Enrique, and Arias-Gago, Ana Rosa. 2022. "Enseñanza de las Matemáticas en Invidentes desde la Revisión Sistemática." *Perfiles Educativos* 44 (176): 169–84. https://doi.org/10.22201/iisue.24486167e.2022.176.57947.

INEGI. 2024. Encuesta Nacional de la Dinámica Demográfica (ENADID) 2023. Comunicado de Prensa Número 305/24. https://www.inegi.org.mx/contenidos/saladeprensa/boletines/2024/ENADID/ENADID2023.pdf (consultado 13 de febrero de 2025).

International Organization for Standardization. 2018. Ergonomics of Human-System Interaction — Part 11: Usability: Definitions and Concepts. https://www.iso.org/obp/ui/#iso:std:iso:9241:-11:ed-2:v1:en.

Kamaghe, Juliana. 2021. "Enhanced M-Learning Assistive Technology to Support Visually Impaired Learners in Tanzania: The Case of Higher Learning Institution." Doctoral dissertation, NM-AIST.

Laeeq, Kashif, and Memon, Zulfiqar Ali. 2019. "Scavenge: An Intelligent Multi-Agent Based Voice-Enabled Virtual Assistant for LMS." *Interactive Learning Environments* 29 (6): 954–72. https://doi.org/10.1080/10494820.2019.1614634.

Lewis, James R., & Sauro, Jeff. 2021. Usability and user experience: Design and evaluation. In G. Salvendy & W. Karwowski (Eds.), *Handbook of Human Factors and Ergonomics* (5th ed., pp. 972–1015). Wiley.

Maboe, Motlhabane Jacobus, Eloff, Mariki and Schoeman, Marthie. 2019. "Comparison of Students Using Electronic Learning Website of University of South Africa." In *2019 Conference on Information Communications Technology and Society (ICTAS)*, 1–6. IEEE. doi: 10.1109/ICTAS.2019.8703622.

Mateus-Nieves, Enrique, and Devia, Harold. 2021. "Desarrollo de Habilidades del Pensamiento Matemático desde la Formulación y Resolución de Problemas de Enunciado Verbal." *Acta Scientiae* 23 (1): 30–52. https://doi.org/10.17648/acta.scientiae.5845

Nahar, Latfun, Sulaiman, Riza and Jaafar, Azizah. 2021. "'Bangla Braille Learning Application' in Smartphones for Visually Impaired Students in Bangladesh." *Interactive Learning Environments* 29 (5): 821–34. https://doi.org/10.1080/10494820.2019.1619588

Ojeda-Castelo, Juan Jesus, Piedra-Fernandez, Jose A. and Iribarne, Luis. 2021. "A Device-Interaction Model for Users with Special Needs." *Multimedia Tools and Applications* 80 (5): 6675–710. https://doi.org/10.1007/s11042-020-10026-0

Sahin Kölemen, Cansu, and Akgün, Ögretim Üyesi Ergün. 2022. "Designing, Development, Implementation and Assessment of the Accessible Mass Open Learning Platforms for the Visually Impaired Individuals." *Turkish Online Journal of Educational Technology (TOJET)* 21 (3): 15–32.

Sauro, Jeff, & Kindlund, Erika. (2005). A method to standardize usability metrics into a single score. In G. C. van der Veer & C. Gale (Eds.), *Proceedings of the SIGCHI Conference on Human Factors in Computing Systems* (pp. 401–409). ACM.

Screpnik, Claudia, Negre-Bennasar, Francisca and Salinas, Jesús. 2024. "Educación Inclusiva con Juegos Serios y Diseño Centrado en el Usuario: Explorando la Intersección de Accesibilidad y Usabilidad." *RELATEC* 23: 27–50. https://doi.org/10.17398/1695-288X.23.2.27.

World Health Organization and World Bank. 2011. *World Report on Disability*. Geneva: World Health Organization. https://apps.who.int/iris/handle/10665/75356.

Xie, Iris, Babu, Rakesh, Lee, Tae Hee, Castillo, Melissa DAvey, You, Sukjin and Hanlon, Ann M. 2020. "Enhancing Usability of Digital Libraries: Designing Help Features to Support Blind and Visually Impaired Users." *Information Processing & Management* 57 (3): 102110.

Xie, I., R. Babu, T. H. Lee, M. D. Castillo, S. You, and A. M. Hanlon. 2020. "Enhancing Usability of Digital Libraries: Designing Help Features to Support Blind and Visually Impaired Users." *Information Processing & Management* 57 (3): 102110.

7 Usability in Product Design

Critical Factors and Regional Research Trends

Mayra Ivette Peña Ontiveros,
Cesar Omar Balderrama-Armendariz, David Cortés Sáenz,
Aide Aracely Maldonado-Macías, and
Juan Luis Hernández Arellano

7.1 INTRODUCTION

Ergonomics encompasses the interactions between humans and elements of a system, applying theory, principles, data, and methods to design to optimize well-being. This practice covers diverse factors such as physical, cognitive, social, and usability factors, which are related to physical activity when using an object (Bonilla 2017), in addition to being considered a scientific discipline that focuses on understanding the complexity of work systems and how the interaction between people and the elements of a system can influence its performance (Aceves-González et al. 2021).

As mentioned above, when considering ergonomics and its relationship with human factors, it is essential to understand the term usability and its practice; therefore, the following literature review is focused on usability, taking these and other related factors into account, as well as the impact that the development of this type of research has in Latin America.

In 1994, the International Organization for Standardization (ISO) defined usability as ease of use. In product design, usability is a highly important element since a product's usability design and evaluation are the features that allow a user to utilize it in a more effective, efficient, and satisfactory way (ISO 2018).

Despite the fact that usability provides a common framework for a variety of fields such as the following: product design, WEB design, computer engineering, human–computer interaction, and human factors, professionals of each take different approaches and consider different factors when evaluating usability; thus, it is common for them to use their own tools and methods to obtain specific results. The bibliography shows a great variety of tools in relation to the fields of human–computer interaction and human factors. In the case of product design, the variety is smaller, so knowledge about usability management and application during product process is of essence (Audoux et al. 2018; Borsci et al. 2019; Ferré 2005).

Usability can be understood as a methodology with a strong component for design evaluation; this includes a set of processes and methods that ensure the fulfillment of three factors essentially: first is functionality, which consists of a product structure showing the functions and skill of a person when performing a task or action, in addition to identifying the positive aspects of person–product interaction (William 2018). The second factor is utility, which consists of the satisfaction aspect (Huerte 2010) if a product is not used; it lacks utility (Pibernat 2008). Likewise, the objective of the product is to reach the goal and, thus, to be valid for its purpose (Pentti 2015). Third, there is compliance with design requirements, which is made up of a set of useful recommendations for product development. Because such factors identify the conditions that must be met to reach a solution, they turn the definition of the problem into measurable and verifiable objectives

DOI: 10.1201/9781003582014-9

(Medina 2013). Consequently, when usability is evaluated, these three factors must be considered, the stated objectives must be met, and a design that meets the user's needs must be developed (Hassan et al. 2004).

Regardless of a stable follow-up to ensure that a design process meets a set specifications (in addition to usability evaluations), it is also important to develop a design strategy since the objective is to facilitate the designer's work by providing the necessary tools and procedures to carry out the development of the product and to control its process (Fernández et al. 2010). A methodology is the study, structure, and application of methods based on their development, their explanation, and their justification in order to understand the research process and not only the results obtained (Maya 2014). In addition to being an orderly process through which it is sought to reach a result, for this reason, the hierarchy of tools, techniques, and operations is required in the process of an investigation (Sierra and Silva 2013). There are a great variety of methodologies to implement at each stage of the design process, but the important thing is the correct selection of which methodology to use, and for this, you need to recognize the market's perception regarding the idea and functionality of the product; in this way, determine the basis for collecting customer expectations and needs to start with an analysis and determine which would be the most appropriate with respect to the process of creating a product; each methodology serves in some way so that the final product is in the condition to satisfy the needs and expectations of what is to be achieved (Martínez et al. 2017).

There are several proposals for usability methods, which use certain means and techniques in order to measure different related aspects; it is important for product designers to understand the knowledge and the type of proposals that exist and have been developed over time, which are presented in this research. This way, they can have a broad view of the study, application, evaluation, and justification of usability to understand and implement it within the research, design, and development processes of a specific product (Maya 2014).

The aim of this research is to provide a better understanding of the terms, showing usability theoretical proposals and their applications over time, so that they can be implemented by product designers in process stages to achieve a better interaction with users.

7.2 METHODOLOGY

7.2.1 Materials and Methods

A linear and analytical methodology was followed to generate a sequence formed by means of phase. First, keywords were determined to extract information on usability in product design. From this point, a timeline is developed by detecting the year of publications and type of proposals (e. g., theories, principles, norms, tools, tests, methods, or techniques). Later, a description of each of the proposals by the author was presented. Then, a theoretical analysis was carried out through the classification of common factors to detect where and how usability has been worked on and how to introduce it in product development. Finally, a graphical analysis has been cited to observe the usability trend. Figure 7.1 shows a four-phase diagram to describe the used methodology.

Phases 1 and 2 correspond to the development on the selection and review of literature. Phases 3 and 4 are the results obtained from the analysis of each selected proposal. The development and the data obtained in each of the phases are described below.

7.3 DATA ANALYSIS

7.3.1 Timeline: Usability in Product Design

In phase 1, a list of keywords related to the usability of the product was developed. This research was carried out using popular databases to find publications on usability in product design, such as

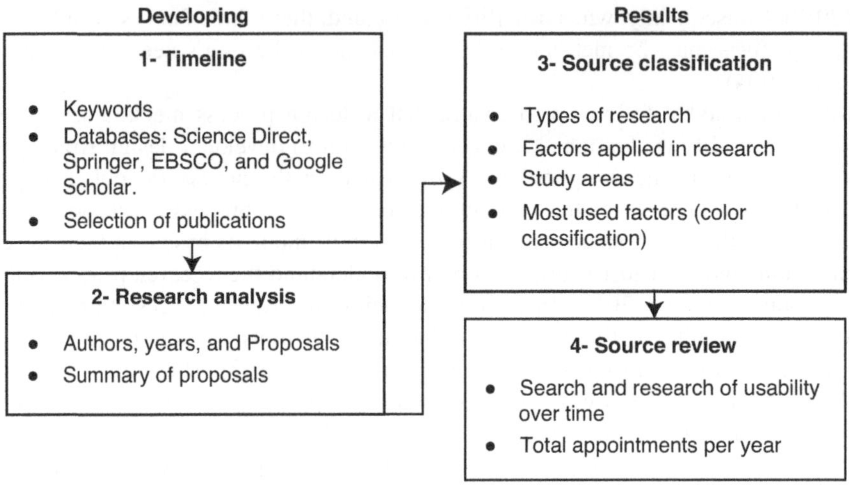

FIGURE 7.1 Research methodological structure.

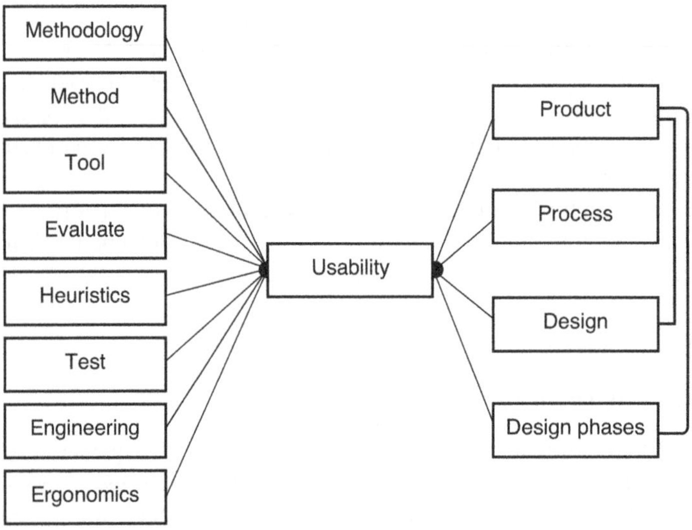

FIGURE 7.2 Search for keywords on usability in product design.

Science Direct, Springer, and EBSCO, as well as documents linked through Google Scholar search application. Combinations of English and Spanish words were used to cover the major sources of information, as seen in Figure 7.2.

A timeline is presented in Figure 7.3 based on the sources searched. These documents include publications related to usability and are considered relevant due to their contributions to sources such as books, book chapters, articles, and conferences, from 1954 to 2024. The products found were filtered to focus on the usability employed in product design. The databases generated a total of 50,400 results. About 250 publications were selected, discarding those unrelated to usability in design and those that did not fully meet the relevant keywords for the search, as well as those that did not address theories, techniques, or methodologies. Finally, 42 publications that met the established criteria were selected and color-coded to differentiate the types of sources: methods and techniques, tests, regulations and tools, and theories and principles.

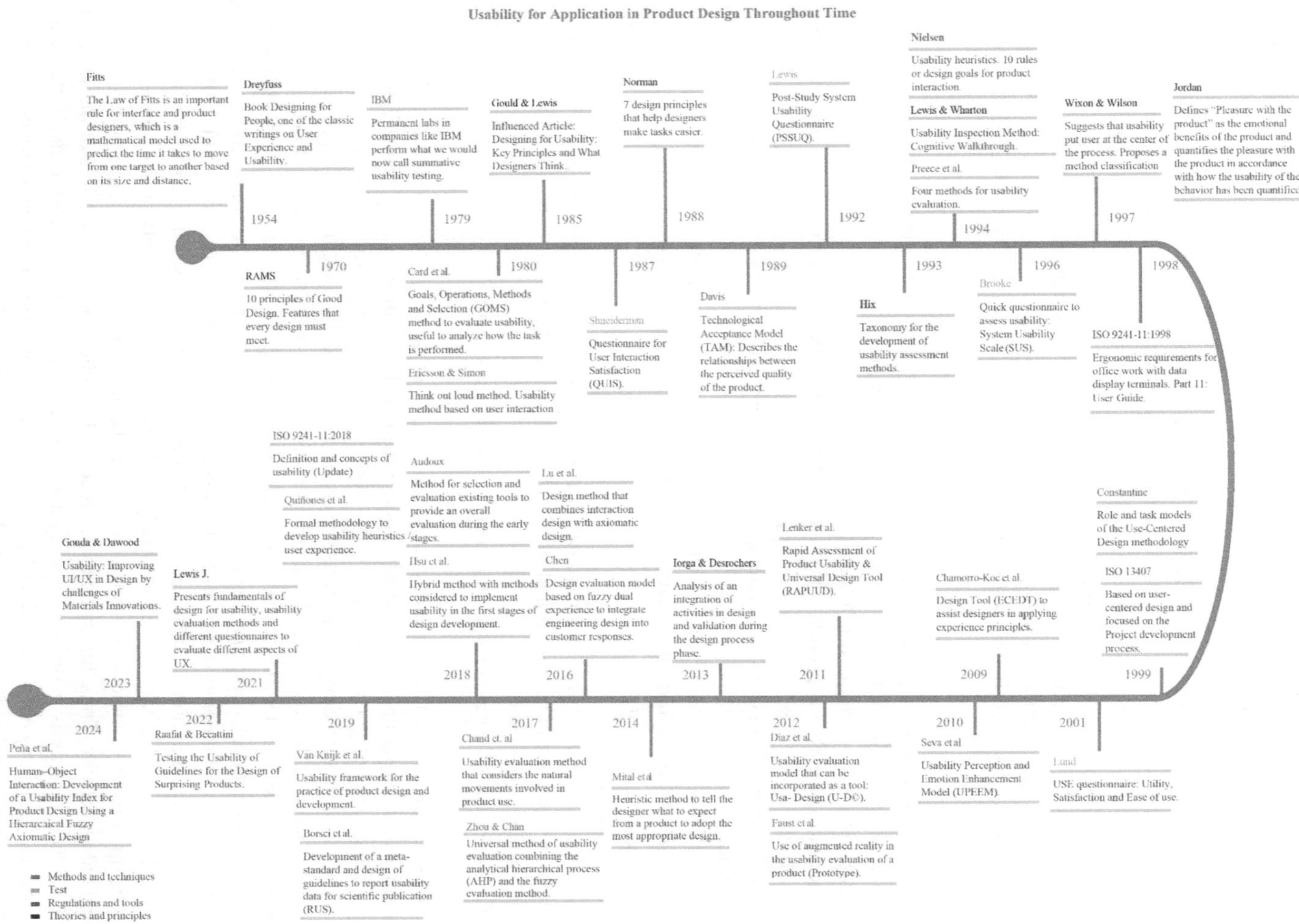

FIGURE 7.3 Usability for application in product design throughout time.

7.3.2 Research Analysis

Based on the publications featured in the timeline, a summary of each was created to highlight the objectives and provide a broader understanding of the authors' context. The data collected are presented chronologically below using the terminology in Figure 7.2.

1954: Paul Fitts publishes the Fitts Law, in which he describes a usability-focused model using a mathematical means to measure the time it takes for an object to interact. This law is applicable to human movement in fast and precise movements. In addition, it has had application in automobile design in terms of the distance of pedals, steering wheel, and brakes, among others. Therefore, the smaller and further away an object is, the longer the activity takes. Thus, this law affects the usability, efficiency, and experience of both the user and the product (Devin 2017; Luna 2013).

1970: Dieter Rams, one of the most influential designers in industrial design history, sets out ten principles of good design, in which he emphasizes the dynamic link between product, usability, consumption, and design creation, a relation that is essential in application during design development. Each principle is presented below, with a brief explanation of its use (Morales 2016; Palazuelos 2015; Vitsoe 2019):

1. *Good design is innovative*: Design innovation uses the opportunities offered by technology.
2. *Good design makes a product useful*: Design must be practical.
3. *Good design is aesthetic*: It is attractive to the user.
4. Good design makes a product understandable: Its function is expressed clearly through intuition.
5. *Good design is honest*: It never attempts to lie about the actual value and innovation of the product.
6. *Good design is discreet*: Every product and its design must be neutral and sober.
7. *Good design has a long life*: It creates valuable and timeless products.
8. *Good design is consistent in its details*: It must be cared for and designed with meticulous attention to detail.
9. *Good design respects the environment*: It contributes to the preservation of the environment by conserving resources and minimizing physical and visual contamination during the product's life cycle.
10. *Good design is design in its simplest expression*: Less is more.

1979: In its laboratory, the IBM company (International Business Machines Corporation) developed the summative usability test, which consists of a method called Plural Usability Run. This method involves three types of participants: users, developers, and usability experts, and it is useful for a general review of product design. The tests are carried out through prototypes on paper, where each participant interprets the role of the device user. Then, each of them should write down the actions to perform the task in as detailed a way as possible; this is to start a debate on the operation among users, then among developers, and ultimately among usability experts to, thus, collect data and detect errors and to make improvements to the product (Matos 2013; Vera and Giraldo 2011).

1980: Ericsson and Simon presented an influential work, which consists of a protocol called Think Out Loud. This method is one of the most widely used to evaluate the usability of a system or product. It is made up of methodological instruments based on the action of a user who thinks out loud while carrying out an activity. Thus, his thoughts are recorded to reflect on the objectives of the work and also to conduct an analysis in order to identify and solve errors or make improvements in the design so that the user can carry out the activities by using the service or product correctly (Armengol 2007).

1987: Shneiderman developed a software that consists of a questionnaire used to evaluate usability according to four attributes: efficiency, effectiveness, ease of learning, and product or service satisfaction (Barajas et al. 2017). This software is called as follows: questionnaire for the satisfaction of the interaction with the user (QUIS, for its acronym in English). In addition, this is considered as a quantitative method used to make changes and quantify the magnitude of design improvements (Chin et al. 1988).

1988: Norman raised design principles based on psychology, the nature of human cognition, emotion, action, and interaction with the world. These aspects must be applied and considered by design to obtain a better quality of life for the product. These principles focus on making improvements in usability or on detecting errors and developing a solution in the product so that the user has full interaction with the system or product. A list of the aforementioned principles is given as follows (Norman 2013):

1. *Equality of use*: The design must be easy to use and suitable for all people, regardless of their abilities and skills.
2. *Flexibility*: The design accommodates a wide range and variety of individual capabilities.
3. *Simple and functional use*: The operation of the design must be simple to understand, regardless of the experience or knowledge. It should eliminate unnecessary complexity.
4. *Understandable information*: The design communicates the necessary information to the user.
5. *Error tolerance*: The design minimizes the dangers and adverse consequences of accidental or involuntary actions.
6. *Low physical effort*: The design can be used efficiently and comfortably with minimum physical fatigue.
7. *Space and size for approach and use*: It is necessary to have spaces of adequate sizes to approach, reach, manipulate, and use the product, regardless of the user's size, posture, or mobility.

1989: Davis describes the relationships in the quality perceived in a technology-acceptance model (TAM), which was designed to assess usability by focusing on acceptance, the usefulness perceived to increase the effectiveness of a task and the ease of use that is referred to by the user's perception, and on the effort required for its use. The advantage of this model is the simplicity, adaptation, and solidity of use, as well as the fulfillment of what is expected when evaluating the product or system (Sánchez et al. 2015).

1992: To evaluate usability, Lewis introduces a questionnaire which is divided into three phases and seeks to obtain an immediate response from the user. The first part is administered after a task in a usability test (*After Scenario Questionnaire*). The second one is used for the evaluation of usability tests (*Post Study System Usability Questionnaire*). The third part addresses the field studies (*Computer System Usability Questionnaire*). The goals of these three phases are to assess user interaction according to ease of performance and completion of a specific task and to collect the opinion and assessment of the effectiveness and capacity of the information generated in the activity (Keinonen 1998).

1993: In this year, Hix, along with Hartson, describes two types of usability evaluations to be applied to a product or service, which depend on the characteristics of the methods used during their development but are also based on the objectives and requirements raised along the way. The methods are classified as follows (Hurtado and Castrillón 2013):

- Analytical evaluation, based on an analysis of the design characteristics, through revision, on the presentation of the product.
- Empirical evaluation based on the observation of users while they are using the product.

1995: Preece proposes four methods for the usability evaluation of a product or system. They are listed below (Preece et al. 2002):

- *Expert evaluation*: Also known as heuristic evaluation, it is normally carried out by experts in design. It requires a description of the problems or failures that they consider are faced by users, as well as suggestions for solutions.
- *Observational evaluation*: It enables data collection, which provides information about what users are doing when they interact with the product.
- *Research evaluation*: It uses questionnaires and interviews to know users' opinions or to understand their preferences about a potential or existing product.
- *Experimental evaluation*: In this type of assessment, an evaluator can manipulate and study its effects on user performance as they interact with the product.

1995: Nielsen developed usability heuristics, which consist of ten rules or design goals on interaction. They are used to evaluate the system or product. Therefore, they must be kept in mind before, during, and after the development of the project, as well as throughout the design process, from creation to changes in the project, to avoid usability problems when the user interacts with the product. Those ten rules are listed below. A brief description of how they are implemented and how they relate to the user product is also included so that they can be better understood (Helander 2006):

1. *Visibility*: It keeps the user informed of what is happening.
2. *Relationship with the system and reality*: It uses the language of the user that it addresses and organizes the information in a natural and logical order.
3. *Control and freedom*: It offers redo and undo functions, which allow users to have free control of their interactions.
4. *Consistency and standards*: It means establishing rules, situations, and logical actions and keeping them at all times.
5. *Error prevention*: It involves including error messages to prevent problems.
6. *Recognition instead of remembering*: It entails making objects, actions, and options visible to avoid having to remember information.
7. *Flexibility and efficiency in use*: It allows the system to be adapted to frequent users. Thus, it designs the accomplishment of advanced tasks in a smooth and efficient way.
8. *Aesthetic and minimalist design*: It shows only what is necessary and relevant in each situation.
9. It helps the user to recognize, diagnose, and recover from errors by signaling the specific problem that is occurring and suggesting constructive solutions.
10. *Help and documentation*: It should include helpful information that is brief, concise, easy to find, and focused on user tasks.

1996: Brooke carried out a questionnaire, which she called system usability scale (SUS), where the ISO 9241-11 standard is supported by indicators of effectiveness, efficiency, and degree of satisfaction. This questionnaire has a 0 to 4 scoring system, where Sections 1, 3, 5, 7, and 9 are rated, while Sections 2, 4, 6, 8, and 10 are rated with a section of five answers: excellent, remarkable, good, normal, and poor. In the end, a resulting analysis is generated from all the data obtained and from what is being assessed in the product; then a conclusion is reached as to whether the product is properly functional (Lilliam et al. 2015).

1997: Jordan defines "product pleasure" as the emotional benefits of the product. In light of this, he identifies such benefits through four dimensions:

- Physical pleasure is related to the touch toward the products.
- Social pleasure refers to the social relationships and communication that are obtained through the product.

- The pleasure of the psyche is achieved when products assist users in achieving tasks or activities.
- The pleasure of the idea refers to the values that the products and their use represent.

Jordan suggests the need to quantify product pleasure in relationship to the usability of the behavior toward the user. This allows for the generation of importance and value of pleasure in the product specifications (Keinonen 1998).

1998: The International Organization for Standardization develops the ISO 9241-11: 1998 standard called as follows: Ergonomic requirements for office work with data display terminals. Part 11: User Guide (1994 update). This standard explains the benefits of evaluating the usability of a product to satisfy the user. Also, it proposes obtaining the design, development, evaluation, and spread of information regarding a product's usefulness. However, it focuses solely on the office workspace with data display screen scenario. Nonetheless, in 2018, the regulations were updated to the current ISO 9241-11: 2018 Ergonomics of human–system interaction. Part 11: Usability: definitions and concepts. This change provides a framework to understand usability and apply it to systems, products, and services to achieve ease of use. In addition, it identifies the foundations of usability and explains the concept's application (ISO 2018).

1999: Constantine and Lockwood developed a method called usage-centered design. This method resulted from usability, so it is a component in the relation between product functionality and user. Likewise, the method works in such a way that attention is directed toward the product's design, its objectives, the tasks associated with its performance, and the skills required for the specific jobs or activities that the designer suggests. Therefore, focus on the user's tasks is all that is needed to understand the user and the context for which the product will be designed (Constantine et al. 2003).

2001: Lund developed a questionnaire to measure usability, called USE (usability, satisfaction, and ease of use). This questionnaire can be adapted according to the program, service, or product to be evaluated; thus, the data to be obtained are divided according to usefulness, satisfaction, ease of learning, and ease of use (Lund 2001).

2009: Chamorro-Koc, Popovic, and Emmison present an investigation on how human experience influences both people and the understanding of a product's usability. The results of the experiment led the authors to identify the relationships between experience, knowledge, and context of use, which influence designers and users. Such relationships were translated into design principles, which allowed experience to inform the activity in order to understand the use of a product. Designers used those results to develop an Experience-Context Enquiry Design Tool (ECEDT) to assist designers in applying these principles. Furthermore, this tool includes and interprets the findings in a way that informs designers about the aspects of experience that must be addressed by product design (Chamorro-Koc et al. 2009).

2010: Seva, Gosaico, Santos, and Pangilinan conducted a study on product quality and ease of use. In such research they developed a model of use to satisfy the user, called Usability Perception and Emotion Enhancement Model (UPEEM), which consists of four phases to follow, namely, product selection, attribute identification, design alternative generation, and design alternative evaluation. In this model, the phases are related to the perception of a product's esthetics to improve ease of design and product use (Seva et al. 2010).

2011: Lenker, Nasarwanji, Paquet, and Feathers developed a tool for the Rapid Assessment of Product Usability & Universal Design (RAPUUD). It consists of the development and psychometric tests of the evaluation of a product. This reporting tool is made up of 12 elements, seven of which are the principles of universal design and are applied to the user. However, the instrument is still being improved and could become a pragmatic tool for product designers to identify usability problems (Lenker et al. 2011).

2012: Díaz, Teixeira, Petry, Díaz, and Amaral propose a usability evaluation model with a methodological approach consisting of two stages. First, there is a literature review focused specifically on usability and product development. Then, it presents the proposed model, called *Usa-Design*

(U-D ©); this model consists of four phases: the first one is the compression of the product's context of use; the second is the evaluation of usability on the efficiency, effectiveness, and satisfaction of the product; the third stage evaluates the principles and results of usability; and finally, the U-D © characteristics of modularity and flexibility are obtained, allowing principles to be changed according to the needs and scenario of each situation. Thus, the authors use qualitative and quantitative measurement scales for the understanding and application of the model, where the results are obtained for the entire product–development process (Merino et al. 2012).

2013: Iorga and Desrochers conduct an analysis of an integration of design and usability validation activities for each phase of the design process, where they show certain elements that stem from the contributions of proposed methodologies in the academic and industrial fields of method design formalization. This allows for the finding of one or more parameter combinations that maximize or minimize a design criterion, while validation activities provide feedback to designers to verify the accuracy of the calculations and the achievement of all the design criteria. The proposed methodologies are useful to generate communication between designers, managers, and users (Iorga and Desrochers 2013).

2014: Mital, Desai, and Subramanian publish a book on product development. In Chapter 10, Design for Usability, they mention that it is the designer's responsibility to develop a highly usable product. This same chapter also focuses on a verification checklist based on a heuristic method, the purpose of which is to guide the designers' expectations for a product and their adoption of the most appropriate design as well as to relate the manufacturing processes to the users' requirements. The cases presented in this chapter were developed and approved through questionnaires for each of the elements focused on usability. The authors consider that this may be a future area on which product designers can focus and thus increase the overall attractiveness of a product (Mital et al. 2014).

2016: Lu, Feng, Zheng, and Tan publish a method where they combine interaction and axiomatic design. This method primarily uses interaction design to acquire the user's needs, in relation to three aspects: user, products, and environment. Later, several authors adopted the axiomatic design to complete conceptual design. The aim of the analysis is to demonstrate and validate the usability of a particular study, such as the study that the researchers presented about a children's bicycle, whose results show that the proposed method improves the experience of the users significantly, in addition to satisfying users' needs, especially in terms of spiritual and psychological needs (Lu et al. 2016).

2017: Zhou and Chan present a universal model for usability evaluation, where they combine the Analytical Hierarchical Process (AHP) and the fuzzy evaluation method. This method seeks to obtain a hierarchically structured index with terms based on three usability components: effectiveness, efficiency, and user satisfaction of a product or system. This form is a useful option to compare the advantages and disadvantages of different products or different versions of a product during its life cycle (Zhou and Chan 2017).

2018: Quiñones, Rusu, and Rusu publish an article where the issue of heuristic usability is addressed with the aim of evaluating application domains and specific characteristics related to usability. In addition, this research shows several sets of heuristics, which were shaped to develop a methodology to evaluate different usability aspects related to user experience. The methodology is based on three steps (Quiñones et al. 2018).

1. *Systematic literature review*: Examines the processes followed in product development and specifies usability heuristics and methodologies.
2. *Experimental results*: Several interviews and surveys are conducted to determine the perception of the evaluators who used Nielsen heuristics.
3. *Rusu methodology*: Application of the proposal and process analysis to create heuristics and processes in the design phases.

2019: Borsci, Federici, Maliza, and De Filippis have published a writing where a debate on the concept of usability, the lack of application of this knowledge, and harmonization was developed.

Thus, they developed a constructive way to practice usability in the different fields of knowledge, proposing the development of a meta-standard to generate a unified framework between the fields; in this manner, they create a connection and communication about how usability is characterized, represented, and operationalized in other fields. In addition, they design guidelines to report usability data for scientific publications (RUS), which can reduce replicability problems. However, the success of such guidelines requires international action by the usability communities and the support of publishers to demand compliance, as well as a unified way of summarizing and presenting data. Actions by the scientific community are needed to accomplish this and to implement these mitigation strategies and harmonize usability practice (Borsci et al. 2019).

2021: James Lewis and Jeff Sauro published a chapter called *Usability and user experience: design and evaluation*, where they present the terms of usability and user experience and the importance of the application of science, art, and craft focused on the principles of design, in addition to showing evaluation methods focusing on usability tests and the different standardized questionnaires for the evaluation of different aspects of UX (Lewis and Sauro 2021).

2022: Raafat and Becattini developed guidelines that can be used to create concept ideas for pop-up and wow-effect products and test their usability. The guidelines should allow different users with training in product design (mechanical engineers and industrial designers) to use them. These will be aimed at exploring and expanding the creativity of designers and motivating them to reach a higher level to generate ideas and capture the wonders of the creative mind in a step-by-step procedure (Raafat and Becattini 2022).

2023: Mahmoud Gouda and Mina Dawood carry out research where they propose a relationship between product design and interface design, in addition to finding out how important it is to evaluate usability from the entire design process to the interaction of the interface, incorporating new materials and technologies. This is to support the designer's decisions to achieve innovative materials that fit the parts of the product to be designed and provide functional solutions and treatments for user interaction interfaces and achieve an element of sustainability of the products in general (Gouda and Dawood 2023).

2024: Peña, Balderrama, and researcher colleagues conducted a study, where they designed and developed a methodology to measure the usability index in product design. Gathering proposals and concepts of usability focused on product design; this helps detect and group attributes through a series of questionnaires and a five-stage process, where attributes were collected and selected and were classified in a degree of importance; fuzzy values were assigned to obtain their numerical representation of use (usability index) and assign a qualitative category (Level of usability) through fuzzy, hierarchical, and axiomatic tools (Peña-Ontiveros et al. 2024).

7.4 SOURCE CLASSIFICATION

In this phase 4, we develop four classifications to understand what type of proposal is presented and where and how to use them; in this way, the designer can select and adapt it according to his objective. The procedure used to make these classifications was based on the articles presented by the authors; through readings, the elements, factors, or variables implemented were detected as well as a count of how many times these elements were used by each author. Also, what type of proposal is it, in what design phase is it applicable, and what types of results are obtained are also assessed.

7.4.1 First Classification: Type of Research and Proposals

Table 7.1 shows the first classification, which consists of identifying the type of methodological proposal addressed by the source to separate its general use in product design. In addition to identifying the type of proposal, it also presents the color code and the number of publications viewed in Figure 7.3 for a precise relationship. Furthermore, it can be determined that most of the publications analyzed are methods and techniques. Also, it can be shown that only four investigations

TABLE 7.1
First Classification of the Timeline

First Classification			
Color Code	**Classification**	**Number of Publications**	**Publications in Latin America**
Red	Methods and techniques	22	4
Yellow	Tests	4	0
Gray	Regulations and tools	6	0
Blue	Theories and principles	11	0

TABLE 7.2
Second Classification of Application of Factors and Attributes

Second Classification	
Interaction	Context of use
User experience	Method combination
User-centered design	Multi-attribute methods
Perception	Attributes – qualitative
Heuristics	Quantitative
Design esthetics	Initial phase
Anthropometry/biomechanics	Validation

TABLE 7.3
Classification by Fields of Study

Third Classification
Experience, emotions, usefulness, and satisfaction
Ergonomics
Conjugation
Measurement
Design stages

were found and analyzed in Latin America and are included in the classification of methods and techniques, leaving a shortage and an area of opportunity to develop new alternatives in each of the classifications.

7.4.2 Second Classification: Factors and Attributes Applied in Research

As shown in Table 7.2, the second classification offers the application purposes, which were identified based on the common areas most worked on in the timeline shown in Section 7.2. The value of this classification is the identification of the fields of research and their use in product design.

7.4.3 Third Classification: Fields of Study

This classification helps identify the area where the attributes and applications used by the authors in their research, shown in Table 7.2, are categorized. This aims to improve the scope and organization of each attribute that was identified in the second classification, as well as to determine and visually access the grouping of the attributes, which are presented in Table 7.3.

7.4.4 Fourth Classification: Frequency of Use

Using color-coding, this classification shows the most common attribute used by the authors. The most widely used attributes are shown in black; attributes with an average score are shown in gray; and attributes with low use are shown in white. This coding helps identify the fields of study in which the lowest and highest levels of research on product usability are found. For greater clarity, Table 7.4 provides a complete visualization in which each of the previous classifications was incorporated.

For further appreciation, Table 7.4 shows a complete visualization of the four classifications. Its purpose is to understand how the research conducted over the years is being implemented and to determine the attributes and factors involved in the development of new proposals in the hope of obtaining an organized perspective on usability implementation in product design. The most relevant observations are offered in the next section.

From the first classification, it can be gathered that over the years, more "methods and techniques" have been developed to evaluate product usability than to test it since only four publications for "tests" were selected.

The second classification focuses on the areas most worked on in publications. These were as follows: the use of "interaction" and the generation of qualitative results to obtain "attributes" to make usability applicable during the design phase of "validation." These two were the factors holding the greatest interest by authors when developing proposals for their research. The factors found to have a low application are the use of "cognitive perception," "heuristics," "design aesthetics," the product's "ergonomic" evaluation, and the development of "multi attribute methods."

The third classification shows the factor groupings. As can be seen, the group "experience, emotions, usefulness and satisfaction" has more factors repeated by authors, whereas the "ergonomic" and "conjugation" groups each have two as no other variables which could be incorporated into these groups were found in the publications. The group labeled as "design stages" shows that these can be applied in the first or last stages of design or during the design development process.

The fourth classification used three colors to show the high, medium, or low frequency levels. The following groupings resulted in the following: "interaction", "attributes," and "validation" formed the grouping with the highest level of frequency. An average frequency level was obtained by the factors of "user experience," "user-centered design," "context of use," "combination of methods," "quantitative results," and "initial design stage." Finally, the lowest level was obtained by the factors of "cognitive perception," "heuristics," "design aesthetics," "ergonomics," and "multi-attributes."

Finally, it was noticed that as time passes (1954–2024), either new attributes, factors, and methods are developed and added to existing proposals or new alternatives emerge. There is also an adaptation of methods to various fields of research. Figure 7.4 includes a quantification of data that shows the percentage of each factor and attribute used in the investigations for this study. The graph shows that the "validation" factor and the "qualitative attributes" rank highest, with a percentage of 17%, while "multi-attribute methods" and "anthropometry/biomechanics" rank lowest, with 1%. Similarly, "design aesthetics" and "heuristics" feature 2% of applications among the authors; "context of use," 3%; "cognitive perception," 4%; and combination of methods," 5% are considered low percentages. These factors and attributes create an opportunity for a tool to extend the usability evaluation in the design of the product.

7.5 SOURCE REVIEW

In phase 5, an analysis of the study and research of usability in product design is presented to determine the topic of increase over the years and detect where there is a weakness in this area of research.

To show interest in addressing these issues over time, citations were searched from the first publication in 1954. Figure 7.5 shows the results of the data obtained in this research, which were

TABLE 7.4
Classification of Research Proposals for Product Usability throughout Time

Classification of Authors for Usability Applied to Produce Design

	Third Classification		Experience, Emotions, Usefulness and Satisfaction					Ergonomics	Conjugation			Measurement		Design Stages	
First Classification	**Second Classification Author**	**Interaction**	**User Experience**	**User-Centered Design Cognitive**	**Cognitive Perception**	**Heuristics**	**Design Aesthetics**	**Biomechanical Anthropometry**	**Context of Use**	**Method Combination**	**Multi-Attribute Methods**	**Attributes**	**Quantitative**	**Initial Stage**	**Validation**
Methods and techniques	Card et al. (1980)	x	x									x			x
	Ericsson and Simon (1980)	x	x	x								x			x
	Davis (1989)				x							x			x
	Nielsen (1994)	x		x		x						x		x	x
	Preece et al. (1994)	x				x						x			x
	Lewis and Wharton (1994)				x							x		x	x
	Wizon and Wilson (1997)	x										x	x	x	
	Constantine and Lockwood (1999)	x	x						x			x		x	x
	Chamorro-Koc et al. (2009)		x	x	x				x			x		x	x
	Seva et al. (2010)		x		x		x					x		x	x
	Lenker et al. (2011)	x								x		x	x		x

(*Continued*)

TABLE 7.4 (*Continued*)
Classification of Research Proposals for Product Usability throughout Time

Classification of Authors for Usability Applied to Produce Design

	Third Classification		Experience, Emotions, Usefulness and Satisfaction					Ergonomics	Conjugation			Measurement		Design Stages	
First Classification	Second Classification Author	Interaction	User Experience	User-Centered Design Cognitive	Cognitive Perception	Heuristics	Design Aesthetics	Biomechanical Anthropometry	Context of Use	Method Combination	Multi-Attribute Methods	Attributes	Quantitative	Initial Stage	Validation
	Diaz et al. (2012)								x			x	x	x	x
	Faust et al. (2012)						x			x		x	x	x	
	Mital et al. (2014)					x	x					x		x	x
	Lu et al. (2016)	x	x	x			x			x	x		x	x	
	Rui-Yang Chen (2016)	x	x		x					x	x		x	x	x
	Zhou and Chan (2017)									x	x	x	x		x
	Chand et al. (2017)	x										x	x		x
	Hsu et al. (2018)		x				x			x		x		x	
	Audoux et al. (2018)									x		x	x	x	
	Quiñones et al. (2018)				x	x				x		x		x	x
	Van Kuijk et al. (2019)	x	x									x		x	x

(Continued)

TABLE 7.4 (*Continued*)
Classification of Research Proposals for Product Usability throughout Time

Classification of Authors for Usability Applied to Produce Design

		Third Classification	Experience, Emotions, Usefulness and Satisfaction					Ergonomics	Conjugation			Measurement		Design Stages	
First Classification	Second Classification Author	Interaction	User Experience	User-Centered Design Cognitive	Cognitive Perception	Heuristics	Design Aesthetics	Biomechanical Anthropometry	Context of Use	Method Combination	Multi-Attribute Methods	Attributes	Quantitative	Initial Stage	Validation
Tests	Shneiderman (1987)	x										x		x	
	James Lewis (1992)	x										x	x	x	x
	Brooke (1996)	x								x		x	x		x
	Lund (2001)							x				x	x		x
Regulations and tools	Hix (1993)											x	x		x
	ISO 13407 (1999)	x	x	x										x	x
	ISO 9241-11 (1998)	x	x											x	x
	ISO 9241-11 (2018)	x	x											x	x
	Borsci et al. (2019)		x									x		x	x
	Raafat and Becattini (2022)	x	x						x					x	x
Theories and principles	Dreyfuss (1954)		x									x		x	x
	Fitts (1954)	x		x									x		x
	Jordan (1997)	x	x									x			x

(*Continued*)

TABLE 7.4 (*Continued*)
Classification of Research Proposals for Product Usability throughout Time

Classification of Authors for Usability Applied to Produce Design

	Third Classification	Experience, Emotions, Usefulness and Satisfaction						Ergonomics	Conjugation			Measurement		Design Stages	
First Classification	Second Classification Author	Interaction	User Experience	User-Centered Design Cognitive	Cognitive Perception	Heuristics	Design Aesthetics	Biomechanical Anthropometry	Context of Use	Method Combination	Multi-Attribute Methods	Attributes	Quantitative	Initial Stage	Validation
	Gould and Lewis (1985)	x										x		x	x
	Norman (1988)	x	x	x	x							x		x	x
	Iorga et al. (2013)	x		x				x	x			x		x	x
	Lewis and Sauro (2021)	x	x	x		x			x	x					
	Gouda and Dawood (2023)	x	x							x				x	x
	Peña at el. (2024)	x	x			x			x	x	x	x	x	x	x

Fourth Classification

High
Medium
Low

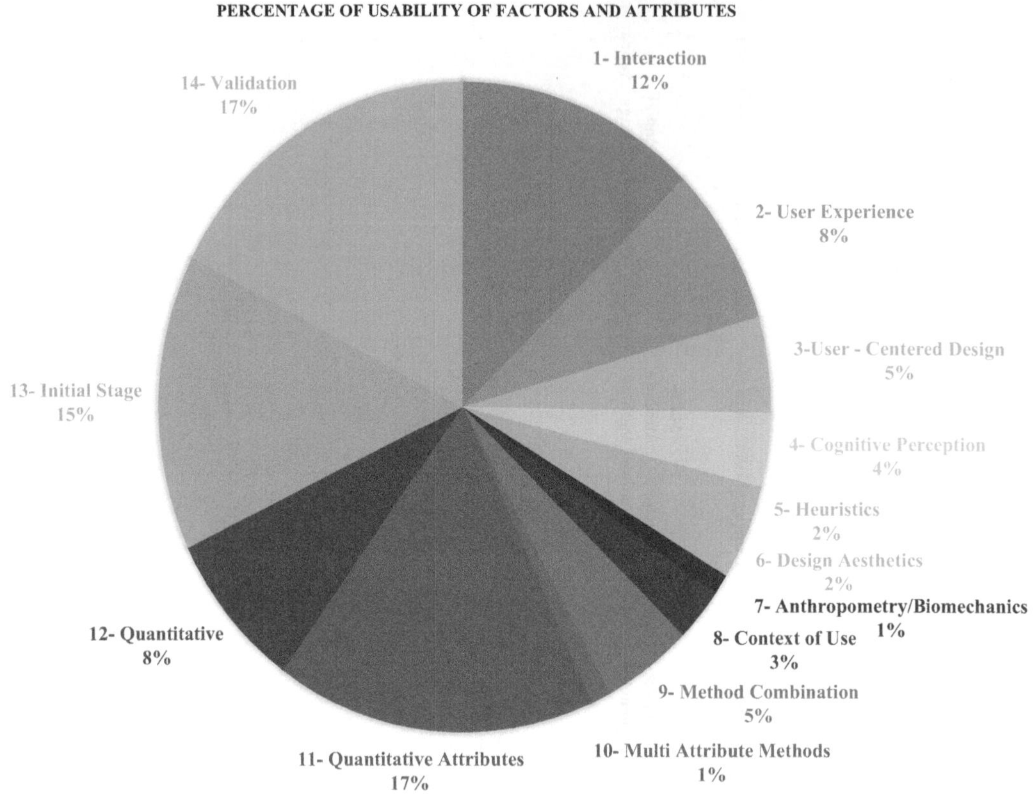

FIGURE 7.4 Usability factor percentage classification.

classified for this analysis by "author" and "article," including the number of citations in logarithmic form up to 2024. The data shown were consulted in each of the author's Google Scholar profiles and in the database where the number of citations for each article is provided.

The trend line in Figure 7.5 goes up and down. A notable increment is present from the early 1980s to 1988; after that, citations by author show a decrement until the present time. After an increment, as can be seen, the most recent authors have the fewest citations, despite the abundance of contributions around usability in the last decade. A case that stands out is that of Fred Davis and Norman, one of the most widely read authors for usability. Both Norman and Nielsen have developed proposals to implement during the design process of a system, product, or service, whereas current authors with low citations develop ideas based on the existing principles or proposals, and thus, in their research, they look to authors who have made a change and have contributed to the history of usability. This does not mean, however, that some authors are better than others, but rather that the usability proposals published today are not as well known and have not had as much impact as those made between 1954 and 2001. One of the causes of this phenomenon could be the lack of knowledge and harmony between the fields of specialization on concepts, principles, theories, tools, and methods, among others, regarding usability (Borsci et al. 2019).

Figure 7.6 shows a second analysis, which consists of a study of the total citations in the publications per year. This is to get a visual idea of how usability in product design has been researched over time. The data indicate that the first four citations were recorded in 1980 and that the highest number of citations occurred in 2024, reaching a total of 61,174. This indicates that the interest in investigating usability in product design increases as time goes by. From this clear and consistent

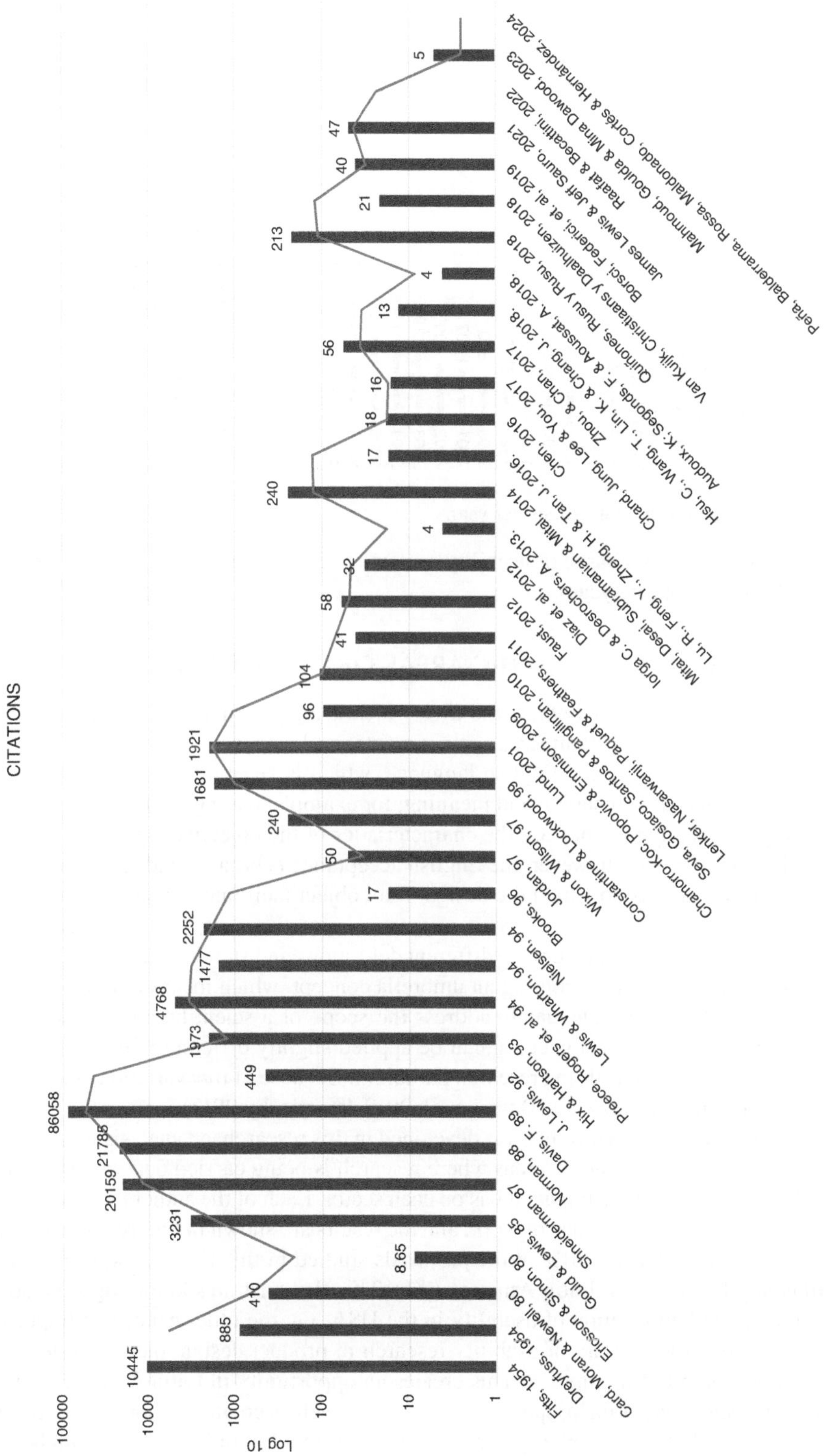

FIGURE 7.5 Citations per author.

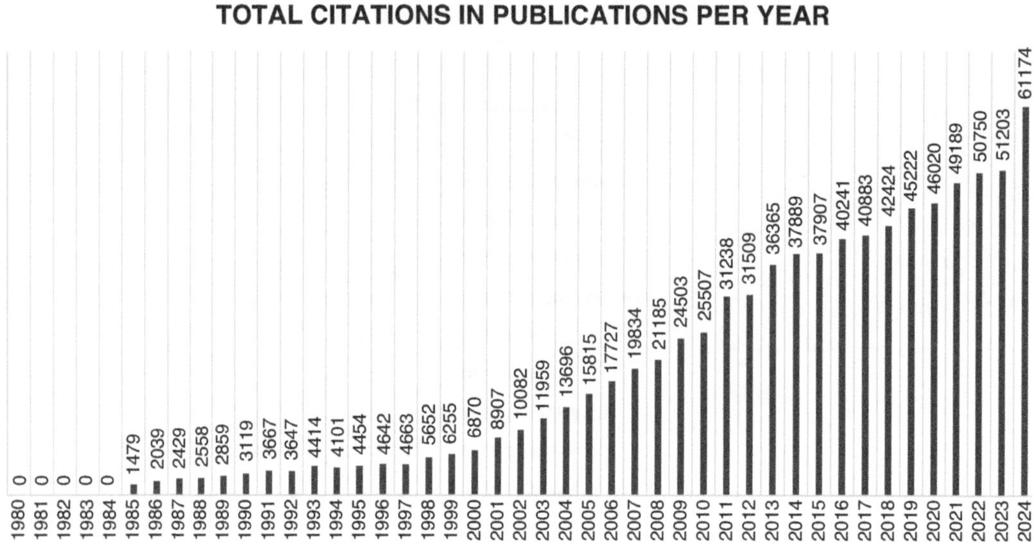

FIGURE 7.6 Total citations in publications per year.

trend, it can be implied that the issue of usability remains relevant in academic publications, with a positive outlook for the coming years.

7.6 PERSPECTIVE ON GEOGRAPHIC AREAS OF USABILITY RESEARCH

The term usability was defined in 1998 by the ISO, which is a global network where standards are developed that are required for commerce, governments, and society (Bonner et al. 2010). This concept is derived directly from the English language, which is "usability," and thus, as in many geographical areas, it has its translation and meaning; for example, having a Latin American focus, it means training for use, that is, they are the characteristics of the objectives specifically designed for the use of those that are not; however, the English acceptance is broader and refers to the ease or level of use, that is, the degree to which the design of an object facilitates or hinders its manipulation (Grau 2007).

This definition aims to unite and connect different definitions and standards of interaction design (ISO 2018); however, usability is becoming an umbrella concept, which means that these are terms that are too broad and weak to adequately address the scope of a scientific community. In other words, due to the breadth of this concept, it can be applied slightly differently in one area than in another and, therefore, lacks a unifying research paradigm. Because of this variation, it is important to harmonize this fragmented concept (Borsci et al. 2019; Tractinsky 2017).

According to the literature review that was developed in this research, an analysis was carried out to gain a perception of the geographic areas where research is being carried out, where usability is being worked on the most and the impact it has on each sector. Each of the authors was investigated, detecting their nationality and/or place of work, and the results are shown in Figure 7.7, where a high percentage in the United States of 51% of the proposals studied in this research was observed and with a significantly low result in Latin America (7%). This demonstrates knowledge and strength in the field of study and application of usability in the USA. On the other hand, it is important to emphasize that, despite the scarcity of usability research in product design, there is little research and application of this in Latin America. This creates an opportunity in Latin America to develop new studies in this field, generating proposals to validate, as well as create harmony on usability, not only in Latin America but also in other places in the world, in addition to designing high-quality, interactive, functional, and user-friendly products.

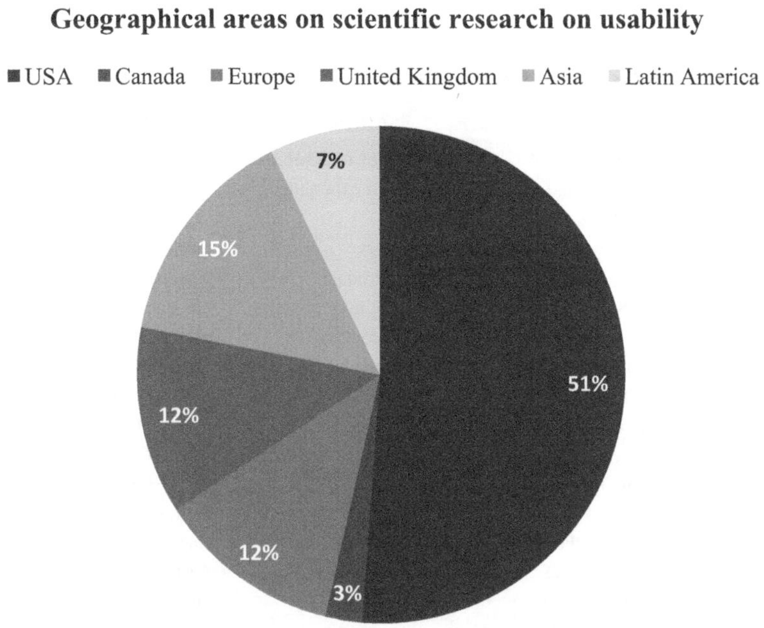

FIGURE 7.7 Percentages of scientific research on usability in geographic areas.

7.6.1 Latin America Perspective

As mentioned above, 7% of the 42 investigations analyzed for this research were found in Latin America, which are as follows: Brazil, Chile, and México. In Chile, research by Quiñones, Rusu, and Rusu was carried out, where they propose a process with three phases to evaluate usability, which consists of a systematic review of the literature, obtaining experimental results, and proposing the Rusu Methodology for applying the proposal and analyzing processes, which, in each phase, considers both usability and Nielsen heuristics (Quiñones, Rusu, and Rusu 2018); On the other hand, in Latin America, in Brazil, Diaz Merino's research team published an article where they developed a model called Usa-Design Model (U-D) ©, which is made up of four phases, to determine the understanding of the product's usage context, carry out usability evaluation, evaluate the principles, and obtain the modularity and flexibility characteristics of the U-D © and in this way obtain attributes and evaluate both the environment and the usability of the product (Merino et al. 2012). And, in México, a methodological tool was developed to determine the usability index of the product by Peña, Balderrama, and collaborators, where they carried out an analysis to detect those attributes that have a high degree of importance in usability, which are implemented during the methodological process that consists of a series of questionnaires to evaluate what is expected of the product and what the design should have, and in this way determine the weight to obtain the level and usability index of the product (Peña-Ontiveros et al. 2024).

As can be seen in Table 7.5, each of these investigations proposes methodologies and methods to evaluate the usability of the product, where they consider attributes that are linked and must be considered when evaluating usability. Both in Latin America and in other regions, they try to apply these attributes when developing a method or tool and evaluating the usability of the product design without leaving aside these factors that are essential for its validation.

7.7 DISCUSSION

The linear methodology that was proposed served to have an orderly follow-up, showing the parts by which this investigation was submitted, as well as what came from each search, analysis, and results that were extracted.

TABLE 7.5
Selected Proposals on Usability from Latin America

Geographical Area	Authors	Article	Attributes	Type of Proposal
Chile	Daniela Quiñonesa, Cristian Rusu, and Virginica Rusu	A methodology to develop usability/user experience heuristics	Heuristics User Experience Cognitive perception Qualitative	Methodology
Brazil	Giselle Díaz Merino, Clarissa Stefani Teixeira, Rodrigo Petry Schoenardie, Eugenio Diáz Merino, and Leila Amaral Gontijo	Usability in product design – The importance and need for systematic assessment models in product development – Usa-Design Model (U-D) ©	Context of use Satisfaction Effectiveness Efficiency Functionality Quantitative Qualitative	Method
México	Mayra Peña-Ontiveros, Cesar Omar Balderrama, Alberto Rossa-Sierra, Aide Maldonado-Macias, David Cortés, and Juan Luis Hernández	Human–Object Interaction: Development of a Usability Index for Product Design Using a Hierarchical Fuzzy Axiomatic Design	Context of use User experience Quality Aesthetics Functionality Satisfaction Effectiveness Efficiency Quantitative	Methodology

During the search through the databases, it was detected that more than 60% of the proposals focused on interface design, leaving other fields with few articles published worldwide, such as biomedical and product design. One of the main objectives of this work is to make known that there are a variety of proposals to evaluate the usability of the product and, in addition, to detect an approach to guide the development of new proposals and/or to generate improvements to the existing ones.

The term "usability" is imposed by the English language, which is what happens with many terms in the field of research. Therefore, it is essential to define the concept of usability to establish a uniform relationship with other disciplines, for example, that it has the same meaning in the use and application of usability tools both in the United States and in Latin America since this term is directly derived from English. In Spanish, it means "capacity of use," that is, the characteristic that distinguishes objects designed to be used from those that are not. But, in English, it is broader, which refers to the ease of, as mentioned above, the degree to which the design of an object facilitates or hinders its manipulation (Grau 2007).

Many opinions have arisen around the subject, and the results of this paper agree with the point of view of Borsci et al. (2019), who raise a debate on the concept of usability, questioning the opinion of the author Tractinsky who proposed that usability reached a dead end, that is, an insufficiently solid concept and a construction incapable of providing stable results and unifying scientific knowledge; to differ from this, it is said that usability is not dead, but it is not sufficiently known in research fields. Borsci and colleagues propose harmony between fields to study and evaluate usability in different projects; for example, the concept is implemented in the same way in the interface design as in the product design, as well as in different geographical areas, to expand the use of usability in the world. And in this way, we expand the knowledge about usability and the application of existing tools.

The analysis on usability in product design through the time span covered in this study is one way to present the studies that have been carried out as well as the evolution that the topic has undergone, presented through classifications to separate the type of proposals and study areas in which they are focused, as well as the most used factors and attributes. Thus, this research is a means for understanding, comparing, and identifying an opportunity to develop new usability tools, methods, or methodologies, in addition to knowing where and how to implement them by professionals and students.

The aim of the data collected in this work is to inform on the interest that the term *usability* has drawn, as well as to identify common factors that can guide designers and like specialties to help offer them a more complete perspective for the selection and use of methodological tools. Also, it is important to create an understanding of the meaning of usability, both in the field of research and in the geographical area.

7.7.1 Usability and Interaction

It is important to consider usability when designing products to improve their ease of use, just as it is essential to consider interactions since these generate how people communicate and establish relationships in their daily and work-life. The two terms develop a connection to create a compelling user experience (Preece et al. 2015). Interaction design concerns how a system must respond to a user's actions (Soler-Adillon 2012). It is related to helping the user reach their goal through fluid and pleasant interactions with a product (Preece et al. 2015).

Buchanan defines interaction design as to how human beings relate to other human beings through the mediating influence of products, services, and systems (Villagómez 2019). In addition to physical representation, there are experiences, activities, and solutions, which are integrated into a new understanding of what a product is or could be. This definition provides the basis for interaction in design, which evolved from interests in the product's impact and social interaction in use (Niedderer 2007).

People have many notions about design, which refers only to decoration or style (i.e., visual appearance). While there is nothing wrong with an attractive esthetic, the design can be more than that; integrating graphic communication and industrial design provides working methods where interaction design is integrated to evaluate the final product. Although we experiment with examples of good and bad designs every day, interaction design as a discipline is complex to define. However, it's part of interdisciplinary roots: involving industrial design, communication, human factors, and human–computer interaction (Saffer 2010).

The concept of usability is applied to situations involving systems (including environments, engineering, software, and simulation), products (including industrial and consumer products), and services (including technical and personal), with the purpose of making them interactive and more helpful by focusing on the use and application of techniques and methods of human factors and ergonomics (ISO 2018).

In other words, interaction is the physical relationship that the user establishes with a product; therefore, various knowledge branches such as ergonomics, user experience, and usability must be considered and studied to understand the interactive process better (Puyuelo Cazorla et al. 2011). The IxDA (Interaction Design Association) says that: "Interaction design defines the structure and behavior of interactive systems; designers seek to create meaningful relationships between people, their environment and products" (IxDA 2005). Interaction design, usability, and other disciplines help us plan and develop an excellent product to offer a good user experience (Cantú 2017).

As previously presented, it isn't easy to separate interaction and usability. For this reason, it is crucial to introduce these types of studies to develop better designs considering methods, techniques, and principles to generate fully functional, attractive, and safe products with an effective interactive to meet user needs.

7.8 CONCLUSIONS

The authors' proposals presented in this research represent an alternative to a study guide for product designers to generate their own selection of methods and develop a functional structure in their methodology; with this, it is possible to evaluate the usability of the product and, likewise, solve the customer needs and obtain a quality and highly usable product, in addition to promoting study and application in the design process.

Three main observations were obtained, first, usability validation was the most used factor in bibliography, which shows that authors focus more on evaluating the usability of the product; second, most of the proposals obtain qualitative values for the evaluation; and third, the researchers also focus on the interaction of the product; however, they leave factors aside as biomechanics and esthetics are alternatives that can be considered in future research. In addition, it can be observed that as time has passed, new proposals continue to be investigated and developed to evaluate usability by introducing new technologies in this professional field. The number of citations found in the initial data suggest that the works that have sparked the greatest interest from the academy took place in the 1980s. On the other hand, the number of authors who have contributed knowledge to the topic has increased over the last decade. Thus, it can be concluded that the interest in usability has observed an upward slope since 1980 and that the subject remains quite current in academic publications. The percentage of the geographic area gives a perspective on the applications of this concept in the world, generating the idea that usability has a strong development and that studies in the United States compared to Latin America are minimal, so it is important to develop and enrich the concept of usability and its application both in Latin America and worldwide.

This type of research is important for designers to develop or provide feedback for their knowledge about this usability term, which is little known or used in the evaluation of a product. This research offers a review regarding the context of usability and that it is necessary to consider at the time of its application or its study.

REFERENCES

Aceves-González, Carlos, Landa-Ávila, Irma Cecilia, Carvalho, Fernando, Ortega-Ruiz, Bertha Alicia and Jun, Gyuchan Thomas. 2021. "Ergonomía En Los Sistemas De Salud De América Latina: Revisión Sistemática De La Situación Actual, Necesidades Y Desafíos Futuros." *EID. Ergonomía, Investigación y Desarrollo* 3 (2): 10–27.

Armengol, Lurdes. 2007. "Los Protocolos de Pensamiento En Voz Alta Como Instrumento Para Analizar El Proceso de Escritura." *Revista Española de Lingüística Aplicada* 20 (20): 27–36.

Audoux, Kévin, Segonds, Frédéric, Kerbrat, Olivier, and Aoussat, Améziane. 2018. "Selection Method for Multiple Performances Evaluation during Early Design Stages." *Procedia CIRP* 70: 204–10. doi:10.1016/j.procir.2018.03.295.

Barajas, Manuel Alejandro, Reyes, Rosa Maria, Riva, Jorge & Maldonado, Aide. 2017. "Estudio Comparativo de Cuestionarios Para La Evaluación de La Usabilidad En Software." *Revista Ingeniantes* 1 (2): 1–7.

Bonilla, Enrique. 2017. "La Ergonomía: Factores Humanos En El Diseño." *Espacio Diseño* 252: 17–21.

Bonner, P., Inklaar, A., & Holloway, G. (2010). Organismos nacionales de normalización en países en desarrollo. *Progresar Rápidamente*. pp. 1–88

Brooke, John. "SUS: A quick and dirty usability scale," In P. W. Jordan, B. Thomas, B. A. Weerdmeester, and I. L. McClelland (Eds.), *Usability evaluation in industry*, pp. 189–194, Taylor & Francis, 1996.

Borsci, Simone, Federici, Stefano, Malizia, Alessio and De Filippis, Maria Laura. 2019. "Shaking the Usability Tree: Why Usability Is Not a Dead End, and a Constructive Way Forward." *Behaviour and Information Technology* 38 (5): 519–32. doi:10.1080/0144929X.2018.1541255.

Cantú, Alberto. 2017. "Intuitivamente." https://blog.acantu.com/que-es-diseno-interaccion/.

Card, Stuart K., Moran, Thomas P., and Newell, Allen, 1980. "The keystroke-level model for user performance time with interactive systems," *Communications of the ACM*, 23 (7): 396–410.

Chamorro-Koc, Marianella, Popovic, Vesna, and Emmison, Michael. 2009. "Human Experience and Product Usability : Principles to Assist the Design of User – Product Interactions." *Applied Ergonomics* 40 (4): 648–56. doi:10.1016/j.apergo.2008.05.004.

Chand, Joonho, et al. 2017. Development of a usability evaluation method using natural product-use motion. *Applied Ergonomics* 60: 171–182.

Chin, John Patrick, Diehl Virginia Ann, and Norman, Kent Lynn. 1988. "Development of an Instrument Measuring User Satisfaction of the Human-Computer Interface." Proceedings of SIGCHI 1: 213–18.

Constantine and Lockwood, Constantine, Larry L., and Lockwood, Lucy A. D. 1999. *Software for use: A practical guide to the models and methods of usage-centered design*, ACM Press/Addison-Wesley.

Constantine, Larry L., Biddle, Robert, & Noble, James. 2003. "Usage-Centered Design and Software Engineering: Models for Integration." In *Conference: Proceedings of ICSE 2003 Workshop on Bridging the Gaps Between Software Engineering and Human-Computer Interaction*. Portland, OR.

Davis, Fred D. "Perceived usefulness, perceived ease of use, and user acceptance of information technology," *MIS Quarterly*, 13, no. 3, pp. 319–340, 1989, doi: 10.2307/249008

Devin, Fabio. 2017. "Historia de La Experiencia de Usuario – Primera Parte: Ergonomía y Psicología." *UXpanol*. https://uxpanol.com/historia-de-la-experiencia-de-usuario/historia-de-la-experiencia-de-usuario-primera-parte-ergonomia-y-psicologia/.

Diaz, Giselle, et al. 2012. "Usability in product design—the importance and need for systematic approach," *Work*, 41 (1): 3251–3254.

Dreyfuss, Henry. *Designing for people*. Skyhorse Publishing Inc., 2003.

Ericsson, K. Anders, and Simon, Herbert A. 1980. *Protocol analysis: Verbal reports as data*, MIT Press.

Faust, F., Roepke, G., Catecati, T., Araujo, F., Ferreira, M. G. G. and Albertazzi, D. "Use of augmented reality in the usability evaluation of products," *Work*, vol. 41, pp. 1164–1167, 2012.

Fernández, Jorge, López, Víctor, Frenández, Carlos. & Viladás, Xènia. 2010. *Diseño Estratégico; Gupia Metodológica*. Fundación Prodintec.

Ferré, Grau Xavier. 2005. "*Marco de Integración de La Usabilidad En El Proceso de Desarrollo Software*." Madrid: Universidad Politécnica de Madrid.

Fitts, Paul M. 1954. "The information capacity of the human motor system in controlling the amplitude of movement," *Journal of Experimental Psychology*, 47 (6): 381–391.

Gouda, Marwa & Dawood, Mai. 2023. "Usability: Improving UI/UX in Design by Challenges of Materials Innovations." *International Design Journal* 13 (1): 37–56. doi:10.21608/idj.2023.276010.

Gould, John. D., and Lewis, Clayton. 1985. "Designing for usability: Key principles and what we know about them," *Communications of the ACM* 28 (4), 300–311.

Grau, Por Jordi. 2007. "Pensando En El Usuario : La Usabilidad Contrataban." *Anuario ThinkEPI*, 1, 172–77.

Hassan, Yusef, Martín, Fransisco J. & Iazza, Ghislaine. 2004. "Diseño Web Centrado En El Usuario: Usabilidad y Arquitectura de La Información." *Hipertext.Net*. https://www.hipertext.net/web/pag206.htm.

Helander, Martin. 2006. "7.4 Usability." In *A Guide to Human Factors and Ergonomics*, 2nd ed., 128–143. Boca Raton, FL: CRC Press.

Hix, Deborah, Hartson, H. Rex. *Developing user interfaces: ensuring usability through product & process*. John Wiley & Sons, Inc., 1993.

Huerte, Felipe. 2010. "El Concepto de Utilidad Según John Rawls." *Revista de Filosofía, Derecho y Política* 11: 127–42.

Hurtado, Luini Leonardo, and Castrillón, Omar Danilo. 2013. "Una Metodología Automatizada Para La Evaluación de Usabilidad de Interfaces de Supervisión Industrial an Automated Methodology for Usability Evaluation of Interfaces of Industrial Supervision" 24: 95–105. doi:10.4067/S0718-07642013000400011.

Hsu, Chin Chao, Wang, Tzone. I., Lin, Ke Jie, and Chang, Jia Wei, 2018. "The effects of the alternate writing and sketching brainstorming method on the creativity of undergraduate industrial design students in Taiwan," *Thinking Skills and Creativity*, 29: 131–141.

Iorga, Constantine, and Desrochers, Alain. 2013. "Generic Product Design & Validation Methodologies at the Detailed Design Stage." *European Scientific Journal* 9 (1): 1–20. https://www.eujournal.org/index.php/esj/article/view/689.

ISO. 2018. Ergonomics of Human-System Interaction – Part 11: Usability: Definitions and Concepts. https://www.iso.org/standard/63500.html.

ISO 9241-11. 1998, ISO. "Ergonomic requirements for office work with visual display terminals (VDTs)—Part 11: Guidance on usability," *ISO Standard* No. 9241-11, 1998.

ISO 13407. 1999. ISO. "Human-centred design processes for interactive systems," *ISO Standard* No. 13407, 1999.

Ix DA. 2005. "IxDA. About & History." https://ixda.org/ixda-global/about-history/.

Jordan (1997), Jordan, Patrick W. *An introduction to usability*, Taylor & Francis, 1998.

Keinonen, Turkka. 1998. *One-Dimensional Usability: Influence of Usability on Consumer's Product Preference*. Helsinki: University of Art and Design Helsinki UIAH.

Lenker, James A., Nasarwanji, Mahiyar, Paquet, Victor and Feathers, David. 2011. "A Tool for Rapid Assessment of Product Usability and Universal Design: Development and Preliminary Psychometric Testing." *Work* 39 (2): 141–50. doi:10.3233/WOR-2011-1160.

Lewis, James R. "Psychometric evaluation of the post-study system usability questionnaire: The PSSUQ," *Proceedings of the Human Factors and Ergonomics Society Annual Meeting,* vol. 36, no. 16, pp. 1269–1273, 1992.

Lewis and Wharton (1994), Wharton, Cathleen, Rieman, John, Lewis, Clayton, and Polson, Peter. "The cognitive walkthrough method: A practitioner's guide," In *Usability inspection methods*, pp. 105–140, John Wiley & Sons, 1994.

Lewis, James R., and Sauro, Jeff. 2021. "Usability and User Experience: Design and Evaluation." In *Handbook of Human Factors and Ergonomics*, 972–1015. doi:10.1002/9781119636113.ch38.

Perurena, Lilliam, Moráguez, Mercedes, Lirola, Francisco J., and Pérez, Adolfina. 2015. "La Usabilidad Percibida y El Grado de Satisfacción En La Plataforma Moodle de La UIB a Partir Del Cuestionario SUS." *Edutec 2015* 24 (2008): 176–94. doi:10.13140/RG.2.1.3813.8329.

Lu, Run Jie, Feng, Yi Xiong, Zheng, Hao and Tan, Jian Rong. 2016. "A Product Design Based on Interaction Design and Axiomatic Design Theory." *Procedia CIRP* 53: 125–29. doi:10.1016/j.procir.2016.07.061.

Luna, Carlos. 2013. Sé Más Persuasivo: Aumenta Tu Capacidad de Influir En Los Demás. *ESIC Edito*. https://books.google.com.mx/books?id=IdCkAQAAQBAJ&printsec=frontcover&dq=Sé+más+persuasivo&hl=en&sa=X&ved=0ahUKEwjVq6j5qYnpAhUCc60KHdsVAzEQ6AEIKzAA#v=onepage&q=Sémás persuasivo&f=false.

Lund, Arnold M. 2001. "Measuring Usability with the USE Questionnaire." *Usability Interface* 8 (2): 3–6. doi:10.1177/1078087402250360.

Martínez, Lina, Pérez, Leidis, Vizcaíno Martha, and Naranjo, Grisel. 2017. "Metodología Para El Diseño y Desarrollo de Productos: Caso de Creación Muebles Optimizadores Con Pop Up." *Revista I+D En TIC* 2 (8): 6–11.

Matos Mareño, Rosalba. 2013. La usabilidad como factor de calidad de páginas web. La Plata: Universidad Nacional de La Plata, Facultad de Informática. https://sedici.unlp.edu.ar/bitstream/handle/10915/33335/Documento_completo__.pdf?sequence=1.

Maya, Esther. 2014. Métodos y técnicas de investigación: Una propuesta ágil para la presentación de trabajos científicos en las áreas de arquitectura, urbanismo y disciplinas afines. Mexico City: Universidad Nacional Autónoma de México, Facultad de Arquitectura.

Medina, Luis J. 2013. "Requerimiento de Diseño." *Prezi*. https://prezi.com/rdf-lgblvq6a/requerimientos-de-diseno/.

Merino, Giselle Schmidt Alves Díaz., Teixeira, Clarissa Stefani., Schoenardie, Rodrigo Petry., Díaz-Merino, Eugenio Andrés, and Gontijo, Leila Amaral. 2012. "Usability in Product Design - The Importance and Need for Systematic Assessment Models in Product Development - Usa-Design Model (U-D) ©." *Work* 41 (Suppl. 1): 1045–52. doi:10.3233/WOR-2012-1011-1045.

Mital, Anil, Anoop Desai, Anand Subramanian, and Aashi Mital. 2014. "Design for Usability." In *Product Development: A Structured Approach to Consumer Product Development, Design, and Manufacture*, 2nd ed., 538. Amsterdam: Elsevier.

Morales, Sofía. 2016. *Dieter Rams, Menos Pero Mejor*. Vol. 1. Universidad de Palermo, Buenos Aires.

Niedderer, Kristina. 2007. "Designing Mindful Interaction: The Category of Performative Object." *Design Issues* 23 (1): 3–17. doi:10.1162/desi.2007.23.1.3.

Nielsen, Jakob. "Enhancing the explanatory power of usability heuristics," *Proceedings of the SIGCHI Conference on Human Factors in Computing Systems,* pp. 152–158, 1994.

Norman, Donald A. 2013. *The Design of Everyday Things*. Revised and expanded ed. New York, NY: Basic Books.

Palazuelos, Félix. 2015. "Los Principios Del 'Buen Diseño' de Dieter Rams." *Hipertextual.Com*.

Peña-Ontiveros, Mayra Ivette, Balderrama-Armendariz, Cesar Omar, Rossa-Sierra, Alberto, Maldonado-Macías, Aide, Cortés-Sáenz, David, and Hernández-Arellano, Juan Luis. 2024. "Human–Object Interaction: Development of a Usability Index for Product Design Using a Hierarchical Fuzzy Axiomatic Design." *Computation* 12 (6). doi:10.3390/computation12060130.

Pentti, Ruotsala. 2015. Usabilidad de los productos: Aspectos de utilidad. Arteología: La ciencia de los artefactos. *Universidad de Aalto. uiah.fi*

Pibernat, Oriol. 2008. *La Utilidad En El Diseño*. Círculo de Bellas Artes: Madrid, Spain.

Preece, Jennifer, Rogers, Yvonne & Sharp, Helen. 2002. *Interaction Design Beyond Human-Computer Interaction*. New York: Wiley.

Preece, Jenny, Rogers, Yvonne, Sharp, Helen, Benyon, David, Holland, Simon, and Carey, Tom. 1994. *Human-computer interaction*, Addison-Wesley.

Preece, Jennifer, Rogers, Yvonne, & Sharp, Helen. (2015). *Interaction design: Beyond human-computer interaction* (4th ed.). Chichester, West Sussex: John Wiley & Sons Ltd.

Puyuelo-Cazorla, Marina, Val-Fiel, Mónica, and Miralles, Fransisco Felip. 2011. "Gathering Interaction, Interface and Aesthetics Considerations in Product Design: Analyzing Devices Related to the Accessibility of Heritage." In *ACHI 2011-4th International Conference on Advances in Computer-Human Interactions*, 164–69. https://www.scopus.com/inward/record.url?eid=2-s2.0-84883061933&partnerID=tZOtx3y1.

Quiñones, Daniela, Rusu, Cristian and Rusu, Virginica. 2018. "A Methodology to Develop Usability/User Experience Heuristics." *Computer Standards and Interfaces* 59 (March): 109–29. doi:10.1016/j.csi.2018.03.002.

Raafat, Omar, and Becattini, Niccolò. 2022. "Testing the Usability of Guidelines for the Design of Surprising Products." *Proceedings of the Design Society* 2: 931–40. doi:10.1017/pds.2022.95.

Saffer, Daniel. 2010. *Designing for Interaction: Creating Innovative Applications and Devices*. 2nd ed. Berkeley, CA: New Riders.

Sánchez, José C., Olmos, Susana. & García, Fransisco J. 2015. "Intención de Uso de Tecnologías Móviles Entre Los Profesores En Formación. Aplicación de Un Modelo de Adopción Tecnológica Basado En TAM Con Los Constructos Compatibilidad y Resistencia Al Cambio." In *Conferencia: XVII Simposio Internacional de Informática Educativa*, SIIE'15, 9. Portugal.

Seva, Rosemary R., Gosiaco, Katherine Grace T., Crea, Ma, Santos, Eurice D. and Pangilinan, Denise Mae L. 2010. "Product Design Enhancement Using Apparent Usability and Affective Quality." *Applied Ergonomics* 42 (3): 511–17. doi:10.1016/j.apergo.2010.09.009.

Shneiderman, Ben. 1987. *Designing the user interface: Strategies for effective human-computer interaction*, Addison-Wesley.

Sierra, Alejandro and Silva, Juan. 2013. "Metodologías Para El Diseño." *Revista Digital de Diseño* 7 (13): 1–6.

Soler-Adillon, Joan. (2012). *Principios de diseño de interacción para sistemas interactivos*. Barcelona: Universitat Pompeu Fabra.

Tractinsky, Noam. 2017. "The Usability Construct: A Dead End?" *Human–Computer Interaction* 33 (2): 131–177. doi:10.1080/07370024.2017.1298038.

van Kuijk, Jasper, Daalhuizen, Jaap and Christiaans, Henri. "Drivers of usability in product design practice: Induction of a framework through a case study of three product development projects," *Design Studies*, 2019. *60*, 139–179

Vera, Javier Mauricio Reyes and Libreros Giraldo,Freddy Alejandro. 2011. "Método Para La Evaluación Integral De La Usabilidad En Sistemas E-Learning." *Integral Evaluation Method of Usability for E-Learning Systems. (English)* (12): 69–79.

Villagómez, Cynthia. 2019. "Análisis de La Investigación En El Diseño de Acuerdo a Richard Buchanan." In *Revista Interiorgráfico de La División de Arquitectura, Arte y Diseño de La Universidad de Guanajuato*, Décima Nov. https://www.interiorgrafico.com/edicion/decima-novena-edicion-agosto-2019/analisis-de-la-investigacion-en-el-diseno-de-acuerdo-richard-buchanan.

Vitsoe. 2019. "The Power of Good Design. Dieter Rams's Ideology." In *Vitsœ, Design Dieter Rams*. https://www.vitsoe.com/gb/about/good-design.

William, James. 2018. "Análisis Del Producto : Concepto, Atributo y Beneficios. Productos de Consumo y Productos Organizacionales." *UNID (Universidad Interamericana Para El Desarrollo)* 1: 15–16.

Wixon, Dennis, and Wilson, Charles. "The usability-engineering framework for product design and evaluation," In *Handbook of human-computer interaction* (2nd ed.), pp. 653–688, Elsevier, 1997,

Zhou, Ronggang, and Chan, Alan H.S. 2017. "Using a Fuzzy Comprehensive Evaluation Method to Determine Product Usability: A Proposed Theoretical Framework." *Work* 56 (1): 9–19.

Section 3

Physical Ergonomics and Workplace Health

8 Impact of Prolonged Standing on Musculoskeletal Health in Colombian Workers

Jonathan Osorio-Vasco and Yordán Rodríguez

8.1 INTRODUCTION

Standing work is a work posture that is observed in various jobs: food service workers, welders, cutters, retail salespeople, receptionists, electricians, pharmacists, teachers, daycare workers, physical therapists, assembly line workers, machine operators, security personnel, engineers, library assistants, hairdressers, laboratory technicians, nurses, and caregivers (Kees and Nicolien 2021; Coenen et al. 2017; Konz 2006).

The number of workers who perform their activities while standing is a global phenomenon, as statistics from several countries indicate. In the European Union, around 2010, 69% of people reported standing or walking for at least a quarter of their working day, and it was estimated that by 2019, 20% of workers performed their work standing up (Kees and Nicolien 2021). On the other hand, in countries such as Spain, Portugal, Romania, and North Macedonia, the proportion of workers who perform their duties standing exceeded the proportion of those who work sitting down (Kees and Nicolien 2021). In other countries, such as Australia, 62% of a sample of workers reported performing their tasks while standing up (Coenen et al., 2016), while in Canada, 55.3% of the working population worked while standing up (Antle et al., 2013).

In contrast, in Latin America, no consolidated figures on the magnitude of standing work were found, and there are limited studies on assessing the adverse effects of standing work. Colombia is no exception, and as of 2016, there were no official figures on this subject (López Pabón and Osorio Vasco 2016). To date, a characterization has been carried out with the support of the Dirección de Vigilancia y Control del Ministerio del Trabajo de Colombia, which reported characteristics of standing work such as posture, time, work surface, type of footwear used, and type of work (Osorio-Vasco, Barera-Causil, and Rodríguez 2025). On the other hand, some studies evaluated the adverse effects on the standing working population in Colombia (Yordan Rodríguez et al. 2021; Rodríguez, Osorio-Vasco, and Pérez 2025; Osorio-Vasco and Rodríguez 2021b, 2021a; Osorio-Vasco and Rodríguez 2022; Osorio-Vasco 2020).

The purpose of this chapter is to present and analyze the results of various studies conducted between 2018 and 2022 on the effects of standing work on workers in the manufacturing, health, and floriculture sectors.

8.1.1 Standing Work: Definition

There is no unified criterion for what is considered prolonged standing work, and in this regard, several definitions can be found. Some definitions related to standing work are presented in Table 8.1.

Based on these definitions, the authors of this chapter define prolonged standing work as follows: "Any work activity performed in a static or dynamic bipedal posture that is maintained for continuous periods of more than 1 hour or accumulated for four or more hours per day without frequent opportunity to sit."

DOI: 10.1201/9781003582014-11

TABLE 8.1
Definitions of Standing Work

Definition of Standing Work	Source
Standing consists of not being able to move at least 20cm to either side, forward, or backward and also not being able to walk or sit for at least 2.5 hours or more per day (LASI, 2009).	German Occupational Safety and Health Committee
Standing is a position in which the body rests on the legs and in which the legs cannot move more than within a circle with a radius of 1 m in relation to the original location. In addition, when standing for more than one continuous hour or more than 4 hours in total, it is considered prolonged (Health Council of the Netherlands 2011).	Health Council of the Netherlands
Prolonged standing at work occurs when standing for more than 5 hours a day without the opportunity to sit down regularly (Direktion für Arbeit – Arbeitsbedingungen 2016).	Swiss Federal Department of Economic Affairs, Education, and Research
Prolonged standing work is defined as follows: standing for 4 hours or more per day and/or standing continuously for more than 1 hour within these 4 hours (Waters and Dick 2015).	National Institute for Occupational Safety and Health (NIOSH) of the United States

8.1.2 Effects of Standing Work

There is evidence to suggest that symptoms such as musculoskeletal pain, discomfort, and fatigue in the lower extremities are linked to standing for long periods of time, either continuously or cumulatively, during the working day (Meijsen and Knibbe 2007). Furthermore, standing for continuous periods of time reduces blood circulation in the lower legs, causing inflammation or swelling in the legs during the workday (Antle et al. 2018; Antle, Vézina, and Côté 2015; Zander, King, and Ezenwa 2004). The decrease in blood flow causes blood to accumulate in the lower extremities as a result of reduced venous pumping activity. This mechanism depends on alternating muscle contractions in the lower leg to facilitate the return of blood to the heart (Antle et al. 2013; Antle and Côté 2013; Antle, Vézina, and Côté 2015; Antle et al. 2018). When blood supply to tense muscles is limited, muscle fatigue and pain increase due to the accumulation of metabolites, which can lead to muscle hypersensitivity (Waters and Dick 2015; Zander, King, and Ezenwa 2004).

The adverse effects of standing work have been confirmed by several literature reviews. Waters and Dick (2015) confirmed the adverse effects of cardiovascular problems, leg edema, musculoskeletal pain, and muscle fatigue in the lower extremities and lower back (Waters and Dick 2015). Coenen et al. (2016) concluded through a meta-analysis that there is an association between musculoskeletal pain symptoms and standing work in the lower back. Additionally, there is inconclusive evidence regarding musculoskeletal pain symptoms in the lower extremities and no association with the upper extremities (Coenen et al. 2016). On the other hand, Coenen et al. (2017) determined, after conducting a meta-analysis of experimental studies, that there is an association between musculoskeletal pain symptoms in the lower back and lower extremities (Coenen et al., 2017).

8.2 METHODOLOGY

8.2.1 Study Design

A cross-sectional study was conducted in nine Colombian companies belonging to the manufacturing (five), health (two), and floriculture (two) sectors.

8.2.2 Study Participants

Participants were selected for convenience. To carry out the study, written authorization was requested from the companies to enter the areas of the company where the workers would take part in the study. Informed consent was requested and signed by each participant who voluntarily wanted to be part of the study. The only conditions for participating in the study were the expression of willingness to participate, being present on the day of data collection, and remaining on their feet for at least 80% of the workday. A total of 177 workers from nine companies participated, distributed and consolidated as follows: manufacturing 93 (52.5%), health 47 (26.6%), and floriculture 37 (20.9%).

8.2.3 Assessment of Musculoskeletal Pain Intensity

To assess musculoskeletal pain symptoms, workers were asked to indicate their pain level using the Visual Analog Scale (VAS) (Price et al. 1983), represented on a line from 0 mm to 100 mm. The left end of the line indicated "no pain, " and the right end indicated "maximum pain." Nine body areas were assessed using the body map from ISO/TS 20646 (ISO 2014) as a reference. The body areas analyzed were as follows: right ankle–foot, right leg, right knee, right thigh–hip, left ankle–foot, left leg, left knee, left thigh–hip, and lower back.

Additionally, three grayscale ranges were established to indicate the final pain intensity levels by body area: mild pain (≤10 mm, white), moderate pain (10 mm < pain ≤20 mm, gray), and severe pain (>20 mm, black). Based on the percentage change in musculoskeletal pain, three ranges were established: mild pain (≤25%, white), moderate pain (>25% and ≤50%, gray), and severe pain (>50%, black). The Ergonomic Assessment Platform developed by Ergoyes (Ergoyes 2025) was used for the graphical representation of musculoskeletal pain and the generation of body map figures.

8.2.4 Measurement of Leg Circumference and Calculation of Volume

To measure the circumference of the right and left legs, the Gulick II measuring tape was used, specifically designed to standardize measurements and minimize errors through the constant application of tension (Country Technology Inc. 1999). This feature allows uniform force to be applied when adjusting the tape, increasing the accuracy and reliability of the measurements obtained (Zander, King, and Ezenwa 2004; Lin, Chen, and Cho 2012; Karimi et al. 2016). Therefore, it is a highly accurate instrument (Labs et al. 2000) and helpful in measuring the legs (Mayrovitz, Sims, and Macdonald 2000).

Leg measurements were taken with participants seated in a chair and stretching their legs out on a chair of equal height. A wooden board divided into six segments of 4 cm each, together with a permanent ink marker, was used to mark reference points on the leg and ensure repeatability of measurements; the board was placed from the lateral malleolus, starting at the ankle, to 20 cm along the longitudinal axis of the malleolus, where the calf has the maximum circumference. In total, there were five segments of the leg. To calculate the volume of the legs, based on the circumference measurements, the truncated cone formula shown in Equation 8.1 was used (Zander, King, and Ezenwa 2004; Lin, Chen, and Cho 2012; Karimi et al. 2016):

$$V = \left[\Sigma\left(X^2 + Y^2 + XY\right)\right]/3\pi \tag{8.1}$$

where "V" is the volume of the leg, "X" is the lower circumference of the segment, and "Y" is the upper circumference of the segment 4 cm away from "X." The truncated cone formula was used to calculate the volume at the start of the workday ($V_{(1)}$) and the volume at the end of the workday ($V_{(2)}$) (Karimi et al. 2016).

8.2.5 Assessment of Perceived Physical Exertion

The Borg CR-10 scale (references) was used to assess the level of perceived exertion. This is a modified version of the exertion perception scale designed to assess feelings of intensity during physical exertion (Borg 1998). To do this, the following question was asked: How intense is the physical effort you are perceiving at this point in the workday? This question was asked during the first hour (start) and the last hour (end) of the workday.

8.2.6 Data Processing

The data were processed using SPSS statistical software version 23. Averages were used to consolidate the measurements of musculoskeletal pain, leg volume, and perceived physical effort at the beginning and end of the workday. The differences between the initial and final averages were calculated, along with the corresponding percentages of change. To determine whether there were significant differences between the beginning and end of the working day, the nonparametric Wilcoxon rank test was applied. Additionally, the Kruskal–Wallis test was used to identify differences in volume change across each economic sector.

8.3 RESULTS OBTAINED

The results are presented consolidated by the three economic sectors. The averages for age, height, weight, and body mass index (BMI), together with the percentages for the dominant foot, are shown in Table 8.2. The results of the musculoskeletal pain assessment are shown by economic sector in Table 8.3. The results of the volume calculation are shown in Table 8.4. The comparison of leg volume by economic sector is shown in Table 8.5. The results of perceived exertion are shown in Table 8.6. Additionally, the average intensity of musculoskeletal pain is shown in Figure 8.1 and the percentage increase by body area in Figure 8.2. The percentage by category of perceived physical effort is presented in Figure 8.3, and the trend in perceived physical effort is presented in Figure 8.4.

TABLE 8.2
Averages of the Individual Characteristics of the Participants

	Gender						Dominant Side	
Economic Sector	**M (%)**	**F (%)**	**Age ($\bar{X}-SD$)**	**Height ($\bar{X}-SD$)**	**Weight ($\bar{X}-DS$)**	**BMI, $\bar{X}-SD$**	**Right (%)**	**Left (%)**
Manufacturing (n=93)	96.8	3.2	38.70 (13.13)	170.07 (6.9)	72.13 (10.93)	28.82 (33.75)	84.9	15.1
Health (n=47)	27.7	72.3	33.17 (8.4)	164.02 (8.52)	68.61 (15.26)	25.37 (4.58)	91.5	8.5
Floriculture (n=37)	62.2	37.8	33.46 (8.9)	163.18 (8.23)	63.25 (9.4)	23.97 (3.37)	86.5	13.5

8.3.1 Results of the Assessment of Musculoskeletal Pain Intensity in the Lower Extremities

TABLE 8.3
Average Results of Musculoskeletal Pain Assessment at the Beginning and End of the Working Day, Including the Difference, Percentage Change, and Statistical Significance of the Variation

Economic Sector	Body Area	Side of the Body	Pain Initial $\bar{X}$	Pain Final $\bar{X}$	$\bar{X}\,\Delta$	$\bar{X}\,\%\Delta D$	Test Wilcoxon Z	P
Manufacturing	Lower back	–	4, 494	7, 957	3, 463	77, 058	–2, 874	0.004*
(n=93)	Thigh–hip	Right	5, 849	11, 607	5, 758	98, 444	–4, 588	0
		Left	5, 354	10, 795	5, 441	101, 625	–2, 960	0.003*
	Right knee	Right	10, 795	11, 935	1, 140	10, 560	–3, 291	0.001*
		Left	9, 139	13, 655	4, 516	49, 415	–2, 995	0.003*
	Left	Right	7, 311	15, 548	8, 237	112, 666	–4, 693	0.000*
		Left	6, 989	14, 279	7, 290	104, 307	–4, 130	0
	Ankle–foot	Right	6, 602	13, 623	7, 021	106, 347	–4, 364	0.000*
		Left	4, 774	12, 494	7, 720	161, 709	–4, 059	0
Health (n=47)	Lower back	–	13, 043	16, 149	3, 106	23, 817	–1, 477	0.140
	Thigh–hip	Right	10, 574	17, 936	7, 362	69, 624	–2, 719	0.007*
		Left	9, 744	14, 361	4, 617	47, 383	–2, 614	0.009*
	Right knee	Right	15, 404	18, 782	3, 378	21, 929	–2, 031	0.042*
		Left	11, 148	15, 383	4, 235	37, 989	–2, 228	0.026
	Left	Right	13, 404	21, 191	7, 787	58, 095	–2, 800	0.005*
		Left	12, 127	16, 446	4, 319	35, 615	–2, 072	0.038*
	Ankle–foot	Right	7, 978	17, 638	9, 660	121, 083	–3, 731	0.000*
		Left	6, 659	15, 170	8, 511	127, 815	–3, 212	0.001*
Floriculture	Lower back	–	4, 405	17, 189	12, 784	290, 216	–4, 544	0
(n=37)	Thigh–hip	Right	2, 648	12, 594	9, 946	375, 604	–2, 985	0.003*
		Left	2, 621	11, 189	8, 568	326, 898	–3, 127	0.002*
	Right knee	Right	5, 594	10, 600	5, 006	89, 489	–2, 055	0.040*
		Left	4, 270	9, 945	5, 675	132, 904	–2, 018	0.044*
	Leg	Right	2, 945	14, 756	11, 811	401, 053	–4, 088	0.000*
		Left	4, 675	12, 243	7, 568	161, 882	–3, 470	0.001*
	Ankle–foot	Right	3, 027	17, 810	14, 783	488, 371	–3, 990	0.000*
		Left	4, 648	14, 189	9, 541	205, 271	–2, 558	0.011*

*P<0.05.

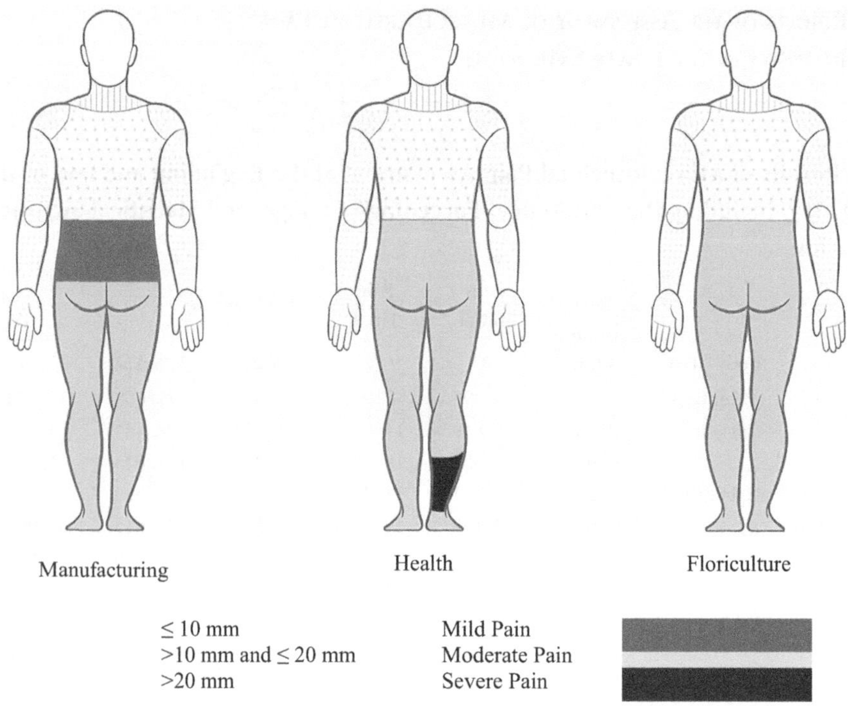

FIGURE 8.1 Average intensity of musculoskeletal pain reported by workers at the end of the workday in the sectors studied—body maps generated with the Ergoyes platform: www.ergoyes.com.

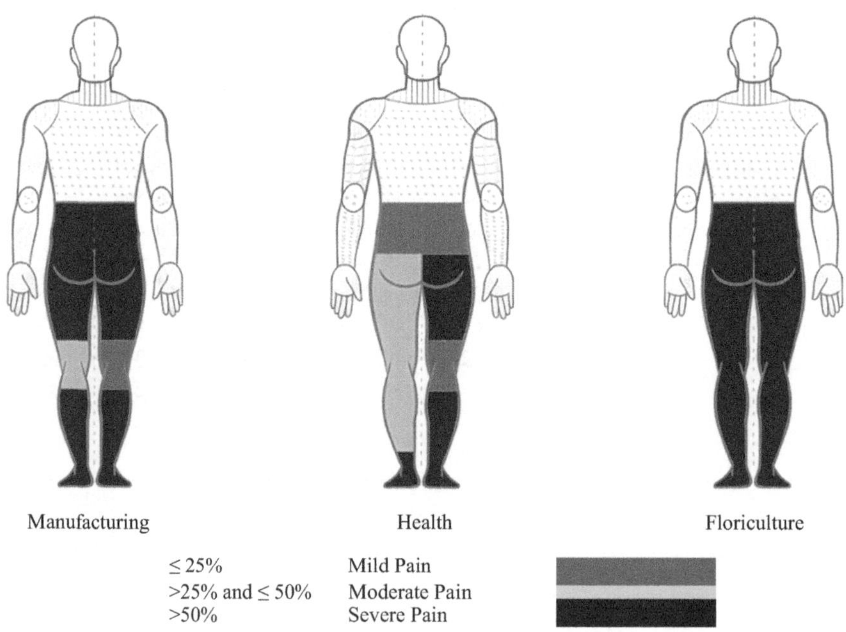

FIGURE 8.2 Average percentage change in musculoskeletal pain reported by workers at the end of the workday in the sectors studied—body maps generated with the Ergoyes platform: www.ergoyes.com.

8.3.2 Results of Leg Circumference Measurements and Volume Calculations

TABLE 8.4
Average Results of Leg Volume at the Beginning and End of the Workday, Difference, Percentage Change, and Significant Difference

		Volume Initial	Volume Final	$\bar{X}\Delta$		Test Wilcoxon	
Economic Sector	Leg	$\bar{X}$(cm³)	$\bar{X}$ (cm³)	(cm³)	$\bar{X}\,\Delta\%$	Z	P
Manufacturing (n=93)	Law	1, 304, 486	1, 345, 588	41, 102	3, 151	–3, 119	0.002*
	Left	1, 297, 638	1, 335, 025	37, 387	2, 881	–2, 698	0.007*
Health (n=47)	Law	1, 419, 165	1, 449, 308	30, 143	2, 124	–3, 642	0.000*
	Left	1, 422, 890	1, 444, 386	21, 496	1, 511	–5, 093	0
Floriculture (n=37)	Law	1, 437, 832	1, 468, 879	31, 047	2, 159	–4, 549	0.000*
	Left	1, 439, 902	1, 485, 487	45, 585	3, 166	–4, 187	0

*P<0.05.

TABLE 8.5
Comparison of Volume between Economic Sectors

Economic Sector	Manufacturing	Health	Floriculture	P
Right leg	92.10	85.03	86.26	0.695
Left leg	89.98	80.45	97.39	0.311

*P<0.05.

8.3.3 Results of Perceived Exertion

TABLE 8.6
Average Results of Perceived Physical Effort at the Beginning and End of the Workday, Difference, Percentage Change, and Significant Difference

	Physical Effort Initial	Physical Effort Final			Initial Test Wilcoxon	
Economic Sector	$\bar{X}$	$\bar{X}$	$\bar{X}\,\Delta$	$\bar{X}\,\%\Delta D$	Z	P
Manufacturing (n=93)	1, 683	3, 038	1, 355	80, 511	–6, 067	0, 000*
Health (n=47)	1, 309	3, 149	1, 840	140, 565	–4, 766	0, 000*
Floriculture (n=37)	1, 149	3, 689	2, 540	221, 062	–5, 001	0, 000*

*P<0.05.

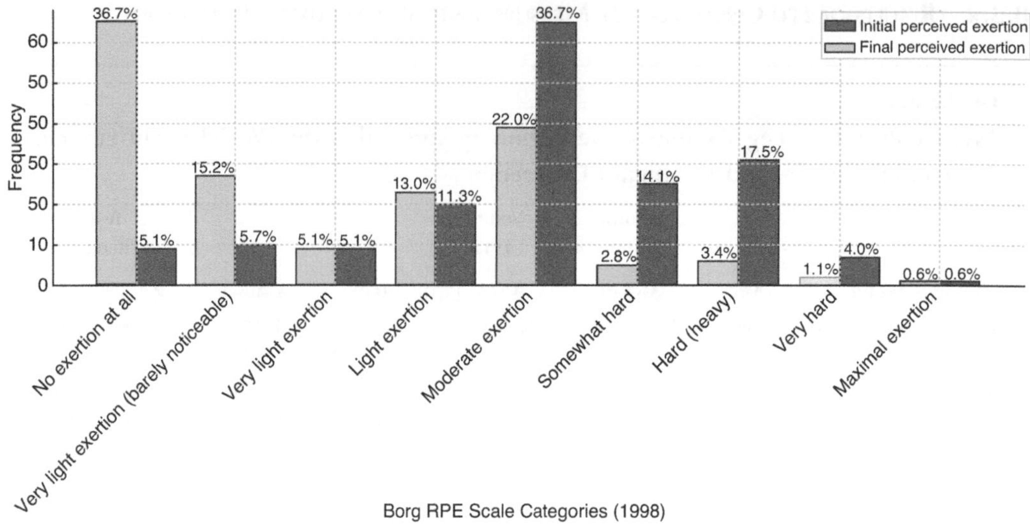

FIGURE 8.3 Frequency of perceived physical exertion at the beginning and end of the workday.

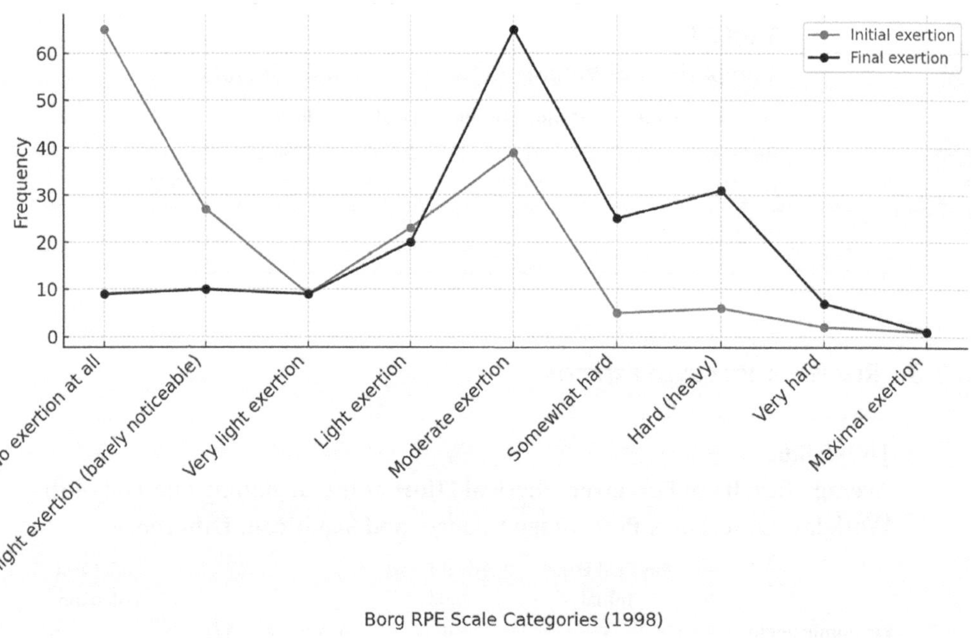

FIGURE 8.4 Trend in perceived physical exertion between the start and end of the workday for each category of the Borg scale.

8.4 ANALYSIS OF FINDINGS

8.4.1 Musculoskeletal Pain in the Lower Extremities

The results presented in this chapter show that, in the manufacturing, healthcare, and floriculture sectors, working while standing during the workday can have adverse effects, increasing musculoskeletal pain in the lower extremities and lower back. Except for the lower back in the healthcare sector, all other areas of the body in the three economic sectors showed significant changes (Table 8.3). These results are consistent with those reported by Coenen et al. (2017), who, in their literature review with a meta-analysis of laboratory studies, demonstrated that working while standing for prolonged periods is associated with an adverse effect on pain in the lower back and lower extremities (Coenen et al., 2017). Similarly, the findings of these studies, which focused on Latin American workers, are consistent with European guidelines on standing work, which also report that one of the adverse effects of standing work on the lower limbs and lower back is musculoskeletal pain. (Kees and Nicolien 2021). Therefore, the symptoms of musculoskeletal pain caused by standing work should promote ergonomic intervention actions in workplaces where workers stand during work, seeking to reduce the symptom of musculoskeletal pain.

The floriculture sector is the economic sector with the most significant differences in musculoskeletal pain. The areas of the lower back (12, 784), right leg (11, 811), and right ankle and foot (14, 783) were the areas of the body that showed most increases in musculoskeletal pain among the other sectors (Table 8.3). This could be explained by the multiple activities that flower growers must perform in their work, such as planting and packing flowers and fumigating and watering crops, most of which must be carried out on unstable and slippery surfaces. These results are relevant for the floriculture sector in general, especially for Colombia, which is the second largest flower exporter in the world (Ministry of Agriculture and Rural Development 2023). For this reason, greater attention should be paid in this economic sector to the care of workers and their standing posture during work as the evidence presented in this chapter indicates that workers may develop future musculoskeletal disorders.

The average intensity of musculoskeletal pain increases at the end of the working day in healthcare workers owing to the greater number of body areas affected by moderate and severe musculoskeletal pain (Figure 8.1). At the same time, in the other two economic sectors, there was at least one area with mild musculoskeletal pain. In terms of the percentage change in pain, the floriculture sector shows a severe percentage change in all body areas evaluated (Figure 8.2); this reaffirms that, in the floriculture sector, the lower extremities and lower back are affected by the activities involved in flower production.

8.4.2 Leg Volume

Leg volume increased significantly in both legs in all three economic sectors (Table 8.4). These results, derived from a direct work context, are consistent with those reported by Antle et al. (2013), who, in an experimental study on the performance of tasks in a standing position, determined that blood flow in the lower extremities increases within 32 minutes (Antle et al. 2013). On the other hand, Waters and Dick (2015) consolidated a series of studies that showed an increase in leg volume during long periods of work.

The results of this chapter show an increase in leg volume in the workers studied. In this regard, the increase in leg volume has been associated with the appearance of varicose veins and, additionally, with cardiovascular problems (Correa Posada et al. 2022; Bahk et al. 2012). Therefore, we believe that interventions should be made to reduce the risk of increasing leg volume while working in a standing position, such as wearing shoes with insoles (Anderson, Williams, and Nester 2021, 2017), working on soft surfaces (Zander, King, and Ezenwa 2004; Lin, Chen, and Cho 2012), and wearing compression stockings (Waters and Dick 2015; Kim, Won, and Ko 2022).

The volume of both legs increased in the three economic sectors by more than 1.5%, with the left leg in floriculture (3.166%–45.585 cm^3) and the right leg in manufacturing (3.151%–41.102 cm^3) achieving the highest growth rates and the greatest average difference[3] and the right leg in the manufacturing sector (3.151%–41.102 cm^3) which obtained the highest growth percentage and the greatest average difference (Table 8.4). It is noteworthy that, in the floriculture sector, the difference in volume in the left leg was greater than the difference in volume in the right leg (1%), with the dominant leg being more frequent (right leg –86.5%) (Table 8.1). This finding is similar to that reported by Osorio and Rodríguez (2021b), who conducted a similar assessment of leg volume in healthcare workers with a smaller population, finding that the volume of the left leg increased by 0.1% more than the volume of the right leg (Osorio-Vasco and Rodríguez 2021b). The increase in volume in one leg or the other may be determined by the amount of time spent standing on one of the two legs while working. It should be clarified that the increase in leg volume is not only related to working on one's feet for long periods of time but also to specific individual characteristics such as age, gender, and tobacco use. (Waters and Dick 2015; Bahk et al. 2012).

There was no difference in the increase in leg volume across the three economic sectors (Table 8.5). Although there are significant visible changes between the start and end of the working day in each economic sector (Table 8.4), there was no evidence that leg volume was greater in any particular economic sector. Therefore, the adverse effect of increased leg volume is the same across the three economic sectors.

8.4.3 Perceived Physical Effort

The physical effort perceived in the three economic sectors increased significantly between the beginning and end of the workday (Table 8.6). The greatest difference in perceived physical effort was found in the floriculture sector, which had a difference of 2, 540 between the beginning and end (Table 8.6). The greatest perceived physical effort may be present in this economic sector because most of the activities carried out are mainly outdoors, usually on uneven and slippery ground, and sometimes, some activities during the workday must be performed in positions other than standing, such as squatting, kneeling, or leaning on one knee.

The most frequent perceived physical effort at the beginning of the workday was no effort, at 36.7%, followed by moderate effort, at 22% (Figure 8.3). At the end of the workday, the greatest effort was moderate, at 36.7%, followed by heavy effort, at 17.5%; no effort was reduced to 5.1% at the end of the workday (Figure 8.3). It should be noted that the trend of physical effort increases in each category reported at the end of the workday, starting with moderate effort (Figure 8.4).

8.4.4 Limitations and Future Work

From a methodological perspective, the sample of 177 workers from three economic sectors is appropriate for characterizing the effects of standing work in the context studied. This is particularly relevant, given that data collection in real working conditions can cause interruptions in work activities and affect worker performance. However, in future research, it would be advisable to expand the sample to a larger number of participants and include more economic sectors. In addition, more representative sampling strategies could be explored in order to generalize the results.

The studies compiled in this chapter were cross-sectional, conducted during a typical 8-hour workday. This methodology limits the possibility of generalizing the effects of standing work to more extended periods, such as a week, a month, or a year. To address this limitation, the design of longitudinal studies that allow for the evaluation of impact over time is recommended. However, such studies present significant methodological challenges as they require greater resources, and their implementation in real work environments involves difficulties in controlling variables and collecting data.

A key aspect of the studies conducted was the collection of data in real working conditions, which provides empirical information on the effects generated by multiple factors present in work activities. However, this methodology makes it difficult to establish a direct causal relationship between the risk factors of standing work and its effects on health. To address this limitation, future research could consider the development of experimental or quasi-experimental studies, which enable better control of variables and the identification of potential causal relationships.

8.5 CONCLUSIONS

The results presented in this chapter provide evidence on the effects of standing work on the musculoskeletal health of a Colombian population of workers. For future research, it is recommended to expand the samples to workers from various economic sectors in Latin America and to adopt experimental methodological designs. This approach would allow for a more accurate establishment of the causal relationships between the risk factors of standing work and its effects on health.

Finally, it is important to highlight the importance of continuing to investigate the issue of standing work in the Latin American context, as well as the need to develop ergonomic methods and tools to assess risk and implement preventive measures in the workplace from an ergonomic perspective.

Appendix 8.1 presents guidelines that can serve as a reference for implementing ergonomic interventions to reduce musculoskeletal pain, leg inflammation, and perceived physical exertion among workers who perform their activities while standing for a significant part of their working day.

Appendix 8.1

A8.1 GENERAL GUIDELINES FOR STANDING WORK

The following guidelines are recommendations based on the findings of this research and the scientific literature that has supported the research carried out.

A8.1.1 Time Spent Working While Standing

It is recommended to avoid continuous and static standing work for 1 continuous hour and a total of 4 hours during the working day (Meijsen and Knibbe 2007). Short rest periods after 40 minutes of continuous and static standing (Coenen et al. 2017), with at least 10 minutes of rest (Rahman et al. 2019), can reduce the manifestations of adverse effects. Ideally, choose periods of standing and rest in 30-minute intervals (Van Dieën and Oude Vrielink 1998). It is also possible to alternate between static posture and short periods of movement or dynamic standing posture, which can help reduce adverse effects (Balasubramanian, Adalarasu, and Regulapati 2009; Jin et al. 2022; Junge et al. 2022).

A8.1.2 Workstation

If occupational tasks are inherently static, it is recommended to avoid hard or rigid support surfaces, such as iron or concrete (Winberg et al., 2022; Lin, Chen, and Cho, 2012). It is recommended to use soft support surfaces such as padded or rubber mats (Cham and Redfern 2001). Additionally, the workstation design should allow the worker to move or make short movements of more than 20 cm to the sides and forward and backward (LASI 2009). Furthermore, the height of the work surface must be adjustable according to the worker's anthropometry. In workstations where workers remain in spaces less than 20 cm to the sides and forward and backward, footrests or foot bars must be provided (Konz 2018).

A8.1.3 Personal Protective Equipment

Footwear with unstable soles is recommended. This sole prevents blood pooling and improves blood flow (Karimi et al. 2016; Anderson, Nester, and Williams 2018). In addition, the use of silicone insoles inserted into footwear helps reduce negative symptoms in the lower extremities and improves foot comfort (Anderson, Williams, and Nester 2017). The use of compression stockings for the legs is also recommended. This improves blood flow in the legs (Garcia et al. 2023).

ACKNOWLEDGMENTS

We want to thank all the workers who voluntarily participated in the studies, the company managers who granted us access to the facilities, and the students who assisted in collecting the information.

REFERENCES

Anderson, Jennifer, Williams, Anita E. and Nester, Chris. 2021. "Musculoskeletal Disorders, Foot Health and Footwear Choice in Occupations Involving Prolonged Standing." *International Journal of Industrial Ergonomics* 81: 103079. doi:10.1016/j.ergon.2020.103079.

Anderson, Jennifer, Williams, Anita E. and Nester, Christopher J. 2017. "A Narrative Review of Musculoskeletal Problems of the Lower Extremity and Back Associated with the Interface between Occupational Tasks, Feet, Footwear and Flooring." *Musculoskeletal Care* 15 (4): 304–15. doi:10.1002/msc.1174.

Anderson, Jennifer, Nester, Christopher and Williams, Anita. 2018. "Prolonged Occupational Standing: The Impact of Time and Footwear." *Footwear Science* 10 (3): 189–201.

Antle, David M., and Côté, Julie N. 2013. "Relationships between Lower Limb and Trunk Discomfort and Vascular, Muscular and Kinetic Outcomes during Stationary Standing Work." *Gait and Posture* 37 (4): 615–19. doi:10.1016/j.gaitpost.2012.10.004.

Antle, David M., Cormier, Lauren, Findlay, Megan, Miller, Linda L. and Julie N. Côté. 2018. "Lower Limb Blood Flow and Mean Arterial Pressure during Standing and Seated Work: Implications for Workplace Posture Recommendations." *Preventive Medicine Reports* 10 (March): 117–22. doi:10.1016/j.pmedr.2018.02.016.

Antle, David M., Vézina, Nicole, and Julie N. Côté. 2015. "Comparing Standing Posture and Use of a Sit-Stand Stool: Analysis of Vascular, Muscular and Discomfort Outcomes during Simulated Industrial Work." *International Journal of Industrial Ergonomics* 45: 98–106. doi:10.1016/j.ergon.2014.12.009.

Antle, David M., Vézina, Nicole, Messing, Karen and Julie N. Côté. 2013. "Development of Discomfort and Vascular and Muscular Changes during a Prolonged Standing Task." *Occupational Ergonomics* 11 (1): 21–33. doi:10.3233/OER-130205.

Bahk, Jin Wook, Kim, Hyunjoo, Jung-Choi, Kyunghee, Jung, Myung Chul, and Lee, Inseok. 2012. "Relationship between Prolonged Standing and Symptoms of Varicose Veins and Nocturnal Leg Cramps among Women and Men." *Ergonomics* 55 (2): 133–39. doi:10.1080/00140139.2011.582957.

Balasubramanian, Venkatesh, Adalarasu, K and Regulapati, Rahul. 2009. "Comparing Dynamic and Stationary Standing Postures in an Assembly Task." *International Journal of Industrial Ergonomics* 39 (5): 649–54. doi:10.1016/j.ergon.2008.10.017.

Borg, Gunnar. 1998. *Borg's Perceived Exertion and Pain Scales*. Champaign, IL: Human Kinetics.

Cham, Rakié, and Redfern, Mark S. 2001. "Effect of Flooring on Standing Comfort and Fatigue." *Human Factors* 43 (3): 381–91. doi:10.1518/001872001775898205.

Coenen, Pieter, Willenberg, Lisa, Parry, Sharon, Shi, Joyce W., Romero, Lorena, Blackwood, Diana M., Maher, Christopher G., Healy, Genevieve N., Dunstan, D avid W. and Straker, Leon M. 2016. "Associations of Occupational Standing with Musculoskeletal Symptoms: A Systematic Review with Meta-Analysis." *British Journal of Sports Medicine*, 1–10. doi:10.1136/bjsports-2016-096795.

Coenen, Pieter, Parry, Sharon, Willenberg, Lisa, Shi, Joyce W., Romero, Lorena, Blackwood, Diana M. Healy, Genevieve N., Dunstan, David W. and Straker, Leon M.. 2017. "Associations of Prolonged Standing with Musculoskeletal Symptoms—A Systematic Review of Laboratory Studies." *Gait and Posture* 58 (August): 310–18. doi:10.1016/j.gaitpost.2017.08.024.

Correa Posada, Ofelia, Martha, Contreras Correa, Laura Maria and García Vélez, John Fernando. 2022. "Factors Associated with Chronic Venous Disease: Study of 1, 136 Patients Treated for Varicose Veins of the Lower Limbs at a Specialized Clinic in Colombia." *Jornal Vascular Brasileiro* 21: 1–8. doi:10.1590/1677-5449.202200513.

Country Technology Inc. 1999. "Gulick II Plus Tape Measure (Model 67019)." https://www.fitnessmart.com/collections/gulick-ii-tape-measures-calibrated-for-accuracy-and-repeatability/products/gulick-ii-plus-tape-measure?variant=29123527047.

Direktion für Arbeit - Arbeitsbedingungen. 2016. "Stehen Bei Der Arbeit." Bern. https://www.seco.admin.ch/seco/de/home/Publikationen_Dienstleistungen/Publikationen_und_Formulare/Arbeit/Arbeitsbedingungen/Broschuren/stehen-bei-der-arbeit.html.

Ergoyes, Ergonomics Community. 2025. "Body Map." https://ergoyes.com/method/details/category/1/method/16.

Garcia, Maria Gabriela, Roman, Maria Gloria, Davila, Andrea., and Martin, Bernard J. 2023. "Comparison of Physiological Effects Induced by Two Compression Stockings and Regular Socks During Prolonged Standing Work." *Human Factors* 65 (4): 562–74. doi:10.1177/00187208211022126.

Health Council of the Netherlands. 2011. "Standing, Kneeling and Squatting Work." 2011/41. The Hague: 2011/41. https://www.gezondheidsraad.nl/documenten/adviezen/2011/12/23/staand-geknield-en-gehurkt-werken.

ISO. TS 20646:2014. Ergonomics guidelines for the optimization of musculoskeletal workload. Geneva, Switzerland: International Organization for Standardization.

Jin, Xu, Dong, Yidan, Wang, Fujiang, Jiang, Ping, Zhang, Zhongbin, He, Lihua, Forsman, Mikael and Yang, Liyun. 2022. "Prevalence and Associated Factors of Lower Extremity Musculoskeletal Disorders among Manufacturing Workers: A Cross-Sectional Study in China." *BMJ Open* 12 (2). doi:10.1136/bmjopen-2021-054969.

Junge, Florian, Konschake, Wolfgang, Haase, Hermann, Vollmer, Marcus and Jünger, Michael. 2022. "Walking Instead of Standing." *Vasa* 51 (2): 78–84. doi:10.1024/0301-1526/a000990.

Karimi, Zanyar, Allahyari, Teimour, Azghani, Mahmood Reza and Khalkhali, Hamidreza. 2016. "Influence of Unstable Footwear on Lower Leg Muscle Activity, Volume Change and Subjective Discomfort during Prolonged Standing." *Applied Ergonomics* 53: 95–102. doi:10.1016/j.apergo.2015.09.003.

Kees, Peereboom, and de Langen, Nicolien. 2021. Prolonged Constrained Standing Postures Health Effects and Good Practice. Luxembourg: European Agency for Safety and Health at Work. doi:10.2802/91149.

Kim, Da Sol, Won, Yu Hui and Ko, Myoung Hwan. 2022. "Comparison of Intermittent Pneumatic Compression Device and Compression Stockings for Workers with Leg Edema and Pain after Prolonged Standing: A Prospective Crossover Clinical Trial." *BMC Musculoskeletal Disorders* 23 (1): 1–14. doi:10.1186/s12891-022-05975-6.

Konz, Stephan. 2006. "Standing Work." In *International Encyclopedia of Ergonomics and Human Factors*, 929–31.Boca Raton, FL: CRC Press.

Konz, Stephan. 2018. *Work Design: Occupational Ergonomics*. 7th Edition. Tylor & Francis Group. doi:10.1201/9780203733714.

Labs, Karl-Heinz, Tschoepl, Martin, Gamba, Giacomo, Aschwanden, Markus, and Jaeger, Kurt Andreas. 2000. "The Reliability of Leg Circumference Assessment: A Comparison of Spring Tape Measurements and Optoelectronic Volumetry." *Vascular Medicine* 5 (2): 69–74. doi:10.1177/1358836X0000500202.

LASI. 2009. "Bewegungsergonomische Gestaltung von Andauernder Steharbeit." https://lasi-info.com/uploads/media/lv50_01.pdf.

Lin, Yen Hui, Chen, Chih Yong and Cho, Min Hsien. 2012. "Influence of Shoe/Floor Conditions on Lower Leg Circumference and Subjective Discomfort during Prolonged Standing." *Applied Ergonomics* 43 (5): 965–70. doi:10.1016/j.apergo.2012.01.006.

López Pabón, Luisa, and Osorio Vasco, Jonathan. 2016. "Initial Development of a Tool to Assess the Level of Risk in Standing Work [Specialization Thesis]." University of Antioquia. https://hdl.handle.net/10495/5537.

Mayrovitz, Harvey N., Sims, Nancy and Macdonald, John M, 2000. "Assessment of Limb Volume by Manual and Automated Methods in Patients with Limb Edema or Lymphedema." *Advances in Skin and Wound Care* 13 (6): 272–76.

Meijsen, Paul, and Knibbe, Hanneke. J J. 2007. "Prolonged Standing in the OR: A Dutch Research Study." *AORN Journal* 86 (3). doi:10.1016/j.aorn.2007.08.007.

Ministry of Agriculture and Rural Development. 2023. "Flowers from the 'Country of Beauty' Conquer the World with Quality." https://www.minagricultura.gov.co/noticias/Paginas/Las-flores-del-'País-de-la-Belleza'-conquistan-el-mundo-con-calidad.aspx#:~:text=Colombia%252C%2520donde%2520existen%2520m%25C3%25A1s%2520de%252050%2520especies, crecimiento%2520frente%2520al%25202021%2520.

Osorio-Vasco, Jonathan, and Rodríguez, Yordán. 2021a. "Changes in Lower Leg Volume Among Health Care Workers During a Working Day." In Lecture Notes in Networks and Systems. Vol. 222, LNNS. Springer International Publishing. doi:10.1007/978-3-030-74611-7_55.

Osorio-Vasco, Jonathan, and Rodríguez, Yordán. 2021b. "Effects of Standing Work on Healthcare Workers." *Cuidarte Magazine*. doi:10.15649/cuidarte.1790.

Osorio-Vasco, Jonathan, and Rodríguez, Yordán. 2022. "Analysis of Musculoskeletal Pain in the Lower Limbs and Lower Back in Manufacturing Workers." *Atacama Journal of Health Sciences* 1. https://www.salud.uda.cl/ajhs/index.php/ajhs/article/view/59.

Osorio-Vasco, Jonathan, Barera-Causil, Carlos and Rodríguez, Yordán. 2025. "Characteristics of Standing Work in the Colombian Working Population." In *Proceedings of the 22nd Congress of the International Ergonomics Association, Volume* 3, edited by Sangeun Jin, Jeong Ho Kim, Yong-Ku Kong, Jaehyun Park, and Myung Hwan Yun. Singapore: Springer Nature Singapore. doi:10.1007/978-981-96-9330-6_18.

Osorio-Vasco, Jonathan. 2020. "Analysis of Signs, Symptoms, and Risk Factors Associated with the Lower Extremities and Lower Back During the Workday in Workers at a Hospital in Medellín." National School of Public Health, University of Antioquia. https://hdl.handle.net/10495/29741.

Price, Donald D., McGrath, Patricia A., Rafii, Amir and Buckingham, Barbara. 1983. "The Validation of Visual Analogue Scales as Ratio Scale Measures for Chronic and Experimental Pain." *Pain* 17 (1): 45–56. doi: 10.1016/0304-3959(83)90126-4.

Rahman, Ismail Abdul, Mohamad, Norlizaa, Rohani, Jafri Mohd and Zein, Raemy Md. 2019. "The Impact of Work Rest Scheduling for Prolonged Standing Activity." *Industrial Health* 56 (6): 492–99. doi:10.2486/indhealth.2018-0043.

Rodríguez, Yordán, Osorio-Vasco, Jonathan and Pérez, Elizabeth. 2025. "Analysis of Musculoskeletal Pain Among Colombian Flower Growers." In *BT - Proceedings of the 22nd Congress of the International Ergonomics Association, Volume* 1, edited by Sangeun Jin, Jeong Ho Kim, Yong-Ku Kong, Jaehyun Park, and Myung Hwan Yun, 169–74. Singapore: Springer Nature Singapore.

Rodríguez, Yordán, Osorio-Vasco, Jonathan, Zuluaga, Ivonne and Múnera, Ana. 2021. *Lecture Notes in Networks and Systems*. Vol. 221, LNNS. Springer International Publishing. doi:10.1007/978-3-030-74608-7_33.

DIEEN, JAAP H. VAN. 1998. "Evaluación de los horarios de trabajo y descanso con respecto a los efectos de la carga de trabajo postural en el trabajo de pie". *Ergonomics* 41 (12): 1832–44. doi:10.1080/001401398186009.

Van Dieën, Jaap H., & Oude Vrielink, Henny H. E. 1998. Evaluation of work-rest schedules with respect to the effects of postural workload in standing work. *Ergonomics*, 41(12), 1832–1844. https://doi.org/10.1080/001401398186009

Waters, Thomas R., and Dick, Robert B. 2015. "Evidence of Health Risks Associated with Prolonged Standing at Work and Intervention Effectiveness." *Rehabilitation Nursing* 40 (3): 148–65. doi:10.1002/rnj.166.

Winberg, Taylor B, Glinka, Michal N., Gallagher, Kaitlin M., Weaver, Tyler B., Laing, Andrew C. and Callaghan, Jack P. 2022. "Anti-Fatigue Mats Can Reduce Low Back Discomfort in Transient Pain Developers." *Applied Ergonomics* 100: 103661. doi:10.1016/j.apergo.2021.103661.

Zander, Joann E., King, Phyllis M. and Ezenwa, Bertram N. 2004. "Influence of Flooring Conditions on Lower Leg Volume Following Prolonged Standing." *International Journal of Industrial Ergonomics* 34 (4): 279–88. doi:10.1016/j.ergon.2004.04.014.

9 Determination of Hand Anthropometric Equivalence and Grip Strength between Campeche and Yucatán Populations

Mayra Pacheco-Cardín and Juan Luis Hernández Arellano

9.1 INTRODUCTION

The study of anthropometric dimensions and their relationship to grip strength is a topic of relevance in ergonomics and biomechanics since hand strength is a key indicator of upper-extremity functionality and performance in various work and daily activities. Previous research has shown that grip strength is influenced by factors such as age, gender, and anthropometric variables such as hand length, palm width, and wrist circumference (Wey et al. 2023). Likewise, studies have reported that differences between populations may be related to environmental and occupational factors rather than geographic location (Muthiah & Lee 2022).

Although the relationship between anthropometric dimensions and hand strength has been extensively documented in the classical ergonomics literature—for example, in the works of Grandjean, Kroemer, Helander, and Pheasant—most of this evidence has been generated in non-Mexican populations. Therefore, there remains limited reference information about these associations in the regional context of Mexico, particularly when comparing subpopulations with distinct occupational and cultural environments such as Campeche and Yucatán.

The importance of establishing reference values for grip strength has been highlighted in recent studies, which have identified that this measure is useful not only as an indicator of physical performance but also as a predictor of nutritional status and body composition (Lucio et al. 2020). Furthermore, grip strength has been found to vary across the lifespan, with a progressive decline at older ages, underlining its relevance as a marker of aging and functional health (Mehmet et al. 2020).

Given that previous studies have suggested that ergonomic adaptation should be based on specific population characteristics (Kang et al. 2024), the findings of this study will allow us to determine whether grip strength data can be used in a generalized manner in both regions without the need for differential adjustments. The ability to generalize grip strength data across regions has implications for the design of tools and work equipment, contributing to the optimization of working conditions and the prevention of musculoskeletal disorders in workers from both populations.

In this context, the present research aims to compare handgrip strength between the economically active populations of Campeche and Yucatán, two regions with similar sociocultural characteristics but with different work environments. The analysis aims to determine whether significant differences in grip strength exist between the two populations after adjusting for key anthropometric variables.

DOI: 10.1201/9781003582014-12

9.2 MATERIALS AND METHODS

9.2.1 Study Design

This study adopted an analytical and comparative approach to evaluate the equivalence of handgrip strength between the economically active populations of Campeche and Yucatán. A cross-sectional design was employed in which anthropometric and hand strength data were collected in both populations. Comparability between the groups was determined by statistical tests that evaluated the distribution of the samples and the possible influence of anthropometric variables on handgrip strength.

9.2.2 Sample

The Campeche sample was a stratified sample by convenience, composed of 382 subjects (198 men and 184 women) residing in the municipalities of Calkiní, Hecelchakán, and Tenabo in the state of Campeche, Mexico. The sample included men and women of economically active age between 15 and 65 years, excluding pregnant women and people with deformities or pathologies that altered the anthropometric characteristics. This sample selection was based on data from the last National Population Survey, following Economic Active Polulation (EAP) distributions from the most recent national survey to approximate population structure.

The Yucatán sample was obtained from an industrial group made up of 11 companies, in which measurements were made on all personnel, comprising 127 workers (78 men and 49 women). The study population included economically active men and women within a similar age range to that of the Campeche sample. This group provided a representative reference of workers in an industrial setting, allowing us to evaluate the applicability of the findings in the ergonomic and anthropometric contexts of the region. The same inclusion and exclusion criteria applied in Campeche were used for the Yucatán group: only economically active men and women between 15 and 65 years of age were considered, excluding pregnant women and individuals with deformities or pathologies that could alter anthropometric characteristics. The Campeche sample was initially larger than the Yucatán sample. To minimize bias in the comparative analyses, the Campeche dataset was randomly subsampled—preserving the age-by-sex strata—by assigning random numbers to eliminate excess cases within each stratum until matching the Yucatán sample size, yielding two region-specific samples of 127 participants each (78 men and 49 women). This procedure ensured statistical equivalence between the samples and allowed for robust comparative analysis. The sample was distributed, as shown in Table 9.1.

9.2.3 Materials

Anthropometric measurements were performed using a new Ergomeasure® anthropometric kit, which includes a stadiometer and short and long anthropometers, with a reported accuracy of ±1 mm.

TABLE 9.1
Final Sample Composition by Region, Sex, and Age Group

Age Group	Campeche Men	Campeche Women	Yucatán Men	Yucatán Women
15–24 years	12	10	12	10
25–34 years	20	14	20	14
35–44 years	17	11	17	11
45–54 years	18	9	18	9
55–65 years	11	5	11	5
Total	**78**	**49**	**78**	**49**

Grip strength was evaluated with a Camry® dynamometer (capacity 90 kg, accuracy ±0.5 kg). Since the instruments were new, it was assumed that they complied with the manufacturer's reported error margins.

All anthropometric measurements were performed by a single trained observer following standardized protocols. Each dimension was measured twice, and a third measurement was taken when discrepancies exceeded the accepted tolerance, recording the mean of the consistent values. Although a formal calculation of the technical error of measurement was not conducted, this procedure minimized intra-observer variability and ensured the reliability of the data.

The anthropometric dimensions considered in the analysis included hand length, palm length, hand width, palm width, wrist width, wrist circumference, and grip circumference. These measurements were selected because of their relevance in previous studies on grip strength and their influence on hand biomechanics.

9.2.4 Data Processing and Analysis

The dataset was reviewed to identify missing values, outliers, or inconsistencies. Outliers were evaluated using boxplot visualization and standardized z-scores. No extreme values were identified that required exclusion, so all data points were retained, ensuring that the analyses reflect the natural variability of the populations studied. Data cleaning was performed using statistical tools in SPSS and Excel, ensuring the integrity of the dataset. Subsequently, descriptive statistics such as mean and standard deviation were calculated for anthropometric variables and manual strength variables.

For the comparison of grip strength between the two populations, several statistical tests were used. A chi-square test was used to evaluate the distribution of age and gender strata between the samples. Then, a t-test for independent samples was applied to determine if there were differences in mean age between the two populations, and a Levene's test was used to verify equality of variances.

For the analysis of grip strength, three main procedures were performed. First, an Analysis of Covariance (ANCOVA) was used to assess whether region had a significant effect on right- and left-hand grip strength after adjusting for anthropometric dimensions. Model assumptions were systematically verified. Homoscedasticity was assessed using White's test and the Breusch–Pagan test. Scatterplots of residuals versus fitted values were used to visually confirm the homogeneity of regression slopes and the absence of systematic patterns. Collinearity among covariates was evaluated through pairwise correlations and variance inflation factors (VIFs).

Subsequently, *t*-tests for independent samples were applied in cases where the data met the assumption of normality. In contrast, in those cases where the distribution of the data was not normal, Mann–Whitney *U* tests were used.

The tests were performed separately for men and women, analyzing both hands independently. Equality of variances was verified before applying the mean comparison tests, ensuring the validity of the analyses. The results obtained made it possible to determine whether the populations could be considered equivalent in terms of grip strength or whether there were significant differences between them.

The selection of statistical tests was based on the fulfillment of normality and homogeneity assumptions. Normality was evaluated using the Kolmogorov–Smirnov and Shapiro–Wilk tests, while equality of variances was verified with Levene's test. Independent-sample t-tests were applied when the assumptions of normality and homogeneity of variances were satisfied. In cases where the data did not meet these assumptions, the nonparametric Mann–Whitney *U* test was used as an alternative. This procedure ensured that the comparisons between groups were conducted using the most appropriate test according to the distributional properties of the data.

Additionally, effect sizes (Cohen's *d*) and their 95% confidence intervals were calculated to provide information on the magnitude and precision of observed differences between groups, beyond statistical significance.

TABLE 9.2
Distribution of Age Strata in the Campeche and Yucatán Populations

Stratum	Campeche	Yucatán	Total
Men 18–29 years	27	27	54
Men 30–49 years	34	34	68
Men 50–70 years	17	17	34
Women 18–29 years	14	14	28
Women 30–49 years	24	24	48
Women 50–70 years	11	11	22
Total	**127**	**127**	**254**

9.3 RESULTS

9.3.1 Validation of Sample Adjustment

To ensure that comparisons between the Campeche and Yucatán populations were valid, preliminary statistical analyses were carried out to evaluate the equivalence of the samples in terms of age and gender distribution.

The first test performed was the chi-square homogeneity test, which made it possible to evaluate the distribution of the age strata in both populations. The result obtained ($\chi^2=0.000$, $p=1.000$) indicates that there are no significant differences in the age distribution between the two samples. The absence of differences suggests that both populations share a similar age composition, which supports the validity of subsequent comparisons of grip strength. The detailed distribution of age strata is shown in Table 9.2, which reports the absolute counts of participants by sex, age group, and region, presenting the full original sample from Yucatán and the adjusted subsample from Campeche used for subsequent analyses.

A t-test for independent samples was then performed to compare the mean age between the two populations. The results showed that the mean age in Campeche population was 37.49 years (SD=13.80), while in Yucatán population, it was 37.99 years (SD=13.24). Statistical analysis (t (252)=0.297, p=0.767) confirmed that there are no significant differences in mean age between the two populations.

In addition, Levene's test for equality of variances was applied, which showed a result of $F=0.239$ with a p-value=0.626. The outcome of this test indicates that the variances of age can be considered homogeneous between the two populations, further reinforcing the validity of the comparison.

Based on these analyses, it was concluded that the Campeche and Yucatán populations are statistically equivalent in terms of age and gender, which allowed us to continue with the grip strength analyses with confidence in the validity of the results.

9.3.2 Tests of Normality

Once the equivalence of the samples was confirmed, the next step was to assess the distribution of grip strength data to determine the most appropriate statistical tests for group comparisons.

Tests of normality (Kolmogorov–Smirnov and Shapiro–Wilk) are summarized in Table 9.3. The results indicated that grip strength values for men in both regions, as well as for women in Campeche, did not deviate significantly from a normal distribution ($p>0.05$). In contrast, grip strength data for women in Yucatán showed significant departures from normality ($p<0.05$). Based on these results, independent-sample t-tests were applied for comparisons in groups with normal distributions. At the same time, Mann–Whitney U tests were used for women in Yucatán, where the assumption of normality was not satisfied.

TABLE 9.3
Tests of Normality (Kolmogorov–Smirnov and Shapiro–Wilk)

Group	Region	Kolmogorov–Smirnov (*p*)	Shapiro–Wilk (*p*)	Normality Assumption
Men—Right hand	Campeche	0.200	0.317	Yes
	Yucatán	0.200	0.359	Yes
Men—Left hand	Campeche	0.200	0.729	Yes
	Yucatán	0.200	0.906	Yes
Women—Right hand	Campeche	0.200	0.366	Yes
	Yucatán	0.025	0.001	No
Women—Left hand	Campeche	0.200	0.441	Yes
	Yucatán	0.036	0.000	No

TABLE 9.4
Descriptive Statistics of Hand Anthropometric Variables in Campeche and Yucatán Populations

Variable	Region	Min	Max	Mean	SD
Hand length (cm)	Campeche	15.3	20.4	17.71	1.18
	Yucatán	14.7	20.7	17.73	1.31
Palm length (cm)	Campeche	8.5	11.9	10.25	0.73
	Yucatán	8.3	12.2	10.31	0.72
Hand width (cm)	Campeche	6.4	11.0	7.93	0.75
	Yucatán	7.8	11.6	9.53	0.93
Palm width (cm)	Campeche	7.8	11.0	9.47	0.76
	Yucatán	5.4	10.6	7.90	0.78
Wrist width (cm)	Campeche	3.3	5.6	4.47	0.45
	Yucatán	5.0	7.6	6.01	0.54
Wrist circumference (cm)	Campeche	14.6	22.5	18.18	1.42
	Yucatán	13.8	22.4	17.10	1.58
Grip circumference (cm)	Campeche	33.0	48.0	39.65	3.59
	Yucatán	36.0	60.0	46.74	4.24

The Mann–Whitney *U* test was selected because it represents the most appropriate nonparametric alternative to the independent-samples *t*-test. This procedure allows for the comparison of two independent groups without assuming normality and focuses on differences in central tendency (medians) rather than means. In addition, it provides robust results even with moderate sample sizes, making it suitable for the subgroup of women analyzed in this study.

9.3.3 Descriptive Statistics of Anthropometric Covariates

After evaluating normality, descriptive statistics of the anthropometric covariates were examined to establish the empirical basis for their inclusion in the subsequent ANCOVA models.

Before conducting ANCOVA analyses, the descriptive statistics of the seven hand anthropometric dimensions included as covariates were examined. Table 9.4 presents the mean, standard deviation, and range (minimum and maximum) of each dimension for both populations. These results provide an overview of how hand anthropometry is compared between Campeche and Yucatán, and establish the empirical basis for their inclusion as covariates in the subsequent models.

As shown in Table 9.4, participants from Yucatán exhibited higher values in hand width and grip circumference, while Campeche participants showed a slightly larger wrist circumference.

Other dimensions were broadly similar across groups. These differences justified the inclusion of anthropometric variables as covariates in the ANCOVA analyses.

9.3.4 Analysis of Covariance ANCOVA

After validating the distributional assumptions and characterizing the anthropometric variables, ANCOVA analyses were conducted to evaluate regional differences in grip strength after adjusting for covariates.

ANCOVA was used to determine if grip strength in both hands presented significant differences between the Campeche and Yucatán populations after adjusting for relevant anthropometric variables. Prior to conducting ANCOVA, model assumptions were evaluated. For men, both the White and Breusch–Pagan tests indicated no evidence of heteroscedasticity in right- or left-hand models ($p>0.35$). In contrast, in women, significant heteroscedasticity was detected in both hand models (Breusch–Pagan $p<0.001$; White $p<0.01$ for left hand). Nevertheless, visual inspection of residual plots showed no significant deviations, and ANCOVA results are reported with caution for these subgroups.

When evaluating right-hand grip strength, Levene's test ($F=0.033$, $p=0.856$) indicated that there are no significant differences in the variances of grip strength between both populations, fulfilling the assumption of homogeneity of variances. Additionally, pairwise correlations and VIF values indicated no problematic collinearity among anthropometric covariates, confirming the adequacy of the model.

Subsequently, ANCOVA revealed that there are no significant differences in right-hand grip strength between Campeche and Yucatán populations after adjusting for anthropometric covariates (F (1,147)$=0.141$, $p=0.708$). The results of this analysis indicate that geographic location does not have a significant impact on grip strength when anthropometric characteristics are considered. Figures 9.1 and 9.2 show the boxplots of right-hand grip strength in Campeche and Yucatán, highlighting the similarity in the distribution of the data. The descriptive statistics of right-hand grip strength for men and women in both regions are presented in Table 9.5.

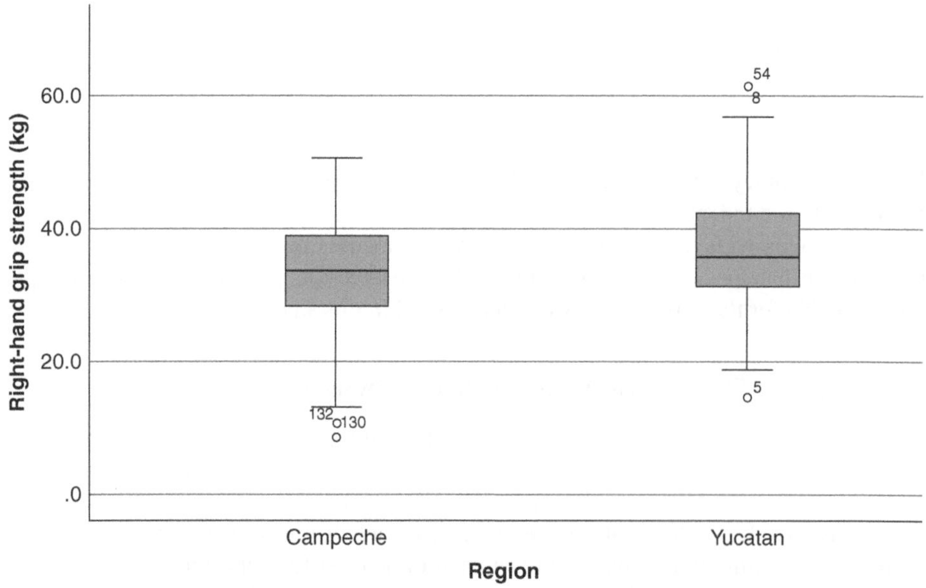

FIGURE 9.1 Boxplots of right-hand grip strength in Campeche and Yucatán men.

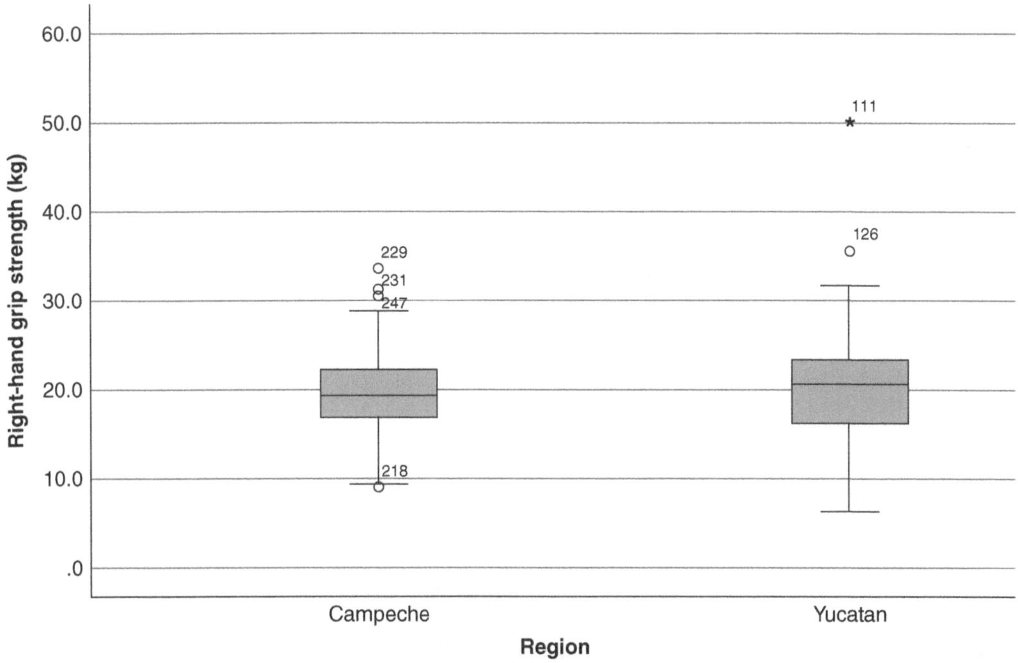

FIGURE 9.2 Boxplots of right-hand grip strength in Campeche and Yucatán women.

TABLE 9.5
Descriptive Statistics of Right-Hand Grip Strength by Region

Sex	Hand	Region	Mean (kg)	SD	*N*
Men	Right	Campeche	32.96	8.93	78
		Yucatán	36.87	8.94	78
		Total	**34.92**	**9.12**	**156**
Women	Right	Campeche	19.26	5.81	49
		Yucatán	19.81	7.28	48
		Total	**19.54**	**6.55**	**97**

Note: Values correspond to unadjusted descriptive means (Mean) with their respective standard deviations (SD) and sample sizes (*N*) for each subgroup. Adjusted means derived from ANCOVA are reported and discussed in the text.

When assessing left-hand grip strength, Levene's test confirmed equality of variances, and ANCOVA results showed that there were also no significant differences in left-hand grip strength between the two populations after adjustment for anthropometric covariates ($F(1,147)=0.141$, $p=0.973$). These findings reinforce the equivalence between the populations in terms of hand strength. Figures 9.3 and 9.4 show the boxplots of left-hand grip strength in Campeche and Yucatán populations, highlighting the similarity in the distribution of the data. The descriptive statistics of left-hand grip strength for men and women in both regions are presented in Table 9.6.

These results reinforce the hypothesis of equivalence between the populations in terms of hand strength, suggesting that the observed differences in grip strength are not determined by geographic location, but by individual factors. Furthermore, anthropometric covariates did not show a significant effect on grip strength variability, supporting the applicability of ergonomic and anthropometric data from both regions in future studies.

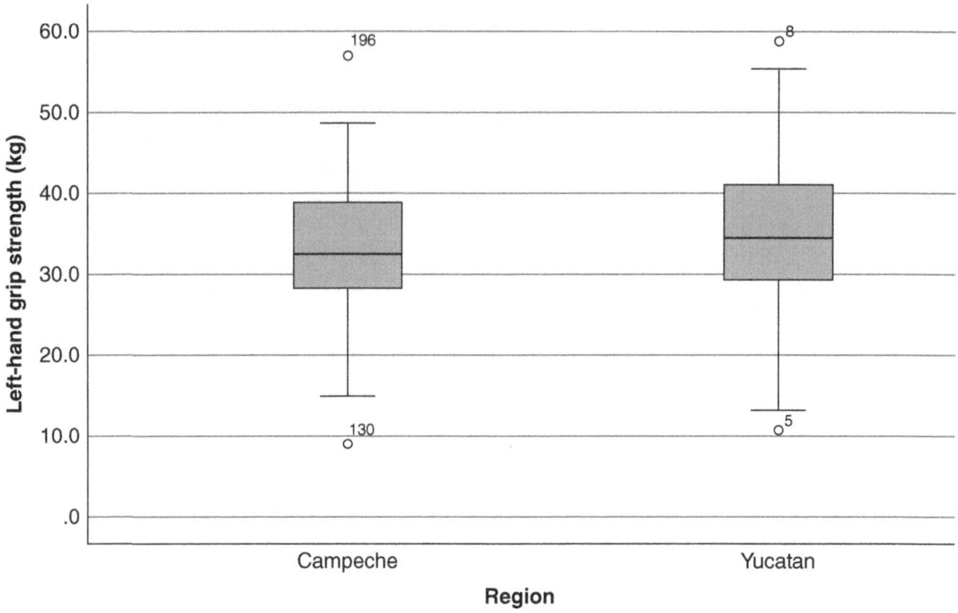

FIGURE 9.3 Boxplots of left-hand grip strength in Campeche and Yucatán men.

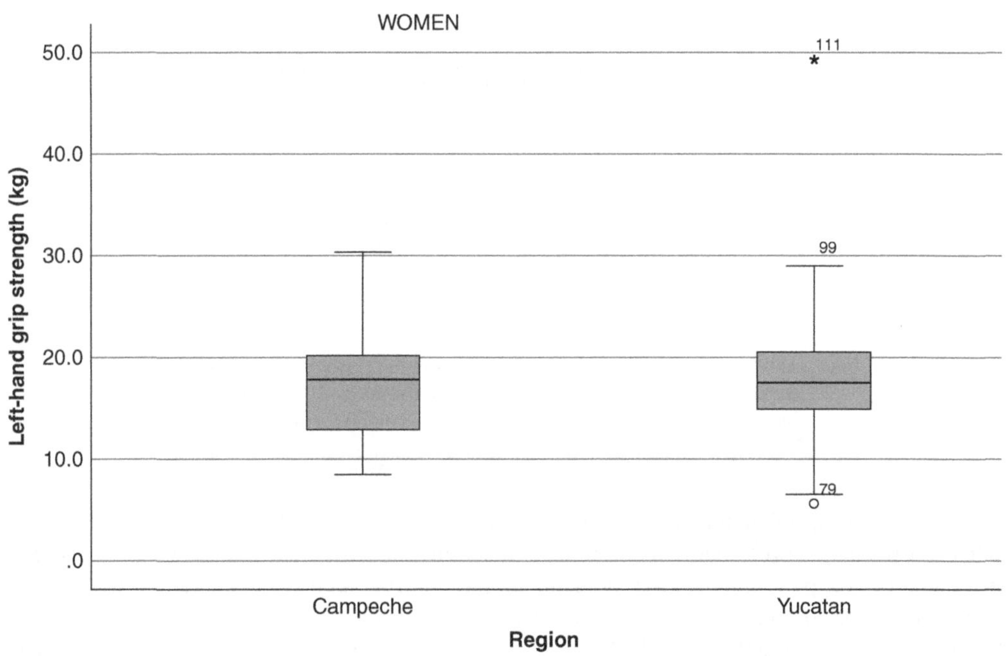

FIGURE 9.4 Boxplots of left-hand grip strength in Campeche and Yucatán women.

TABLE 9.6
Descriptive Statistics of Left-Hand Grip Strength by Region

Sex	Hand	Region	Mean (kg)	SD	*N*
Men	Left	Campeche	32.51	8.90	78
		Yucatán	34.96	9.24	78
		Total	**33.74**	**9.13**	**156**
Women	Left	Campeche	17.52	5.17	49
		Yucatán	18.12	6.98	48
		Total	**17.82**	**6.10**	**97**

Note: Values correspond to unadjusted descriptive means (Mean) with their respective standard deviations (SD) and sample sizes (*N*) for each subgroup. Adjusted means derived from ANCOVA are reported and discussed in the text.

9.3.5 t-Tests and Mann–Whitney *U* Tests for Comparison of Grip Force

Finally, to complement the ANCOVA results and provide a direct comparison of group differences, t-tests and Mann–Whitney U tests were performed according to the normality of the data.

For the comparison of grip strength between the Campeche and Yucatán populations, t-tests for independent samples and Mann–Whitney *U* tests were performed when the assumption of normality was not met.

For comparison of right-hand grip strength in men, the t-test showed a significant difference between the Yucatán and Campeche populations ($t(154)=2.730$, $p=0.007$), with a higher mean value in Yucatán (36.87 kg) compared to Campeche (32.96 kg) populations. Levene's test indicated that the variances were homogeneous ($F=0.046$, $p=0.830$), so the results were valid under this condition.

For comparison of left-hand grip strength in men, the t-test found no significant differences between regions ($t(154)=1.688$, $p=0.093$), with an average value of 34.96 kg in Yucatán and 32.51 kg in Campeche populations. Levene's test indicated that variances were homogeneous ($F=0.186$, $p=0.667$).

In women, because the data did not meet normality in the Kolmogorov–Smirnov and Shapiro–Wilk tests for right-hand grip strength, the Mann–Whitney *U* test was used for comparison. The results indicated that there were no significant differences in right-hand grip strength between women from Campeche and Yucatán populations ($U=0.532$, $p=0.532$). For left-hand grip strength, the Mann–Whitney U test also showed no significant differences between the two populations ($U=0.614$, $p=0.614$). It is worth noting that tests for heteroscedasticity (White and Breusch–Pagan) indicated nonconstant variance in women's models. However, the nonparametric results converged with ANCOVA findings, supporting the conclusion that no regional differences were present in women.

To complement the significance testing, effect sizes (Cohen's *d*) with 95% confidence intervals were calculated for the comparisons between regions. These values provide additional insights into the magnitude and practical relevance of the observed differences. Results indicated minor effect sizes for most comparisons, except for men's right-hand grip strength, which showed a moderate effect size, consistent with the statistically significant difference detected by the t-test, as shown in Table 9.7.

TABLE 9.7
Effect Sizes (Cohen's *d*) and 95% Confidence Intervals for Grip Strength Comparisons between Campeche and Yucatán Populations

Group	Hand	Cohen's *d*	95% CI	Interpretation
Men	Right	0.44	[0.12, 0.75]	Moderate
Men	Left	0.20	[−0.03, 0.43]	Small (ns)
Women	Right	0.08	[−0.31, 0.47]	Negligible (ns)
Women	Left	0.09	[−0.30, 0.48]	Negligible (ns)

9.4 DISCUSSION

The results of this study indicate that, after statistically controlling for the effect of anthropometric variables such as hand length, palm width, and wrist circumference using ANCOVA, no significant differences in hand grip strength were found between the Campeche and Yucatán populations. These results are consistent with those of previous studies that have analyzed anthropometric comparisons between geographic regions and ethnic groups, concluding that differences in hand strength are usually more related to environmental and occupational factors than to geographic location (Muthiah & Lee, 2022).

A relevant observation was found in men's right-hand grip strength. While the independent-samples t-test indicated a significant regional difference favoring Yucatán population, this effect disappeared once anthropometric covariates were included in the ANCOVA model. This shift suggests that the apparent regional difference was at least partly explained by underlying hand dimensions rather than by geographic location itself. In other words, when hand length, palm dimensions, wrist measures, and grip circumference were accounted for, the strength advantage initially detected in Yucatán men was no longer significant. This finding underscores the importance of adjusting for anthropometric variability to avoid confounding effects when comparing physical performance between populations.

This finding, although statistically significant (d=0.44, moderate effect), disappeared after adjusting for anthropometric covariates. This indicates that the observed difference likely reflects individual variability in hand dimensions rather than regional disparities and therefore has limited ergonomic relevance.

In absolute terms, the mean difference in men's right-hand grip strength was 3.9 kg (36.9 kg in Yucatán vs. 32.9 kg in Campeche), equivalent to approximately 11.8% of the overall mean. This magnitude, while statistically detectable, represents a relatively small functional difference. Tasks requiring sustained or repetitive use of the dominant hand could, in theory, be influenced by such a variation; however, once individual hand dimensions were accounted for, the effect was no longer significant. This reinforces that morphological variability explains a larger proportion of the variance in grip strength (adjusted R^2 between 0.25 and 0.32 in predictive models) than geographic origin. From an ergonomic perspective, these results imply that although minor differences may emerge in raw comparisons, they are insufficient to warrant differentiated design standards across regions. Instead, the emphasis should be placed on individual adaptation of tools and work protocols, particularly in occupations with high manual demands.

The effect size analysis further clarified this pattern. Although men's right-hand grip strength initially suggested a regional advantage for Yucatán, the moderate effect size (Cohen's d=0.44) indicates that the difference was of limited practical significance and disappeared once anthropometric covariates were considered. All other comparisons reflected negligible or minor effects, reinforcing the overall equivalence between populations and supporting a condition of functional equivalence between the Campeche and Yucatán groups for ergonomic applications and design purposes.

A comparative analysis of different population groups in Asia showed that there are significant variations in anthropometric dimensions due to racial and environmental differences, suggesting that body structure may be modified by geography and lifestyle (Kang et al. 2024). However, in the present study, regional differences do not appear to significantly influence grip strength, suggesting that the Campeche and Yucatán populations maintain anthropometric homogeneity, possibly due to a shared ethnic origin.

The main contribution of this work is referential: it provides validated anthropometric and grip strength data for two underrepresented Mexican populations, showing that they can be treated as equivalent for ergonomic applications. This finding supports the use of unified regional standards in the Yucatán Peninsula, facilitating the design of tools, equipment, and workspaces adapted to local workers. Beyond conceptual novelty, the value of this chapter lies in establishing a reliable reference dataset for ergonomic practice in a context where such information is scarce.

The impact of anthropometric variables on grip strength has been documented in several studies. Characteristics such as palm width and wrist circumference have been shown to have a direct correlation with grip strength, reinforcing the importance of statistical fit in the analysis (Wey et al. 2023). In addition, studies have suggested that grip strength may be an indicator of nutritional condition and body composition, highlighting its importance as a functional measure in occupationally active populations (Lucio et al. 2020)

Although the present study focused on inter-regional comparisons, potential intra-group variability within each population was not specifically analyzed. Factors such as age distribution, occupational demands, or habitual physical activity levels may contribute to differences in grip strength among subgroups, and their influence could provide additional insights beyond the regional comparison.

From an ergonomic perspective, it has been identified that anthropometric data can be used to design tools and workspaces that optimize performance and reduce the risk of injury (Kang et al. 2024). The absence of significant differences between Campeche and Yucatán populations suggests that the same ergonomic strategies can be implemented in both regions without the need for differential adjustments, given that the equivalence in grip strength suggests a functional similarity between the two populations.

These findings may be especially relevant in occupational contexts such as manufacturing, retail trade, and service industries, which are prevalent in both regions. For example, the design of hand tools, workstations, or load-handling protocols can be standardized without compromising safety or efficiency. Moreover, the results provide a basis for preventive ergonomic interventions aimed at reducing the risk of musculoskeletal disorders in jobs with high manual demands.

On the other hand, studies in older populations have shown that grip strength decreases with age and that women present lower grip strength than men (Mehmet et al. 2020). These findings coincide with the results obtained in this study and reinforce the importance of considering age as a key factor in future research.

It is also important to consider that socioeconomic and occupational factors may influence grip strength outcomes. According to the 2019 Economic Census (INEGI 2019), in the northern region of Campeche—covering the municipalities of Calkiní, Hecelchakán, and Tenabo—29.3% of the economically active population is employed in manufacturing industries, 35.6% in retail trade, and 21.1% in service-related activities. These occupational profiles suggest that physical demands associated with prevalent economic sectors could partly explain the similarities observed between the Campeche and Yucatán populations. Furthermore, other contextual elements such as nutritional status and habitual physical activity, although not directly assessed in this study, may also play a role and should be addressed in future research.

Finally, comparison with international reference values allows contextualizing the findings. A study conducted in Brazil has established reference values for grip strength in adults and older people, providing a framework to assess functionality in different populations (Amaral et al. 2019).

In future research, it would be advisable to expand the sample in Yucatán to strengthen the analyses and explore other factors that may influence grip strength, such as work activity and fitness level.

9.5 LIMITATIONS

This study has certain limitations. First, although the original Campeche sample was larger, a random subsampling procedure was applied to equalize both regional samples to 127 participants each (78 men and 49 women), while preserving the age-by-sex strata. While this approach ensured equivalence in the analyses, future research should aim to expand the sample size in Yucatán rather than reduce the Campeche data, in order to preserve representativeness. In addition, the Yucatán sample was drawn exclusively from an industrial group, which may limit its generalizability to the broader economically active population of the region, as workers in industrial settings could present specific physical demands that influence grip strength.

Second, the study design was cross-sectional, which restricts the possibility of establishing causal relationships or evaluating changes in grip strength and anthropometric dimensions over time. Longitudinal designs would provide valuable insights into the evolution of these characteristics across age groups, lifestyle changes, and occupational trajectories. Incorporating these elements in future studies would allow a more dynamic and comprehensive understanding of how grip strength develops and declines throughout the life course and under different working conditions.

Third, only a limited set of anthropometric variables (hand length, palm width, and wrist circumference) was considered as covariates. While these dimensions are strongly related to grip strength, other relevant factors—such as body composition, forearm length, or functional reach—were not included. Incorporating these variables in future studies would provide a more comprehensive perspective and allow for a more precise interpretation of regional differences in grip strength.

Fourth, non-anthropometric factors, including occupational activity at the individual level, nutritional status, and habitual physical activity, were not assessed. However, they may play an important role in grip strength outcomes. Addressing these aspects in future studies would strengthen the generalizability and interpretability of the findings.

Fifth, although heteroscedasticity was detected in the models for women's grip strength, the application of nonparametric tests provided consistent results with ANCOVA, supporting the robustness of the conclusions. Future research should consider larger female samples and alternative modeling approaches (e.g., generalized linear models or robust ANCOVA) to confirm these findings further.

Sixth, measurements were conducted at different times of the day, depending on the logistics of data collection in each region. Time of day was not considered as a factor in the analyses, which could have introduced uncontrolled variability. Future studies may address this aspect to evaluate its potential influence on grip strength outcomes.

Finally, intra-group variability was not examined in detail. Differences between subgroups defined by age, occupation, or lifestyle factors also influence grip strength outcomes, and future studies should incorporate these aspects to improve the interpretability of results. Expanding the Yucatán sample and including a wider range of anthropometric and contextual variables would further enhance the external validity of future analyses.

9.6 CONCLUSIONS

The results obtained in this study indicate that the Campeche and Yucatán populations are statistically equivalent in terms of age and gender, which allowed comparisons of grip strength with a high level of confidence. Validation of the sample fit using the chi-square test, the t-test for independent samples, and Levene's test confirmed that there are no significant differences in age distribution or mean age between the two populations. These validation results support the applicability of subsequent analyses and reduce the possibility of bias in the interpretation of the data.

ANCOVA allowed us to evaluate the impact of anthropometric differences on grip strength. The results showed that, after controlling dimensions such as hand length, palm width, and wrist circumference, the geographic variable had no significant effect on grip strength in either hand. These findings suggest that the observed differences in hand strength are not determined by geographic location, but by individual factors.

Independent-samples t-tests and Mann–Whitney U tests reinforced these findings by showing that, except for right-hand grip strength in men, there were no significant differences between the two populations. The difference detected in the right-hand grip strength of Yucatán men in the initial t-test was not significant after adjustment for covariates in the ANCOVA, indicating that the observed variability may be more related to individual anthropometric characteristics than to the region of origin.

From an ergonomic perspective, these findings are relevant because they suggest that anthropometric and grip strength data can be used interchangeably between the Campeche and Yucatán populations without the need for differential adjustments. Such results have implications for the design of work tools and equipment, ensuring that ergonomic standards applied in one region can be equally valid in the other.

Finally, comparison with previous studies reaffirms that differences in hand grip strength are mainly influenced by factors such as body composition, age, and occupational activity rather than by geographic region. In future research, it is recommended to expand the sample in the population of Yucatán, as well as to evaluate the influence of other factors such as occupational activity and physical condition on hand grip strength, to gain a more comprehensive understanding of the variability in the physical capacities of these populations.

ACKNOWLEDGMENTS

We thank the Consejo Nacional de Humanidades Ciencias y Tecnologías (CONAHCYT) for the funding granted for the development of this research, through the award of the Beca Nacional Tradicional 2022-1, with application number 2022-000002O1NACF-02496.

REFERENCES

Amaral, Cledir Araújo et al. 2019. "Hand Grip Strength: Reference Values for Adults and Elderly People of Rio Branco, Acre, Brazil". *PLoS ONE* 14 (1): 1–13. doi:10.1371/journal.pone.0211452.

INEGI. 2019. "Censos Económicos 2019". *Programas de Información*. https://www.inegi.org.mx/programas/ce/2019/#Tabulados.

Kang, Jiyeong et al. 2024. "Cross-Cultural Comparison of Body Dimensions and Trends: An Analysis of ANSUR and Size Korea Datasets for Automotive Seating Design". *Ergonomics* 67 (1): 123–134. doi:10.1080/00140139.2023.2206073.

Lucio, Ana Gabriela Palos et al. 2020. "Handgrip Strength as a Predictor of Body Composition in Female College Students". *Revista Chilena de Nutricion* 47 (4): 604–611. doi:10.4067/S0717-75182020000400604.

Mehmet, Hanife; Yang, Angela W.H. & Robinson, Stephen R. 2020. "Measurement of Hand Grip Strength in the Elderly: A Scoping Review with Recommendations". *Journal of Bodywork and Movement Therapies* 24 (1): 235–243. doi:10.1016/j.jbmt.2019.05.029. https://doi.org/10.1016/j.jbmt.2019.05.029.

Muthiah, Arunachalam & Lee, Yu Chi. 2022. "Comparative Analysis of Male Cyclist Population in Four Asia Countries for Anthropometric Measurements". *International Journal of Environmental Research and Public Health* 19 (16). doi:10.3390/ijerph191610078.

Wey, Alysia Koh Xin et al. 2023. "Effect of Hand Anthropometry Indices on the Measurement of Hand Grip Strength Using a Handheld Dynamometer in Young Asian Males". *European Journal of Physical Education and Sport Science* 9 (5): 82–93. doi:10.46827/ejpe.v9i5.4763.

10 Comparative Ergonomic Analysis of the Mechanical Demand Imposed by the Order-Picking Activity in Distribution Centers with Different Levels of Mechanization

Fernanda Maradei, Shyrle Berrio, Oscar Bernal, and Leonardo Quintana

10.1 INTRODUCTION

According to the Colombian Ministry of Labor, the disease rate resulting from activities associated with the storage and transportation of goods is 95.37 per 100,000 people affiliated with the labor risk system (Consejo Colombiano de Seguridad, 2022). The most frequent occupational diseases diagnosed in Colombian workers in 2011 were carpal tunnel syndrome, painful shoulder rotation syndrome, epicondylitis, intervertebral disk displacement, unspecified back pain, lumbago, synovitis, and tenosynovitis (Fasecolda, 2013). Another report suggested that the causes of occupational diseases in Colombia are related to repetitive hand and/or arm movements (31.4%), awkward postures (25.5%), and load-lifting throughout the workday (7.6%) (Ministerio de la Protección Social, 2004). This situation leads to absenteeism and work-related disabilities, generating high annual costs. According to a study conducted in the general services sector, absenteeism and occupational disabilities affect organizational budgets and influence decisions to introduce ergonomic adjustments to reduce potential musculoskeletal injuries (Kumar & Kumar, 2008).

At the international level, upper-limb discomfort has become a hallmark of manual loading activities (Hanvold et al., 2014; Sterud et al., 2014; Luime et al., 2004). To address this issue, electronic or mechanized systems have been incorporated into these processes.

However, it has been generally observed that, depending on the type of mechanization system employed in a given distribution center (DC), specific conditions affect different body parts, particularly the arms and trunk (Braam et al., 1996). Furthermore, regardless of the level of mechanization, there are activities common to all DCs (Posada, 2011).

The present study assessed the order-picking activity in two DCs with different mechanization conditions, focusing on the impact that the mechanical demands of each task have on workers. The study was carried out in a food distribution company, whose largest proportion of employees work in the two studied DC types. Therefore, this population is ideal for observing the results yielded by the proposed methodology regarding the physical and postural effects of order-picking (Susic et al., 2010).

 DOI: 10.1201/9781003582014-13

For this purpose, the ergonomic interventions aimed to benefit workers in their daily activities. The study used electromyography (EMG) and electrogoniometry (GNM), based on previous similar works and methodologies, following the European EMG guidelines (Portnoy et al., 2015; Forsman et al., 2002; Takala & Toivonen, 2013; Susic et al., 2010; Chen et al., 2010).

Additionally, descriptive and quantitative records were kept of the segmentation of the studied tasks over time, a procedure known as activity chronicles (Rasoulifar et al., 2009). A quantitative integration of this tool with EMG and GNM was used to analyze the results obtained under different mechanization conditions affecting the mechanical demands on the back and upper limbs of workers. Finally, this approach supported decision-making aimed at improving the ergonomic conditions of the order-picking activity (Chen et al., 2010).

10.2 METHODOLOGY

10.2.1 Participants

Fifteen participants conducted the real picking activity in a moderately mechanized (Mm) DC, separating the products of each customer's order in corresponding containers and placing them on a roller conveyor. The other 15 participants working in a slightly mechanized (Sm) DC palletized and classified the ordered products by client and distribution area or route. The demographic characteristics of the sample and the environmental conditions of the facilities are shown in Table 10.1.

The difference in the gender distribution of the sample ($n = 15$) between the two DCs is explained by the fact that the SmDC had a total of 15 male workers only, whereas the MmDC had a larger workforce, of which only 9 were males. Therefore, it was necessary to include a minimum of 15 workers for each DC, regardless of gender.

10.2.2 Experimental Design and Procedure

The objective of this research was to study the physical workload of the picking activity, comparing the influence of the mechanization level of the DC and how the biomechanical demands affect the back and upper limbs of the worker population. This research was conducted under real operation conditions in two types of DCs. In regard to the environmental conditions, the MmDC, which is located in the city of Bogotá, had an average temperature of 15.9°C (SD 1.75°C) and a

TABLE 10.1
Demographic and Environmental Data

	Mm ($N = 15$)	Sm ($N = 15$)
Age (years)	29.33 (5.56)	26.28 (4.63)
Height (cm)	165 (8)	172.2 (6)
Weight (kg)	62.5 (9.2)	71.4 (9.5)
Experience (months)	56.6 (40.4)	40 (25.8)
Gender	60% M – 40% F	100% M
Dominance	93% right-handed	86% right-handed
Room temperature (°C)	15.9 (1.75)	24 (2.84)
Relative humidity (%)	58% (9)	66% (7)

Source: Mean (Standard Deviation). Authors' own creation.

relative humidity of 58% (SD 9%). The SmDC, located in the city of Bucaramanga, had an average temperature of 24°C (SD 1.84°C) and relative humidity of 66% (SD 7%). The measurements were performed simultaneously on each of the workers at the beginning of the workday, thus controlling for thermal stress.

10.2.2.1 Physical Workload

Physical load, posture, and muscle activity data were obtained as inputs for further analysis. Postural measurements were taken using flexible biaxial electro-goniometers (Biometrics Ltd., UK) placed on the joints under study. The evaluated parameters and corresponding biometric sensors were position of the hand relative to the forearm, flexion–extension, and ulnar–radial deviation (SG65, Biometrics Ltd., UK); shoulder adduction–abduction (SG150, Biometrics Ltd., UK); and back posture, including flexion–extension and lateral inclination or flexion (SG110, Biometrics Ltd., UK).

Muscle activity was evaluated using eight surface EMG sensors (SX230, Biometrics Ltd., UK) placed on the muscle belly of the extensor carpi radialis (ECR), extensor carpi ulnaris (ECU), flexor carpi radialis (FCR), flexor carpi ulnaris (FCU), middle deltoid (mD), and on the dominant side of the upper trapezius (uT). For the back muscles, sensors were placed on the left erector (lE) and right erector (rE) muscles. Participants were prepared following standard procedures, including skin preparation, sensor placement, and measurement protocols (Perotto et al., 2011; Freriks et al., 1999). The band-pass filter was set to 20–460 Hz, signal noise was <5 μV, and input impedance exceeded 1,000 Ω.

10.2.2.2 Workload

A series of parameters (tasks, work halts, problem-solving, attention to task, and demand for labor) were analyzed through "activity chronicles," which consist of the appraisal of the mentioned parameters in a synchronized chronological order with the physical load data. The basic information was obtained through 45-minute video recordings and analyzed in an Actogram Kronos 2 software package (France) by qualified observers. The tasks identified in the MmDC were radiofrequency control, product handling, container manipulation, container pushing, label rebonding, standing on the rack, box opening, container placement on the conveyor, and product search. In the SmDC, the identified tasks were pallet truck manipulation, radiofrequency control, product handling, product search, pallet marking, box opening, and pallet handling. Also, the Work Demand variable common to both DCs was classified as Physical, Mental, Both, or None.

10.2.3 Analysis of Results

All activity chronicle records were synchronized with the evaluated physical workload data. Surface electromyography signals recorded in American Standard Code for Information Interchange (ASCII) were processed using algorithms developed in MatLab R 2013a (USA). The root mean square (RMS) signal was estimated from a moving window of 200 ms (Berrio et al., 2012) (Lee et al., 2009) and normalized with each muscle's maximum voluntary contraction (MVC) record. P10 (static level), P50 (medium level), and P90 (dynamic level) were estimated in the Amplitude Probability Distribution Function (APDF). The posture information signal was processed by the data link software of the equipment (Biometrics Ltd., UK) and recorded in ASCII (after filtering) and was measured in degrees. Angles of interest were associated with corresponding APDF analyses, particularly emphasizing average, P10, P50, and P90 data. All records were processed and analyzed in SPSS 22.0. Given a particular task, prioritization took into account if the muscles exceeded 20% average MVC and if percentiles P10 and P90 of the body segments fell out of the neutral ranges.

10.3 RESULTS

10.3.1 Muscular Activity

The most intense muscular activity was evident in the extensors (radial and ulnar) and the right and left erector spinae. In all tasks, these four muscles exceeded 20% of the MVC during dynamic activity (P90). The exception was the ECR in the tasks called label rebonding and product search. In the MmDC, box opening showed the highest muscular activity, mainly concentrated in the upper limbs, with ECU muscle demands of up to 39.8%. The tasks of standing on the rack and product handling were particularly demanding for back muscular activity, with levels of 36.5% for the rE and 35.8% for the lE muscle. Figure 10.1 shows P90 of MVC, as assessed for each muscle across MmDC tasks.

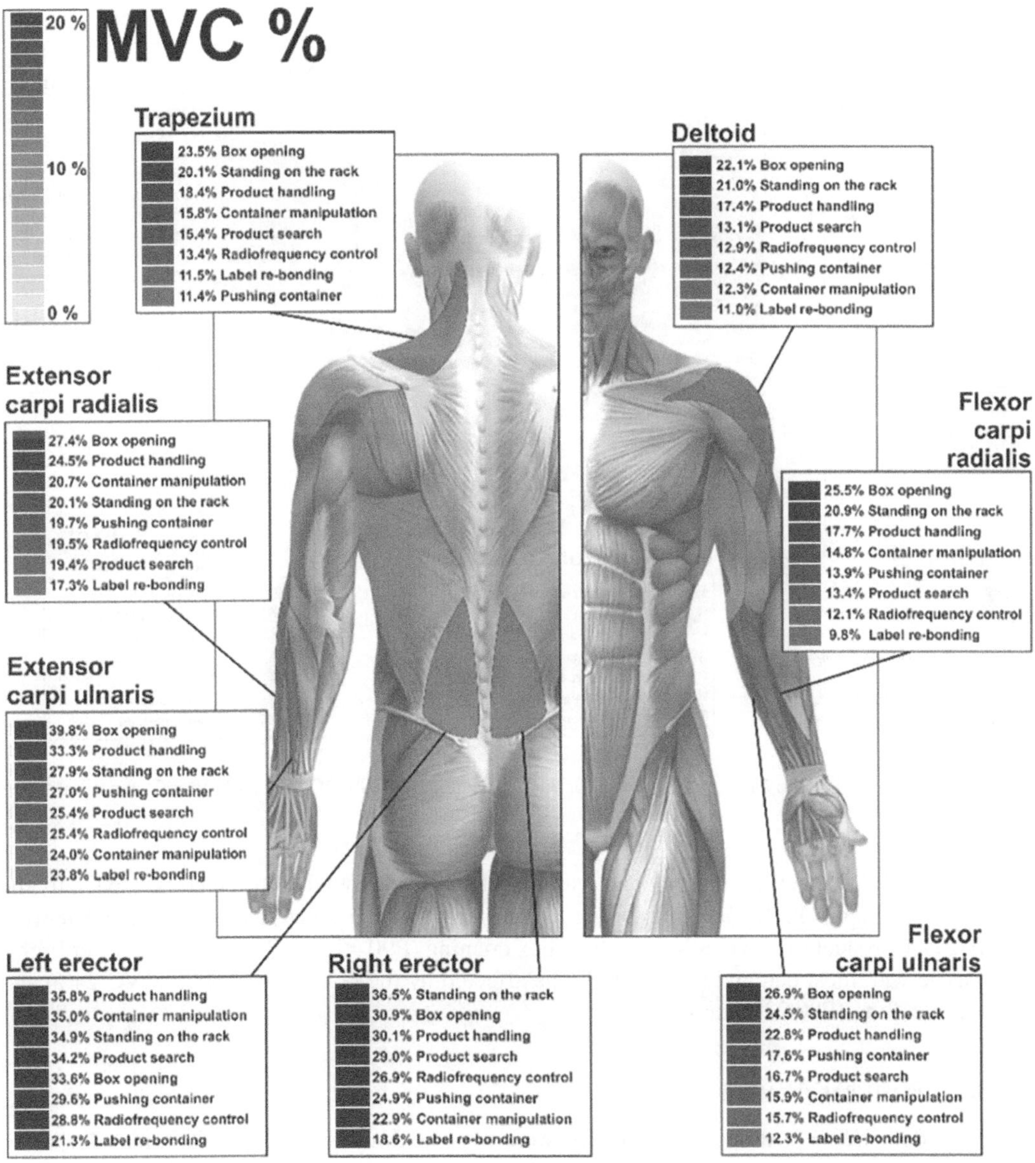

FIGURE 10.1 P90 of the MVC, as assessed for each muscle across MmDC tasks. (Authors' own creation.)

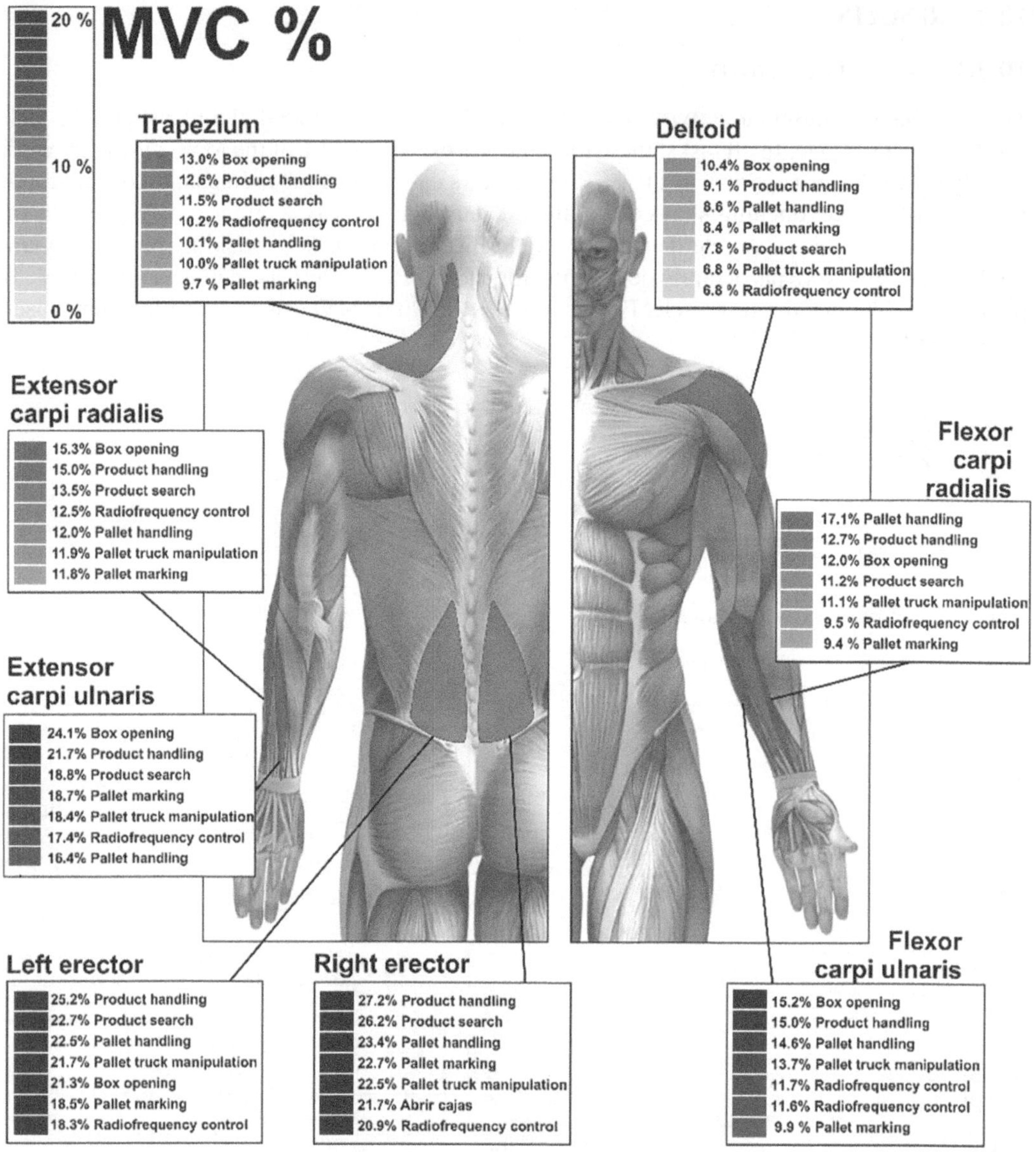

FIGURE 10.2 P90 of the MVC, as assessed for each muscle across SmDC tasks. (Authors' own creation.)

In the SmDC, the highest muscular activity (P90 MVC) was found in the erectors, especially the right one (see Figure 10.2). Tasks exhibiting the most dynamic back activity were product handling (P90=27%), product search (P90=26%), and box opening (P90=23%). Regarding the upper limbs, the muscle with most demand was the ECU, as affected by the box opening (P90 MVC 24%) and product handling (P90 22%) tasks.

10.3.2 Postural Analysis

The most demanding postures in the MmDC were wrist flexion–extension, radial–ulnar deviation, and back flexion–extension. All tasks registered over 45° wrist extensions, with up to 22.9% of the time in this range of motion. In the SmDC, activities were carried out with wrist extensions above

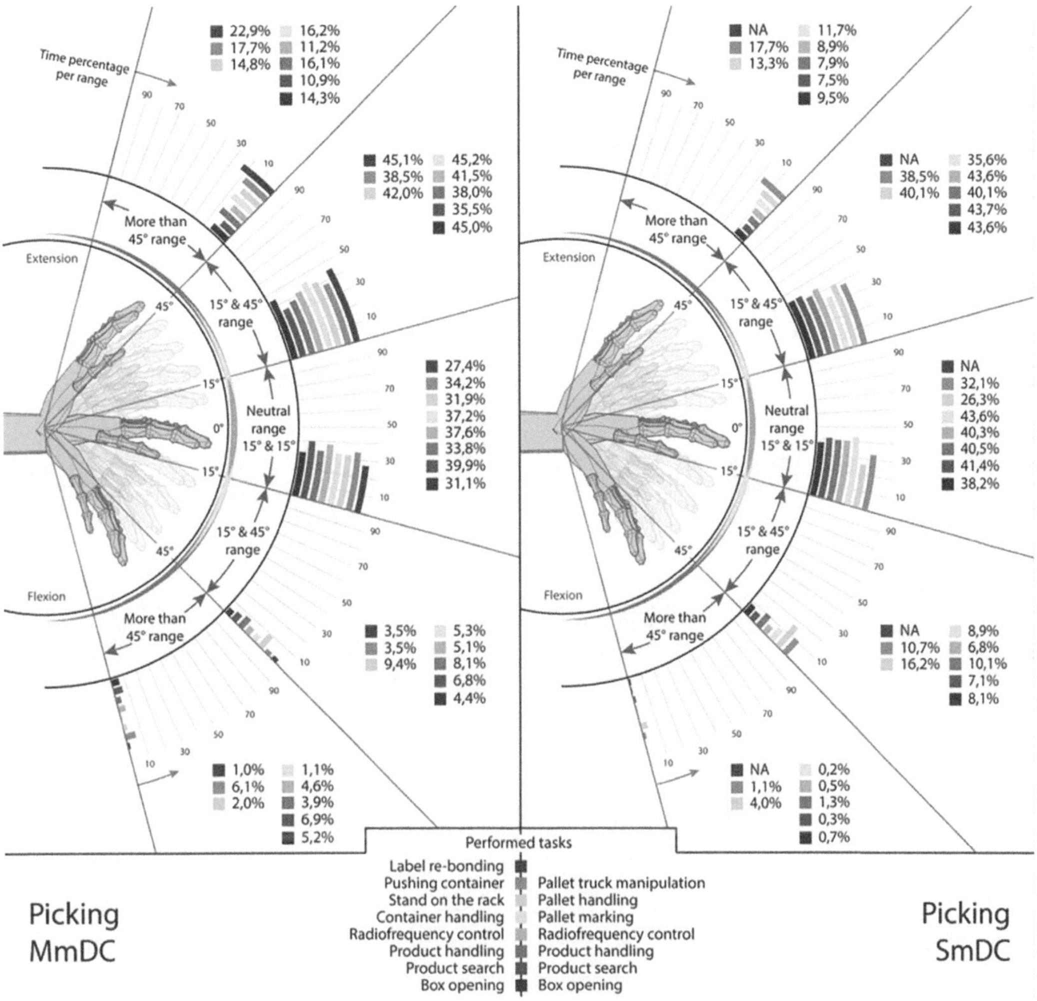

FIGURE 10.3 Percentage of time at hand flexion–extension angles across DC types. (Authors' own creation.)

45° during 17.7% of the time, especially in tasks like pallet truck manipulation and box opening, with wrist extension postures of up to 54.9° and 44.3°, as seen in Figure 10.3.

Regarding wrist lateral deviation, picking activity at the MmDC exhibited up to 34° ulnar deviation and 17.9° radial deviation. In tasks like box opening and product handling, the radial deviation in the wrist reached ranges higher than 25° during 5% of the time. Postures of up to 23.9° ulnar deviation and 21.4° radial deviation were observed in the SmDC, with ulnar deviation reaching over 40° during up to 6.9% of the time, as shown in Figure 10.4.

In the SmDC, the picking activity required postures that fell out of the neutral angle range. Back extension angles between 0° and 30° occurred during 19.8% of the time, while back flexion angles between 20° and 60° were observed during 27.9% of the time (see Figure 10.5). These postures were especially found in tasks such as pallet truck manipulation and box opening, with respective back flexion angles of up to 26.2° and 31.1°. In turn, the MmDC demanded back extension angles ranging from 0° to 30° during 19.2% of the time, while 20° to 60° back flexions were observed for 53.1% of the time.

On average, back lateral deviations showed neutral movement ranges for 97.4% of the time in the MmDC and 97.6% of the time in the SmDC. Left lateral deviations were dominant at the MmDC,

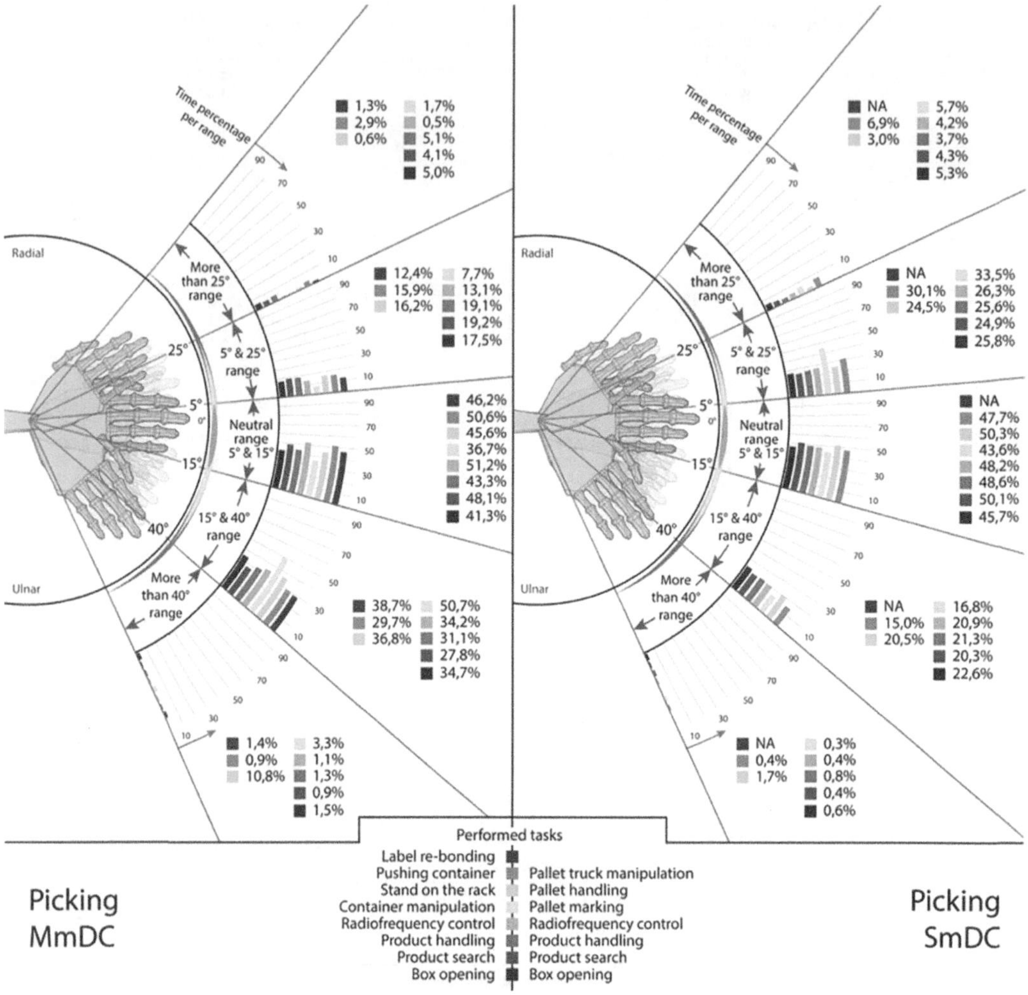

FIGURE 10.4 Percentage of time at wrist radial and ulnar deviation angles, as assessed across DC types. (Authors' own creation.)

taking 80.9% of the time, particularly in the container handling task, which involved taking the container from the conveyor belt and putting it on the floor. In turn, right lateral deviations were dominant at the SmDC during 61.2% of the time, especially in the pallet handling task.

Shoulder postures ranged between 10° of adduction and 90° of abduction, averaging 90.9% and 97.1% of the time in the MmDC and the SmDC, respectively. However, in the MmDC, 75% of the box opening and container handling tasks exceeded 10° of adduction, which reached adduction angles of 15.2° and 13.8°, respectively.

10.3.3 Physical Demand of DC Tasks of Interest

Table 10.2 shows the myoelectric activity of the muscles, as well as the postural behavior measured through GNM. Values highlighted in red indicate those that exceed the appropriate thresholds for ergonomic work and, therefore, represent a potential risk of developing subsequent pathologies. Both the box-opening and standing-on-the-shelf tasks showed a high level of muscle activity above 20% of the MVC. Likewise, forced postures were found mainly in joints such as the wrist and arm.

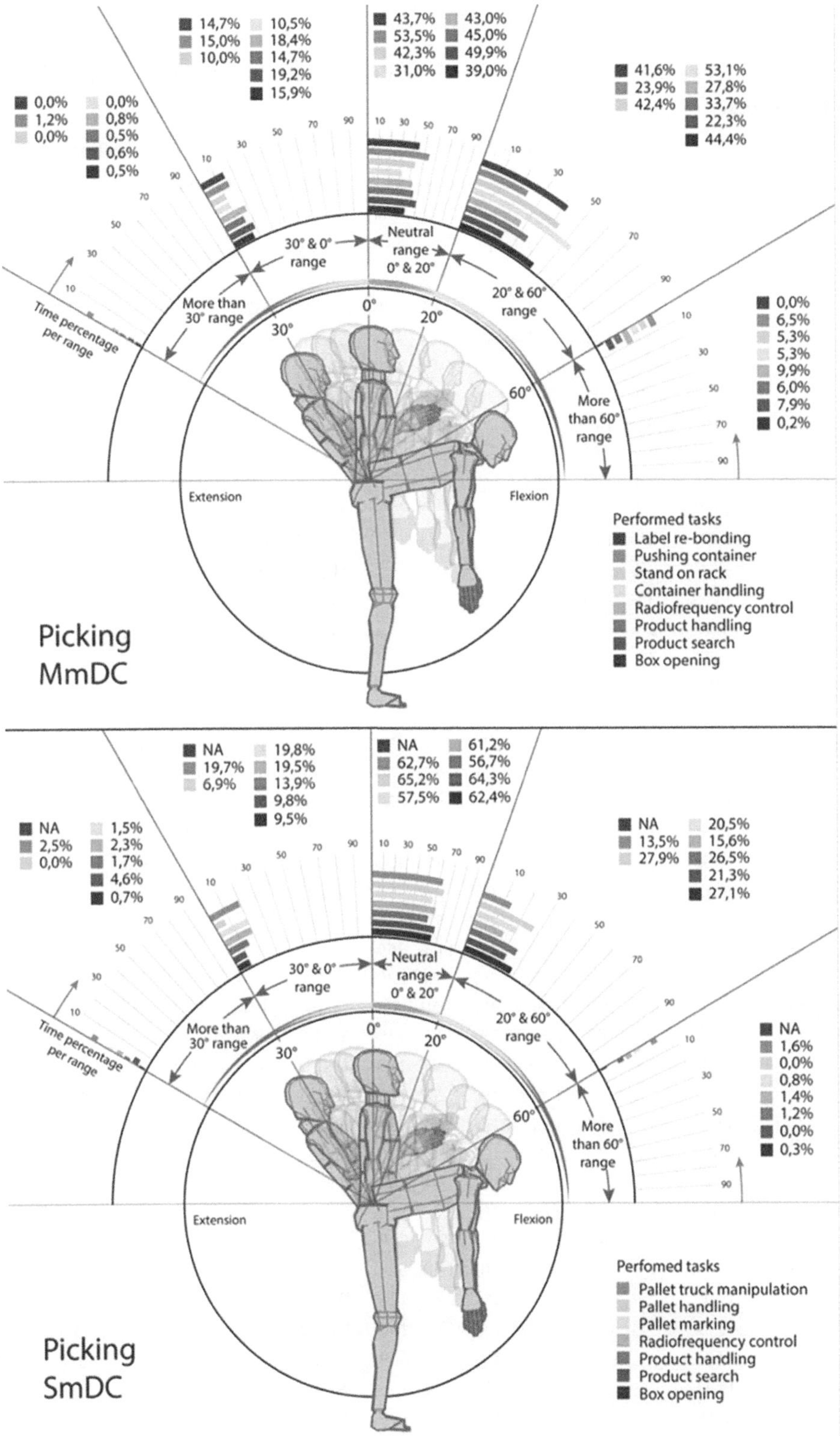

FIGURE 10.5 Percentage of time at back flexion–extension angles across DC types. (Authors' own creation.)

TABLE 10.2
Summary of the Results According to DC Task Prioritization—Physical Demand of DC Tasks of Interest

Tasks of Interest	Muscles	EMG						Joint	GNM					
		Mm			Sm				Mm			Sm		
Box opening		P10	P50	P90	P10	P50	P90		P10	P50	P90	P10	P50	P90
	ECR	3%	12%	27%	1%	7%	15%	W_FE	−50	−12	14.1	−44.3	−15.9	12.6
	ECU	3%	17%	40%	2%	10%	24	UR_D	−16.4	7.6	29	−16.4	2.2	23.9
	FCR	1%	8%	26%	1%	4%	12%	AD_AB	−37.8	−11.1	15.2	−33.1	−14.4	0.6
	FCU	3%	12%	27%	1%	5%	15%	S_LF	−16.8	−2.3	15.7	−13.9	−3	10.1
	mD	1%	7%	22%	1%	4%	10%	B_FE	−1.8	16.6	40.8	0	13.4	31.1
	uT	2%	9%	23%	2%	7%	13%							
	IE	4%	14%	34%	2%	9%	19%							
	rE	5%	14%	31%	3%	10%	22%							
Standing on the shelf		P10	P50	P90		NA.			P10	P50	P90		NA.	
	ECR	2%	9%	20%				W_FE	−54	−14.8	21.1			
	ECU	3%	14%	28%				UR_D	−10.8	9.3	27.2			
	FCR	1%	7%	21%				AD_AB	−39.9	−14.6	11.8			
	FCU	3%	12%	24%				S_LF	−18	−6.4	8.5			
	mD	2%	8%	21%				B_FE	0.2	25.2	35.2			
	uT	3%	11%	20%										
	IE	6%	17%	35%										
	rE	6%	19%	36%										
Product handling		P10	P50	P90	P10	P50	P90		P10	P50	P90	P10	P50	P90
	ECR	2%	9%	25%	1%	6%	15%	W_FE	−53.4	−12.8	19.9	−41.7	−13.1	17.5
	ECU	2%	12%	33%	1%	8%	22%	UR_D	−17.9	6.4	28.5	−15.9	2.8	23.1
	FCR	1%	5%	18%	1%	3%	13%	AD_AB	−34.1	−8.2	12.1	−29.4	−11.8	3.1
	FCU	2%	8%	23%	1%	4%	15%	S_LF	−16.8	−2.4	12.6	−9.3	1.5	12.9
	mD	1%	5%	17%	1%	3%	9%	B_FE	−0.9	24.2	42.6	−3.6	12.7	33.8
	uT	2%	7%	18%	2%	5%	13%							
	IE	4%	13%	36%	3%	11%	25%							
	rE	4%	12%	30%	3%	10%	27%							
Radiofrequency control		P10	P50	P90	P10	P50	P90		P10	P50	P90	P10	P50	P90
	ECR	2%	8%	20%	1%	4%	13%	W_FE	−46.7	−10.3	14.3	−43.2	−16.2	10.3
	ECU	1%	9%	25%	1%	5%	17%	UR_D	−7.5	10.2	28	−16.8	1.9	22
	FCR	1%	3%	12%	1%	2%	10%	AD_AB	−32.1	−7.2	10.8	−25.4	−10.9	2
	FCU	2%	5%	16%	1%	3%	12%	S_LF	−15.7	−1.2	14.8	−21	−1.4	10.6
	mD	1%	3%	13%	1%	2%	7%	B_FE	−3.2	27.9	55.3	−5.8	8.4	25.6
	uT	2%	5%	13%	2%	5%	10%							
	IE	3%	10%	29%	2%	8%	21%							
	rE	4%	8%	27%	2%	8%	23%							
Pallet handling			NA.		P10	P50	P90			NA.		P10	P50	P90
	ECR				2%	15%	12%	W_FE				−49	−12.7	34.5
	ECU				2%	9%	16%	UR_D				−15.7	3.4	22.9
	FCR				1%	6%	17%	AD_AB				−26.8	−13.7	1.2
	FCU				1%	8%	15%	S_LF				−9.6	2.2	10.3
	mD				1%	4%	9%	B_FE				1.3	14.5	28.7
	uT				2%	5%	10%							
	IE				3%	14%	23%							
	rE				4%	18%	21%							

(Continued)

TABLE 10.2 (*Continued*)
Summary of the Results According to DC Task Prioritization—Physical Demand of DC Tasks of Interest

Tasks of Interest	Muscles	EMG Mm	EMG Sm			Joint	GNM Mm	GNM Sm		
Pallet truck		NA.	P10	P50	P90		NA.	P10	P50	P90
manipulation	ECR		1%	4%	12%	W_FE		–54.9	–20.3	18.4
	ECU		1%	6%	18%	UR_D		–21.4	–0.9	19.4
	FCR		1%	3%	11%	AD_AB		–31.8	–12.8	1.3
	FCU		1%	4%	14%	S_LF		–10.7	1.4	13.9
	mD		1%	2%	7%	B_FE		–5	8.6	26.2
	uT		2%	5%	10%	.			.	
	lE		2%	9%	22%					
	rE		3%	9%	23%					

Source: Authors' own creation.
W_FE, wrist flexion (+) extension (–); UR_D, ulnar (+) radial (–) deviation; AD_AB, shoulder adduction (+) abduction (–); S_LF, right (+) left (–) lateral flexion of the spine; B_FE, back flexion (+) extension (–); mD, middle deltoid; uT, upper trapezius; lE, left erector; rE, right erector.

10.3.4 Simultaneous Task Demand

The mental requirement of the tasks imposed the highest objective demand on the part of the picking activity at the MmDC (39.5%) and the SmDC (42%). This is an expectable result because this factor involves counting units and reviewing codes, references, and other data issued by the radiofrequency tool. In the SmDC, the physical burden represented 33.8% of the picking activity, while in the MmDC, this (Figure 10.6) was 22%. In the latter, the highest demand resulted from combined physical and mental demand (26.5%) (see Figure 10.6).

At the MmDC, the tasks with greater mental demands were label rebonding (83.22%) and moving the container to the next station (62.88%). The task with the greatest physical demand was standing on the rack (62.50%). It is worthwhile noting how the product handling task exhibited a combined (physical and mental) demand of 40.15%.

At the SmDC, the highest mental demand (57.22%) was imposed by the radiofrequency control task. Product search imposed a mental demand of 53.98%. Additionally, pallet handling showed an elevated physical demand (72.73%). No task exhibited combined (physical and mental) demand above 15%.

10.4 DISCUSSION

The research and development of techniques and methods for the optimization of order picking have focused primarily on the efficiency of the costs of the logistics processes and to a lesser extent on the engineering of human factors (Grosse et al., 2015). This may be one of the reasons why order-picking is still an eminently manual process despite the wide possibility of its mechanization and automation. And this in turn can be one of the main determinants of the important physical load attributed to this work activity that, by its nature, involves risks such as demanding postures, repetitive tasks, and manual handling of loads. Logically, there are also different methods and techniques for evaluating these risks. Thus, frequency measurements and tools such as GNM and analysis techniques such as video recording have been used a lot in terms of the study of repetitiveness

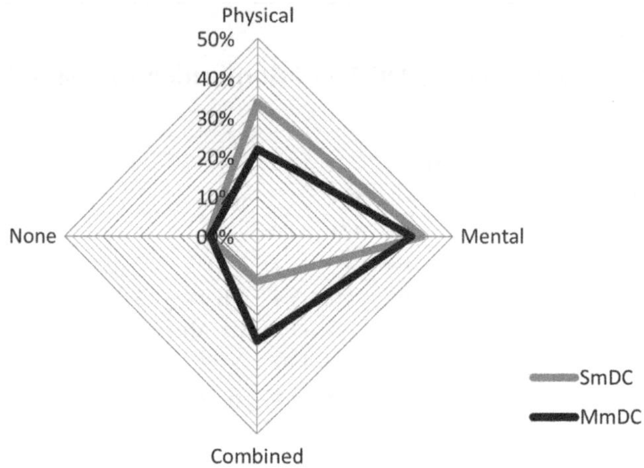

FIGURE 10.6 Activity demand at the MmDC (left) and SmDC (right). (Authors' own creation.)

(You & Kwon, 2005). While the evaluation of the posture is torn between the specificity and generality of the methods proposed for this purpose, ranging from some widely known and validated, as is the case of Rapid Upper-Limb Assessment (RULA) and Rapid Entire Body Assessment (REBA), to mention just two among many others, including novel proposals such as the PERA method for the ergonomic evaluation of postural risk (Chander & Cavatorta, 2017), such methods are based on the observation of posture with the help of photographic and video records and risk assessment mainly according to the amplitude of the joint movement and the duration of the posture.

Thus, different studies conducted in industrial sectors related to the risk of physical load exposure show the implementation of postural assessment methods such as RULA and REBA. These have been recognized as useful and relevant tools for evaluating postural behavior in workers who perform their tasks both in a seated position and while standing (Ansari & Sheikh, 2014; Boulila et al., 2018). However, compared to the RULA method, the REBA method covers the entire human body, making it the preferred choice when a more comprehensive assessment of load or postural behavior is desired (Kee & Karwowski, 2015). It should be noted that these methods are observational; therefore, their assessment is considered subjective, depending on the expertise of the researcher applying the method. In this regard, other authors highlight the importance of direct measurement assessment (Village et al., 2005; Dai, Jin, Ning, & Mirka, 2010; Dempsey, McGorry, & O'Brien, 2004) as it allows for obtaining accurate results, which in turn provides confidence and reliability to the study (Boulila et al., 2018). Therefore, in seeking to compare different mechanization conditions and their influence on the musculoskeletal demands of workers, this study employed EMG and GNM as they are objective recordings free from perception biases, with the aim of obtaining reliable results.

Undoubtedly, more accurate indicators are required to assess the physical workload, as is the case of physiological indicators, for example, the valuation of muscle work by surface electromyography. This resource not only has great applications for the direct evaluation of the biomechanical demand imposed by the work activity, for example, in the manual handling of loads, but it is also useful to assess the application of indirect methods for the evaluation of the manual handling of charges (Ranavolo et al., 2018). A substantial amount of research conducted in various industrial processes suggests a relationship between work activity and the possible effects on health from exposure to physical workload. However, fewer studies cover the comparison between physical load and mental load.

In this research, a method was developed to assess critical points in the activities conducted in a supply chain operation. Specifically, we sought to determine whether DC mechanization modalities improve working conditions, taking into account the angles adopted by the most frequently involved body segments and the performance of the corresponding muscles across different tasks. Likewise, the methodological strategy adopted allowed qualifying the mental demand associated with the work activity. In this measure, an approximation to the composition of the tasks studied is achieved according to their physical and mental demands. The evaluation of the intrinsic physical and mental requirements of these tasks that make up the picking process in two DCs with different levels of mechanization provides important information to control possible risks to workers' health.

An important part of the interventions on the conditions and the working environment that are implemented from ergonomic studies is based on mechanical solutions, either the incorporation of mechanical aids for the accomplishment of a specific task or the partial or total mechanization of a work process, although it must be taken into account that this mechanization also results from the need to optimize the productive processes. In any case, the results of the present study suggest that a higher level of mechanization does not necessarily imply less physical workload. The overall analytical results suggest that mechanization in a DC implies an elevated muscular demand combined with prolonged wrist and back exposures at extreme postural ranges. This indicates the need for mechanization to require prior engineering of human factors in order to positively impact risk conditions.

In this sense, it is important to note that the present study cannot be considered conclusive due to its limitations, especially with regard to the population subject to study. However, the methodology used can serve as a reference for the accomplishment of similar studies that allow greater inference about the relation that exists between the level of mechanization of the DCs, the physical and mental loads imposed by the tasks of picking, and possible damage to health.

10.5 CONCLUSION

The present results reflect measurements obtained under the real operation conditions and constraints of the picking activity in two DCs. The study details the tasks involved in the picking activity of two DC types by assessing the muscular physical load and body segment posture with specialized equipment. In the MmDC, the tasks exceeding 20% of the MVC were box opening, standing on the rack, and product handling. In the SmDC, the tasks exceeding 20% of the MVC were pallet truck manipulation and pallet handling. In this way, the current findings complement those previously reported in "The workload of warehouse workers in three different working systems" (Braam et al., 1996). The study allows justified recommendations about changes in the physical structure of the DC. In the MmDC, it is advisable to place heavier and higher turnover products in the middle of the rack in order to decrease trunk flexion and activity of the erector muscles when lifting heavy loads. To reduce muscle activity in the arm extensors, the implementation of a box-opening tool is suggested. In the SmDC, preventive maintenance of the mechanical pallet trucks—the manipulation of which affects the erector spinae muscles—is recommended since it reduces the pushing effort.

It is clear that the mechanization of the MmDC does not prevent the physical risks to which workers are exposed when performing logistics tasks. It cannot be assured that mechanization, the benefits of which have already been proven in other studies, is the likely generator of the elevated MVCs observed in the MmDC. However, prior application of human factor engineering can be taken into consideration to generate a positive impact on the risky conditions of the workers. In addition, self-care was observed to be greater in the SmDC than in the MmDC. For further research, it is important to consider the time pressure factors involved in the picking operation in connection with its physical and postural load.

Finally, it is important to consider that the results of this study are universal in nature, meaning they can be transferred to other logistical contexts with similar conditions. However, it is essential to analyze the work activity beforehand, as established by the concept of ergonomics. The demands of an activity are specific and are framed within specific work contexts at a given time.

REFERENCES

Ansari, Nafis, & Sheikh, Mohd. (2014). Evaluation of work posture by RULA and REBA: A case study. *IOSR Journal of Mechanical and Civil Engineering,* 11(4), 18–23.

Berrio, Shyrle, Barrero, Lope, & Quintana, Leonardo Augusto. (2012). A field experiment comparing mechanical demands of two pruners for flower cutting. *Work,* 41, 1342–1345. https://doi.org/10.3233/WOR-2012-0321-1342

Boulila, Atef, Ayadi, Mahfoudh, & Mrabet, Khadija. (2018). Ergonomics study and analysis of workstations in Tunisian mechanical manufacturing. *Human Factors and Ergonomics in Manufacturing,* 28(4), 166–185. https://doi.org/10.1002/hfm.20732

Braam, Ingrid, Dormolen, Max van, & Frings-Dresen, Momique H. (1996). The workload of warehouse workers in three different working systems. *International Journal of Industrial Ergonomics,* 17, 469–480. https://doi.org/10.1016/0169-8141(95)00008-9

Chander, Divyaksh Subhash, & Cavatorta, Maria Pia. (2017). An observational method for postural ergonomic risk assessment (PERA). *International Journal of Industrial Ergonomics,* 57, 32–41. https://doi.org/10.1016/j.ergon.2016.11.007

Chen, Hsieh-Ching, Chang, Cha-Mei, Liu, Yung-Ping, & Chen, Chih-Yong. (2010). Ergonomic risk factors for the wrists of hairdressers. *Applied Ergonomics,* 41, 98–105. https://doi.org/10.1016/j.apergo.2009.05.001

Consejo Colombiano de Seguridad. (2022, 8 de marzo). Accidentes de trabajo y enfermedades laborales en Colombia 2021. Disponible online: https://ccs.org.co/atel-col-2021/

Dai, Boyi, Jin, Sangeun, Ning, Xiaopeng, & Mirka, Gary. (2010). The effects of horizontal load speed and lifting frequency on lifting technique and biomechanics. *Ergonomics,* 53(8), 1024–1032. https://doi.org/10.1080/00140139.2010.493957

Dempsey, Patrick, Mcgorry, Raymond, & O'Brien, Niall. (2004). The effects of work height, workpiece orientation, gender, and screwdriver type on productivity and wrist deviation. *Industrial Ergonomics,* 33, 339–346. https://doi.org/10.1016/j.ergon.2003.10.006

Hanvold, Therese Nordberg, Wærsted, Morten, Mengshoel, Anne Marit, Bjertness, Espen, & Veiersted, Kaj Bo. (2014). Work with prolonged arm elevation as a risk factor for shoulder pain: A longitudinal study among young adults. *Applied Ergonomics,* 47, 43–51. https://doi.org/10.1016/j.apergo.2014.08.019

Fasecolda. (2013, abril). La enfermedad laboral en Colombia. Disponible online: https://fasecolda.com/cms/wp-content/uploads/2019/08/enfermedad-laboral-colombia-2013.pdf

Forsman, Mikael, Hansson, Gert Ake, Medbo, L., Asterland, P., & Engström, T. (2002). A method for evaluation of manual work using synchronized video recordings and physiological measurements. *Applied Ergonomics,* 33, 533–540. https://doi.org/10.1016/S0003-6870(02)00070-4

Freriks, Bart, Hermens, Hermie, Disselhorst-Klug, Cathenie, & Rau, Günter. (1999). The recommendations for sensors and sensor placement procedures for surface electromyography. *Roessingh research and development,* *8*(2), 13–54.

Grosse, Eric H., Glock, Christoph H., & Neumann, W. Patrick. (2015). Human factors in order picking: A content analysis of the literature. *International Journal of Production Research,* 53(8), 2341–2358. https://doi.org/10.1080/00207543.2016.1186296

Kee, Dohyung, & Karwowski, Waldemar (2015). A Comparison of three observational techniques for assessing postural loads in industry a comparison of three observational techniques for assessing postural loads in industry. *International Journal of Occupational Safety and Ergonomics,* 13(1), 3–14. https://doi.org/10.1080/10803548.2007.11076704

Kumar, Rupesh, & Kumar, Shrawan. (2008). Musculoskeletal risk factors in cleaning occupation—A literature review. *International Journal of Industrial Ergonomics,* 38, 158–170. https://doi.org/10.1016/j.ergon.2006.04.004

Lee, David L., Kuo, Po.-Ling, Jindrich, Devin L., & Dennerlein, Jack T. (2009). Computer keyswitch force–displacement characteristics affect muscle activity patterns during index finger tapping. *Journal of Electromyography and Kinesiology*, 810–820. https://doi.org/10.1016/j.jelekin.2008.03.011

Luime, Jolanda, Koes, Bart, Hendriksen, Ingrid, Burdorf, Alex, Verhagen, Arianne, Miedema, Harald, & Verhaar, Jan. (2004). Prevalence and incidence of shoulder pain in the general population: A systematic review. *Scandinavian Journal of Rheumatology,* 33, 73–81. https://doi.org/10.1080/03009740310004667

Portnoy, Sigal, Rosenberg, Limor, Alazraki, Tal, Elyakim, Esti, & Friedman, Jason. (2015). Differences in muscle activity patterns and graphical product quality in children copying and tracing activities on horizontal or vertical surfaces. *Journal of Electromyography and Kinesiology,* 25, 540–547. https://doi.org/10.1016/j.jelekin.2015.01.011

Perotto, Aldo, Delagi, Edwar, Iazzetti, John, & Morrison, Daniel. (2011). *Anatomical Guide for the Electromyographer: The Limbs and Trunk*. Springfield, IL: Charles C Thomas Publisher, Ltd.

Posada, Juan Gregorio. (2011). Aspectos a considerar para una buena gestión en los almacenes de las empresas (Centros de Distribución, CEDIS). *Journal of Economics, Finance and Administrative Science,* 16(30), 83–96.

Ranavolo, Alberto, Varrecchia, Tiwana, Iavicoli, Sergio, Marchesi, Agnese, Rinaldi, Martina, Serrao, Mariano, Conforto, Silvia, Cesarelli, Mario, & Draicchio, Francesco. (2018). Surface electromyography for risk assessment in work activities designed using the "revised NIOSH lifting equation". *International Journal of Industrial Ergonomics,* 68, 34–45. https://doi.org/10.1016/j.ergon.2018.06.003

Sterud, Tom, Johannessen, Håkon A., & Tynes, Tore. (2014). Work-related psychosocial and mechanical risk factors for neck/shoulder pain: A 3-year follow-up study of the general working population in Norway. *International Archives of Occupational and Environmental Health,* 87, 471–481. https://doi.org/10.1007/s00420-013-0886-5

Susic, Aleksandar, Lulic, Tanja J., & Veljovic, Fikret. (2010). Ergonomic evaluation of task execution: Surface electromyography in muscular activity screening. *Periodicum Biologorum,* 122, 33–38. https://hrcak.srce.hr/52689

Rasoulifar, Rahi, Meillon, Brigitte, Thomann, Guillaume, & Villeneuve, Francois. (2009). Observation, annotation and analysis of design activities: How to find an appropriate tool? In *DS 58-5: Proceedings of ICED 09, the 17th International Conference on Engineering Design, Vol. 5, Design Methods and Tools (pt. 1),* Palo Alto, CA, USA, 24.-27.08. 2009.

Takala, Esa-Pekka, & Toivonen, Risto. (2013). Placement of forearm surface EMG electrodes in the assessment of hand loading in manual tasks. *Ergonomics,* 56, 1159–1166. https://doi.org/10.1080/00140139.2013.799235

Village, J., Frazer, M., Cohen, M., Leyland, A., Park, I., & Yassi, Annalee. (2005). Electromyography as a measure of peak and cumulative workload in intermediate care and its relationship to musculoskeletal injury: An exploratory ergonomic study. *Applied Ergonomics,* 36(5), 609–618. https://doi.org/10.1016/j.apergo.2005.01.019

You, Heecheon, & Kwon, Ochae. (2005). A survey of repetitiveness assessment methodologies for hand-intensive tasks. *International Journal of Industrial Ergonomics,* 35(4), 353–360. https://doi.org/10.1016/j.ergon.2004.07.006

11 Taxonomy, Applications, and Future of Passive Exoskeletons
A Comprehensive Analysis of Opportunities and Challenges

Pablo Antonio Muñoz Gutiérrez,
Juan Luis Hernández Arellano, and David Cortés Sáenz

11.1 INTRODUCTION

Passive exoskeletons have emerged as an innovative solution in today's industry to improve ergonomic conditions and reduce fatigue during physically demanding tasks (Preethichandra et al. 2024; Halim et al. 2022; Lévy 2013). Unlike active exoskeletons, which require an external power source, passive exoskeletons operate without any external energy by leveraging mechanical materials and mechanisms to provide support in activities that demand physical effort, such as lifting heavy loads, maintaining prolonged postures, and performing repetitive tasks (Fox 2020; Yang et al. 2008; Lowe, Billotte, and Peterson 2019; ERGOIBV 2024). These devices have found significant applications in sectors such as manufacturing, construction, and logistics, where workers are exposed to high physical loads and the risk of musculoskeletal injuries. Their adoption contributes to occupational safety and well-being, improving both productivity and worker satisfaction by reducing physical strain (Kuber and Rashedi 2021; Kim et al. 2019; Riccò et al. 2021; Upasani et al. 2019).

11.2 PURPOSE OF THE STUDY

The aim of this study is to develop a clear and detailed taxonomy of passive exoskeletons, classifying them according to functional criteria, application sectors, and mechanisms of action. The intention is to create a reference framework that facilitates a better understanding of their applications in various industrial environments, while also identifying areas for innovation and potential challenges in their future implementation.

11.3 IMPORTANCE OF THE TAXONOMY

A structured taxonomy facilitates research and the development of new technologies in passive exoskeletons by providing a guide for designers and engineers in choosing features according to workplace needs. Moreover, it enables companies to make informed decisions regarding which type of exoskeleton best suits their requirements, thereby optimizing the design and functionality of these devices to improve industrial ergonomics.

11.4 LITERATURE REVIEW

11.4.1 Review Methodology

To address the lack of an integrated taxonomy and to ensure rigor and transparency, a systematic literature review was conducted following the PRISMA (Preferred Reporting Items for Systematic

DOI: 10.1201/9781003582014-14

Reviews and Meta-Analyses) protocol guidelines. The review protocol focused on identifying and extracting data related to the following aspects:

Inclusion criteria: Published studies, articles, and use cases examining passive exoskeletons in occupational settings were selected. The focus was on devices designed to reduce physical fatigue and improve ergonomics in sectors such as manufacturing, construction, healthcare, and logistics.

Exclusion criteria: Studies focused exclusively on active (motorized) exoskeletons or those designed primarily for medical rehabilitation without direct occupational application were excluded.

Analysis of results: The review identified key patterns in the literature, including the most commonly assisted body areas (functionality), the industrial sectors with the highest adoption (application sector), the mechanical principles used (mechanism of action), and the types of movements or efforts assisted. These findings laid the foundation for the development of the proposed taxonomy, which seeks to integrate these multiple dimensions into a coherent framework, thereby facilitating better understanding and decision-making.

11.4.2 Definition of Passive Exoskeletons

Passive exoskeletons are external devices that provide physical support to the user without the need for electrical energy or additional motorized sources (Sefton et al. 2025; Siedl and Mara 2021; McFarland and Fischer 2019; Steinhilber et al. 2020). Instead of relying on motors or batteries, these devices employ mechanical principles such as springs, tension mechanisms, and rigid or semi-rigid structures to reduce the physical load on the human body (Okunola et al. 2024; Arachchige et al. 2024; Govaerts et al. 2023; Bosch et al. 2016; Koopman et al. 2019). Their primary objective is to alleviate stress in specific areas of the body such as the back, shoulders, or lower extremities—making them useful tools for tasks that involve forced postures or repetitive movements (Mahmud et al. 2022; Zelik et al. 2022; Carrasquillo et al. 2025; Garosi et al. 2022; Omoniyi et al. 2020). In contrast to active exoskeletons, which offer greater control and precision through motorized systems, passive exoskeletons have limitations in adaptability and the amount of force they can provide. Nevertheless, they offer advantages such as lower weight, reduced cost, and simplicity of use.

11.4.3 Current State and Applications

Current literature evidences a growing use of passive exoskeletons in various industries (Omoniyi et al. 2020; Thamsuwan et al. 2020; Di Natali et al. 2025; Mendoza n.d.; Erden and Rainey 2024). In the manufacturing sector, passive exoskeletons are employed to reduce muscular fatigue during repetitive tasks, assisting workers who perform continuous movements on assembly lines (Spada et al. 2017; Reese 2024). In construction, these devices offer support in activities that involve heavy lifting and prolonged postures, such as material handling or work on scaffolding, thereby contributing to injury prevention (Behjati Ashtiani n.d.; Ojha et al. 2024; Zhu, Dutta, and Dai 2021; Kim et al. 2019). In logistics, passive exoskeletons facilitate loading and unloading tasks, enhancing efficiency while protecting the worker (Clement Thevenot and Mornieux 2025). They have also been explored in the healthcare sector, where they can be used by medical personnel to alleviate strain on the lumbar region and limbs during prolonged physical assistance tasks, such as patient mobilization (Baltrusch et al. 2018).

11.4.4 Existing Classifications

Several classification proposals exist in the literature; however, most are partial and limited to specific criteria (Crea et al. 2021; Howard et al. 2020). Some authors have classified passive exoskeletons based on the body area they support (e.g., exoskeletons for the back, legs, or arms) (Schwerha et al. 2021; Nussbaum et al. 2019). Other studies have grouped them according to their application sectors, such as those oriented toward industry or healthcare (Fournier and Yung 2023; Burks et al.

2021; Steinhilber et al. 2020). Although these classifications are useful, they do not provide an integrated view that encompasses functional, mechanical, and application aspects simultaneously. The absence of an exhaustive taxonomy limits the ability to compare and evaluate devices consistently and hampers the identification of opportunities for design and functional improvements.

11.4.5 Current Innovations and Challenges

The current trends in passive exoskeleton design focus on creating devices that are lighter, more ergonomic, and adaptable (Kuber and Rashedi 2021; Zelik et al. 2022; Khamaisi et al. 2024; Hefferle et al. 2020; De Bock et al. 2022). Passive exoskeletons have been developed using advanced materials that combine strength and flexibility, offering greater comfort and freedom of movement. Additionally, adjustable mechanisms are being incorporated to allow the level of support to be personalized according to the task and the user's body type (Sefton et al. 2025; Del Ferraro et al. 2020; Kim et al. 2018; Proud et al. 2022). However, significant challenges remain, including limitations in adjustment and adaptation for different body types, as well as issues related to comfort during prolonged use (Kirkwood, Wilkenfeld, and Dunbar 2022; Theurel and Desbrosses 2019). Another obstacle is cost as some models—even though passive—require specialized materials that increase the price and hinder widespread implementation. These challenges underscore the need for ongoing research to enhance the effectiveness and accessibility of passive exoskeletons in the workplace.

11.5 DEVELOPMENT OF THE TAXONOMY

To comprehensively classify passive exoskeletons, criteria were established based on their functional characteristics and practical applications. The proposed taxonomy is organized into five main dimensions (Figure 11.1).

11.5.1 Functionality

This criterion focuses on the body region that receives support, allowing the differentiation of devices according to the area of assistance:

- *Lumbar support:* Designed specifically to alleviate load in the lower back, these exoskeletons are intended for tasks involving heavy lifting or maintaining prolonged postures, thereby reducing the risk of lumbar injuries. See Figure 11.2 for the EVO-C exoskeleton, a mobile passive device that combines support for the back and hips, designed for industrial applications, warehouses, and cold rooms. This low-cost system, weighing 3 lb, significantly contributes to the reduction of work-related fatigue.
- *Lower extremities:* These devices provide support for the legs and are especially useful in tasks requiring squatting or prolonged standing. By better distributing the load on the lower limbs, these devices aim to prevent fatigue and minimize joint impact. See Figure 11.3 for the EVO-I exoskeleton from EVO, a rigid harness that activates at any time to lock in position, effectively transforming the exoskeleton into a chair. This innovative design allows rapid transitions between standing, sitting, and walking without altering the workstation, enabling the worker to adopt a temporary ergonomic posture that alleviates leg strain and promotes a healthy, upright stance.
- *Upper extremities:* Designed to assist the arms and shoulders, these exoskeletons are applied in contexts where overhead lifting or repetitive movements are common, thereby reducing muscular tension and enhancing task precision. See Figure 11.4 for the EVO-S exoskeleton, specifically designed for tasks that require work with the arms at or above chest level. Its purpose is to alleviate back and shoulder pain generated by awkward postures, improving ergonomics and safety during elevated work tasks.

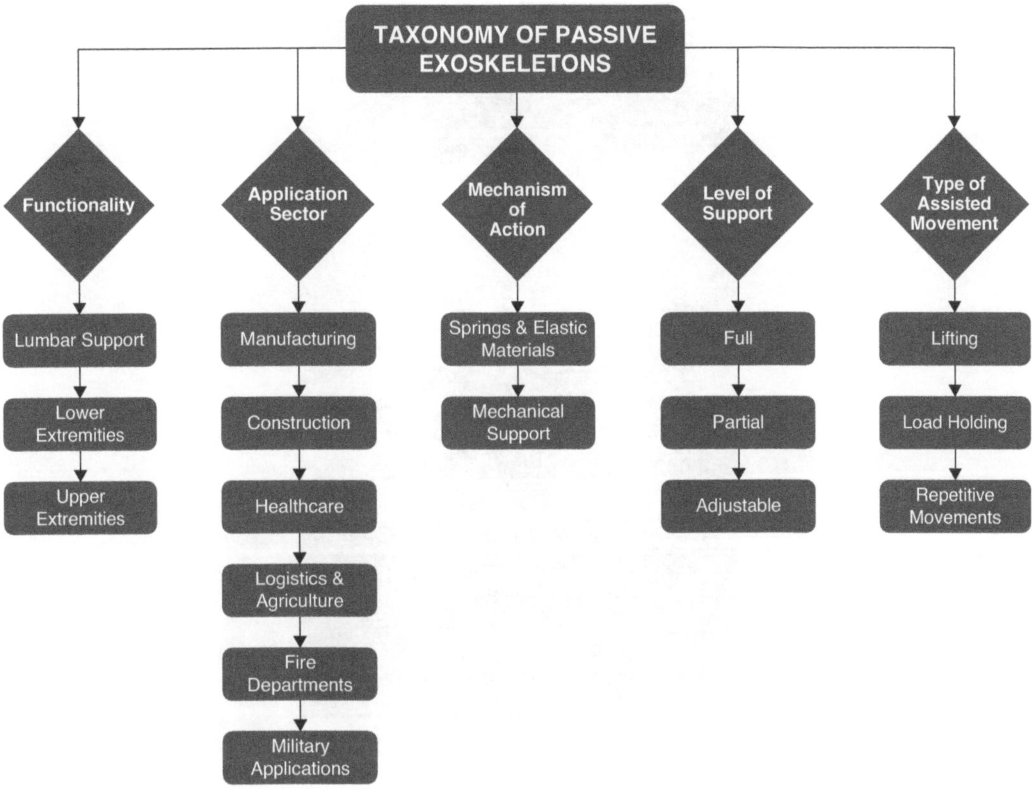

FIGURE 11.1 Taxonomy.

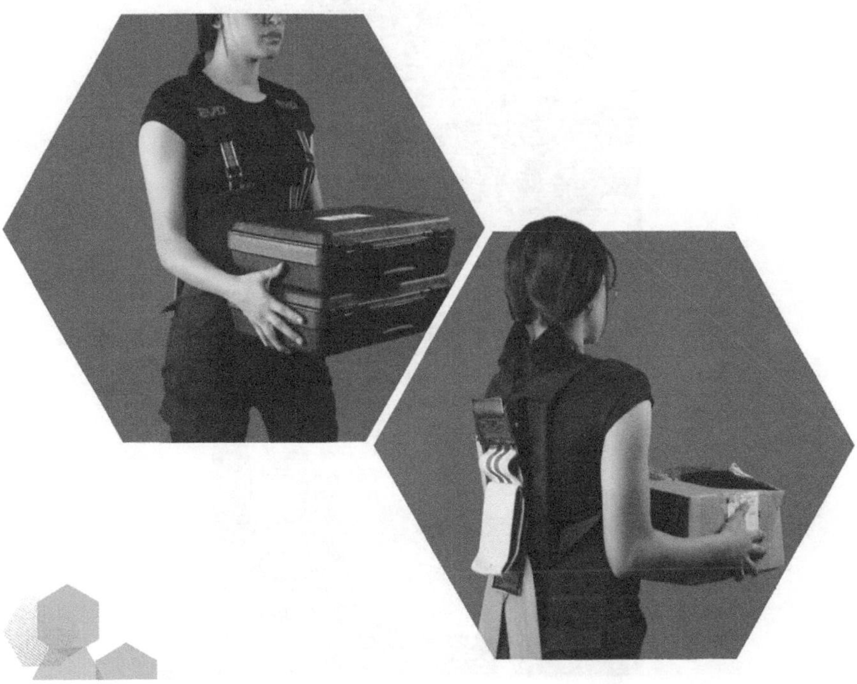

FIGURE 11.2 *EVO*-C exoskeleton.

FIGURE 11.3 *EVO*-I exoskeleton.

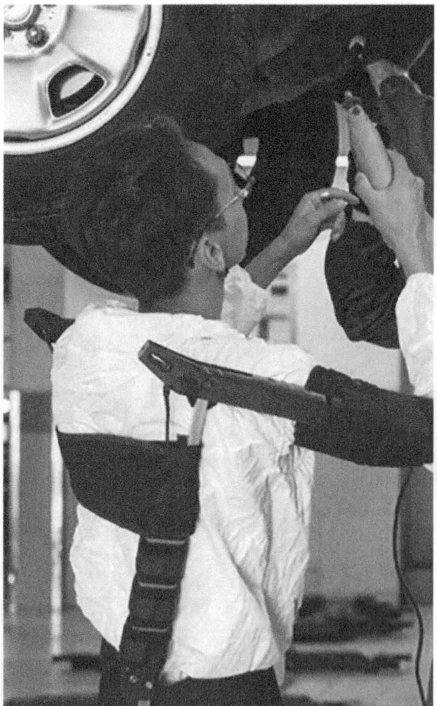

FIGURE 11.4 *EVO*-C exoskeleton.

- *Postural:* These exoskeletons offer support during prolonged static positions, such as standing or squatting for long periods. Their goal is to reduce muscle load and the fatigue associated with maintaining strained postures, and they may include devices that function as temporary seats for the worker. Many exoskeletons function either as purely postural supports or as part of mixed systems.

11.5.2 Application Sector

The main sectors where the implementation of passive exoskeletons has demonstrated promising results have been identified and categorized as follows:

- *Manufacturing:* Passive exoskeletons have been deployed in assembly lines and production processes where repetitive movements can lead to cumulative fatigue among workers. These devices help mitigate physical effort during precise and continuous assembly tasks, reducing the incidence of musculoskeletal injuries. Additionally, they facilitate ergonomics by evenly distributing the load, which can translate into increased operational efficiency and reduced downtime due to fatigue or pain (Schwerha et al. 2021; Todorovic, Constantinescu, and Popescu 2018; Madinei et al. 2020; Iranzo et al. 2020).
- *Construction:* In the construction sector, passive exoskeletons provide essential support in activities that involve heavy lifting and material handling. Their use enables workers to maintain safe postures and reduces the impact of repetitive movements on the spine and limbs. Implementing these devices helps diminish the risk of injuries by alleviating tension in critical areas and facilitating the performance of tasks in high-demand physical environments (Okunola et al. 2024; Behjati Ashtiani n.d.; Zhu, Dutta, and Dai 2021; Kim et al. 2019; Mahmud et al. 2022; Gutierrez et al. 2024; Gonsalves et al. 2024; Afolabi, Yusuf, and Akanmu 2025; Daniel and Olaoye 2025).
- *Healthcare:* Hospitals and medical care facilities have begun to incorporate passive exoskeletons to assist staff in tasks that require prolonged physical effort, such as patient transfers and equipment mobilization. These devices not only improve the ergonomics of healthcare professionals but also reduce the likelihood of injuries associated with manual handling, resulting in a safer work environment and enhanced patient care quality (Kim et al. 2019; Baltrusch et al. 2018; Pinelli, Manetti, and Lettieri 2023; Baptista et al. 2022; Kinge et al. 2015).
- *Logistics and agriculture:* In the sectors of logistics and agriculture—where handling loads and performing physically demanding tasks are common—passive exoskeletons have become valuable tools. In logistics, they facilitate the loading and unloading of goods by providing stability and support during lifting, thereby enhancing the operational efficiency. In agriculture, exoskeletons help workers endure long shifts under demanding conditions, facilitating tasks such as fertilizer application, handling of heavy tools, and other labor-intensive activities, which leads to reduced physical strain and increased productivity (Upasani et al. 2019; Siedl and Mara 2021; Arachchige et al. 2024; Clement Thevenot and Mornieux 2025; Thamsuwan et al. n.d.; Jakob et al. 2023; Renner 2018; Borisov et al. 2021).
- *Fire departments:* In firefighting, exoskeletons assist in transporting and operating heavy equipment, a crucial factor during emergencies. Additionally, these devices offer potential enhancements, such as integrated cooling systems or protective mechanisms, which can be decisive in high-risk situations like structural collapses (Duffus 2019; Schnieders et al. 2017; Anitha et al. 2024).
- *Military applications:* Within the military domain, passive exoskeletons have been adopted to support soldiers during extended marches, facilitating the transport of heavy loads and improving the precision of weapon handling. These devices contribute to reducing physical fatigue and optimizing performance in combat conditions, representing a significant

advantage in tactical operations (Proud et al. 2022; Scataglini, Andreoni, and Gallant 2015; Mudie et al. 2018).

- *Rehabilitation:* Although this study focuses on occupational applications, it is important to recognize that passive exoskeletons also play a role in rehabilitation. In this sector, passive exoskeletons are used to help patients regain mobility, provide support during physical therapy, and facilitate the performance of therapeutic exercises with less assistance from a therapist.

11.5.3 Mechanism of Action

This criterion describes the way in which the device operates to provide support and is divided into the following:

- *Springs and elastic materials:* These mechanisms exploit the mechanical energy stored in components such as springs or elastomers to assist the user without the need for external energy sources, thereby enabling a rapid and adaptable response during movement.
- *Mechanical support without additional energy:* This approach is based on the use of rigid or articulated structures that passively redistribute the load. Such a mechanism is fundamental in applications where simplicity and robustness are essential for the device's continuous operation.

11.5.4 Level of Support

This refers to the degree of assistance provided by the exoskeleton, which is crucial for adapting the tool to various tasks:

- *Total, partial, and adjustable:* Total support implies that the device assumes most of the load throughout the workday, while partial support is activated at specific moments or for particular movements. The adjustable option allows the user to modify the level of assistance according to individual needs, combining flexibility with ergonomic efficiency.

11.5.5 Type of Assisted Movement

This aspect classifies exoskeletons according to the nature of the movement they facilitate:

- *Lifting:* Designed to assist in the elevation of heavy loads, thereby minimizing strain on the back and arms.
- *Load holding:* Focused on maintaining a stable position while sustaining a load over an extended period, thus reducing fatigue in the stabilizing muscles.
- *Repetitive movements:* These devices are targeted at tasks involving constant and repetitive movements, helping to prevent cumulative fatigue and improving the efficiency of continuous operations.

11.5.6 Examples of Classification

To illustrate the application of this taxonomy, several practical examples can be highlighted:

- *Lumbar support exoskeleton in manufacturing with partial support for load holding:* These are exoskeletons designed with adjustable springs that allow for the adaptation of the support level to the specific needs of workers on assembly lines, thereby reducing fatigue and preventing lumbar injuries.

- *Upper extremity exoskeleton in construction with adjustable support for repetitive movements:* These devices focus on providing assistance to the arms and shoulders during heavy lifting and material handling, thus optimizing ergonomics in environments with high physical demands.
- *Healthcare and logistics exoskeleton for lower extremities with partial support for lifting:* Exemplified by exoskeletons that facilitate patient mobilization in hospital settings or load handling in distribution centers, these devices are designed to offer adequate support for the legs, enhancing stability and reducing effort during prolonged work periods.

11.6 IMPLEMENTATION LIMITATIONS AND CHALLENGES

Despite the demonstrated benefits in reducing fatigue and improving ergonomics, the implementation of passive exoskeletons faces several challenges that must be addressed to achieve widespread and effective adoption:

- *User acceptance:* Resistance to change is one of the main obstacles. Many workers may perceive exoskeletons as uncomfortable or restrictive devices that limit natural movement (Elprama, Vanderborght, and Jacobs 2022; Jakob et al. 2023; Luciani et al. 2023; Wioland et al. 2025; Gilotta et al. 2019; del Rio Carral et al. 2022; de Looze, Krause, and O'Sullivan 2017; Baltrusch et al. 2021; Andrade and Nathan-Roberts 2022). Studies have shown that the perception of discomfort and the sensation of rigidity can reduce the motivation for continuous use. To counteract this barrier, it is essential to involve employees in the selection and adjustment process and to implement awareness campaigns that emphasize the health and productivity benefits (Elprama et al. 2023; Moyon, Poirson, and Petiot 2019; Schwerha et al. 2022).
- *Fit and adaptation:* Since body morphologies vary widely among users, exoskeletons must be highly adjustable. An inadequate fit not only reduces the device's effectiveness but may also cause discomfort or even exacerbate musculoskeletal issues (Garosi et al. 2022; Lowe, Billotte, and Peterson 2019; Sylla et al. 2014; Harrison et al. 2019). The device engineering must consider modular or personalized adjustment systems, based on precise anthropometric measurements, to ensure optimal contact and a balanced load distribution. This adaptation is essential to maximize ergonomic benefits and prevent adverse effects on user health. A significant challenge is to ensure that passive exoskeletons are comfortable enough for prolonged use. Discomfort may lead to rejection by the user, especially if the device restricts movement or causes chafing. Designing exoskeletons that naturally conform to the user's physiology while providing effective support without interfering with tasks is crucial to improve usability and acceptance.
- *Required training:* The proper use of a passive exoskeleton is not intuitive for all workers and requires specific training. It is necessary to establish training programs that include correct device donning, adjustments according to the task, and safety practices to avoid improper use (Schnieders et al. 2017; Poggensee and Collins 2021; Kuber and Rashedi 2024). Lack of training can lead to errors in device use, thereby reducing its effectiveness and even increasing the risk of injuries. Consequently, implementation should be accompanied by regular training sessions and supervised use protocols during the initial adoption phase.
- *Costs:* Although passive exoskeletons generally have a lower cost than active devices, the initial investment for acquisition, implementation, and maintenance can be significant, especially for small- and medium-sized enterprises (Manetti, Turchetti, and Fusco 2020). Moreover, the development of solutions that offer a high degree of personalization and adjustment can increase manufacturing costs. This economic barrier limits mass adoption and requires strategies such as financing, subsidies, or leasing models to facilitate access to these technologies for a wider range of organizations.

11.7 ADOPTION AND ORGANIZATIONAL CULTURE

The acceptance and integration of passive exoskeletons in the workplace largely depend on organizational culture. The following factors are critical in the process of integrating these devices:

- *Perception and resistance to change:* Workers may be skeptical about using exoskeletons, viewing them as an additional burden or a restriction on their natural movement. Furthermore, if they are not fully informed about the benefits of these devices, resistance to change is likely (Mahmud et al. 2022; Zhu, Dutta, and Dai 2021; Upasani et al. 2019; Kim et al. 2019). To counteract this, organizations must invest in awareness and training programs that explain how exoskeletons can improve ergonomics, reduce fatigue, and prevent injuries.
- *Training and instruction:* A lack of knowledge regarding the proper use of exoskeletons can lead to ineffective or even unsafe usage. Therefore, training is crucial to ensure that workers understand how to adjust the device and fully benefit from its support without risk (Crea et al. 2021; Karvouniari et al. 2018; Pacifico et al. 2022; Dufraisse et al. 2025). Organizations should implement specific training programs to teach users best practices and optimize exoskeleton usage.

11.8 FUTURE PERSPECTIVES

Passive exoskeletons will evolve alongside technological trends and the needs of the labor market. Some of the expected innovations and developments for the future include the following:

- *Integration of smart technologies:* Although passive exoskeletons do not rely on motorized components or external power sources, integrating smart technologies can open new possibilities. For instance, pressure sensors and accelerometers integrated into the device could enable real-time monitoring of the user's load and posture. These data could be transmitted to a digital platform that, through artificial intelligence, provides personalized recommendations for adjustment and safe use, thereby helping to reduce fatigue and the risk of injuries.
- *Advanced personalization:* Customized design is one of the trends in industrial ergonomics, and passive exoskeletons are no exception. In the future, devices could be manufactured using three-dimensional scans of the user to achieve a perfect fit or even incorporate interchangeable modules that adapt to different contexts and specific tasks. This advanced personalization would facilitate a better fit to the user's body structure and allow for configurations with different support levels.
- *High-tech and sustainable materials:* The search for lighter, stronger, and more sustainable materials will be a priority in the design of passive exoskeletons. Materials such as graphene or shape-memory alloys can provide strength without adding extra weight, while being more environmentally friendly. Additionally, the use of recyclable materials or those with a low carbon footprint would help reduce the environmental impact of these devices, aligning them with corporate sustainability policies.
- *Global standards and regulations:* The increasing adoption of exoskeletons in various industries will generate a need for global standards and regulations that govern their safe and effective use. These standards will establish minimum requirements for safety, ergonomics, and sustainability, allowing passive exoskeletons to be harmoniously integrated into diverse work environments and ensuring that companies can rely on their quality and functionality.

These trends and perspectives underscore a promising future for passive exoskeletons, with improvements in comfort, functionality, and adaptability. A collaboration among researchers, manufacturers, and companies will be key to overcoming current challenges and capitalizing on opportunities for the evolution of these devices.

11.9 STUDY LIMITATIONS

Despite the efforts to provide a comprehensive analysis, this study has certain limitations that should be considered:

- *Focus on passive exoskeletons:* The analysis focuses exclusively on passive exoskeletons. Therefore, the conclusions and the proposed taxonomy are not directly applicable to active or hybrid exoskeletons, which have distinct characteristics and challenges.
- *Taxonomy validation:* The presented taxonomy is a theoretical proposal based on the current literature. It has not yet undergone an empirical validation process by the scientific or industrial community, which will be a necessary future step to confirm its robustness and practical utility.

11.10 CONCLUSION

11.10.1 Summary of Contributions

This study has presented a detailed taxonomy of passive exoskeletons, classifying them according to criteria such as functionality, application sector, mechanism of action, level of support, and type of assisted movement. This taxonomy provides a structured foundation for understanding and categorizing passive exoskeletons across their various industrial applications, assisting designers, researchers, and companies in selecting and developing devices tailored to their specific needs. Additionally, the analysis of practical applications in sectors such as construction, manufacturing, healthcare, agriculture, and logistics highlights the potential of these devices to improve workplace ergonomics, reduce fatigue, and increase productivity. Conversely, challenges and opportunities for the continuous improvement of these devices have been identified, addressing issues such as comfort, adaptability, and production cost. This information offers both a practical and theoretical guide that supports research and the development of innovative solutions in the field of passive exoskeletons.

Despite significant advances in the design and application of passive exoskeletons, key areas still require further investigation. First, long-term effectiveness studies are recommended to evaluate the impact of these devices on users' physical health and their role in reducing musculoskeletal injuries. Likewise, it is essential to conduct adaptability tests in various industrial environments to optimize design and ensure that exoskeletons meet the specific requirements of each task and sector. Research into new lightweight and durable materials that not only improve user comfort but also reduce manufacturing costs is also fundamental to making these devices more accessible. Finally, exploring the integration of smart technologies for real-time monitoring and advanced personalization is suggested, which could enhance the functionality of passive exoskeletons and allow for optimal ergonomic adjustment for each user.

We invite researchers, designers, and companies to adopt this taxonomy as a robust foundation for identifying areas of improvement and opportunities for innovation in the development of passive exoskeletons. Interdisciplinary collaboration will be essential to overcome current challenges and capitalize on the opportunities for the evolution of these devices. Researchers can use this taxonomy to guide studies that substantiate the ergonomic and safety benefits of passive exoskeletons, while designers can rely on these criteria to create devices that are more adaptable, comfortable, and functional. Likewise, companies can play a crucial role by implementing and evaluating these devices in real-world settings, thereby providing valuable data for continuous improvement. Together, these efforts can drive the evolution of passive exoskeletons as essential tools in enhancing working conditions and promoting worker well-being.

REFERENCES

Afolabi, Afolabi., Yusuf, Anthony and Akanmu, Abiola. "Predicting Mental Workload of Using Exoskeletons for Construction Work: A Deep Learning Approach," *Journal of Information Technology in Construction*, vol. 30, pp. 1–21, 2025, doi: 10.36680/j.itcon.2025.001.

Andrade, Christina. and Nathan-Roberts, Dan. "Occupational Exoskeleton Adoption and Acceptance in Construction and Industrial Work: A Scoping Review," *Proceedings of the Human Factors and Ergonomics Society Annual Meeting*, vol. 66, no. 1, pp. 1325–1329, Sep. 2022, doi: 10.1177/1071181322661415.

Anitha, Dhakshina, Kolangiammal, Sankaradass, Lalitha, Kandasamy, Naga Valli, Vijaya, Ganesh, Eniyanallore Narasimhan, and Muthulekshmi, M."IoT-Enabled Exoskeletons for Firefighters Using Reinforcement Learning for Adaptive Support in Emergency Situations," in *2024 Second International Conference on Intelligent Cyber Physical Systems and Internet of Things (ICoICI)*, 2024, pp. 455–460. doi: 10.1109/ICoICI62503.2024.10696056.

Arachchige, Sanura Danu, Piyathilaka, Lasitha, Sul, Jung-Hoon. and Preethichandra, D. M. G. "A Review of Potential Exoskeletons for the Prevention of Work-Related Musculoskeletal Disorders in Agriculture," *Sensors*, vol. 24, no. 21, p. 7026, Oct. 2024, doi: 10.3390/s24217026.

Baltrusch, Saskia, Houdijk, Euphemia., van Dieën, Jaap H., and Johannes de Kruif, "Passive Trunk Exoskeleton Acceptability and Effects on Self-Efficacy in Employees with Low-Back Pain: A Mixed Method Approach," *Journal of Occupational Rehabilitation*, vol. 31, no. 1, pp. 129–141, Mar. 2021, doi: 10.1007/s10926-020-09891-1.

Baltrusch, Saskia, van Dieën, Jaap, van Bennekom, Coen A. M., and Houdijk, Johannes. "The Effect of a Passive Trunk Exoskeleton on Functional Performance in Healthy Individuals," *Applied Ergonomics*, vol. 72, pp. 94–106, Oct. 2018, doi: 10.1016/j.apergo.2018.04.007.

Baptista, Renato. et al., "Training-Induced Muscle Fatigue with a Powered Lower-Limb Exoskeleton: A Preliminary Study on Healthy Subjects," *Medical Sciences*, vol. 10, no. 4, p. 55, Sep. 2022, doi: 10.3390/medsci10040055.

Behjati Ashtiani, M., "Understanding the Drivers and Barriers to Adopting Exoskeletons in Construction: Results 2." n.d., [Online]. Available: https://orcid.org/0000-0002-1887-8431

Borisov, Alexander, Borisova, Viktoria. L., Konchina, Lyudmila. V., Kulikova, Maria. G. and. Maslova, Kristina. S. "Application of Active Exoskeletons with Touch Sensing, Electric Drives, Energy Recuperators and Gravity Compensators in Agricultural Technologies," in *IOP Conference Series: Earth and Environmental Science*, IOP Publishing Ltd, Mar. 2021. doi: 10.1088/1755-1315/699/1/012019.

Bosch, Tim, van Eck, Jennifer, Knitel, Knitel. and de Looze, Michiel. "The Effects of a Passive Exoskeleton on Muscle Activity, Discomfort and Endurance Time in Forward Bending Work," *Applied Ergonomics*, vol. 54, pp. 212–217, 2016, doi: 10.1016/j.apergo.2015.12.003.

Burks, Garret. et al., "A Framework for Virtual Reality-Based Motor Skills Training for the Use of Exoskeletons," in *Proceedings of the Human Factors and Ergonomics Society*, SAGE Publications Inc., 2021, pp. 277–278. doi: 10.1177/1071181321651170.

Carrasquillo, Carlos, Bajpai, Aakash, Iyengar, Divya, Collins, Killian, Mazumdar, Anirban, and Young, Aaron J. "Enhancing Human Navigation Ability Using Force-Feedback from a Lower-Limb Exoskeleton," *IEEE Transactions on Haptics*, pp. 1–13, 2025, doi: 10.1109/TOH.2025.3533974.

Clement Thevenot, X. P. and Mornieux, G. "Effects of an Occupational Soft-Back Exoskeleton during Order Picking: A Field Study in Logistics," *Ergonomics*, pp. 1–14, 2025, doi: 10.1080/00140139.2024.2447867.

Crea, Simona. et al., "Occupational Exoskeletons: A Roadmap toward Large-Scale Adoption. Methodology and Challenges of Bringing Exoskeletons to Workplaces," *Wearable Technologies*, 2021. doi: 10.1017/wtc.2021.11.

Daniel, Simon. and Olaoye, Godwin. "Analyzing IoT-Driven Solutions for Improving Compliance with Construction Safety Standards," Jan. 27, 2025. doi: 10.20944/preprints202501.2006.v1.

De Bock, Sander. et al., "Benchmarking Occupational Exoskeletons: An Evidence Mapping Systematic Review," Jan. 01, 2022. doi: 10.1016/j.apergo.2021.103582.

de Looze, M. P., F. Krause, and L. W. O'Sullivan, "The Potential and Acceptance of Exoskeletons in Industry," in *Biosystems and Biorobotics*, vol. 16, Springer International Publishing, 2017, pp. 195–199. doi: 10.1007/978-3-319-46532-6_32.

Del Ferraro, Simona, Falcone, Tiziana, Ranavolo, Alberto and Molinaro, Vincenzo. "The Effects of Upper-Body Exoskeletons on Human Metabolic Cost and Thermal Response during Work Tasks—A systematic Review," Oct. 02, 2020. doi: 10.3390/ijerph17207374.

del Rio Carral, Maria, Bourqui, Vanlisa, Vuilleumier, Noémi, Ortieb, Amalric, and Bouri, Mohamed. "Are Functional Measures Sufficient to Capture Acceptance? A Qualitative Study on Lower Limb Exoskeleton Use for Older People," *International Journal of Social Robotics*, vol. 14, no. 3, pp. 603–616, Apr. 2022, doi: 10.1007/s12369-021-00810-9.

Di Natali, Christian, Poliero, Tommaso, Sposito, Matteo, Fanti, Vasco, Leggieri, Sergio and Caldwell, Darwin G. "From the Idea to the User: a Pragmatic Multifaceted Approach to Testing Occupational Exoskeletons," *Wearable Technologies*, vol. 6, p. e5, Jan. 2025, doi: 10.1017/wtc.2024.28.

Duffus, LuAnn McClernan, "Exoskeleton Requirements for Firefighters," 2019. Ohio University, Athens, OH, United States.

Dufraisse, Marc, Cegarra, Julien, Atain Kouadio, Jean-Jacques, Clerc-Urmès, Isabelle. and Wioland, Lien. "From Unknown to Familiar: An Exploratory Longitudinal Field Study on Occupational Exoskeletons Adoption," *Applied Ergonomics*, vol. 122, p. 104393, 2025, doi: 10.1016/j.apergo.2024.104393.

Elprama, Shirley A. et al., "Design and Implementation Requirements for Increased Acceptance of Occupational Exoskeletons in an Industrial Context: A Qualitative Study," *International Journal of Human–Computer Interaction*, pp. 1–16, 2023, doi: 10.1080/10447318.2023.2247597.

Elprama, Shirley A., Vanderborght, Bram and Jacobs, An. "An Industrial Exoskeleton User Acceptance Framework Based on a Literature Review of Empirical Studies," *Applied Ergonomics*, vol. 100, p. 103615, 2022, doi: 10.1016/j.apergo.2021.103615.

Erden, Yesemin J. and Rainey, Stephen. "An Ethical Assessment of Powered Exoskeletons: Implications from Clinical Use to Industry and Military Contexts," Oct. 01, 2024, *John Wiley and Sons Inc.* doi: 10.1111/aor.14822.

ERGOIBV, "Exoesqueletos y ergonomía en el trabajo." Accessed: Nov. 19, 2024. [Online]. Available: https://www.ergoibv.com/es/posts/exoesqueletos-ergonomia-trabajo/

Fournier, Daniel. and Yung, Marcus, Somasundram, Kumara. "Quality, Productivity, and Economic Implications of Exoskeletons for Occupational Use: A Systematic Review," *PLoS One*, vol. 18, no. 6 June, Jun. 2023, doi: 10.1371/journal.pone.0287742.

Fox, Stephen, Aranko, Olli, Heilala, Juhaji, & Vahala, Päivi. (2020). Exoskeletons: Comprehensive, comparative and critical analyses of their potential to improve manufacturing performance. *Journal of Manufacturing Technology Management, 31*(6), 1261–1280. Emerald Publishing Limited. https://doi.org/10.1108/JMTM-01-2019-0023

Garosi, Ehsan. et al., "Design and Ergonomic Assessment of a Passive Head/Neck Supporting Exoskeleton for Overhead Work Use," *Applied Ergonomics*, vol. 101, p. 103699, 2022, doi: 10.1016/j.apergo.2022.103699.

Gilotta, Stefano, Spada, Stefania, Ghibaudo, Luca, Isoardi, Monica. and Mosso, Cristina Ottavia. "Acceptability beyond Usability: A Manufacturing Case Study," in *Advances in Intelligent Systems and Computing*, Springer Verlag, 2019, pp. 922–934. doi: 10.1007/978-3-319-96071-5_95.

Gonsalves, Nihar, Akanmu, Abiola, Shojaei, Alireza, and Agee, Philip. "Factors Influencing the Adoption of Passive Exoskeletons in the Construction Industry: Industry Perspectives," *International Journal of Industrial Ergonomics*, vol. 100, p. 103549, 2024, doi: 10.1016/j.ergon.2024.103549.

Govaerts, Renée. et al., "Work Performance in Industry: The Impact of Mental Fatigue and a Passive Back Exoskeleton on Work Efficiency," *Applied Ergonomics*, vol. 110, Jul. 2023, doi: 10.1016/j.apergo.2023.104026.

Gutierrez, Nancy. et al., "Perceived Benefits, Barriers, Perceptions, and Readiness to Use Exoskeletons in the Construction Industry: Differences by Demographic Characteristics," *Applied Ergonomics*, vol. 116, p. 104199, 2024, doi: 10.1016/j.apergo.2023.104199.

Halim, Isa, et al., "Critical Factors Influencing User Experience on Passive Exoskeleton Application: A Review," *International Journal of Integrated Engineering*, vol. 14, no. 4, pp. 89–115, 2022, doi: 10.30880/ijie.2022.14.04.009.

Hefferle, Michael, Lechner, Maria, Kluth, Karsten and Christian, Marc. "Development of a Standardized Ergonomic Assessment Methodology for Exoskeletons Using Both Subjective and Objective Measurement Techniques," in *Advances in Intelligent Systems and Computing*, Springer Verlag, 2020, pp. 49–59. doi: 10.1007/978-3-030-20467-9_5.

Howard, John, Murashov, Vladimir V. Lowe, Brian D. and Lu, Ming-Lun. "Industrial Exoskeletons: Need for Intervention Effectiveness Research," *American Journal of Industrial Medicine*, vol. 63, no. 3, pp. 201–208, Mar. 2020, doi: 10.1002/ajim.23080.

Iranzo, Sofía, Piedrabuena, Alicia, Iordanov, Daniel, Martinez-Iranzo, Ursula and Belda

Lois, Juan-Manuel. "Ergonomics Assessment of Passive Upper-Limb Exoskeletons in an Automotive Assembly Plant," *Applied Ergonomics*, vol. 87, p. 103120, 2020, doi: 10.1016/j.apergo.2020.103120.

Jakob, Martina, Balaguier, Romain, Park, Hyunjin and Trask, Catherine. "Addressing Exoskeleton Implementation Challenges: Case Studies of Non-Acceptance in Agriculture," *Journal of Agromedicine*, 2023, doi: 10.1080/1059924X.2023.2236605.

Karvouniari, Anna, Michalos, George, Dimitropoulos, Nikos and Makris, Sotirios. "An Approach for Exoskeleton Integration in Manufacturing Lines Using Virtual Reality Techniques," *Procedia CIRP*, vol. 78, pp. 103–108, 2018, doi: 10.1016/j.procir.2018.08.315.

Khamaisi, Riccardo Karim. et al., "A Multi-Facet Approach to Functional and Ergonomic Assessment of Passive Exoskeletons," *Procedia Computer Science*, vol. 232, pp. 584–594, 2024, doi: 10.1016/j.procs.2024.01.058.

Kim, Sunwook. et al., "Potential of Exoskeleton Technologies to Enhance Safety, Health, and Performance in Construction: Industry Perspectives and Future Research Directions," *IISE Transactions on Occupational Ergonomics and Human Factors*, vol. 7, no. 3–4, pp. 185–191, Oct. 2019, doi: 10.1080/24725838.2018.1561557.

Kim, Sunwook, Nussbaum, Maury A. Esfahani, Mohammad, Alemi, Mohammad, Alabdulkarim, Saad, and Rashedi, Ehsan. "Assessing the Influence of a Passive, Upper Extremity Exoskeletal Vest for Tasks Requiring Arm Elevation: Part I – 'Expected' Effects on Discomfort, Shoulder Muscle Activity, and Work Task Performance," *Applied Ergonomics*, vol. 70, pp. 315–322, Oct. 2018, doi: 10.1016/j.apergo.2018.02.025.

Kinge, Jonas Minet, Knudsen, Ann Kristin. Skirbekk, Vergad. and Vollset, Stein Emil. "Musculoskeletal Disorders in Norway: Prevalence of Chronicity and Use of Primary and Specialist Health Care Services," *BMC Musculoskeletal Disorders*, vol. 16, no. 1, 2015, doi: 10.1186/s12891-015-0536-z.

Kirkwood, Gavin, Wilkenfeld, J Nan. and Dunbar, Norah E. "Exoskeletons and the Future of Work: Envisioning Power and Control in a Workforce Without Limits," *Human-Machine Communication*, vol. 4, no. 1, pp. 187–206, 2022, doi: 10.30658/hmc.4.10.

Koopman, Axel S, Kingma, Idsart. Faber, Gert S. de Looze, Michiel P. and van Dieën, Jaap H. "Effects of a Passive Exoskeleton on the Mechanical Loading of the Low Back in Static Holding Tasks," *Journal of Biomechanics*, vol. 83, pp. 97–103, Oct. 2019, doi: 10.1016/j.jbiomech.2018.11.033.

Kuber, Pranav M. and Rashedi, Ehsan. "Product Ergonomics in Industrial Exoskeletons: Potential Enhancements for Workforce Efficiency and Safety," *TheoreTical Issues in Ergonomics Science*, vol. 22, no. 6, pp. 729–752, 2021, doi: 10.1080/1463922X.2020.1850905.

Kuber, Pranav Madhav and Rashedi, Ehsan. "Training and Familiarization with Industrial Exoskeletons: A Review of Considerations, Protocols, and Approaches for Effective Implementation," *Biomimetics*, vol. 9, no. 9, p. 520, Aug. 2024, doi: 10.3390/biomimetics9090520.

Lévy, Pierre-Denis, "Beyond Kansei Engineering : The Emancipation of Kansei Design Citation for Published Version (APA)," *International Journal of Design*, 2013. [Online]. Available: www.tue.nl/taverne

Lowe, Brian D., Billotte, William G. and Peterson, Donald R. "ASTM F48 Formation and Standards for Industrial Exoskeletons and Exosuits," *IISE Transactions on Occupational Ergonomics and Human Factors*, vol. 7, no. 3–4, pp. 230–236, 2019, doi: 10.1080/24725838.2019.1579769.

Luciani, Beatrice, Braghin, Francesco, Pedrocchi, Alessandra Laura and Gandolla, Marta. "Technology Acceptance Model for Exoskeletons for Rehabilitation of the Upper Limbs from Therapists' Perspectives," *Sensors*, vol. 23, no. 3, Feb. 2023, doi: 10.3390/s23031721.

Madinei, Saman, Alemi, Mohammad Mehdi, Kim, Sunwook Srinivasan, Divya, and Nussbaum, Maury A. "Biomechanical Evaluation of Passive Back-Support Exoskeletons in a Precision Manual Assembly Task: 'Expected' Effects on Trunk Muscle Activity, Perceived Exertion, and Task Performance," *Human Factors*, vol. 62, no. 3, pp. 441–457, Oct. 2020, doi: 10.1177/0018720819890966.

Mahmud, Dilruba. et al., "Identifying Facilitators, Barriers, and Potential Solutions of Adopting Exoskeletons and Exosuits in Construction Workplaces," *Sensors*, vol. 22, no. 24, Dec. 2022, doi: 10.3390/s22249987.

Manetti, Sara, G. Turchetti, Guiseppe, and Fusco, Francesco, "Determining the Cost-Effectiveness Requirements of an Exoskeleton Preventing Second Hip Fractures Using Value of Information," *BMC Health Services Research*, vol. 20, no. 1, Dec. 2020, doi: 10.1186/s12913-020-05768-4.

McFarland, Taylor. and Fischer, Steven. "Considerations for Industrial Use: A Systematic Review of the Impact of Active and Passive Upper Limb Exoskeletons on Physical Exposures," *IISE Transactions on Occupational Ergonomics and Human Factors*, vol. 7, no. 3–4, pp. 322–347, Oct. 2019, doi: 10.1080/24725838.2019.1684399.

Mendoza, Franklin, "Advances in Exoskeletons for Military Use," *Athenea Journal*, vol. 4, pp. 43–54, https://athenea.autanabooks.com/index.php/revista/article/view/57

Moyon, Adrien, Poirson, Emilie, and Petiot, Jean-Francois. "Development of an Acceptance Model for Occupational Exoskeletons and Application for a Passive Upper Limb Device," *IISE Transactions on Occupational Ergonomics and Human Factors*, vol. 7, no. 3–4, pp. 291–301, Oct. 2019, doi: 10.1080/24725838.2019.1662516.

Mudie, Kurt. L. et al., "Consensus Paper on Testing and Evaluation of Military Exoskeletons for the Dismounted Combatant," *Journal of Science and Medicine in Sport*, 2018. doi: 10.1016/j.jsams.2018.05.016.

Nussbaum, Maury, Lowe, Brian, de Looze, Michiel, Harris-Adamson, Carisa, and Smets, Michael. "An Introduction to the Special Issue on Occupational Exoskeletons," *IISE Transactions on Occupational Ergonomics and Human Factors*, vol. 7, no. 3–4, pp. 153–162, Oct. 2019, doi: 10.1080/24725838.2019.1709695.

Ojha, Abhishek, Gautam, Yash, Jebelli, Houtan, and Akanmu, Abiola. "Physiological Impact of Powered Back-Support Exoskeletons in Construction: Analyzing Muscle Fatigue, Metabolic Cost, Ergonomic Risks, and Stability," *Automation in Construction*, vol. 168, p. 105742, 2024, doi: 10.1016/j.autcon.2024.105742.

Okunola, Akinwale, Afolabi, Adedeji, Akanmu, Abiola, Jebelli, Houtan, and Simikins, Susan."Facilitators and Barriers to the Adoption of Active Back-Support Exoskeletons in the Construction Industry," *Journal of Safety Research*, vol. 90, pp. 402–415, 2024, doi: 10.1016/j.jsr.2024.05.010.

Omoniyi, Adebayo, Trask, Catherine, Milosavljevic, Stephan, and Thamsuwan, Ornwipa. "Farmers' Perceptions of Exoskeleton Use on Farms: Finding the Right Tool for the Work(er)," *International Journal of Industrial Ergonomics*, vol. 80, Nov. 2020, doi: 10.1016/j.ergon.2020.103036.

Pacifico, Isaac. et al., "Exoskeletons for Workers: A Case Series Study in an Enclosures Production Line," *Applied Ergonomics*, vol. 101, p. 103679, 2022, doi: 10.1016/j.apergo.2022.103679.

Pinelli, Michele, Manetti, Sara, and Lettieri, Emanuele. "Assessing the Social and Environmental Impact of Healthcare Technologies: Towards an Extended Social Return on Investment," *International Journal of Environmental Research and Public Health*, vol. 20, no. 6, Mar. 2023, doi: 10.3390/ijerph20065224.

Poggensee, Katherine. L. and Collins, Steven. H. "How Adaptation, Training, and Customization Contribute to Benefits from Exoskeleton Assistance," *Science Robotics*, vol. 6, no. 58, Sep. 2021, doi:10.1126/scirobotics.abf1078

Preethichandra, Dulitha et al., "Passive and Active Exoskeleton Solutions: Sensors, Actuators, Applications, and Recent Trends," Nov. 01, 2024, *Sensors*. doi: 10.3390/s24217095.

Proud, Joshua. K. et al., "Exoskeleton Application to Military Manual Handling Tasks," vol. 64, no. 3, pp. 527–554, 2022, doi: 10.1177/0018720820957467.

Reese, M., "Understanding and Incorporating Wearable Exoskeleton Design Understanding and Incorporating Wearable Exoskeleton Design Acceptance into the Construction and Manufacturing Industries by Acceptance into the Construction and Manufacturing Industries by Evaluating Key Stakeholders within an Enterprise Organization Evaluating Key Stakeholders within an Enterprise Organization," Thesis. Mississippi State University. 2024. [Online]. Available: https://scholarsjunction.msstate.edu/td/6165/

Renner, Thomas, "Exoskeleton with Lightweight Plastic Components Finds Applications in Medicine, Manufacturing, And Agriculture," *Plastics Engineering*, vol. 74, no. 3, pp. 30–35, Mar. 2018, doi: 10.1002/j.1941-9635.2018.tb01855.x

Riccò, Matteo, Ranzieri, Silvia, Vezzosi, Luigi, Balzarini, Federica, and Bragazzi, Nicola Luigi. "Wearable Exoskeletons on the Workplaces: Knowledge, Attitudes and Perspectives of Health and Safety Managers on the Implementation of Exoskeleton Technology in Northern Italy," *Acta Biomedica*, vol. 92, no. 6, 2021, doi: 10.23750/abm.v92i6.10437.

Scataglini, Sofia, Andreoni, Giuseppe, and Gallant, James. "A Review of Smart Clothing in Military," in *WearSys 2015- Proceedings of the 2015 Workshop on Wearable Systems and Applications*, Association for Computing Machinery, Inc, May 2015, pp. 53–54. doi: 10.1145/2753509.2753520.

Schnieders, Thomas, Stone, Richard, Oviatt, Timothy and Danford-Klein, Erika. "ARCTiC LawE: An Upper-Body Exoskeleton for Firearm Training," *Augmented Human Research*, vol. 2, no. 1, Dec. 2017, doi: 10.1007/s41133-017-0004-4.

Schwerha, Diana, McNamara, Nicjola. Nussbaum, Maury. A. and Kim, Sumwook. "Adoption Potential of Occupational Exoskeletons in Diverse Enterprises Engaged in Manufacturing Tasks," *International Journal of Industrial Ergonomics*, vol. 82, p. 103103, 2021, doi: 10.1016/j.ergon.2021.103103.

Schwerha, Diana, McNamara, Nicholas, Kim, Sumwook, and Nussbaum, Maury. A. "Exploratory Field Testing of Passive Exoskeletons in Several Manufacturing Environments: Perceived Usability and User Acceptance," *IISE Transactions on Occupational Ergonomics and Human Factors*, vol. 10, no. 2, pp. 71–82, Aug. 2022, doi: 10.1080/24725838.2022.2059594.

Sefton, JoEllen et al., "The Impact of a Commercial Lower Extremity Exoskeleton on Metabolic Load, Perceived Exertion, and Physiological Response to a Challenging Military Relevant Task: A Randomized Cross-Over Design Pilot Study," *PLoS One*, vol. 20, no. 1 January, Jan. 2025, doi: 10.1371/journal.pone.0314613.

Siedl, Sarah Maria. and Mara, Martina."Exoskeleton Acceptance and Its Relationship to Self-Efficacy Enhancement, Perceived Usefulness, and Physical Relief: A Field Study among Logistics Workers," *Wearable Technologies*, vol. 2, 2021, doi: 10.1017/wtc.2021.10.

Spada, Stefahia, Ghibaudo, Luca, Gilotta, Stefano, Gastaldi, Laura, and Cavatorta, Maria Pia. "Investigation into the Applicability of a Passive Upper-limb Exoskeleton in Automotive Industry," *Procedia Manuf*, vol. 11, pp. 1255–1262, 2017, doi: 10.1016/j.promfg.2017.07.252.

Steinhilber, Benjamin. et al., "The Use of Exoskeletons in the Occupational Context for Primary, Secondary, and Tertiary Prevention of Work-Related Musculoskeletal Complaints," *IISE Transactions on Occupational Ergonomics and Human Factors*, vol. 8, no. 3, pp. 132–144, 2020, doi: 10.1080/24725838.2020.1844344.

Sylla, Ninon, Bonnet, Vincent, Colledani, Frederic, and Fraisse, Philippe. "Ergonomic Contribution of ABLE Exoskeleton in Automotive Industry," *International Journal of Industrial Ergonomics*, vol. 44, no. 4, pp. 475–481, 2014, doi: 10.1016/j.ergon.2014.03.008.

Thamsuwan, Ornwipa, Srinivasan, Divya, Milosavljevic, Stephan, Bath, Brenna, Crizzle, Alexnader, and Bashiri, Banafsheh. "Ergonomics Evaluation of Exoskeletons in Agriculture: Report to Alberta OHS Futures." n.d., [Online]. Available: www.alberta.ca/ohs-futures-research-grants

Thamsuwan, Ornwipa, Milosavljevic, Stephan, Srinivasan, Divya, and Trask, Catherine, "Potential Exoskeleton Uses for Reducing Low Back Muscular Activity During Farm Tasks," *American Journal of Industrial Medicine*, vol. 63, no. 11, pp. 1017–1028, Nov. 2020, doi: 10.1002/ajim.23180.

Theurel, Jean and Desbrosses, Kevin. "Occupational Exoskeletons: Overview of Their Benefits and Limitations in Preventing Work-Related Musculoskeletal Disorders," *IISE Transactions on Occupational Ergonomics and Human Factors*, vol. 7, no. 3–4, pp. 264–280, 2019, doi: 10.1080/24725838.2019.1638331.

Todorovic, Olija, Constantinescu, Carmelo, and Popescu, Daniela. "*Foundations for Economic Evaluation of Exoskeletons in Manufacturing", Acta Technica Napocensis*. vol. 61. 2018, pp. 221–230.

Upasani, Shaswata, Franco, Roberto, Niewolny, Kim. and Srinivasan, Divya. "The Potential for Exoskeletons to Improve Health and Safety in Agriculture—Perspectives from Service Providers," *IISE Transactions on Occupational Ergonomics and Human Factors*, vol. 7, no. 3–4, pp. 222–229, Oct. 2019, doi: 10.1080/24725838.2019.1575930.

Wioland, Lucas, Atain Kouadio, Jean-Jaques, Bréard, Hugo, Clerc-Urmès, Isabel and Paty, Benjamin. "The Adoption of Occupational Exoskeletons: From Acceptability to Situated Acceptance, Questionnaire Surveys," *International Journal of Human–Computer Interaction*, vol. 41, no. 2, pp. 1446–1458, 2025, doi: 10.1080/10447318.2024.2314359.

Yang, Can Jun, Zhang, Jian Feng, Chen, Ying, Dong, Yan Min. and Zhang, Yu. "A Review of Exoskeleton-Type Systems and Their Key Technologies," Aug. 2008. doi: 10.1243/09544062JMES936.

Zelik, Karl. E., Nurse, Christopher. A., Schall, Mark. C., Sesek, Richard. F., Marino, Matthew. C. and Gallagher, Sean. "An Ergonomic Assessment Tool for Evaluating the Effect of Back Exoskeletons on Injury Risk," *Applied Ergonomics*, vol. 99, p. 103619, 2022, doi: 10.1016/j.apergo.2021.103619.

Zhu, Zhaojun, Dutta, Anirban, and Dai, Fei. "Exoskeletons for Manual Material Handling – A Review and Implication For Construction Applications," *Automation in Construction*, vol. 122, p. 103493, 2021, doi: 10.1016/j.autcon.2020.103493.

Section 4

Psychosocial and Organizational Ergonomics

12 Obsessive Passion at Work and Its Effect on Stress and Burnout

María Marisela Vargas-Salgado, Aurora Irma Máynez-Guaderrama, Erwin Adán Martínez-Gómez, and Virginia Guadalupe López-Torres

12.1 INTRODUCTION

Passion is a valuable asset to explore in the workplace (Bester et al. 2020). It has the potential to enhance motivation; improve commitment, personal well-being, and work performance; provide a sense of meaning to daily activity; and trigger energy at work (Vallerand et al. 2010; Yukhymenko-Lescroart and Sharma 2022). Although passion for work is beneficial, it can also have adverse effects. According to the dualistic model of passion, individuals can be committed to a common goal but may pursue it differently: in harmonious passion, individuals maintain self-control and are able to choose their level of commitment and time spent pursuing the goal; in obsessive passion, control goes beyond individuals, so they feel compelled and pressured to continue pursuing the goal (Lobato et al. 2023). This study investigates obsessive passion based on its potential harm.

Passion can lead to work that drains the person, triggering conflict and burnout (Vallerand et al. 2010). Burnout syndrome is an ongoing and latent problem in organizations, is a "job-related problem", and the number of people suffering from this condition is increasing (Castillo del Toro and Hernández-López 2025; Alshurtan et al. 2024; West et al. 2018). Burnout involves internal or external factors that affect physical and emotional well-being (Choudhary et al. 2022). This syndrome is an occupational disease resulting from chronic stress caused by intense work demands, rotating shifts, role conflict, job instability, lack of social support at work, and conflictual relationships (Babiker et al. 2025; Galanis et al. 2024; Hsieh et al. 2011; Maslach et al. 2001; Moreno-Jiménez et al. 2022). The people most affected by burnout are those who are exposed to higher levels of stress (Thanh et al. 2024).

Stress is a feeling of tension, uneasiness, or worry (Muis et al. 2021). Stress leads employees to perceive their work experience negatively (Saleem et al. 2021). Occupational stress triggers physical and psychological manifestations (Choudhary et al. 2022). Furthermore, stress affects employees' health and organizational performance (Lukenda et al. 2024; Saleem et al. 2021).

The results of the literature review revealed that although previous studies (Colín Flores 2020; Armenta et al. 2018) have studied burnout in Mexican maquila workers, none have explored its association with obsessive passion and stress simultaneously. There is a lack of research on the effect of work passion on stress and burnout syndrome in the export manufacturing industry. This study addresses this research gap and, in that context, its objective was to explore whether obsessive passion directly and indirectly affects burnout through stress mediation.

DOI: 10.1201/9781003582014-16

12.2 THEORETICAL BACKGROUND

12.2.1 Obsessive Passion

Passion is defined as a strong inclination to engage in an activity that the individual enjoys (or even loves), appreciates (and therefore considers important), and thus invests time and energy in it (Vallerand et al. 2003). Unlike other motivational constructs, passion is more stable and less transitory because the activities involved become an integral element of the subject's identity (Pollack et al. 2020). Harmonious and obsessive passions are two types of working passions (Yukhymenko-Lescroart and Sharma 2022); the former has a positive relationship to favorable emotions, while the latter has mixed effects, such as reduced performance and stress levels, and produces both positive and negative emotions (Paquette et al. 2022). These types of passion have different effects on how work activities are balanced with other areas of life (Lobato et al. 2023).

As previously stated, this research delves into obsessive passion. Passion is a strong and positive attitude toward a task that can lead to favorable or unfavorable organizational consequences (Gülbahar and Özkan 2023). Obsessive passion is defined as a strong desire or internal pressure that compels the individual to perform an activity; it is an uncontrollable force that leads the person to engage in the activity in a persistent, inflexible, and rigid manner (Lucidi et al. 2016; Ratelle et al. 2004). In the workplace, people with obsessive passion are driven by an uncontrollable impulse and determination to keep at their work regardless of whether this leads to conflicts in other life activities (Vergauwe et al. 2022; Yukhymenko-Lescroart and Sharma 2022). Moreover, in this kind of passion, sustained and long-term commitment is driven by external pressures and motivations, such as reputation, social acceptance, and monetary rewards, which trigger intense emotions in people (Lee et al. 2024).

12.2.2 Obsessive Passion and Stress

Work-related stress is one of the most common illnesses among the working population (Seijas-Solano 2019). Stress is the body's response to various harmful agents, a psychological phenomenon and physical problem that is affected by stressors that can lead to negative effects (Wu et al. 2020). Stress can be caused by fatigue, mental tiredness, poor self-esteem, excessive working hours, inefficient work processes, and poor interpersonal relationships (West et al. 2018). Moreover, work-related stress can be the result of requirements that exceed the capabilities and resources of the worker, pressures caused by the need to increase productivity, a lack of training, or poorly resolved conflicts (West et al. 2018).

Studies have examined the relationship between obsessive passion and stress, most of which target educational and medical personnel; however, research on industrial workers is limited. Manufacturing employees often face stress because of several challenges in their work environment (Morales and Ronquillo 2021). Production processes and accountability for achieving high production quotas and quality levels are trigger factors, while unrealistic work problems and expectations contribute significantly to their stress (Liao et al. 2024).

Studies have reported that obsessive passion can lead to stress. Obsessive passion hinders a person from completely detaching from their work, and this condition becomes all-consuming and therefore usually has negative outcomes, including stress (Di Santo et al. 2024). In parallel, the feeling of pressure can create personal conflicts and even affect people's health (Pollack et al. 2020), situations in which stress is also likely to occur. Studies conducted in the health (Landay et al. 2022), sports (Lucidi et al. 2016), and commercial (Bélanger et al. 2015) sectors found positive and significant relationships between obsessive passion and work-related stress in nurses, marathon runners, and employees of a car dealership and a hardware company. Therefore, it is postulated that:

H1. Obsessive passion has a positive and significant effect on stress.

12.2.3 Stress and Burnout

Continuous exposure to work-related stress causes exhaustion and ultimately leads to burnout syndrome (Seijas-Solano 2019; Wang et al. 2015). Studies conducted in the education sector in Australia (Carroll et al. 2022) and Pakistan (Ul Ain 2024) have revealed that stress significantly contributes to teacher burnout. Another study with Vietnamese citizens (Thanh et al. 2024) found that stress during the COVID-19 pandemic resulted in burnout. Furthermore, the work by Spányik et al. (2022) with healthcare workers highlighted that both general and work-related stress increase burnout. Therefore, it is proposed that:

H2. Stress has a significant impact on burnout

12.2.4 Obsessive Passion and Burnout

Burnout is a multidimensional variable comprising emotional exhaustion (a feeling of being overloaded and of having no emotional or physical resources left), depersonalization or cynicism (a negative, insensitive, or distant response to different work-related problems), and a feeling of diminished personal accomplishment (Chirico et al. 2021; Maslach et al. 2001). People with burnout are more likely to suffer accidents at work (Hodkinson et al. 2022). Burnout may lead to absenteeism, turnover, and intention to quit (Choudhary et al. 2022; Maslach and Leiter 2016).

Empirical evidence indicates that obsessive passion is associated with the onset of burnout syndrome in various ways (Curran et al. 2013). González-Brignardello and Sánchez-Elvira (2023) found that obsessive passion has a slightly detrimental effect on all burnout dimensions. In their study, they concluded that obsessive passion is less strongly predictive of burnout but is directly related to it, exerting an obstructive role as a factor that hinders well-being. Another study in Italy (Bélanger et al. 2015) reported that obsessive passion relates positively to nurses' burnout experience. Likewise, a study conducted with athletes in the province of Quebec (Lopes and Vallerand 2020) found that obsessive passion was directly and positively associated with burnout. Other studies in the healthcare sector (Zito et al. 2022) and in the education sector (Jiatong et al. 2024) reported that obsessive passion is positively related to burnout in nurses and teachers. Furthermore, a study of sports coaches (Birkeland et al. 2018) found that obsessive passion is positively and significantly related to emotional exhaustion and cynicism. Therefore, it is possible to consider that

H3. Obsessive passion positively and significantly impacts burnout.

12.3 METHODOLOGY

This study employed a quantitative, empirical, cross-sectional, and explanatory research design, employing a nonprobabilistic sample from the maquiladora export industry in Ciudad Juárez, Chihuahua. The inclusion criteria were as follows: being over 18 years of age and working in the sector under study; gender, educational level, and tenure are irrelevant. An electronic questionnaire was sent to the participants in October and November 2024. Participation was voluntary, anonymous, and confidential, and data privacy was ensured.

A total of 390 responses were received, of which only 386 were useful. This sample is adequate because models including endogenous variables receiving two structural paths require at least 130 observations to detect an R2 of 0.10, with a statistical power of 80% and a significance level of 1% (Hair Jr. et al. 2019). The sample consisted of 56.5% females, 42.2% males, and 1.3% other gender participants. Regarding marital status, 71.5% were single, and the majority were between 18 and

29 years old (75.4%). Regarding educational level, 37.6% had a bachelor's degree, and 30.1% had a high school education. Regarding employment, 31.6% worked in companies in the automotive sector, 45.1% worked as assistants, 32.4% worked in the production department, and 70.2% had worked for the company for less than 2 years.

Scales available in the literature were used to measure the variables under study. The scales were adapted because they were not originally designed for the study population (Hati et al. 2021). Obsessive passion was assessed with seven items from the Vallerand et al. (2003) scale; examples of these items are as follows: "I feel that I cannot live without this job" and "I feel a drive for my job so strong that I cannot avoid working". The questions were answered on a five-point Likert scale (from "never" to "always"). The 11-item scale of Cohen et al. (1983) was used to assess perceived stress. Some examples of these items are as follows: "In the last month, I have been bothered by unexpected changes at work (e.g., shift changes, staff cuts, excessive workload, reduction in working hours, more overtime)" and "I have felt that the most important things in my life are getting out of control". This scale assesses stress in general, not just work-related stress. Recent studies indicate that stress is a problem that affects people's lives, activities, and productivity (Garcés-Delgado et al. 2023), and obsessive passion not only triggers work stress but also stress in many other areas of life. Burnout was approached as a second-order construct comprising three dimensions. It was measured with 18 items using the Maslach and Jackson (1981) scale. Examples of these items are as follows: "I feel emotionally fatigued" and "I feel frustrated". Participants rated the items of perceived stress and burnout on a five-point Likert response scale ("strongly disagree" to "strongly agree").

The proposed relationships were assessed using a partial least squares structural equation model (PLS-SEM), due to its usefulness for prediction and theoretical development, as well as its causal approach aimed at maximizing the explained variance of the dependent variables (Hair Jr et al. 2020). SmartPLS version 4.0 software was used.

12.4 RESULTS

Structural modeling involves two steps: the measurement model review and structural model evaluation. The first step in the measurement model revision was an examination of factor loads, reliability, convergent validity, and discriminant validity. Item loadings should be equal to or greater than 0.708, and *t*-values should be statistically significant (Hair Jr. et al. 2020). Of the original 36-item model, 10 were eliminated because they did not meet the recommended cut-off points: 1 for obsessive passion [OP5], 3 for stress [S1, S4, S8], and 6 for burnout [I3, I4, I6, C6, C7, and E3]. Considering that this is a reflective model, in the evaluation of the measurement model, indicators with loadings higher than 0.6 can be maintained since these can provide construct validity (Hair et al., 2012; Sarstedt et al., 2014), but in this case, the deleted items did not affect such validity. According to Hair et al. (2011), only indicators with standardized loadings lower than 0.70 should be excluded if their elimination raises the composite reliability to the critical threshold, which occurred in the case of our study.

Cronbach's alpha, ρ_a, and ρ_c coefficients were reviewed to assess convergent validity. The coefficient values of the model used fall in the recommended ranges (between 0.700 and 0.950), and this is also the case for the average variance extracted (AVE) scores (equal to or greater than 0.500) as they fall between 0.591 and 0.865 (Hair Jr. et al. 2020). The indicators are listed in Table 12.1. Discriminant validity was assessed using the heterotrait–monotrait ratio. This ratio should be lower than 0.850 because the model includes conceptually different constructs (Ringle et al. 2020). The ratios are as follows: Emotional exhaustion and Cynicism (0.727), Inefficacy and Cynicism (0.539), Inefficacy and Emotional exhaustion (0.555), Obsessive passion and Cynicism (0.345), Obsessive passion and Emotional exhaustion (0.389), Obsessive passion and Inefficacy (0.367), Stress and Cynicism (0.643), Stress and Emotional exhaustion (0.669), Stress and Inefficacy (0.545), and Stress and Obsessive passion (0.485). Stress and Emotional exhaustion (0.669), Stress and Inefficacy (0.545), and Stress and Obsessive passion (0.485).

TABLE 12.1
Reliability and Convergent Validity of the Measurement Model

Constructs	Cronbach Alpha	Composite Reliability (ρ_a)	Composite Reliability (ρ_c)	Average Variance Extracted (AVE)
Burnout	0.941	0.943	0.949	0.591
Cynicism	0.930	0.931	0.947	0.783
Emotional exhaustion	0.918	0.920	0.942	0.803
Inefficacy	0.922	0.922	0.951	0.865
Obsessive passion	0.906	0.921	0.927	0.679
Stress	0.935	0.937	0.947	0.690

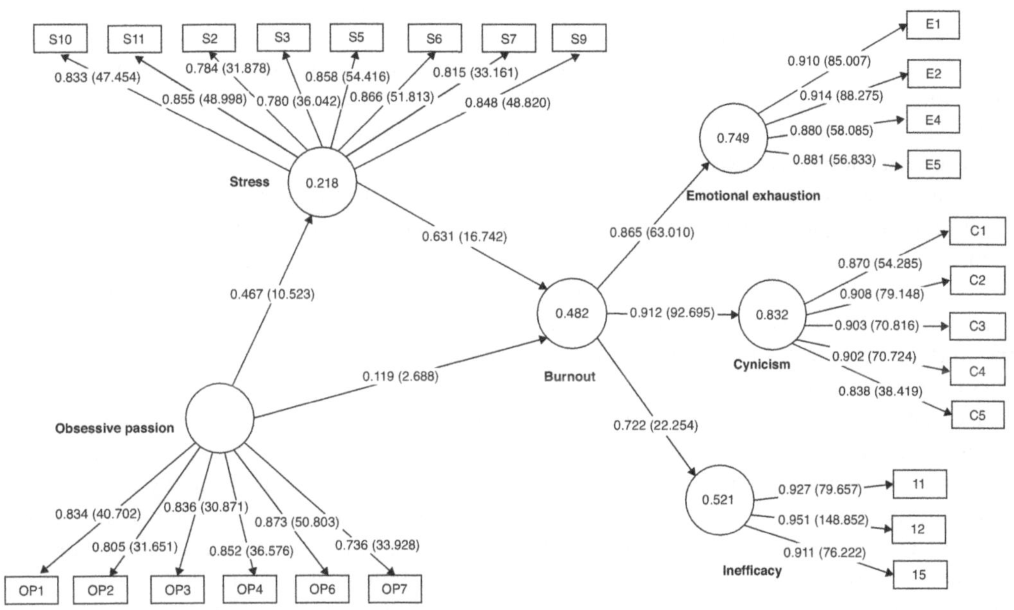

FIGURE 12.1 Final model.

After ensuring the reliability, convergent, and discriminant validity of the measurement model, the structural model was revised. It has been found that obsessive passion generates stress (H1: β=0.467, t=10.523) and that stress triggers exhaustion (H2: β=0.631, t=16.742). Furthermore, obsessive passion leads to burnout (H3: β=0.119, t=2.688).

R^2, f^2, and Q^2 were also evaluated. The model weakly explains stress (R^2=0.218) and moderately explains burnout (R^2=0.482). Moreover, three significant total effects (f^2) were found: Obsessive passion -> Stress (0.467), Obsessive passion -> Burnout (0.413), and Stress -> Burnout (0.631). The Q^2 indicator shows a medium relevance for stress (0.212) and burnout (0.163).

Lastly, the mediating role of stress in the relationship between obsessive passion and burnout was evaluated. It was identified that this construct partially and complementarily mediates this relationship (direct β 0.119, t=2.688; indirect β=0.294, t=8.816; total β=0413, t=8.798) (Cepeda-Carrión et al. 2017). Figure 12.1 shows the final model.

12.5 DISCUSSION

Passion for work implies benefits; however, it can also have adverse effects. This study aimed to explore whether obsessive passion directly and indirectly affects burnout through the mediation of stress in the maquiladora export industry of Ciudad Juárez, Chihuahua.

In the context of the maquiladora export industry, obsessive passion was found to directly and indirectly influence burnout through stress. Additionally, in the study context, obsessive passion is highlighted when employees show obsessive feelings, feel emotional dependence, and cannot imagine living without the work they do. These results coincide with those of previous studies, which highlight that people with obsessive passion have an uncontrollable determination to continue in their work, triggering intense emotions in people (Lee et al. 2024; Vergauwe et al. 2022; Yukhymenko-Lescroart and Sharma 2022), which can generate negative consequences that increase stress (Di Santo et al. 2024). In this study, it is important to highlight that burnout is triggered because people feel unable to contribute effectively in the activities they perform in the company and in solving problems that arise and feel frustrated and emotionally fatigued due to work and also doubt the importance and value of work. This result is different from the findings of Jafari (2019) and Moreno Pellicer (2017), which showed no significant effect of obsessive passion on athletes' exhaustion. This disparity might be attributed to differences between the occupations of the subjects analyzed, considering that those studies examined athletes, whereas the participants in this study were manufacturing workers.

We also confirmed that stress has a positive and significant influence on burnout. These results confirm that people who are exposed to job stress manifest exhaustion and develop burnout (Seijas-Solano 2019; Spányik et al. 2022; Wang et al. 2015). In the manufacturing context studied, stress manifests when people feel that things get out of control; they feel insecure about work problems and have difficulty solving them; on the other hand, they are not aware of things happening in the work environment. In addition, stress was found to mediate the relationship between obsessive passion and burnout.

Passion for work has benefits in the work environment and can increase well-being. However, obsessive passion at work has a detrimental effect on people because such obsessive enthusiasm triggers stress and burnout, and even personal activities can contribute to it. Human Resource managers should develop interventions to reduce stress and burnout by encouraging a more balanced passion at work. This will contribute to employee well-being and a healthier, more committed, and more sustainable work environment. For example, managers can implement policies that promote work–life balance, flexible schedules, and encourage autonomy in decision-making; design comprehensive wellness programs that include physical and emotional health; motivate employees to manage their time effectively, ask for support when needed, and maintain self-care practices. Implementing these strategies allows organizations to foster a harmonious passion and minimize the risks associated with obsessive passion, leading to healthier, more committed, and more productive employees.

Similar to other studies, this research has some limitations. First, it was conducted on a nonprobability sample taken from a specific context, so its results cannot be generalized. It is worth doing more research in other settings and conducting longitudinal studies to advance the knowledge. Second, it is important to consider that the content validity of the constructs may be reduced due to the elimination of some items from the measurement model. Ineffectiveness was the variable most affected in the model. Third, burnout was approached as a general construct to identify the level of risk rather than establishing a categorical diagnosis. This approach represents a crucial first step in identifying employees at risk, who can then be referred for further assessment and support (De Beer et al. 2024). In any case, this limitation highlights the importance of systematically evaluating whether the conclusions presented here can be generalized to other burnout measures and in what way. In further studies, it is worth looking at burnout in a more comprehensive way (including cognitive and physical burnout, as well as disengagement). Contextual variables such as workload, contract type, or organizational culture were not considered in this research. Finally, our model provides a limited explanation of stress (only obsessive passion was used as a predictive variable) and a moderated explanation of burnout (stress and obsessive passion as predictors). Therefore, future studies should consider including additional variables. It would also be interesting to incorporate some predictors of obsessive passion at work, such as supervisor support, organizational justice, psychological capital, leadership style, work climate, and trust.

REFERENCES

Alshurtan, Kareemah Salem, Saad, Yousef Aldhaifi, Khalid, Abdulkarim Alshammari, Othman, Mohammed Alodayli, Khalid, Falah Alqahtani, and Sarah, Yousef Aldhaifi. 2024. "Burnout syndrome among critical care health providers in Saudi Arabia." *Journal of Multidisciplinary Healthcare* 17: 843–854. https://doi.org/10.2147/JMDH.S452294

Armenta, Oziely Daniela, Maldonado Macias, Aidé Aracely, and García Alcaraz, Jorge Luis. 2018. "Determinación del síndrome de burnout en mandos medios y superiores de la industria maquiladora de Ciudad Juárez Chihuahua." Cultura Científica y Tecnológica 63: 330–338. https://erevistas.uacj.mx/ojs/index.php/culcyt/article/view/2213

Babiker, Marwan, Merisalu, Eda, Roja, Ženija, and Kalkis, Henrijs. 2025. "Prospective effects of artificial intelligence on burnout syndrome: Reducing risks and enhancing psychological well-being." *Sigurnost* 67(2): 135–141. https://doi.org/10.31306/s.67.2.4

Bélanger, Jocelyn J, Pierro, Antonio, Kruglanski, Arie W, Vallerand, Robert J, De Carlo, Nicola, and Falco, Alessandra. 2015. "On feeling good at work: The role of regulatory mode and passion in psychological adjustment." *Journal of Applied Social Psychology* 45(6): 319–329. https://doi.org/10.1111/jasp.12298

Bester, Marais S, Coetzee, Melinde, and van Lill, Xander. 2020. "Exploring the factor structure of the Passion Scale: Are the dualistic types of passion relevant for workers in the South African context?." *SA Journal of Industrial Psychology* 46(1): 1–10. https://doi.org/10.4102/sajip.v46i0.1788

Birkeland, Ide Katrine, Richardsen, Astrid M, and Dysvik, Anders. 2018. "The role of passion and support perceptions in changing burnout: A Johnson-Neyman approach." *International Journal of Stress Management* 25(2): 163–180. https://doi.org/10.1037/str0000057

Carroll, Annemaree, Forrest, Kylee, Sanders-O'Connor, Ema, Flynn, Libby, Bower, Julie M, Fynes-Clinton, Samuel, York, Ashley, and Ziaei, Maryam. 2022. "Teacher stress and burnout in Australia: Examining the role of intrapersonal and environmental factors." *Social Psychology of Education* 25: 441–469. https://doi.org/10.1007/s11218-022-09686-7

Castillo del Toro, Juan C, and Hernández-López, Edgar A. 2025. "Prevalence of operational fatigue, a form of occupational burnout syndrome, in flight instructors at the Escuela Militar de Aviación y del Centro de Adiestramiento de Helicópteros de la Fuerza Aérea Mexicana." *Revista de Sanidad Militar* 79(2): 1–29. https://doi.org/10.56443/1ss39s64

Cepeda-Carrión, Gabriel, Nitzl, Christian, and Roldán, José L. (2017). Mediation analyses in partial least squares structural equation modeling: Guidelines and empirical examples. In: Latan, H., Noonan, R. (eds) *Partial Least Squares Path Modeling*. Springer, Cham, 173–195. https://doi.org/10.1007/978-3-319-64069-3_8

Chirico, Franceso, Crescenzo, Pietro, Sacco, Angelo, Ricco, Matteo, Ripa, Serena, Nucera, Gabriella, and Magnavita, Nicola. 2021. "Prevalence of burnout syndrome among Italian volunteers of the Red Cross: A cross-sectional study." *Industrial Health* 59(2): 117–127. https://doi.org/10.2486/indhealth.2020-0246

Choudhary, Mahima, Kumar, Ashok, Pandey, Vandana, Choudhary, Vikas, and Patidar Natwar. 2022. "Burnout syndrome assessment scale for nurses working in intensive care units: Development and validation." *International Journal of Applied and Basic Medical Research* 12(2): 82–86. https://doi.org/10.4103/ijabmr.ijabmr_547_21

Cohen, Sheldon, Kamarck Tom, and Mermelstein Robin. 1983. "A global measure of perceived stress." *Journal of Health and Social Behavior* 24(4): 385–396. https://doi.org/10.2307/2136404.

Colín Flores, Carlos G. 2020. "Capital psicológico como mediador entre adicción al trabajo y burnout en trabajadoras en la industria de maquila en Guanajuato." *Ciencias Administrativas Teoría y Praxis* 16(1):11–29. https://hdl.handle.net/11285/637241

Curran, Thomas, Appleton, Paul R, Hill, Andrew P, and Hall, Howard K. 2013. "The mediating role of psychological need satisfaction in relationships between types of passion for sport and athlete burnout." *Journal of Sports Sciences* 31(6): 597–606. https://dx. doi.org/10.1080/02640414.2012.742956

De Beer, Leon T, Christensen, Marit, Sørengaard, Torhild A, Innstrand, Siw T, and Schaufeli, Wilmar B. 2024. "The psychometric properties of the Burnout Assessment Tool in Norway: A thorough investigation into construct-relevant multidimensionality." *Scandinavian Journal of Psychology* 65(3): 479–489. https://doi.org/10.1111/sjop.12996

Di Santo, Daniela, Tesi, Alessio, Aiello, Antonio, and Pierro, Antonio. 2024. "The job can kill you or make you stronger: The role of passion in job demands, work motivational imbalance, and work-life conflict." *TPM: Testing, Psychometrics, Methodology in Applied Psychology* 31(4): 487–500. https://doi.org/10.4473/TPM31.4.4

Galanis, Petros, Moisoglou, Ioannis, Katsiroumpa, Aglaia, and Sourtzi, Panayota. 2024. "Impact of workplace bullying on job burnout and turnover intention among nursing staff in Greece: Evidence after the COVID-19 pandemic." *AIMS Public Health* 11(2): 614–627. https://doi.org/10.3934/publichealth.2024031

Garcés-Delgado, Yaritza, García-Álvarez, Edgar, López-Aguilar, David. and Álvarez-Pérez, Pedro R. 2023. "Incidencia del género en el estrés laboral y burnout del profesorado universitario." *Revista Iberoamericana sobre Calidad, Eficacia y Cambio en Educación* 21(3): 41–60. https://doi.org/10.15366/reice2023.21.3.003

González-Brignardello, Marcela P, and Sánchez-Elvira Paniagua, Ángeles. 2023. "Contribución De La Pasión Y La Procrastinación Académica Multidimensional Sobre El Burnout en El Ámbito Universitario a Distancia." *Acción Psicológica* 20(1): 131–142. https://doi.org/10.5944/ap.20.1.39246

Gülbahar, Yasemin, and Özkan, Osman S. 2023. "The antecedents and outcomes of obsessive passion in the workplace." *Current Psychology* 42: 21263–21277. https://doi.org/10.1007/s12144-023-04877-7

Hair Jr, Joseph F, Hult, G Tomas M, Ringle, Christian M, Sarstedt, Marko, Castillo Apraiz, Julen, Cepeda Carrión, Gabriel A. and Roldán, José L. (2019). *Manual de Partial Least Squares Structural Equation Modeling (PLS-SEM)*. 2da. ed. OmniaScience.

Hair Jr., Joseph F, Howard, Matt C, and Nitzl, Christian. 2020. "Assessing measurement model quality in PLS-SEM using confirmatory composite analysis." *Journal of Business Research* 109: 101–110. https://doi.org/10.1016/j.jbusres.2019.11.069

Hair, Joe F, Ringle, Christian M, and Sarstedt, Marko. 2011. "PLS-SEM: Indeed a silver bullet." *Journal of Marketing Theory and Practice* 19(2): 139–152. https://doi.org/10.2753/MTP1069-6679190202

Hair, Joe F, Sarstedt, Marko, Ringle, Christian M, and Mena, Jeannette A. 2012. "An assessment of the use of partial least squares structural equation modeling in marketing research." *Journal of the Academy of Marketing Science* 40(3): 414–433. https://doi.org/10.1007/s11747-011-0261-6

Hati, Sri Rahayu H, Zulianti, Ina, Achyar, Adrian, and Safira, Anya. 2021. "Perceptions of nutritional value, sensory appeal, and price influencing customer intention to purchase frozen beef: Evidence from Indonesia." *Meat Science*, 172: 108306. https://doi.org/10.1016/j.meatsci.2020.108306

Hodkinson, Alexander, Zhou, Anli, Johnson, Judith, Geraghty, Keith, Riley, Ruth, Zhou, Andrew, Panagopoulou, Efharis, Chew-Graham, Carolyn, Peters, David, Esmail, Aneez, and Panagioti, Maria. 2022. "Associations of physician burnout with career engagement and quality of patient care: Systematic review and meta-analysis." *Bmj* 378: e070442 https://doi.org/10.1136/bmj-2022-070442

Hsieh, Mei-Lin, Li, Yin-Ming, Chang, En-Ting, Lai, Hui-Ling, Wang, Wan-Hsiang, and Wang, Shu-Chen. 2011. "Sleep disorder in Taiwanese nurses: A random sample survey." *Nursing & Health Sciences* 13(4): 468–474. https://doi.org/10.1111/j.1442-2018.2011.00641.x

Jafari, Zahra M. 2019. "Effect of passion for physical activity on (physical) burnout in student athletes." *Sport TK: revista euroamericana de ciencias del deporte* 8(2): 125–132. https://doi.org/10.6018/sportk.391851

Jiatong, Wang, Dukhaykh, Suad, Jun, Fuba, Madina, Tussupbekova, and Khan, Muhammad T. 2024. "Impact of passion at work on emotional exhaustion: Mediating role of negative emotions." *Current Psychology*, 43: 29133–29142. https://doi.org/10.1007/s12144-024-06516-1

Landay, Karen, Arena Jr, David F, and King, Dennis A. 2022. "Passion in the pit: The effects of harmonious and obsessive passion on nurse burnout." *Journal of Managerial Psychology* 37(3): 192–205. https://doi.org/10.1108/JMP-03-2021-0181

Lee, Sanggeun, Lee, Younggeun, and Park, Kyung M. 2024. "Performance feedback and obsessive passion: The moderating role of human capital." PLOS *ONE*, 19(4): 1–20. https://doi.org/10.1371/journal.pone.0302180

Liao, Ying Y, Soltani, Ebrahim, Iqbal, Abdullah, and van der Meer, Robert. 2024. "The utility of performance review systems: A total quality management perspective." *Strategic Change* 33(4): 287–310. https://doi.org/10.1002/jsc.2580

Lobato, Roberto M, García-Coll, Josep, and Moyano, Manuel. 2023. "Disconnected out of passion: Relationship between social alienation and obsessive passion." *Journal of Interpersonal Violence* 38(1–2): 1950–1969. https://doi.org/10.1177/08862605221094631

Lopes, Maxime, and Vallerand, Robert J. 2020. "The role of passion, need satisfaction, and conflict in athletes' perceptions of burnout." *Psychology of Sport and Exercise* 48: 101674. https://doi.org/10.1016/j.psychsport.2020.101674

Lucidi, Fabio, Pica, G, Mallia, L, Castrucci, E, Manganelli, S, Bélanger, J J, and Pierro, A. 2016. "Running away from stress: How regulatory modes prospectively affect athletes' stress through passion." *Scandinavian Journal of Medicine & Science in Sports* 26(6): 703–711. https://doi.org/10.1111/sms.12496

Lukenda, Kimberly, Sülzenbrück, Sandra, and Sutter, Christine. 2024. "Expressive writing as a practice against work stress: A literature review." *Journal of Workplace Behavioral Health* 39: 1–24. https://doi.org/10.1080/15555240.2023.2240512

Maslach, Christina, and Jackson, Susan E. 1981. "The measurement of experienced burnout." *Journal of Organizational Behavior* 2(2): 99–113. https://doi.org/10.1002/job.4030020205

Maslach, Christina, and Leiter, Michael P. 2016. "Understanding the burnout experience: Recent research and its implications for psychiatry." *World Psychiatry* 15(2): 103–111. https://doi.org/10.1002/wps.20311

Maslach, Christina, Schaufeli, Wilmar B, and Leiter, Michael P. 2001. "Job burnout." *Annual Review of Psychology* 52: 397–422. https://doi.org/10.1146/annurev.psych.52.1.397

Morales Rodríguez, Mario, and Ronquillo Horsten, José. 2021. "Caracterización del estrés laboral y su impacto en la productividad en una industria maquiladora de Reynosa Tamaulipas." *593 Digital Publisher CEIT* 6(2): 183–192. https://doi.org/10.33386/593dp.2021.2.498

Moreno Pellicer, Raquel. 2017. *Estudios de variables motivacionales y de liderazgo y su relación con el bienestar y malestar en fútbol categoría juvenil en la comarca de La Safor*. Doctoral dissertation. Universitat de València. https://dialnet.unirioja.es/servlet/tesis?codigo=157087

Moreno-Jiménez, Jennifer E, Demerouti, Evangelina, Blanco-Donoso, Luis M, Chico-Fernández, Mario, Iglesias-Bouzas, Maria I, and Garrosa, Eva. 2022. "Passionate healthcare workers in demanding intensive care units: Its relationship with daily exhaustion, secondary traumatic stress, empathy, and self-compassion." *Current Psychology* 42(33): 29387–29402. https://doi.org/10.1007/s12144-022-03986-z

Muis, Masyitha, Nai'em, M. Furqaan, Arsin, Andi A, Darwis, A. Muflihah, Thamrin, Yahya, and Hans, Nural Anniza P. (2021). "The effect of multiple role conflicts and work stress on the work performance of female employees." *Gaceta Sanitaria* 35: S90–S93. https://doi.org/10.1016/j.gaceta.2020.12.025

Paquette, Virginie, Vallerand, Robert J, Houlfort, Nathalie, and Fredrickson, Barara L. 2022. "Thriving through adversity: The role of passion and emotions in the resilience process." *Journal of Personality* 91(3): 789–805. https://doi.org/10.1111/jopy.12774

Pollack, Jeffrey M, Ho, Violet T, O'Boyle, Ernest H, and Kirkman, Bradley L. 2020. "Passion at work: A meta-analysis of individual work outcomes." *Journal of Organizational Behavior* 41(4): 311–331. https://doi.org/10.1002/job.2434

Ratelle, Catherine F, Vallerand, Robert J, Mageau, Genevieve A, Rousseau, François L, and Provencher, Pierre. 2004. "When passion leads to problematic outcomes: A look at gambling." *Journal of Gambling Studies* 20(2): 105–119. https://doi.org/10.1023/B:JOGS.0000022304.96042.e6

Ringle, Christian M, Sarstedt, Marko, Mitchell, Rebecca, and Gudergan, Siegfried P. 2020. "Partial least squares structural equation modeling in HRM research." *The International Journal of Human Resource Management* 31(12): 1617–1643. https://doi.org/10.1080/09585192.2017.1416655

Saleem, Farida, Malik, Muhammad I, and Qureshi, Saiqa S. 2021. "Work stress hampering employee performance during COVID-19: Is safety culture needed?" *Frontiers in Psychology* 12: 655839. https://doi.org/10.3389/fpsyg.2021.655839

Sarstedt, Marko, Ringle, Christian M, Smith, Donna, Reams, Russell, and Hair Jr, Josep F. 2014. "Partial least squares structural equation modeling (PLS-SEM): A useful tool for family business researchers." *Journal of Family Business Strategy* 5(1): 105–115. https://doi.org/10.1016/j.jfbs.2014.01.002

Seijas-Solano, David E. (2019). "Riesgos psicosociales, estrés laboral y síndrome burnout en trabajadores universitarios de una escuela de bioanálisis." *Revista de Salud Pública* 21(1): 102–108. https://doi.org/10.15446/rsap.V21n1.71907

Spányik, András, Simon, Dávid, Rigó, Adrien, Griffiths, Mark D, and Demetrovics, Zsolt. 2022. "Subjective COVID-19-related work factors predict stress, burnout, and depression among healthcare workers during the COVID-19 pandemic but not objective factors." *PloS One* 17(8): e0270156. https://doi.org/10.1371/journal.pone.0270156

Thanh, Tran Le, Hoang, Nguyen T, Mai Lan, Nguyen T, Trang, Vu T, Ha, Le Vu, Hoa, N. T, Xuan Lieu, Cao. and Minh Chinh, Ha Thi. 2024. "Stress in relationship with burnout due to Covid-19: The moderating role of coping strategies." *ASEAN Journal of Psychiatry* 25(3): https://doi.org/10.54615/2231-7805.47347

Ul Ain, Syeda Noor. 2024. "Association between Perceived Work Stress, Emotional Regulation, and Burnout among teachers." *International Journal of Social Science Archives (IJSSA)* 7(3): 1170–1177.

Vallerand, Robert J, Blanchard, Celine, Mageau, Genevieve A, Koestner, Richard, Ratelle, Catherine, Léonard, Maude, Gagné, Marylene, and Marsolais, Josée. 2003. "Les passions de l'âme: On obsessive and harmonious passion." *Journal of Personality and Social Psychology* 85(4): 756–767. https://doi.org/10.1037/0022-3514.85.4.756

Vallerand, Robert J, Paquet, Yvan, Philippe, Frederick L, and Charest, Julie. 2010. "On the role of passion for work in burnout: A process model." *Journal of personality* 78(1): 289–312. https://doi.org/10.1111/j.1467-6494.2009.00616.x

Vergauwe, Jasmine, Wille, Bart, De Caluwé, Elien, and De Fruyt, Filip. 2022. "Passion for work: Relationships with general and maladaptive personality traits and work-related outcomes." *Personality and Individual Differences* 185: 111306. https://doi.org/10.1016/j.paid.2021.111306

Wang, Hui, Hall, Nathan C, and Rahimi, Sonia. 2015. "Self-efficacy and causal attributions in teachers: Effects on burnout, job satisfaction, illness, and quitting intentions." *Teaching and Teacher Education* 47: 120–130. https://doi.org/10.1016/j.tate.2014.12.005

West, Colin P, Dyrbye, Lotte N, and Shanafelt, Tait D. 2018. "Physician burnout: Contributors, consequences and solutions." *Journal of Internal Medicine* 283(6): 516–529. https://doi.org/10.1111/joim.12752

Wu, Wenzhi, Zhang, Yan, Wang, Pu, Zhang, Li, Wang, Guixiang, Lei, Guanghui, Xiao, Qiang, Cao, Xiaochen, Bian, Yueran, Xie, Simiao, Huang, Fei, Luo, Na, Zhan, Jingyuan, and Luo, Mingyan. 2020. "Psychological stress of medical staffs during outbreak of COVID-19 and adjustment strategy." *Journal of Medical Virology* 92: 1–9. https://doi.org/10.1002/jmv.25914

Yukhymenko-Lescroart, Mariya A, and Sharma, Gitima. 2022. "Passion for work and well-being of working adults." *Journal of Career Development* 49(3): 505–518. https://doi.org/10.1177/0894845320946398

Zito, Margherita, Emanuel, Federica, Bertola, Lara, Russo, Vincenzo, and Colombo, Lara. 2022. "Passion and flow at work for the reduction of exhaustion at work in nursing Staff." *Sage Open* 12(2). https://doi.org/10.1177/21582440221095009

13 Employee Well-Being through Leadership 4.0 and Empowerment in the Mexican Industry 4.0

Berenice García-Vázquez, Aurora Irma Máynez-Guaderrama, and Gabriela Jacobo-Galicia

13.1 INTRODUCTION

Well-being is a broad concept that spans a range of aspects of life (Danna and Griffin 1999). In the workplace, well-being of employees is closely tied to their job (Yukhymenko-Lescroart and Sharma 2022) and is evidenced in their perception of and level of satisfaction with their experience of employment (Danna and Griffin 1999). Given that occupational well-being is a dimension of general well-being, efforts should be made to ensure that work is meaningful for the individual and contributes to the fulfillment of his or her life; that is, it should contribute to his or her well-being (Kan et al. 2022).

Industry 4.0 represents not merely a working environment but a paradigm shift in production systems, widely recognized as the Fourth Industrial Revolution. Initially coined in Germany as part of a national strategy to modernize manufacturing, the concept has since evolved into a global framework adopted across industries and governments worldwide. This paradigm is characterized by the integration of cyber-physical systems (CPS), automation, digitalization, and real-time data exchange throughout the value chain (Ghobakhloo et al. 2021). The implementations of this industry (such as robotics, cyber-physical systems (CPS), Big Data, and the Internet of Things (IoT)) lead organizations to new production processes that allow them to manufacture products of high-quality standards, personalized according to customer demand (Javaid et al. 2022; Xu, Xu, and Li 2018).

Beyond technological adoption, Industry 4.0 also implies transformations in organizational structures, workforce competencies, and sociotechnical systems. In this context, it becomes essential to safeguard employee well-being. In these types of organizations, new relationships are generated between human beings and the production infrastructure (Kwiotkowska et al. 2021). The transition to new ways of working and the adoption of new technologies could lead to a decline in workers' well-being due to stressors such as robotization, reduced human involvement in the work, and the perception of a lack of skills to perform adequately in these new conditions (Kadir and Broberg 2020; Koropets and Plutova 2021).

Empowerment and leadership styles influence occupational well-being. In applied psychology, it is considered that empowerment promotes well-being (Zhou et al. 2023) as it is related to greater occupational satisfaction and affective commitment, lower levels of occupational anxiety, and better organizational performance (García-Juan, Escrig-Tena, and Roca-Puig 2023). Psychological well-being can be particularly affected in contexts requiring high resource consumption and risk-taking. In such environments, empowerment has been found to act as a moderating variable between high demands and worker well-being (Kim 2024). The academic literature (Saldaña et al.

DOI: 10.1201/9781003582014-17

2020; Walsh and Arnold 2020) emphasizes the importance of deploying leadership styles in companies that promote the psychological well-being of employees, such as transformational leadership, a style that leads to leadership 4.0.

Leadership 4.0 is a style developed for Industry 4.0, considering aspects such as digitalization, technology, innovation, the environment, and the behaviors of transformational leadership: this style integrates transformational behaviors and the encouragement of the use of technology (Magesa and Jonathan 2019; Mihardjo et al. 2019). Leadership 4.0 encourages collaboration, is open to new ideas, uses digital media to disseminate information, evaluates the digital environment to select appropriate technologies for organizations, builds environments of trust and experimentation with new technologies, and develops a vision to prepare its followers to work with new technologies (Guzmán et al. 2020).

Although some academic studies associate well-being at work with leadership characteristics and autonomy in export manufacturing companies (Vargas-Salgado and Máynez-Guaderrama 2019), no evidence of this relationship exists in Industry 4.0 companies. For this reason, this work explores this knowledge gap.

Grounded in the theoretical perspectives of organizational behavior and Industry 4.0 leadership, the objective of this work is to examine whether leadership 4.0 and empowerment have a positive and significant effect on employees' occupational well-being within companies operating in an Industry 4.0 environment in Ciudad Juárez.

Considering the notion of leadership 4.0 and the importance of preserving employee well-being in Industry 4.0, through leadership styles that have a positive impact on this factor, and the fact that there are studies that relate well-being at work to leadership (Vargas-Salgado and Máynez-Guaderrama 2019), the following research hypothesis is proposed:

13.1.1 Leadership 4.0 Has a Positive and Significant Effect on Employee Psychological Well-Being in Ciudad Juárez's Industry 4.0

An additional characteristic of leadership 4.0 is the tendency to establish flat hierarchies in which decision-making is flexible and decentralized (Helming et al. 2019). Industry 4.0 requires a structure characterized by autonomous decision-making that empowers workers and allows them to act by their own initiative so that they can solve problems in a prompt, flexible, and strategic way (Kaasinen et al. 2020; Elnadi and Abdallah 2024). This is significant because, in the transition to a more digitalized environment, employees' favorable perception of empowerment is important because it increases their involvement in the process, motivation, and willingness to change (Gfrerer et al. 2021). In view of the above, the following research hypothesis is proposed:

13.1.2 Leadership 4.0 Has a Positive and Significant Effect on Empowerment in the Industry 4.0 in Ciudad Juárez

Moreover, the autonomy granted to employees in their work is related to their well-being. When workers do not have control over the resources they need to do their job and cannot choose how the job is performed, their psychological and mental needs are affected, which can negatively affect their well-being (Neffa 2018). By contrast, the findings of a recent study conducted on teachers suggest that empowerment can predict well-being; moreover, according to the authors, teachers can experience happiness at work because empowerment is usually coupled with positive emotions, which are crucial components of well-being (Yu et al. 2024). Thus, a proactive approach to empowerment can improve employee well-being (Blanco and Moros 2020; Rachmad 2022). Even in highly technical environments, innovation is more feasible when employees are encouraged to participate in decision-making by facilitating interaction and the exchange of information between colleagues so that they can influence and learn from each other more naturally (Guinan, Parise, and Langowitz 2019). Therefore, the following research hypothesis is proposed.

13.2 METHODOLOGY

A nonprobability convenience sample was used. The fieldwork was carried out in March–May 2023. An electronic questionnaire was sent to the employees of the export manufacturing industry in Ciudad Juárez, Mexico. A total of 564 responses were received, of which 535 were usable. This sample is adequate because models with endogenous variables received two structural paths and required a minimum of 130 observations for detecting R^2 values of 0.10, with a statistical power of 80% and a significance level of 1% (Hair, Howard, and Nitzl 2020). The sample comprised 56% males and 44% females. Regarding educational level, 41% attended university and earned a bachelor's degree. The majority of respondents reported that they were between 23 and 35 years old (68.7%). Concerning the company's line of business, 41.3% work in the auto parts industry and 34.4% report working as engineers and technicians.

Leadership 4.0, empowerment, and well-being at work were assessed using validated instruments that have been widely applied in peer-reviewed research, ensuring reliability, construct validity, and theoretical soundness. These established instruments were selected and adapted because they are robust measures and relevant for analyzing organizational dynamics in Industry 4.0 contexts. Leadership 4.0 was measured with 15 items from García-Vázquez et al. (2023); the following are examples of the items: In this company, the leader "Shares necessary information with us, using digital media (email, whatsapp, intranet), to enable us to learn and/or contribute with new ideas", "Identifies the requirements for creating environments to use and experiment with new technologies", and "Encourages us to become more open and adopt new technologies at work". Twelve items from the Menon (2001) and Spreitzer (1995) scales were used to measure empowerment, including the following: "The company in which I work inspires me to work toward its goals," "I think that I have the ability to perform my job well," and "In this company I can decide about how to do my job". Well-being at work was measured using eight items using the scale developed by Pradhan and Hati (2019), examples of which are as follows: "I am very satisfied with my job", "I am motivated by my work achievements", and "My job offers me broad opportunities for career growth". Respondents rated items on a five-point Likert scale (1 = strongly disagree; 5 = strongly agree). Construct scores were obtained through the partial least squares (PLS) algorithm in SmartPLS, which computes latent variable scores based on outer loadings, instead of relying on raw item sums or averages. To ensure that there were no common method bias problems, the variance inflation factor (VIF) values of the constructs were examined. Their results were less than 3.3, suggesting that collinearity problems were unlikely to exist (Kock 2015). This methodological approach ensures transparency in the operationalization of constructs and reinforces the consistency and validity of the measurement model.

13.3 RESULTS

The hypotheses were tested using the partial least squares structural equation modeling (PLS-SEM) with SmartPLS version 4. In this technique, the measurement model is first evaluated, and then the theoretical model tested is reviewed. In the measurement model evaluation, the first step is to examine the factor loadings, reliability, convergent validity, and discriminant validity. Items must exhibit loadings greater than 0.708 with statistically significant t-values (Hair, Howard, and Nitzl 2020). The initial model included 35 items, of which seven were eliminated due to high collinearity with items of their latent variable [L4.4, L4.6, L4.8, L4.14, L4.15, and E10]. Convergent validity was then assessed by evaluating Cronbach's alpha, ρ_a, and ρ_c coefficients. Table 13.1 presents the values of these coefficients, which range between 0.931 and 0.954, and fulfill recommendations regarding their minimum and maximum values; moreover, the average variance extracted (AVE) scores are higher than the minimum recommended cut-off point of 0.500 (Hair, Howard, and Nitzl 2020). Discriminant validity was assessed using the heterotrait–monotrait relationship (HTMT). The maximum cut-off point was 0.850 because the model constructs were conceptually distinct

TABLE 13.1
Reliability and Convergent Validity of the Measurement Model

Constructs	Cronbach's Alpha	Composite Reliability (ρ_a)	Composite Reliability (ρ_C)	Average Variance Extracted (AVE)
Empowerment	0.940	0.942	0.948	0.605
Leadership 4.0	0.946	0.947	0.954	0.673
Well-being at work	0.931	0.936	0.944	0.677

(Ringle et al. 2020). The HTMT ratios are as follows: Leadership 4.0 – Wellbeing at work (0.733), Leadership 4.0 – Empowerment (0.795), and Empowerment-Wellbeing at work (0.829).

After ensuring the reliability, convergent validity, and discriminant validity of the measurement model, the structural model was reviewed. It was found that leadership 4.0 has a positive and significant effect on empowerment (H1: β=0.751, t=29.092) and work well-being (H2: β=0.230, t=4.752). Likewise, empowerment had a significant positive effect on occupational well-being (H3: β=6.10, t=12.130).

The results of R^2, f^2, and Q^2 were subsequently examined. The model moderately explains empowerment (R^2=0.565) and moderately to substantially explains job well-being (R^2=0.636). In addition, significant effects were found among the variables, a small effect on leadership 4.0-> well-being at work (f^2=0.063, t=2.114) and two large effects on leadership 4.0-> empowerment (1.296, t=6.207) and empowerment -> well-being at work (0.445, t=3.802). The Q^2 indicator evidenced a predictive capacity of the model with medium relevance for well-being at work (0.470) and high relevance for empowerment (0.561). Regarding mediation, empowerment's mediating role in the leadership 4.0-> well-being at work relationship was assessed. The results showed that the variable has a complementary partial mediating role (direct β=0.230, t=4.752; indirect β=0.459, t=10.904; total β=0.688, t=20.700) (Carrión, Nitzl, and Roldán 2017). The model tested with SmartPLS that includes the constructs, their items, outer loadings, t-values, and R^2 is shown in Figure 13.1.

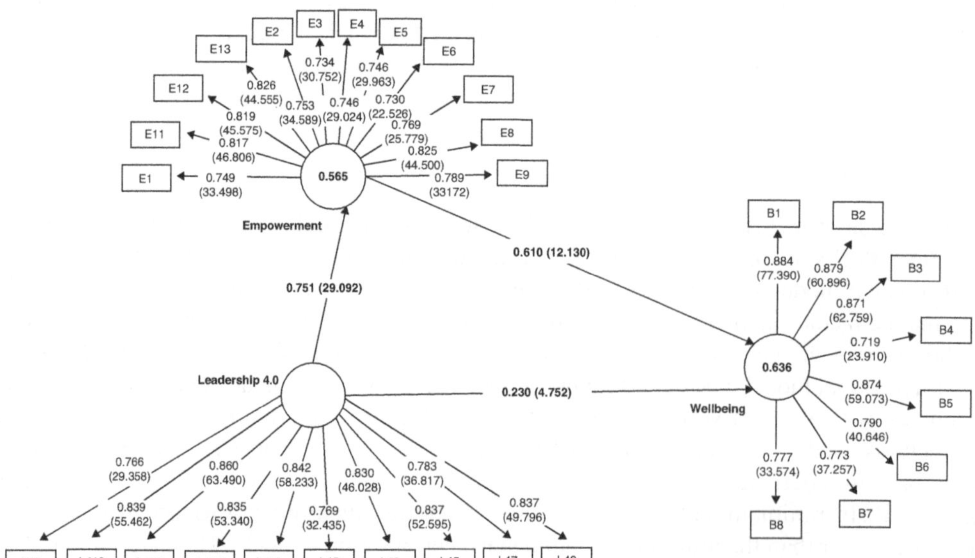

FIGURE 13.1 Tested model. (Image by the authors.)

TABLE 13.2
Hypothesis Tests

Hypothesis	Path Coefficient	*T*-value	Decision
H1. Leadership 4.0 has a positive and significant effect on the psychological well-being of employees in Ciudad Juárez's Industry 4.0.	0.230	4.752***	Do not reject
H2. Leadership 4.0 has a positive and significant effect on empowerment in Industry 4.0 in Ciudad Juárez	0.751	29.092***	Do not reject
H3: Empowerment has a positive and significant effect on the psychological well-being of employees in Industry 4.0 in Ciudad Juárez	0.610	12.130***	Do not reject

***Significant with an alpha of 0.001, if $t > 3.291$ for a two-tailed test.

The results of the statistical analysis indicate that employee well-being is positively affected by leadership 4.0 and empowerment and that Leadership 4.0 has a positive and significant effect on empowerment. In addition, empowerment was found to have a partial mediating effect between leadership 4.0 and well-being. The results of the hypothesis testing are presented in Table 13.2.

13.4 DISCUSSION

In the context of Industry 4.0 in Ciudad Juárez, it was found that there is a positive and statistically significant relationship between Leadership Style 4.0 and employee well-being. The literature suggests that in highly technological environments, a leadership style that promotes employee well-being should be developed (Saldaña et al. 2020; Walsh and Arnold 2020). In this sense, leadership 4.0 has a positive impact on well-being through its social attributes, in which behaviors that generate agreements, trust, collaboration, and participation stand out (Guzmán et al. 2020). Thus, participants reported feeling satisfied with their work, which contributed to their perception of well-being.

It was also confirmed that there is a positive and statistically significant relationship between Leadership Style 4.0 and empowerment in this context. The surveyed employees reported that their manager recognizes the requirements for creating an environment for using and experimenting with new technologies, is able to evaluate the company's environment to select the most favorable communication platforms, and intends that, in the future, everyone in the company will participate, collaborate, contribute new ideas, and make decisions. The behavior of the leader perceived by his subordinates in this study coincides with that recommended by the literature, which suggests that in the transition to a more digitized environment, the positive perception of employee empowerment is important because it increases their motivation, their willingness to change, and the level at which they are involved in the process (Kaasinen et al. 2020; Elnadi and Abdallah 2024).

The literature indicates that empowerment is a critical factor for workers to achieve an optimal level of well-being (Rachmad 2022). This study confirms this assertion by revealing a positive and statistically significant effect of empowerment on employee well-being. In fact, empowerment has been associated with positive feelings that promote employee happiness and thus employee well-being (Yu et al. 2024). In this study, this is mainly reflected in the feeling of inspiration that the organization creates in the employee so that he/she strives to achieve his/her goals, to get where he/she wants to go, and to feel that his/her work is important to him/her.

13.5 CONCLUSIONS

This study proposed and tested a research model examining the relationships among leadership 4.0, empowerment, and employee well-being in the context of Industry 4.0 in Ciudad Juárez. These relationships were transformed into statistical hypotheses and validated through PLS-SEM.

The present work provides an analysis of the impact of leadership 4.0 and empowerment on employee well-being in the context of Industry 4.0 in Ciudad Juárez. It was demonstrated that in this context, leadership 4.0 and empowerment have a positive impact on employee well-being, as well as showing that there is a mediating effect of empowerment between this leadership style and well-being. Given the literature's emphasis on the importance of promoting employee well-being in these environments, it is recommended that leadership styles that promote this factor through empowerment, as well as behaviors that encourage the adoption of technology through its use and experimentation, be established in organizations to reduce the stress that may arise in the transition to Industry 4.0. Likewise, research should be conducted to explore other variables that affect and promote employee well-being in this context.

However, an important conceptual consideration arises regarding the construct of leadership 4.0. Although the items used reflect characteristics of transformational leadership, the construct of leadership 4.0 is precisely built upon the combination of this style with an emphasis on encouraging followers to use technology and on creating digital environments that foster learning and experimentation with such technologies. It is recommended that future research works further analyze and measure this construct to strengthen its theoretical contribution and practical applicability.

REFERENCES

Blanco, Yaima, and Moros, Helena. 2020. Empoderamiento organizacional: Factor protector del bienestar laboral. *Revista Cubana de Salud y Trabajo* 21 (2): 60–65.

Carrión, Gabriel, Nitzl, Christian. and Roldán, José L. 2017. Mediation analyses in partial least squares structural equation modeling: Guidelines and empirical examples. In *Partial least squares path modeling: Basic concepts, methodological issues and applications*, edited by H. Latan and R. Noonan, 173–95. Cham: Springer International Publishing. https://doi.org/10.1007/978-3-319-64069-3_8

Danna, Karen, and Griffin, Ricky W. 1999. Health and well-being in the workplace: A review and synthesis of the literature. *Journal of Management* 25 (3): 357–84. https://doi.org/10.1177/014920639902500305

Elnadi, Moustafa, and Abdallah, Yasser Omar. 2024. Industry 4.0: Critical investigations and synthesis of key findings. *Management Review Quarterly* 74 (2): 711–44. https://doi.org/10.1007/s11301-022-00314-4

García-Juan, Beatriz, Escrig-Tena, Ana B. and Roca-Puig, Vicente. 2023. Structural empowerment and organisational performance: The mediating role of employees' well-being in Spanish local governments. *International Journal of Human Resource Management* 34 (10): 1907–39. https://doi.org/10.1080/09585192.2020.1859581

García-Vázquez, Berenice, Vargas-Salgado, María Maricela and Máynez-Guaderrama, Aurora I. 2023. Validez de contenido de un instrumento para medir liderazgo 4.0. *DYNA* 90 (229). 170–74. https://doi.org/10.15446/dyna.v90n229.108410

Gfrerer, Anne, Hutter, Katja. Füller, Johann and Ströhle, Thomas. 2021. Ready or not: Managers' and employees' different perceptions of digital readiness. *California Management Review* 63 (2): 23–48. https://doi.org/10.1177/0008125620977487

Ghobakhloo, Morteza, Iranmanesh, Mohammad, Grybauskas, Andrius, Vilkas, Mantas and Petraitė, Monika. 2021. Industry 4.0, innovation, and sustainable development: A systematic review and a roadmap to sustainable innovation. *Business Strategy and the Environment* 30 (8): 4237–57. https://doi.org/10.1002/bse.2867

Guinan, Patricia J., Parise, Salvatore and Langowitz, Nan. 2019. Creating an innovative digital project team: Levers to enable digital transformation. *Business Horizons* 62 (6): 717–27. https://doi.org/10.1016/j.bushor.2019.07.005

Guzmán, Valeria E., Muschard, Bernd, Gerolamo, Mateus, Kohl, Holger and Rozenfeld, Henrique. 2020. Characteristics and skills of leadership in the context of Industry 4.0. *Procedia Manufacturing* 43: 543–50. https://doi.org/10.1016/j.promfg.2020.02.167

Hair, Joe, Howard, Matt and Nitzl, Christian. 2020. Assessing measurement model quality in PLS-SEM using confirmatory composite analysis. *Journal of Business Research* 109: 101–10. https://doi.org/10.1016/j.jbusres.2019.11.069

Helming, Sina, Ungermann, Florian, Hierath, Natalie, Stricker, Nicole and Lanza, Gisela. 2019. Development of a training concept for leadership 4.0 in production environments. *Procedia Manufacturing* 31: 38–44. https://doi.org/10.1016/j.promfg.2019.03.007

Javaid, Mohd, Haleem, Abid, Singh, Ravi Pratap, Suman, Rajiv and Gonzalez, Ernesto. 2022. Understanding the adoption of Industry 4.0 technologies in improving environmental sustainability. *Sustainable Operations and Computers* 3: 203–17. https://doi.org/10.1016/j.susoc.2022.01.008

Kaasinen, Eija, Schmalfuß, Franziska, Özturk, Cemalettin, Aromaa, Susanna, Boubekeur, Menouer Heilala, Juhani, Heikkilä, Paivi, Kuula, Timo, Liinasuo, Marja, and Mach, Sebastian. 2020. Empowering and engaging industrial workers with Operator 4.0 solutions. *Computers & Industrial Engineering* 139: 105678. https://doi.org/10.1016/j.cie.2019.01.001

Kadir, Bzhwen A., and Broberg, Ole. 2020. Human well-being and system performance in the transition to Industry 4.0. *International Journal of Industrial Ergonomics* 76: 102936. https://doi.org/10.1016/j.ergon.2020.102936

Kan, Chiemi, Koremura, Yuka, Nobuhiro, Masaki, Kobayashi, Yusuke, Asano, Yuki, Yasui, Nobuyuki, Shimizu, Shoken and Hojo, Reiko. 2022. Study on well-being of workers under corporative worksite at an automobile manufacturing. In *2022 IEEE 4th Global Conference on Life Sciences and Technologies (LifeTech)*, 506–7. IEEE. https://doi.org/10.1109/LifeTech53646.2022.9754867

Kim, Cheol Young. 2024. Innovative behavior, empowerment, and work engagement: Joint curvilinear effect on well-being. *Social Behavior and Personality: An International Journal* 52 (5): 13095E–3108E. https://doi.org/10.2224/sbp.13095

Kock, Ned. 2015. Common method bias in PLS-SEM: A full collinearity assessment approach. *International Journal of E-Collaboration* 11 (4): 1–10. https://doi.org/10.4018/ijec.2015100101

Koropets, O., and Plutova, M. 2021. Well-being of workers and career strategies in the conditions of Industry 4.0. In *Proceeding of the International Science and Technology Conference FarEastCon 2020*, 649–55. Springer. https://doi.org/10.1007/978-3-030-68014-5_63

Kwiotkowska, Anna, Gajdzik, Bozena, Wolniak, Radoslaw, Vveinhardt, Jolita and Gębczyńska, Magdalena. 2021. Leadership competencies in making Industry 4.0 effective: The case of Polish heat and power industry. *Energies* 14 (14): 4338. https://doi.org/10.3390/en14144338

Magesa, Mwita, and Jonathan, Joan 2019. Digital leadership for digital transformation. *Electronic Scientific Journal* 10 (4): 2082–677. https://en.wspolczesnagospodarka.pl/

Menon, Sanjay. 2001. Employee empowerment: An integrative psychological approach. *Applied Psychology* 50 (1): 153–80. https://doi.org/10.1111/1464-0597.00052

Mihardjo, Leonardus, Sasmoko, Sasmoko, Alamsjah, Firdaus and Elidjen, Elidjen. 2019. Digital leadership role in developing business model innovation and customer experience orientation in Industry 4.0. *Management Science Letters* 9 (11): 1749–62. https://doi.org/10.5267/j.msl.2019.6.015

Neffa, Julio César. 2018. Los riesgos psicosociales en el trabajo. *Boletín del Consejo Académico de* Ética *en Medicina* 14. https://ri.conicet.gov.ar/bitstream/handle/11336/266695/CONICET_Digital_Nro.8c2f180f-44cc-427c-b963-89bfee18de39_B.pdf?sequence=2

Pradhan, Rabindra Kumar, and Hati, Lopamudra. 2019. The measurement of employee well-being: Development and validation of a scale. *Global Business Review*. https://doi.org/10.1177/0972150919859101

Rachmad, Yoesoep Edhie. 2022. *Empowerment theory*. Cáceres Conquistadores Publicaciones Internacionales, Edición Especial 2022.

Ringle, Christian M., Sarstedt, Marko, Mitchell, Rebecca and Gudergan, Siegfried P. 2020. Partial least squares structural equation modeling in HRM research. *International Journal of Human Resource Management* 31 (12): 1617–43. https://doi.org/10.1080/09585192.2017.1416655

Saldaña, Claudia, Polo-Vargas, Jean David Gutiérrez-Carvajal, Oscar Iván and Madrigal, Berta Ermila 2020. Bienestar psicológico, estrés y factores psicosociales en trabajadores de instituciones gubernamentales de Jalisco-México. *Revista de Ciencias Sociales* 26 (1): 25–37.

Spreitzer, Gretchen M. 1995. Psychological empowerment in the workplace: Dimensions, measurement, and validation. *Academy of Management Journal* 38 (5): 1442–65. https://doi.org/10.2307/256865

Vargas-Salgado, María Marisela, and Máynez-Guaderrama, A. I. 2019. Liderazgo, contrato psicológico y empoderamiento: Sus vínculos en organizaciones fronterizas mexicanas. *Acta Universitaria* 29: 1–13. https://doi.org/10.15174/au.2019.2047

Walsh, Megan M, and Arnold, Kara A. 2020. The bright and dark sides of employee mindfulness: Leadership style and employee well-being. *Stress and Health* 36 (3): 287–98. https://doi.org/10.1002/smi.2926

Xu, Li Da, Xu, Eric L., and Li, Ling. 2018. Industry 4.0: State of the art and future trends. *International Journal of Production Research* 56 (8): 2941–62. https://doi.org/10.1080/00207543.2018.1444806

Yu, Xiajun, Lin, Xue, Xue, Danni and Zhou, Hui. 2024. Impact of work engagement on teachers' workplace well-being: A serial mediation model of perceived organizational support and psychological empowerment. *SAGE Open* 14 (4): 1–14. https://doi.org/10.1177/21582440241291344

Yukhymenko-Lescroart, Mariya A., and Sharma, Gitima. 2022. Passion for work and well-being of working adults. *Journal of Career Development* 49 (3): 505–18. https://doi.org/10.1177/0894845320946398

Zhou, Xiaofeng, Sun, Ninghao, Zhao, Xu, Wang, Leyao and Tong, Ruipeng. 2023. Revealing the practical strategies for promoting Chinese migrant workers' health and well-being: Organizational empowerment perspective. *Applied Psychology: Health and Well-Being* 15 (3): 1085–109. https://doi.org/10.1111/aphw.12387

14 Effects of Mental Workload on University Faculty Members at a Higher Education Institution in Mexico

Arturo Realyvásquez-Vargas, José Ángel Díaz-Alcaraz, Guadalupe Hernández-Escobedo, Aide Aracely Maldonado-Macías, Jorge Luis García-Alcaraz, and David Antonio Manjarrez-Zazueta

14.1 INTRODUCTION

Mental workload (MWL) is defined as the cognitive demands of a task (Mohammadian et al. 2022). According to several authors, human MWL has the highest demand among multidimensional constructs in Ergonomics and Human Factors science (Longo et al. 2022). Among the professions with the highest demand for the MWL are physicians (Gyorffy, Dweik, and Girasek 2016), nurse educators (Vauhkonen et al. 2024), engineers (Rachmuddin, Dewi, and Dewi 2021), and university faculty members (Akca and Küçükoğlu 2022).

Regarding university faculty members, several studies have been conducted that make clear the causes and consequences of high levels of MWL. For example, Akca and Küçükoğlu (2022) investigated the relationship between MWL, burnout, and job performance. To do so, these authors collected data from 144 academics in Turkey. Their results revealed that MWL increases burnout and decreases university faculty members' job performance.

On the other hand, Pace et al. (2021) explored the level of MWL in 252 university faculty members from 10 institutions in Europe. In addition, they also assessed psychosocial risks related to MWL, such as workaholism and psychological distress. They found that MWL has a significant effect on university faculty members' mental health as well as on their workaholism. In another research, Sakarji et al. (2024) conducted systematic literature reviews to investigate the salient factors influencing the mental health of academics in higher education. Their results revealed that job demand significantly impacts the mental health of academics in higher education.

On the other hand, Longo and Orrú (2022) calculated the MWL levels of university faculty members with three different methods: NASA Task Load Index (NASA-TLX), Workload Profile, and Rating Scale Mental Effort. In all three methods, the mean MWL had a value greater than 40.

Currently, MWL is one of the main reasons why university faculty members leave their profession (Heffernan et al. 2022). This background reveals that there is a worrying pattern of increasing work pressures and frequency of mental health problems among academics. Therefore, this chapter aims to measure the levels of MWL to which university faculty members at an institution in Mexico are exposed to and understand the effects of MWL on them, such as stress, mental fatigue, headache, irritability, depression, anxiety, and increased body weight.

DOI: 10.1201/9781003582014-18

14.1.1 Research Context

This study was conducted in a higher education institution in Tijuana, Mexico. This institution offers 19 undergraduate programmes and eight graduate programmes. By the end of 2022, more than 12,000 undergraduate, master's, and doctoral students were enrolled. Industrial engineering is one of the undergraduate programmes offered.

This programme lasts 10 semesters and offers students solid training in different areas such as mathematics, physics, statistics, computer science, economics, process engineering, production systems, and administration. In addition, the industrial engineering department has laboratories for work-study, metrology, manufacturing processes, simulation, ergonomics, Six Sigma, advanced manufacturing, physics, and chemistry. These laboratories allow students to apply their theoretical and practical skills.

The Industrial Engineering Department's organizational chart comprises one head, two secretaries, three coordinators, one head of research projects, two heads of laboratories, two heads of teaching, two heads of liaison, and teaching staff.

14.1.2 Problem Statement

The educational programme under study has 66 professors, whether tenured, temporary, interim, or hired on a fee basis. The average age of the academic staff is 43. However, more than 50% are over 55, candidates for retirement, and more than 55% correspond to professors hired on a fee basis, some of whom are not recurrent.

On the other hand, the educational programme has more than 1,863 students enrolled. In other words, the ratio is 28.22 students per university faculty member, which may increase the MWL faced by most (Tarraya 2023). Sometimes, university faculty members must deal with groups of up to 40 students. Attending to, reviewing assignments, grading, and capturing the qualifications of a relatively large number of students increase the MWL for each university faculty member.

In addition, faculty members must advise five to eight students per semester on their residency projects. Sometimes, the number of students to be advised accumulates semester after semester. During the advising sessions, the professors must check that the residency project document complies with the structure and that the content is correct and legible. This also favours the increase of MWL among the university faculty members.

Some professors must assist graduate students in their subjects and advise and review their thesis projects, as if that were not enough. Additionally, some must carry out high-level research projects to remain in the Mexican National System of Researchers. University faculty members of the department under study are evaluated in two modalities. The first is a departmental evaluation carried out by the head of the department, and in turn, the university faculty member performs a self-evaluation. The teaching, tutoring, liaison, and research items are evaluated in this modality.

The second modality is an evaluation carried out by the students. Through this modality, the university faculty member is evaluated academically regarding knowledge capacity, audiovisual media and technologies used, ease of expression, punctuality, responsibility, commitment to content, planning, and evaluating the subject.

It is well known that, to teach at the university level, a university professor must have knowledge and strategy competencies, ingenuity, creativity, and professionalism. However, according to a survey, some students have commented that university faculty members, in general, have been showing some emotional instability, difficulty in mastering technology, and little or no responsibility in the fulfilment of plans and programmes, as well as in the planning of the course. Some students highlight the ability of university faculty members to teach.

However, it is also evident that several of them show high MWL due to their multiple administrative, liaison, and research occupations, which increase their mental load and impact their performance in the classroom and generally within the institution.

Despite the above, no studies have been conducted to determine simultaneously the effects of MWL that may impact the work performance of university faculty members in the department under study and their physical and emotional health. More precisely, at least in the last 5 years, no research has been conducted in Mexico that simultaneously analyses the effects of MWL on stress, headaches, irritability, mental fatigue, depression, anxiety, and weight gain. That is the contribution of this research.

14.1.3 Objectives

This research aims to identify the effects of MWL on university faculty members of the academic department under study. In addition, there are the following specific objectives:

- To determine the MWL levels of university faculty members by applying the NASA-TLX method.
- To determine which dimensions of the NASA-TLX method contribute most to MWL.
- To identify which conditions are associated with MWL concerning the type of contracting through mediation analysis.

14.2 LITERATURE REVIEW

14.2.1 NASA-TLX Method

MWL refers to the mental effort required to perform a specific task (Wolpe, Holton, and Fletcher 2024). Excessive MWL can negatively affect individuals' performance and health. In environments such as the workplace, MWL can be an essential factor to consider to ensure the efficiency and well-being of employees. In this context, the assessment of MWL becomes crucial.

The NASA-TLX method is a commonly used tool to assess the MWL perceived by employees in various environments (Putra, Pratama, and Farahdiansari 2023). NASA developed it in the 1980s as part of research on human–computer interaction in advanced aviation systems (Kosch et al. 2023). More specifically, as Braarud (2021) mentioned, Hart and Staveland (1988) developed the NASA-TLX method. In the book *Human Mental Workload*, they published the chapter entitled *Development of NASA-TLX (Task Load Index): Results of Empirical and Theoretical Research*. Therefore, the NASA-TLX method focuses on the subjective measurement of MWL. Since then, the method has become popular for assessing MWL in various fields, including user interface design, psychology, medicine, and ergonomics (Realyvásquez-Vargas et al. 2023).

This method is based on the premise that workload is multidimensional and can be assessed most effectively by considering several factors (Virtanen et al. 2022). These factors include mental demand, physical demand, temporal demand, self-performance, perceived effort, and perceived frustration (Rodio-Trevisan et al. 2024). Each of these factors contributes to the overall MWL perceived by an individual and may vary in importance depending on the specific task and context (Dethier, Stevens, and Boden 2024). Each of these factors is defined below.

Mental demand refers to the mental effort and attention capacity required to perform a specific task (Martínez-García 2021). This dimension evaluates the task's cognitive complexity and the mental resources required to complete it successfully. Mental demands can range from simple tasks that require little concentration to highly complex tasks that involve multiple simultaneous cognitive demands. Mental demand is one of the key factors contributing to an individual's perceived overall workload (Knisely, Joyner, and Vaughn-Cooke 2021). Similarly, physical demand refers to the physical effort required to perform a specific task (Baumgartner et al. 2025). This dimension assesses the intensity and type of physical activity required to complete the task successfully. Physical demands range from sedentary tasks involving little or no physical exertion to highly physically demanding tasks (van As et al. 2022). Regarding the dimension of physical demand, as mentioned by DiDomenico and Nussbaum (2011), physically demanding work may impact MWL by impairing mental processing or decreasing performance. Furthermore, DiDomenico and Nussbaum (2008) mention that although physical demand is included in the NASA-TLX, the index is not intended to measure physical workload but instead recognizes the potential influence of physical activity on the perception of MWL.

For its part, time demand refers to the amount of time available to complete a task and the degree of haste or urgency associated with it (Comezaquira-Reay, Terán-Ángel, and Pedro José Quijada-Martínez 2021). This dimension assesses the time pressure experienced by an individual when performing the task and how it influences perceived workload (Pooladvand and Hasanzadeh 2022). The fourth factor is self-performance, which refers to an individual's judgement of his/her performance on a specific task (Mendoza, Yan, and King 2023). This dimension assesses personal satisfaction regarding performance and perceived effectiveness in completing the task. It must be remembered that self-performance is a subjective assessment and may not objectively reflect actual performance on the task (Wick 2021). The fifth factor is effort. This refers to the physical and mental effort associated with a specific task (Hakim et al. 2022). This dimension assesses the amount of mental and physical work required to complete the task and the amount of difficulties encountered in the process (Pagnotta et al. 2022). Finally, the perceived frustration factor refers to dissatisfaction and displeasure associated with a specific task (Ghosn et al. 2023). This dimension assesses the amount of stress and anxiety experienced by an individual when performing the task and how it influences the perceived workload (Brunzini et al. 2021).

When assessing MWL using the NASA-TLX method, participants are asked to assign values to each factor based on their experience with a specific task (Nur, Iskandar, and Ade 2020). Values are assigned on a scale of 1 to 20, where 1 represents no load and 20 represents the highest possible load (Bolton, Biltekoff, and Humphrey 2023) (for the self-performance factor only, 1 = Good and 20=Poor). This value is represented as d_i. Subsequently, the worker performs 15 pairwise comparisons of the six factors. From these 15 pairs, the worker selects the factor that has contributed most to the MWL. Subsequently, each factor is rated according to the total number of times the worker has selected it. This rating is designated as w_i and comprises a scale from 0 (not relevant) to 5 (more important than any other factor) (Virtanen et al. 2022).

Subsequently, these values are used to calculate an overall MWL index that quantitatively measures perceived MWL (Hoogendoorn et al. 2020). For this purpose, Equation (14.1) is applied (Realyvásquez-Vargas et al. 2023):

$$\text{NASA-TLX} = \frac{1}{15}\sum_{1}^{6} d_i w_i \tag{14.1}$$

The NASA-TLX method offers the following advantages (Virtanen et al. 2022):

1. The NASA-TLX offers a rapid and straightforward method for assessing operator workload.
2. The NASA-TLX approach has exhibited sensitivity to demand modifications in various flying studies.
3. The NASA-TLX has undergone extensive testing and has been the focus of several validation studies.
4. Numerous studies have demonstrated its superiority compared to the Subjective Workload Assessment Technique (SWAT).
5. The NASA-TLX is accessible in a pen-and-paper version for individuals lacking computers.
6. It is likely the most prevalent method for assessing MWL.
7. The NASA-TLX employs a multidimensional methodology for evaluating MWL.
8. Implementing the NASA-TLX in the software package alleviates most of the analyst's workload, leading to an expedited and straightforward process.
9. The method is non-intrusive to primary task performance when applied after the trial.

10. The NASA-TLX sub-scales are universal, allowing the technique to be utilized across several domains.

However, it also has the following disadvantages (Malakoutikhah et al. 2021):

1. When the method is administered retrospectively, individuals may have overlooked the high workload elements of the activity.
2. Workload assessments may correlate with task performance; individuals who do poorly on the core task may assess their workload as excessively high, and vice versa.
3. The sub-scale weighting process is arduous and prolongs the procedure.
4. When administered online, the NASA-TLX may disrupt primary task performance.

The NASA-TLX method is considered a valid and reliable tool for assessing perceived workload in various contexts (Hutson et al. 2024).

14.2.2 Effects of Mental Workload on University Faculty Members

Stress is a natural physiological and psychological response to the demands of daily life. Although it is a regular part of life, chronic or overwhelming stress can be detrimental (Valent and De La Serna 2021). Several studies mention that excessive MWL can increase stress, negatively impacting job satisfaction and overall well-being (Kirana, Lukitaningsih, and Londin 2021). For instance, a study examining the impact of MWL on stress among Australian university faculty members found that excessive MWL can contribute significantly to overall stress levels (Carroll et al. 2022). Other research has shown that tasks requiring high concentration and managing multiple responsibilities increase perceived stress (Agyapong et al. 2022).

Based on the above, we propose the following hypothesis, H_1.

H_1: MWL has a direct and positive effect on the stress level of university faculty members.

Another effect that MWL can have on people is headaches. For instance, one study mentions that higher levels of MWL were associated with an increased risk of developing tension-type headaches among university faculty members (De Carvalho Leite Peres and Oliven 2023). Another study found that higher quantitative demands and role conflict, less control over decisions, control over work intensity, and job satisfaction (i.e., MWL) were related to more severe headaches (Christensen and Knardahl 2012; Cabezas-Heredia et al. 2024). Finally, it has been reported that university faculty members from England have headaches due to heavy MWL (Johnson and Coleman 2025).

From this background, hypothesis H_2 arises.

H_2: MWL directly and positively affects headaches in university faculty members.

Regarding irritability, a third effect of MWL, a literature review on university faculty members' perceived MWL and its impact on their health found that their mental demands translated into emotional symptoms such as irritability (González-Palacios, Ceballos-Vásquez, and Rivera-Rojas 2021). Another study mentions that work overload (i.e., MWL) in university faculty members occurs when the demands of the job exceed the individual's potential to cope with them, which causes irritability (Cadena-Povea et al. 2025). Finally, in a third study, MWL was detected to cause university faculty members' emotional exhaustion syndrome, manifested by irritability (Aigerim and Almazhay 2024).

These studies serve as a basis for proposing hypothesis H_3.

H_3: MWL has a direct and positive effect on the irritability of university faculty members.

In addition, studies are showing that MWL can cause mental fatigue in people. Mental fatigue is a psychobiological state induced by prolonged mental effort (e.g., study and work commitments or physical activity) (Van Cutsem et al. 2022). It is experienced as a feeling of tiredness and lack of energy, a reduction in cognitive performance, or an increase in the effort required to maintain performance (Van Cutsem et al. 2022). Karanfil and Khatami (2021) mention that, in the Turkish context, Pinar (2018) investigated the burnout levels of English as a foreign language university faculty members as a function of some demographic characteristics. The study concluded that heavy workload (i.e., MWL), among other factors, generates burnout in EFL university faculty members, and that, in turn, burnout causes emotional exhaustion (i.e., mental fatigue). Similarly, Van Cutsem et al. (2022) mention that prolonged high MWL can lead to task abandonment by perceiving a task as too difficult or depleting resources and inability to maintain performance.

In another study, Kreuzfeld, Felsing, and Seibt (2022) investigated teachers in Germany. For this purpose, they applied a cross-sectional survey to 6,109 teachers. They determined how long upper secondary school teachers work per week, what influences their working hours, and how different recording methods affect the total working hours. Their results showed that teachers who worked more than 45 hours per week (i.e., more MWL) suffered more frequently from inability to recover (46%) and emotional exhaustion (32%). From this background, hypothesis H_4 emerges.

H_4: MWL directly and positively affects mental fatigue in university faculty members.

Another effect that MWL can cause in people is depression. Depression is defined as a mental health disorder characterized by persistently low mood or loss of interest in activities, which causes significant impairment in daily life (Okeke et al. 2023). Agyapong et al. (2022) conducted a literature review to determine the extent of the current literature on the prevalence of depression among teachers. The results indicated that workload causes depression in teachers because it increases professional burnout, and that when only clinically significant psychological disorders (moderate to severe) among teachers were taken into account, the prevalence of depression ranged from 4% to 77%. Hojo (2021) mentions that, in Japan, significant correlations were found between workload and depressive symptoms in teachers.

Meanwhile, Ozamiz-Etxebarria et al. (2021) conducted a rapid systematic review with meta-analysis based on eight studies of depression among teachers during the COVID-19 pandemic. This review reported overall prevalence rates of depression in teachers of different ages, sexes, countries, and educational sectors. Their results indicate that 19% of teachers reported depression. The authors mention this may be due to teacher overload (MWL). Johansson, Falkstedt, and Almroth (2022) point out that teachers are at particular risk for mental health problems. Therefore, they conducted a study to compare the risk of being diagnosed with depression in different types of teachers in Sweden, both with each other and with the rest of the population. His study included more than 3 million individuals between the ages of 30 and 60, of whom 256,166 were teachers. Their results indicated that primary and special education teachers had a higher risk among men when comparing the different types of teachers with university faculty members. In comparison, primary and secondary education teachers had a lower risk among women. From the above, hypothesis H_5 emerges.

H_5: MWL has a direct and positive effect on depression in university faculty members.

A sixth effect that high MWL can cause is anxiety. Anxiety is characterized by worried thoughts, tension, and feelings of fear, as well as rapid breathing, among other physical changes (Ashipala and Shilunga 2023). Several studies simultaneously analyse MWL and anxiety, but not the relationship between these variables. Some of these works are those conducted by Alsuraykh et al. (2019), Colin et al. (2023), and Tseng, Chuang, and Liu (2022). Desouky and Allam (2017) conducted a cross-sectional study with 568 Egyptian teachers to assess the prevalence of anxiety. Respondents filled out the validated Arabic version of Taylor's Manifest Anxiety Scale to assess anxiety. The results indicated a prevalence of anxiety of 67.5% among the teachers. The scores were significantly higher in teachers with higher workloads. Similarly, Shablova (2024) found that the increased MWL in the teaching profession represents a factor in the development of anxiety disorders. This result is similar to that of Agyapong et al. (2022), who mention that as teachers' workloads increase, their working hours will invariably increase, resulting in increased work demands and, ultimately, anxiety.

From this background, hypothesis H_6 emerges.

H_6: MWL has a direct and positive effect on the anxiety of university faculty members.

Finally, a seventh effect that MWL can cause is body weight gain. Body weight gain is defined as overweight or obesity. It is characterized by an excessive increase in adipose tissue (Sarikaya and Gökçe 2024). It is considered a disease of civilization that has become one of the significant public health problems worldwide (Switala and Leonska-Duniec 2021). There is little literature on the impact of MWL on body weight gain in teachers from 2020 to date. In one study, Silva et al. (2023) mention that teachers have to work long hours, which increases their MWL and, therefore, the probability of presenting obesity problems. In another study, in which MWL was not considered, Ramos-Valencia, Buitron-Gonzalez, Aristizábal-Grisales, and Villaquiran-Hurtado (2023) conducted a study to determine the prevalence and factors associated with obesity and overweight among university faculty members in Colombia, selecting 256 members who were administered a survey on sociodemographic data, knowledge of obesity and overweight, habits, and anthropometric measurements. Their results indicated that 47.7% of the university faculty members were classified as overweight. Similarly, Corbett, Phongsavan, Okely, Peralta, and Bauman (2024) conducted a cross-sectional study examining the lifestyle behaviours of 1,136 teachers in Australia and assessed the association between work-related factors and lifestyle behaviours. Of the total number of teachers, 53% were overweight or obese. The authors found that hours worked and teaching load were associated with obesity.

Based on the above, hypothesis H_7 is proposed.

H_7: MWL directly and positively affects body weight gain in university faculty members.

14.2.3 Effects of Stress on University Faculty Members

Currently, the teaching profession is recognized as the most stressful job in the world (Pau et al. 2022). Numerous studies mention that stress can cause headaches in people. For example, Suleman, Khattak, and Hussain (2021) conducted a study to examine job stress and its symptoms among 101 secondary school principals in Pakistan. Similarly, Khalifa, Khalaf, and Mohammed (2022) conducted a study to determine the prevalence of job stress in 106 teachers and to detect the prevalence of physical symptoms. The results indicated that all teachers complained of stress, mainly of a moderate level (55.7%), and that there was a significant relationship between stress and headaches experienced by teachers. Another study assessed the health status of 519 teachers in the Republic of Moldova regarding working conditions. The results indicated that 507 (97.7%) teachers suffered from headaches (Cheptea 2021). Another similar study investigated whether the high vocal demands specific to the teaching profession are associated with the mental health of 152 female kindergarten teachers in Germany. The results showed that the educators felt stressed by the excessive number of work tasks and children in each class, the noisy working environment, and loud conversations (Darius, Voigt-Zimmermann, and Böckelmann 2023). Muawanah, Muzayanah, Pandin, Alam, and Trisnaningtyas (2023) mention that the COVID-19 pandemic changed many learning processes, such as distance learning. This caused stress among the teachers. Therefore, these authors conducted a study with 1,520 teachers to describe the mental health problems during the COVID-19 pandemic. The results indicated that teachers experienced at least three types of stress at different levels. Physically, online learning methods were more strenuous for teachers than face-to-face ones. The reasons include the requirement to stare at a screen for extended periods, a heavier workload, and unclear distribution of office hours due to the work-from-home policy. This manifested itself in various physical symptoms, such as headaches.

Darius, Voigt-Zimmermann, and Böckelmann (2023) studied the association between job demands in early childhood educators and their mental health. They found that the educators were stressed and suffered from headaches. Similarly, Commey-Mintah et al. (2022) examined the effects of stress on 150 trainee teachers in Ghana. Their results showed that stress caused headaches in teachers.

From this background, hypothesis H_8 emerges.

H_8: Stress directly and positively affects headaches in university faculty members.

Another effect of stress is irritability, that is defined as "the quality or state of being irritable, which includes (a) rapid excitability to annoyance, impatience, or anger: Petulance and (b) abnormal or excessive excitability of an organ or part of the body" (Saatchi, Olshansky, and Fortier 2023). Literature mentions that teaching is a challenging profession for the emotional health of teachers, who experience high levels of job stress, which manifests itself in several mental disorders, including irritability (Emeljanovas et al. 2023).

Ruiz-Ruano García et al. (2025) conducted research in which they performed a cluster analysis of a sample of 257 teachers to find groups that differed in their burnout profiles. Subsequently, the authors determined the stress symptoms for each group. The results showed that irritability was the central symptom of stress in teachers suffering from burnout. Similarly, Herrera Soria, Trávez Osorio, Cangui Basantes, and Jiménez Zambrano (2024) analysed different symptoms of job stress in Ecuadorian teachers, including irritability. The results indicated that irritability had a mean of 3.1 on a range of 1 to 5. With this, the authors concluded that irritability is a common symptom in stress-affected teachers, affecting their professional and personal lives. In another study, Toohey (2020) identified conceptualizations of specific characteristics of irritability. He administered an irritability questionnaire to 402 participants and found that the fourth most common cause of irritability was stress from feeling pressured and being in a hurry.

Hypothesis H_9 follows from the above.

H_9: Stress directly and positively affects irritability in university faculty members.

A third effect of stress may be mental fatigue. As the teaching profession is currently recognized as the most stressful job in the world, it is expected that university faculty members who experience high levels of job stress would negatively affect their mental health (Pau et al. 2022). Agyapong et al. (2022) mention that chronic stress can lead to mental fatigue among teachers. In contrast, Kunasegaran et al. (2023) mention that stress can increase mental fatigue when performing a task of high cognitive and attentional demand. An example of this is the study by Ormiston, Nygaard, and Apgar (2022), who conducted a systematic review of the literature to answer the question: How have compassion fatigue (CF) (i.e., emotional fatigue) and secondary traumatic stress (STS) in teachers been studied? Qualitative analysis of the data led to the conclusion that when teachers serve children with traumatic backgrounds, they are at risk for developing CF and STS. In another similar study, Oberg, Carroll, and Macmahon (2023) mention that in Australia, more and more teachers do not want to remain in the profession, with burnout being one of the contributing factors. These authors also mention that teachers working with students with traumatic backgrounds experience increased levels of CF, STS, and burnout.

In another research, Dilekçi, Kaya, and Çiçek (2025) investigated the effects of job stress, burnout, and change fatigue (i.e., mental and emotional fatigue due to constant organizational changes) on silent job abandonment among 366 teachers. Their results suggested positive and significant relationships between job stress, burnout, change fatigue, and silent quitting. Similarly, Pau et al. (2022) conducted a study to characterize the mental health and well-being of 776 Malaysian secondary school teachers. For this purpose, emotional exhaustion, sources of stress, and their manifestations were assessed. The analysis revealed that most teachers had experienced job stress, which manifested as mental fatigue. Finally, Zhang et al. (2021) applied a potential marker of Rényi entropy to investigate the changes in mental fatigue in a long-term high-level stress condition. Their study showed that mental fatigue accumulates during long-term high-level stress situations.

From this background, hypothesis H_{10} is proposed.

H_{10}: Stress directly and positively affects mental fatigue in university members.

On the other hand, multiple studies relate stress to depression. Agyapong et al. (2022) mention that worldwide, stress is a problem that causes depression in teachers. These authors conducted a study to determine the extent of the current literature on the prevalence and correlations of stress and depression among teachers. When only clinically significant psychological conditions among teachers were considered, the prevalence of stress ranged from 8.3% to 87.1%, and that of depression ranged from 4% to 77%. Similarly, Cao et al. (2023) mention that because teachers are in highly stressful roles, they face increasing depression. Agyapong et al. (2022) note that worldwide, stress remains a problem among teachers, leading to anxiety and depression.

It is well known that, during the COVID-19 pandemic, teachers were forced to suspend their face-to-face classes and replace them with online teaching and homeschooling. These factors, along with concurrent family responsibilities, increased their stress levels and mental health problems (Lacomba-Trejo et al. 2022). A study with 614 Chilean teachers determined the impact of COVID-19-related concerns on teachers' emotional symptoms. The results indicated that teachers showed high levels of stress and depression. Specifically, 46.60% suffered from stress, and 23.20% suffered from depressive symptoms (Lacomba-Trejo et al. 2022). Similarly, Ma et al. (2022) conducted a comprehensive systematic review to determine the prevalence of stress and depression among teachers during the COVID-19 outbreak and to identify factors associated with these domains of teachers' psychological well-being. Their study included 54 studies synthesizing data from 256,896 teachers from 22 countries. Their results showed a higher prevalence of stress (62.6%, 95% confidence interval [CI]: 46.1–76.6) compared to depression (59.9%, 95% CI: 43.4–74.4) among teachers. Finally, Mendes Rodrigues et al. (2020) analysed the presence of stress and signs indicative of depression in 163 teachers. They found that there was a moderate positive correlation between the scales, so teachers are subject to stress and depression.

This background allows the formulation of hypothesis H_{11}.

H_{11}: Stress has a direct and positive effect on depression in university faculty members.

In addition, there are also studies indicating a possible relationship between stress and anxiety. For example, Agyapong et al. (2022) point out that stress is a problem that causes anxiety in teachers worldwide. Similarly, Jakubowski and Sitko-Dominik (2021) state that teaching is a stressful job that involves experiencing negative emotions, such as anxiety. Moreover, Zheng et al. (2022) mention that, during the COVID-19 pandemic, the operation of the technology presented a challenge for professors as few of them knew how to operate it accurately and some needed training, which was not possible due to the coronavirus. This caused university professors to feel stressed and anxious during the COVID-19 pandemic.

Pressley, Ha, and Learn (2021) conducted a study exploring the association between stress and anxiety in 329 elementary teachers across the United States during the COVID-19 pandemic. The results indicated that stress was a significant predictor of increased teacher anxiety, with those in virtual instruction having the most significant increase in anxiety. In another study, Kamal, Eldein Ahmed, Mohamed Wassif, and Ali Awad Greda (2021) measured the frequency of job stress and its correlation with anxiety in 400 public school teachers in Egypt. The results showed significant positive correlations between stress and anxiety. Similarly, Santamaría, Mondragon, Santxo, and Ozamiz-Etxebarria (2021) analysed stress and anxiety levels in 1,633 teachers in northern Spain, from early childhood education to university studies. The results showed that a high percentage of teachers present symptoms of stress and anxiety.

Based on the above, hypothesis H_{12} is proposed.

H_{12}: Stress has a direct and positive effect on anxiety in university faculty members.

A sixth effect that stress can cause is body weight gain. Lestari and Evanov (2021) mention that teaching is one of the professions that has a high risk of work stress due to the many demands of the job, duties, and responsibilities and that such a level of stress favours the appearance of obesity caused by changes in behaviour and metabolism. Therefore, these authors conducted a study to analyse the relationship between stress and obesity in 63 secondary school teachers. The results showed that most respondents presented a moderate stress level with a healthy body mass index

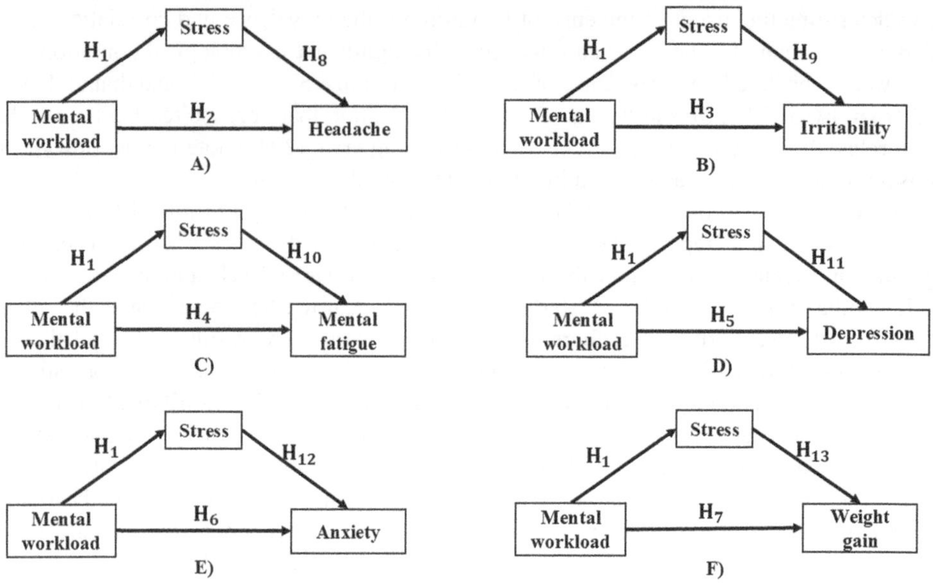

FIGURE 14.1 Hayes' models 4: (a) Model A, (b) Model B, (c) Model C, (d) Model D, (e) Model E, and (f) Model F.

(BMI) category. No significant relationship was found between job stress and obesity in teachers. In another study, Silverman and Wang (2021) mention that school closures during the COVID-19 pandemic led teachers to a more sedentary teaching environment, which, along with pandemic-related stress, could be correlated with weight gain. These authors conducted a study of 129 teachers, of whom 52% reported weight gain, with a higher prevalence observed among kindergarten and elementary school teachers compared to secondary school teachers.

Similarly, Zagkas et al. (2023) conducted a study to demonstrate the relevance of stress in teachers. To do so, they examined data from 786 Greek primary school teachers. The results showed that Greek primary school teachers experience significant stress. In terms of BMI categories, most participants were in the "normal weight" category (57.9%); however, there were significant percentages of teachers in the "overweight" and "obese" categories (28.8% and 11.2%, respectively). Finally, they evaluated whether psychological stress and anthropometric and clinical variables influence resting heart rate variability (HRV) in 65 teachers. Results indicated that parasympathetic parameters of HRV, representing good physical and mental health, are positively associated with stress symptom reduction components.

This background allows us to formulate hypothesis H_{13}.

H_{13}: Stress directly and positively affects weight gain in university faculty members.

Figure 14.1 shows Hayes' model 4 with the corresponding hypotheses.

14.3 MATERIALS AND METHODS

14.3.1 Development and Administration of the NASA-TLX Questionnaire

In this research, in addition to the 21 questions of the NASA-TLX questionnaire, 18 more questions were added to collect sociodemographic information and to associate chronic health problems in university faculty members. The questionnaire was divided into three sections: (a) Sociodemographic data (gender, age, marital status, number of children, among others), (b) Evaluation of MWL with the NASA-TLX method, and (c) Effects of MWL on university faculty members (stress, headache, mental fatigue, irritability, anxiety, and increased body weight).

This last section applied a five-point Likert-type scale (1 = Never, 2 = Seldom, 3 = Sometimes, 4 = Almost always, and 5 = Always).

The questionnaire was applied through a Google form and sent under the direction and instruction of the head of the Industrial Engineering department. The convenience method was used. Participation was voluntary, and the university faculty members had 1 week to answer it.

SPSS Statistics® 25 software was used to validate the questionnaire, which helped evaluate its internal consistency through Cronbach's alpha. For an item to be retained in the questionnaire and subsequent statistical analyses, it had to have a Cronbach's alpha value greater than or equal to 7. The introduction to the questionnaire states that, by answering the questionnaire, the university faculty member permits their responses to be used for research and publication. The questionnaire is in Spanish at the following link: https://forms.gle/R1rDc7oUJPM5pwHDA.

14.3.2 MWL Calculation

Once the questionnaire was validated, the MWL of the university faculty members was evaluated using the NASA-TLX method (Section 2 of the questionnaire). The dimensions of mental demand, physical demand, temporal demand, performance, effort, and frustration were evaluated individually for each respondent. Subsequently, 15 pairwise comparisons between dimensions were performed, and the dimension that contributed most to increasing university faculty members' MWL was selected for each paired comparison. University faculty members' weight for each dimension was obtained from these paired comparisons. This weighting (w_i in Equation (14.1)) equals the number of times the respondent selected each dimension, ranging from 0 to 15.

Subsequently, the score d_i given to each dimension was calculated. This was obtained in the first stage, in which each university faculty member gave a score of 1 to 20 for each dimension (1 = good; 20 = bad). Once these values were calculated, Equation (14.1) was applied to calculate the MWL level for each university faculty member. Table 14.1 was used to categorize the MWL level.

14.3.3 Hypothesis Tests

The mediation analysis of Andrew F. Hayes' Model 4 was used for the hypothesis test. More specifically, the macro process of the SPSS Statistics® 25 software was used. It was decided to use Hayes' mediation analysis because the observed variables can act as both antecedent and consequent variables. Therefore, first, the MWL level of each university faculty member was taken as a basis. In Section 3, it will be sought whether there is a relationship and magnitude between MWL and those variables directly associated with it, which are stress, headache, mental fatigue, irritability, depression, anxiety, and weight gain.

Direct relationships between two observed variables were indicated by direct arrows, where a standardized β value was obtained as a measure of dependence statistically tested. Indirect associations were analysed with a mediator variable M between the antecedent variable X and the consequent variable Y. These associations occurred through two segments or stages. The estimates were interpreted as the expected change in the variable Y when the variable X

TABLE 14.1
MWL Levels Based on NASA-TLX Value

Value of MWL	Level of MWL
NASA-TLX index ≤ 33	Low
33 < NASA-TLX index < 66	Medium
NASA-TLX index ≥ 66	High

experiences a one-unit change through the influence of the mediating variable M, which in turn affects Y.

To assess the validity of each association, we used the bootstrap method, which allows us to obtain robust estimates and perform statistical tests. To do this, 95% bias-corrected, accelerated confidence intervals were created using 10,000 bootstrap samples. All associations will be significant at $p < 0.05$ if zero is not in the 95% confidence interval.

14.4 RESULTS

Table 14.2 shows the descriptive results by sociodemographic groups of the 44 university faculty members surveyed concerning the first section of the questionnaire. As can be seen, more than half of the respondents were men. Regarding age, more than 30% were over 50 years of age. On the other hand, more than 65% turned out to be married; 18.18% had no children at the time, while 15.71% had three to four children. About seniority, 50% had 0–5 years working in the institution, while 11.36% had more than 20 years. It can also be observed that 31.82% of the respondents taught more than 30 hours of class/week and 70.45% had a fee contract. Finally, concerning the modality, it can be observed that 47.73% of the respondents taught face-to-face, while 27.27% taught semi-face-to-face.

These aspects may contribute to university faculty members' MWL levels by impacting mental demand, time demand, effort, frustration, and performance. The base university faculty members had a higher MWL level, ranging from 66 to 75. The interim or honorary university faculty members had an average MWL of 60, ranging between 45 and 65.

Figure 14.2 shows the results of the hypothesis tests. The hypotheses for which a p-value less than or equal to 0.05 was obtained (marked with a continuous arrow) are not rejected, while those for which a p-value greater than 0.05 was obtained (marked with dashed arrows) are rejected. As can be seen, hypotheses H_2, H_3, H_5, H_6, and H_7 are rejected, so MWL only directly affected stress and mental fatigue. The effect of MWL on stress was $\beta=0.0154$, with a p-value$=0.0036$, while the effect on mental fatigue was $\beta=0.0117$, with a p-value$=0.027$.

For its part, stress directly affected all variables. The most significant variable was body weight gain, with a value of $\beta=0.7459$ and a p-value$=0.0019$. The smallest effect of stress was on depression, with $\beta=0.3132$ and a p-value$=0.0261$. All R^2 values were more significant than 0.02, so there is sufficient predictive validity, and the models can be considered appropriate (Díaz-Reza et al. 2024).

Table 14.3 shows the variables with which MWL had a significant indirect association, with stress as the mediating variable.

TABLE 14.2
Descriptive Results

Genre		Age (years)		Marital Status	Number of Children	
Man	25 (56.82%)	25–35	8 (18.18%)	Single 12 (27.27%)	0	8 (18.18%)
Woman	19 (43.18%)	36–40	5 (11.36%)	Married 29 (65.91%)	1	9 (20.45%)
		41–45	12 (27.27%)	Free union 3 (6.82%)	2	20 (45.45%)
		46–50	5 (11.36%)		3–4	7 (15.91%)
		>50 14	(31.82%)			
Seniority (years)		**Class-Hours/Week**		**Type of Contract**	**Modality**	
0–5	22 (50%)	1–10	10 (22.73%)	Base 11 (25%)	On-site 21 (47.73%)	
6–10	8 (18.18%)	11–20	13 (29.54%)	Interim 2 (4.55%)	Semi-face-to-face 11 (25%)	
11–15	9 (20.45%)	21–30	7 (15.91%)	Fees 31 (70.45%)	Both 12 (27.27%)	
16–20	0	>30	14 (31.82%)			
>20	5 (11.36%)					

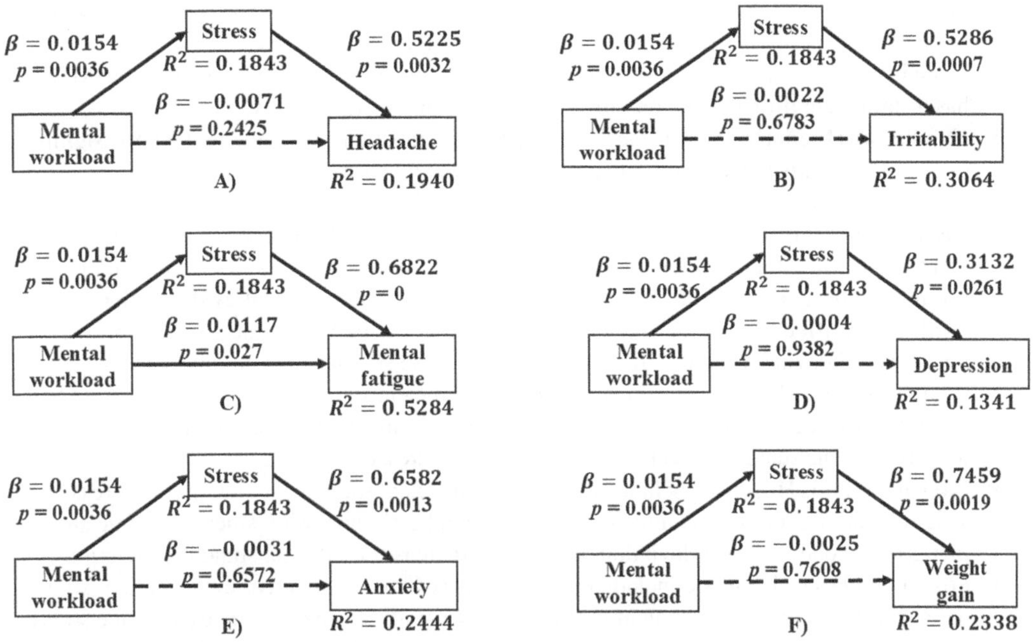

FIGURE 14.2 Results of the hypothesis tests for each model.

TABLE 14.3
Variables with Which MWL was Indirectly Associated

Model	Indirect Association
A	Headache
B	Irritability
C	Mental fatigue
D	Depression
E	Anxiety
F	Weight gain

14.5 CONCLUSIONS AND RECOMMENDATIONS

After analysing the results, we conclude the following:

- From hypothesis H_1, it is concluded that there is sufficient statistical evidence that MWL has a significant direct effect on stress since every time MWL increases by one standard deviation, stress increases by 0.0154 standard deviations.
- From hypothesis H_2, it is concluded that there is not enough statistical evidence that MWL has a significant direct effect on university faculty members' headaches.
- From hypothesis H_3, it is concluded that there is not enough statistical evidence that MWL has a significant direct effect on university faculty members' irritability.
- From hypothesis H_4, it is concluded that there is sufficient statistical evidence that MWL has a significant direct effect on university faculty members' mental fatigue since every time MWL increases by one standard deviation, mental fatigue increases by 0.0117 standard deviations.

- From hypothesis H_5, it is concluded that there is not enough statistical evidence that MWL has a significant direct effect on university faculty members' depression.
- From hypothesis H_6, it is concluded that there is not enough statistical evidence that MWL has a significant direct effect on university faculty members' anxiety.
- Hypothesis H_7 concludes that there is insufficient statistical evidence that MWL significantly affects university faculty members' body weight gain.
- From hypothesis H_8, it is concluded that there is sufficient statistical evidence that stress has a significant direct effect on university faculty members' headaches as each time stress increases by one standard deviation, headache increases by 0.5225 standard deviations.
- From hypothesis H_9, it is concluded that there is sufficient statistical evidence that stress significantly affects university faculty members' irritability. Each time stress increases by one standard deviation, irritability increases by 0.5286 standard deviations.
- From hypothesis H_{10}, it is concluded that there is sufficient statistical evidence that stress has a significant direct effect on university faculty members' mental fatigue since each time stress increases by one standard deviation, mental fatigue increases by 0.6822 standard deviations.
- From hypothesis H_{11}, it is concluded that there is sufficient statistical evidence that stress has a significant direct effect on university faculty members' depression since each time stress increases by one standard deviation, depression increases by 0.3132 standard deviations.
- From hypothesis H_{12}, it is concluded that there is sufficient statistical evidence that stress has a significant direct effect on university faculty members' anxiety since each time stress increases by one standard deviation, anxiety increases by 0.6582 standard deviations.
- From hypothesis H_{13}, it is concluded that there is sufficient statistical evidence that stress has a significant direct effect on university faculty members' body weight gain since each time stress increases by one standard deviation, body weight gain increases by 0.57459 standard deviations.

Based on the results and conclusions, the following is recommended:

- Extend the research to other institutional departments and increase the sample size.
- Periodically monitor and record the mental state of university faculty members, asking them if they suffer from MWL, stress, headache, irritability, anxiety, mental fatigue, depression, or weight gain.
- Implement improvements to reduce MWL, such as having more base university faculty members and giving university faculty members no more than three class groups.
- Implement practices to reduce stress, headache, irritability, anxiety, mental fatigue, depression, or weight gain. Such practices may include regular meditation, physical exercise, promoting water consumption, and other nonwork activities.
- Perform an analysis with structural equation models to include latent variables with several items each.
- Analyse the effect of metal workload on sustainability, especially on the social dimension.
- Include items related to the type of contract (permanent, temporary, and freelance), the time devoted to work (full-time and part-time), and whether or not university faculty members regularly conduct research. The aim is to understand the effect of these conditions on university faculty members' MWL.

14.6 LIMITATIONS

The research has some limitations, which are mentioned below:

1. The study was conducted with a relatively small sample of 44 university faculty members. Therefore, it is recommended that the sample size be increased.

2. The sampling method used was convenience sampling. Therefore, it is recommended that an investigation be carried out using random sampling.
3. The study was conducted with university faculty members from the same department in a single institution. It is recommended that the study be carried out with university faculty members from different departments and different institutions.
4. The hypothetical models include only observed variables. A study with latent variables is recommended.
5. The study does not differentiate between basic, interim, or honorary university faculty members by the number of classes assigned nor between university faculty members conducting research and those not. This leaves the opportunity to research the differentiation of MWL levels among these categories.
6. No distinction is made between the number of university faculty members who perform only undergraduate activities and those who teach graduate classes and conduct research activities in addition to teaching undergraduate classes.
7. In this study, the data for the Depression variable are self-reported. Therefore, it is recommended to be evaluated through a diagnosis, i.e., clinical evaluation.

The results of this study are similar to those obtained by Akca and Küçükoğlu (2022), Pace et al. (2021), and Sakarji et al. (2024), in that all of these studies confirmed that high levels of mental load have adverse effects on university faculty members' well-being.

REFERENCES

Agyapong, Belinda, Gloria Obuobi-Donkor, Lisa Burback, and Yifeng Wei. 2022. "Stress, Burnout, Anxiety and Depression among Teachers: A Scoping Review." *International Journal of Environmental Research and Public Health* 19 (17): 10706. https://doi.org/10.3390/ijerph191710706.

Aigerim, Sharipova, and Yegenissova Almazhay. 2024. "How Does Teacher Emotional Burnout Syndrome Affect Teachers' Development?" *Yessenov Science Journal* 48 (1): 32–38. https://doi.org/10.56525/ugtu5532.

Akca, Meltem, and Mubeyyen Tepe Küçükoğlu. 2022. "Relationships Between Mental Workload, Burnout, and Job Performance: A Research Among Academicians." In *Research Anthology on Changing Dynamics of Diversity and Safety in the Workforce*, edited by Information Resources Management Association (IRMA), 4th ed., 877–97. Hershey, Pennsylvania: IGI Global. https://doi.org/10.4018/978-1-6684-2405-6.CH044.

Alsuraykh, Norah H., Max L. Wilson, Paul Tennent, and Sarah Sharples. 2019. "How Stress and Mental Workload Are Connected." In *PervasiveHealth'19: Proceedings of the 13th EAI International Conference on Pervasive Computing Technologies for Healthcare*, edited by Oscar Mayora, Stefano Forti, Jochen Meyer, and Lena Mamykina, 371–76. Trento, Italy: Association for Computing Machinery. https://doi.org/10.1145/3329189.3329235.

Ashipala, Daniel Opotamutale, and Anna P. K. Shilunga. 2023. "Overview and Management of Anxiety." In *Acceleration of the Biopsychosocial Model in Public Health*, edited by Simon George Taukeni, 1st ed., 194–210. Hershey, Pennsylvania: IGI Global. https://doi.org/10.4018/978-1-6684-6496-0.ch009.

Baumgartner, Neal, Matthew F. Gruse, Erin M. Flerlage, Andrea K. Rhodes, and Kimberly N. Hale. 2025. "A Physical Demands Analysis to Identify and Delineate Critical Physical Tasks for Physically Demanding US Air Force Occupations." *Work* 80 (1): 1–19. https://doi.org/10.3233/WOR-230385.

Bolton, Matthew L., Elliot Biltekoff, and Laura Humphrey. 2023. "The Mathematical Meaninglessness of the NASA Task Load Index: A Level of Measurement Analysis." *IEEE Transactions on Human-Machine Systems* 53 (3): 590–99. https://doi.org/10.1109/THMS.2023.3263482.

Brunzini, Agnese, Margherita Peruzzini, Fabio Grandi, Riccardo Karim Khamaisi, and Marcello Pellicciari. 2021. "A Preliminary Experimental Study on the Workers' Workload Assessment to Design Industrial Products and Processes." *Applied Sciences* 11 (24): 12066. https://doi.org/10.3390/APP112412066.

Cabezas-Heredia, Edmundo, Luis Cabezas-Chávez, Melanie Cabezas-Chávez, Ana Paula Molina-Recalde, Diana Vallejo-Tixi, and Juan Carlos SantillánLima. 2024. "Carga Mental En Docentes Universitarios: Estudio de Caso." *Tesla Revista Científica* 4 (2). https://doi.org/10.55204/TRC.V4I2.E424.

Cadena-Povea, Henry, Marco Hernández-Martínez, Gabriela Bastidas-Amador, and Josué Calderón-Muñoz. 2025. "Perceived Stress: Psychosocial-Sociodemographic Factors as Predictors of Tension, Irritability, and Fatigue Among Ecuadorian University Professors." *International Journal of Environmental Research and Public Health* 22 (1): 107. https://doi.org/10.3390/IJERPH22010107.

Cao, Cui Hong, Xiao Ling Liao, Xing Yong Jiang, Xu Dong Li, I. Hua Chen, and Chung Ying Lin. 2023. "Psychometric Evaluation of the Depression, Anxiety, and Stress Scale-21 (DASS-21) among Chinese Primary and Middle School Teachers." *BMC Psychology* 11 (1): 1–16. https://doi.org/10.1186/S40359-023-01242-Y.

Carroll, Annemaree, Kylee Forrest, Emma Sanders-O'Connor, Libby Flynn, Julie M. Bower, Samuel Fynes-Clinton, Ashley York, and Maryam Ziaei. 2022. "Teacher Stress and Burnout in Australia: Examining the Role of Intrapersonal and Environmental Factors." *Social Psychology of Education* 25 (2–3): 441–69. https://doi.org/10.1007/S11218-022-09686-7.

Cheptea, Dumitru. 2021. "Assessment of Teachers' Health in Relation to Working Conditions." *The Moldovan Medical Journal* 64 (5): 51–55. https://doi.org/10.52418/MOLDOVAN-MED-J.64-5.21.10.

Christensen, Jan Olav, and Stein Knardahl. 2012. "Work and Headache: A Prospective Study of Psychological, Social, and Mechanical Predictors of Headache Severity." *PAIN®* 153 (10): 2119–32. https://doi.org/10.1016/J.PAIN.2012.07.009.

Colin, Catherine, Violaine Prince, Jean-Luc Bensoussan, and Marie-Christine Picot. 2023. "Music Therapy for Health Workers to Reduce Stress, Mental Workload and Anxiety: A Systematic Review." *Journal of Public Health* 45 (3): e532–e541. https://doi.org/10.1093/pubmed/fdad059.

Comezaquira-Reay, Ana Candelaria, Guillermo Terán-Ángel, and Pedro José Pedro José Quijada-Martínez. 2021. "Carga Laboral y Obstáculos de Rendimiento Del Personal de Enfermería En La Unidad de Cuidados Intensivos." *Revista Cubana de Enfermería* 37 (4): e3942–62. https://scielo.sld.cu/pdf/enf/v37n4/1561-2961-enf-37-04-e3942.pdf

Commey-Mintah, Priscilla, Florence Christianah Awoniyi, Millicent Narh-Kert, Cecilia Tutu-Danquah, and Kwaku Darko Amponsah. 2022. "Effects of Stress and Coping Mechanisms: The Case of Pre-Service Teachers in Selected Colleges of Education in Ghana." *American Journal of Educational Research* 10 (8): 491–500.

Corbett, Lucy, Philayrath Phongsavan, Anthony D Okely, Louisa R Peralta, and Adrian Bauman. 2024. "A Cross-Sectional Study of Australian Teachers' Health: Are Work-Related Factors Associated with Lifestyle Behaviours?" *Health Promotion International* 39 (1): 1–12. https://doi.org/10.1093/heapro/daad192.

Darius, Sabine, Susanne Voigt-Zimmermann, and Irina Böckelmann. 2023. "Effects of Occupation-Specific Vocal Stress on the Mental Health of Day Care Teachers." *Folia Phoniatrica et Logopaedica* 75 (5): 306–15. https://doi.org/10.1159/000530283.

De Carvalho Leite Peres, Fabíola, and Ruben George Oliven. 2023. "On the One Hand, Excessive Workload; on the Other, Lack of Concentration: The Relationship between Teachers and Students from Private and Public School in Porto Alegre, Brazil, in Emergency Remote Teaching during the COVID-19 Pandemic." *Artigos* 48 (3): 92–111. https://doi.org/10.4000/AA.11436.

Desouky, Dalia, and Heba Allam. 2017. "Occupational Stress, Anxiety and Depression among Egyptian Teachers." *Journal of Epidemiology and Global Health* 7 (3): 191–98.

Dethier, Erik, Gunnar Stevens, and Alexander Boden. 2024. "The Burden of Administrative Household Labor—Measuring Temporal Workload, Mental Workload, and Satisfaction." *Social Sciences* 13 (8): 404. https://doi.org/10.3390/SOCSCI13080404.

Díaz-Reza, José Roberto, Jorge Luis García-Alcaraz, Cuauhtémoc Sánchez-Ramírez, and Arturo Realyvásquez Vargas. 2024. "Assessing the Impact of Lean Manufacturing on the Social Sustainability through Structural Equation Modeling and System Dynamics." *Jordan Journal of Mechanical and Industrial Engineering* 18 (1): 113–30. https://doi.org/10.59038/jjmie/180109.

DiDomenico, Angela, and Maury A. Nussbaum. 2008. "Interactive Effects of Physical and Mental Workload on Subjective Workload Assessment." *International Journal of Industrial Ergonomics* 38 (11–12): 977–83. https://doi.org/10.1016/J.ERGON.2008.01.012.

DiDomenico, Angela, and Maury A. Nussbaum. 2011. "Effects of Different Physical Workload Parameters on Mental Workload and Performance." *International Journal of Industrial Ergonomics* 41 (3): 255–60. https://doi.org/10.1016/J.ERGON.2011.01.008.

Dilekçi, Ümit, Ayça Kaya, and Ilhan Çiçek. 2025. "Occupational Stress, Burnout, and Change Fatigue as Predictors of Quiet Quitting among Teachers." *Acta Psychologica* 254: 104812–18. https://doi.org/10.1016/j.actpsy.2025.104812.

Emeljanovas, Arūnas, Stanislav Sabaliauskas, Brigita Mežienė, and Natalja Istomina. 2023. "The Relationships between Teachers' Emotional Health and Stress Coping." *Frontiers in Psychology* 14 (November): 1276431. https://doi.org/10.3389/FPSYG.2023.1276431

Ghosn, Farah, Manuel Perea, Rosa Sahuquillo-Leal, Alba Moreno-Giménez, Belén Almansa, Pablo Navalón, Máximo Vento, and Ana García-Blanco. 2023. "The Effects of Reward and Frustration on the Task Performance of Autistic Children and Adolescents." *Research in Developmental Disabilities* 140 (September): 104567. https://doi.org/10.1016/J.RIDD.2023.104567.

González-Palacios, Yanni Lonnette, Paula Andrea Ceballos-Vásquez, and Flérida Rivera-Rojas. 2021. "Mental Workload in Faculty and Consequences in Their Health: An Integrative Review." *Cadernos Brasileiros de Terapia Ocupacional* 29 (September): e2808. https://doi.org/10.1590/2526-8910.CTOAR21232808.

Gyorffy, Zsuzsa, Diana Dweik, and Edmond Girasek. 2016. "Workload, Mental Health and Burnout Indicators among Female Physicians." *Human Resources for Health* 14 (12): 1–10. https://doi.org/10.1186/s12960-016-0108-9.

Hakim, Hassen, Aymen Khemiri, Oussema Gaied Chortane, Samia Boukari, Sabri Gaied Chortane, Antonino Bianco, Santo Marsigliante, Antonino Patti, and Antonella Muscella. 2022. "Mental Fatigue Effects on the Produced Perception of Effort and Its Impact on Subsequent Physical Performances." *International Journal of Environmental Research and Public Health* 19 (17): 10973. https://doi.org/10.3390/IJERPH191710973.

Hart, Sandra G., and Lowell E. Staveland. 1988. "Development of NASA-TLX (Task Load Index): Results of Empirical and Theoretical Research." *Advances in Psychology* 52 (January): 139–83. https://doi.org/10.1016/S0166-4115(08)62386-9.

Heffernan, Amanda, David Bright, Misol Kim, Fiona Longmuir, and Bertalan Magyar. 2022. "'I Cannot Sustain the Workload and the Emotional Toll': Reasons behind Australian Teachers' Intentions to Leave the Profession." *Australian Journal of Education* 66 (2): 196–209. https://doi.org/10.1177/00049441221086654.

Herrera Soria, Ruth Guadalupe, Gladys Mercedes Trávez Osorio, Sonia Verónica Cangui Basantes, and Braulio Antonio Jiménez Zambrano. 2024. "Job Stress in Ecuadorian Teachers: Factors, Symptoms and Mitigation Strategies." *Explorador Digital* 8 (3): 83–100. https://doi.org/10.33262/exploradordigital.v8i3.3081.

Hojo, Masakazu. 2021. "Association between Student-Teacher Ratio and Teachers' Working Hours and Workload Stress: Evidence from a Nationwide Survey in Japan." *BMC Public Health* 21 (1): 1–8. https://doi.org/10.1186/S12889-021-11677-W.

Hoogendoorn, Marga E., Charlotte C. Margadant, Sylvia Brinkman, Jasper J. Haringman, Jan Jaap Spijkstra, and Nicolette F. de Keizer. 2020. "Workload Scoring Systems in the Intensive Care and Their Ability to Quantify the Need for Nursing Time: A Systematic Literature Review." *International Journal of Nursing Studies* 101 (January): 103408. https://doi.org/10.1016/J.IJNURSTU.2019.103408.

Hutson, Jeremy W., Ashley E. Franklin, Beth A. Rogers, and Danielle Walker. 2024. "Psychometric Testing of NASA-TLX to Measure Learners' Cognitive Load in Individual and Group Nursing Simulations." *Clinical Simulation in Nursing* 95 (October): 101607. https://doi.org/10.1016/J.ECNS.2024.101607.

Jakubowski, Tomasz Daniel, and Magdalena Maja Sitko-Dominik. 2021. "Teachers' Mental Health during the First Two Waves of the COVID-19 Pandemic in Poland." *PLoS One* 16 (9): e0257252. https://doi.org/10.1371/JOURNAL.PONE.0257252.

Johansson, Erika, Daniel Falkstedt, and Melody Almroth. 2022. "Depression among Teachers: A Swedish Register-Based Study." *BMC Public Health* 22 (1): 1–10. https://doi.org/10.1186/S12889-022-12758-0.

Johnson, Martin, and Victoria Coleman. 2025. "Teaching in Uncertain Times: Exploring Links between the Pandemic, Assessment Workload, and Teacher Well-Being in England." *Research in Education* 121 (1): 69–92. https://doi.org/10.1177/00345237231195270.

Kamal, Abdelaziz Mohamed, Waleed Salah Eldein Ahmed, Ghada Ossama Mohamed Wassif, and Maha Hussien Ali Awad Greda. 2021. "Work Related Stress, Anxiety and Depression among School Teachers in General Education." *QJM: An International Journal of Medicine* 114 (Supplement_1): i327. https://doi.org/10.1093/QJMED/HCAB118.003.

Karanfil, Ferhat, and Mojtaba Khatami. 2021. "The Correlation between Teachers' Burnout and Workload: The Case of Iranian EFL Teachers." *The Journal of Asia TEFL* 8 (3): 1023–31. https://doi.org/10.18823/asiatefl.2021.18.3.22.1023.

Khalifa, Eman Mohamed,.Khalaf, Ossama Ossama and Mohammed, Rateba Said. 2022. "Prevalence of Occupational Stress and Depression among School Teachers." *Egyptian Journal of Occupational Medicine* 46 (1): 1–16.

Kirana, Kusuma Chandra, Ambar Lukitaningsih, and Dorce Marchiliya Londin. 2021. "Analysis of the Impact of Workload and Work Stress on Job Satisfaction with Motivation as Intervening Variable." *Jurnal Aplikasi Manajemen* 19 (4): 791–803. https://doi.org/10.21776/UB.JAM.2021.019.04.08.

Knisely, Benjamin M., Janell S. Joyner, and Monifa Vaughn-Cooke. 2021. "Cognitive Task Analysis and Workload Classification." *MethodsX* 8 (January): 101235. https://doi.org/10.1016/J.MEX.2021.101235.

Kosch, Thomas, Jakob Karolus, Johannes Zagermann, Harald Reiterer, Albrecht Schmidt, and Paweł W. Woźniak. 2023. "A Survey on Measuring Cognitive Workload in Human-Computer Interaction." *ACM Computing Surveys* 55 (13s): 1–39. https://doi.org/10.1145/3582272.

Kreuzfeld, Steffi, Christoph Felsing, and Reingard Seibt. 2022. "Teachers' Working Time as a Risk Factor for Their Mental Health - Findings from a Cross-Sectional Study at German Upper-Level Secondary Schools." *BMC Public Health* 22 (1): 1–12. https://doi.org/10.1186/s12889-022-12680-5.

Kunasegaran, K, AMH Ismail, S Ramasamy, JV Gnanou, BA Caszo, and PL Chen. 2023. "Understanding Mental Fatigue and Its Detection: A Comparative Analysis of Assessments and Tools." *PeerJ* 11: e15744. https://doi.org/10.7717/peerj.15744.

Lacomba-Trejo, Laura, Konstanze Schoeps, Selene Valero-Moreno, Constanza del Rosario, and Inmaculada Montoya-Castilla. 2022. "Teachers' Response to Stress, Anxiety and Depression During COVID-19 Lockdown: What Have We Learned From the Pandemic?" *Journal of School Health* 92 (9): 864–72. https://doi.org/10.1111/JOSH.13192.

Lestari, Nina Dwi, and Altori Samantha Evanov. 2021. "The Correlation Between Work-Stress Level and Obesity Among Teachers." In *Proceedings of the International Conference on Health and Medical Sciences (AHMS 2020)*, 195–99. Yogyakarta, Indonesia: Atlantis Press. https://doi.org/10.2991/AHSR.K.210127.043.

Longo, Luca, and Giuliano Orrú. 2022. "Evaluating Instructional Designs with Mental Workload Assessments in University Classrooms." *Behaviour & Information Technology* 41 (6): 1199–1229. https://doi.org/10.1080/0144929X.2020.1864019.

Longo, Luca, Christoper D. Wickens, Gabriela M. Hancock, and P. A. Hancock. 2022. "Human Mental Workload: A Survey and a Novel Inclusive Definition." *Frontiers in Psychology* 13 (June): 883321. https://doi.org/10.3389/FPSYG.2022.883321.

Ma, Kang, Luyao Liang, Muhammad Chutiyami, Sandy Nicoll, Teguh Khaerudin, and Xuan Van Ha. 2022. "COVID-19 Pandemic-Related Anxiety, Stress, and Depression among Teachers: A Systematic Review and Meta-Analysis." *Work* 73 (1): 3–27. https://doi.org/10.3233/WOR-220062.

Malakoutikhah, Mahdi, Reza Kazemi, Hadiseh Rabiei, Moslem Alimohammadlou, Asma Zare, and Soheil Hassanipour. 2021. "Comparison of Mental Workload with N-Back Test: A New Design for NASA-Task Load Index Questionnaire." *International Archives of Health Sciences* 8 (1): 13. https://doi.org/10.4103/IAHS.IAHS_126_20.

Martínez-García, Laura Leidys. 2021. "Propuesta de Procedimiento Para Evaluar Puestos de Trabajo Con Elevada Demanda Cognitiva En El Ministeria de Trabajo, Municipio Cárdenas." Matanzas: Universidad de Matanzas. Facultad de Ciencias Empresariales. https://rein.umcc.cu/handle/123456789/938.

Mendes Rodrigues, Louise Tatiana, Eliana Campelo Lago, Camila Aparecida Pinheiro Landim Almeida, Ivonizete Pires Ribeiro, and Gerardo Vasconcelos Mesquita. 2020. "Stress and Depression in Teachers from a Public Education Institution." *Enfermería Global* 57: 232–42. https://doi.org/10.6018/eglobal.19.1.383201.

Mendoza, Norman B., Zi Yan, and Ronnel B. King. 2023. "Supporting Students' Intrinsic Motivation for Online Learning Tasks: The Effect of Need-Supportive Task Instructions on Motivation, Self-Assessment, and Task Performance." *Computers & Education* 193 (February): 104663. https://doi.org/10.1016/J.COMPEDU.2022.104663.

Mohammadian, Mostafa, Hosein Parsaei, Hamidreza Mokarami, and Reza Kazemi. 2022. "Cognitive Demands and Mental Workload: A Field Study of the Mining Control Room Operators." *Heliyon* 8 (2): e08860. https://doi.org/10.1016/j.heliyon.2022.e08860.

Muawanah, Siti, Umi Muzayanah, Moses G. R. Pandin, Mochamad D. S. Alam, and Januari P. N. Trisnaningtyas. 2023. "Stress and Coping Strategies of Madrasah's Teachers on Applying Distance Learning During COVID-19 Pandemic in Indonesia." *Qubahan Academic Journal* 3 (4): 206–18.

Nur, Indira., Iskandar Heri., and Ade, Ryani F. 2020. "The Measurement of Nurses' Mental Workload Using NASA-TLX Method (A Case Study)." *Malaysian Journal of Public Health Medicine* 20 (Special 1): 60–63. https://doi.org/10.37268/MJPHM/VOL.20/NO.SPECIAL1/ART.705.

Oberg, Glenys, Annemaree Carroll, and Stephanie Macmahon. 2023. "Compassion Fatigue and Secondary Traumatic Stress in Teachers: How They Contribute to Burnout and How They Are Related to Trauma-Awareness." *Frontiers in Education* 8 (March): 1128618. https://doi.org/10.3389/FEDUC.2023.1128618.

Okeke, Nzubechukwu Gift, Johncross Ogbonna Okenyi, Eberechi Cyprian Ohanu, and Chidera Okeke. 2023. "Prevalence of Depression among Senior Secondary School Students in Enugu Metropolis." *MedRxiv*. University of Nigeria, Enugu Campus. https://doi.org/10.1101/2023.10.12.23296984.

Ormiston, Heather E., Malena A. Nygaard, and Sophia Apgar. 2022. "A Systematic Review of Secondary Traumatic Stress and Compassion Fatigue in Teachers." *School Mental Health* 14 (4): 802–17. https://doi.org/10.1007/S12310-022-09525-2.

Ozamiz-Etxebarria, Naiara, Nahia Idoiaga Mondragon, Juan Bueno-Notivol, María Pérez-Moreno, and Javier Santabárbara. 2021. "Prevalence of Anxiety, Depression, and Stress among Teachers during the Covid-19 Pandemic: A Rapid Systematic Review with Meta-Analysis." *Brain Sciences* 11 (9): 1172. https://doi.org/10.3390/BRAINSCI11091172.

Pace, Francesco, Giulio D'Urso, Carla Zappulla, and Ugo Pace. 2021. "The Relation between Workload and Personal Well-Being among University Professors." *Current Psychology* 40 (7): 3417–24. https://doi.org/10.1007/S12144-019-00294-X.

Pagnotta, Murillo, David M. Jacobs, Patricia L. de Frutos, Ruben Rodríguez, Jorge Ibáñez-Gijón, and David Travieso. 2022. "Task Difficulty and Physiological Measures of Mental Workload in Air Traffic Control: A Scoping Review." *Ergonomics* 65 (8): 1095–1118. https://doi.org/10.1080/00140139.2021.2016998.

Pau, Kee, Aslina Binti Ahmad, Hsin Ya Tang, Ahmad Jazimin Bin Jusoh, Asma Perveen, and Kong Kwoi Tat. 2022. "Mental Health and Well-being of Secondary School Teachers in Malaysia." *International Journal of Learning, Teaching and Educational Research* 21 (6): 50–70. https://doi.org/10.26803/IJLTER.21.6.4.

Pinar, Hacer. 2018. "Burnout among English Language Teachers in Iğdır Province." Kafkas University Institute of Social Sciences.

Pooladvand, Shiva, and Sogand Hasanzadeh. 2022. "Neurophysiological Evaluation of Workers' Decision Dynamics under Time Pressure and Increased Mental Demand." *Automation in Construction* 141 (September): 104437. https://doi.org/10.1016/J.AUTCON.2022.104437.

Pressley, Tim, Cheyeon Ha, and Emily Learn. 2021. "Teacher Stress and Anxiety During COVID-19: An Empirical Study." *School Psychology* 36 (5): 367–76. https://doi.org/10.1037/SPQ0000468.

Putra, Aditiya Anggara, Rio Chandra Pratama, and Ardana Putri Farahdiansari. 2023. "Analysis Of Mental Workload With NASA- TLX Method On Employees Of Kareb Bojonegoro Cooperative." *Journal of Information System, Technology and Engineering* 1 (3): 95–103. https://doi.org/10.61487/jiste.v1i3.28.

Rachmuddin, Y, D S Dewi, and R S Dewi. 2021. "Workload Analysis Using Modified Full Time Equivalent (M-FTE) and NASA-TLX Methods to Optimize Engineer Headcount in the Engineering Services Department." *IOP Conference Series: Materials Science and Engineering* 1072 (1): 012036. https://doi.org/10.1088/1757-899X/1072/1/012036.

Ramos-Valencia, Omar Andrés, Yuliana Buitron-Gonzalez, Juan Carlos Aristizábal-Grisales, and Andrés Felipe Villaquiran-Hurtado. 2023. "Prevalencia y Factores Asociados a Sobrepeso y Obesidad En Docentes de Una Universidad Pública de Popayán- Colombia En 2021." *Medicas UIS* 36 (1): 21–34. https://doi.org/10.18273/revmed.v36n1-2023002.

Realyvásquez-Vargas, Arturo, Jorge Luis García-Alcaraz, Karina Cecilia Arredondo-Soto, Guadalupe Hernández-Escobedo, and Yolanda Angélica Báez-López. 2023. "Effects of Mental Workload on Manufacturing Systems Employees: A Mediation Causal Model." *Work* Preprint (Preprint): 1–19. https://doi.org/10.3233/WOR-220148.

Rodio-Trevisan, Karen Rayany, Roberto Moraes Cruz, Cyntia Nunes, Daniela Ornellas Ariño, Patricia Dalagasperina, Rafaela Luiza Trevisan, and Doris Cecilia Molina Albarracín. 2024. "Evidence of Validity and Precision of the Instruments Used to Measure Mental Workload: An Integrative Review of the Literature." *Revista Interamericana de Psicología/Interamerican Journal of Psychology* 58 (1): e1894–1915. https://doi.org/10.30849/ripijp.v58i1.1894.

Ruiz-Ruano García, Ana María, Javier Rodríguez Fragoso, Enrique Javier Garcés de los Fayos Ruiz, Dyana Muñoz, Jorge López Puga, and Francisco José Moya-Faz. 2025. "Symptoms Networks in Teachers Suffering from Burnout." *Psychology, Society & Education* 17 (2): 15–25. https://doi.org/10.21071/PSE.V17I2.17810.

Saatchi, Babak, Ellen F. Olshansky, and Michelle A. Fortier. 2023. "Irritability: A Concept Analysis." *International Journal of Mental Health Nursing* 32 (5): 1193–1210. https://doi.org/10.1111/inm.13140.

Sakarji, Siti Rosnita, Abdul Kadir Othman, Bagus Shandy Narmaditya, Raja Mayang Delima Mohd Beta, and Azuati Mahmud. 2024. "Determinants of Academicians' Mental Health in Higher Education Institutions: A Systematic Literature Review." *Information Management and Business Review* 16 (1(I)S): 1–12. https://doi.org/10.22610/imbr.v16i1(I)S.3720.

Santamaría, María Dosil, Nahia Idoiaga Mondragon, Naiara Berasategi Santxo, and Naiara Ozamiz-Etxebarria. 2021. "Teacher Stress, Anxiety and Depression at the Beginning of the Academic Year during the COVID-19 Pandemic." *Global Mental Health* 8: e14. https://doi.org/10.1017/GMH.2021.14.

Sarikaya, Ecenur Esra, and Halil İbrahim Gökçe. 2024. "Investigations of Insulin Resistance in Obese Dogs." *Mediterranean Veterinary Journal* 9 (3): 364–70. https://doi.org/10.24880/meditvetj.1611164.

Shablova, Svetlana. 2024. "Studying the Anxiety Level of Teachers in Schools." *The Annals of the "Ştefan Cel Mare" University of Suceava* 17 (2): 126. https://doi.org/10.4316/SAM.2024.0213.

Silva, Claudiele C.M. da, Victor S. Beretta, Fernanda S. Gil, Leandro D. Delfino, Enrique G.F. Leite, Gerson Ferrari, William R. Tebar, and Diego G.D. Christofaro. 2023. "High Workload Is Related to Lower Quality of Life in Public School Teachers: A Cross-Sectional Study." *Work* 77 (3): 1023–29. https://doi.org/10.3233/WOR-230187.

Silverman, Jill R., and Branden Z. Wang. 2021. "Impact of School Closures, Precipitated by COVID-19, on Weight and Weight-Related Risk Factors among Schoolteachers: A Cross-Sectional Study." *Nutrients* 13 (8): 2723. https://doi.org/10.3390/NU13082723.

Suleman, Qaiser, Alam Zeb Khattak, and Ishtiaq Hussain. 2021. "Occupational Stress: Associated Factors, Related Symptoms, and Coping Strategies Among Secondary School-Heads." *Pakistan Journal of Psychological Research* 31 (4): 529–53. https://doi.org/10.33824/PJPR.2021.36.4.29.

Switala, Katarzyna, and Agata Leonska-Duniec. 2021. "Physical Activity and Gene Association with Human Obesity." *Baltic Journal of Health and Physical Activity* 13 (4): 99–111. https://doi.org/10.29359/BJHPA.13.4.10.

Tarraya, Hilger Ojos. 2023. "Teachers' Workload Policy: Its Impact on Philippine Public School Teachers." *Puissant* 5: 1–16. https://files.eric.ed.gov/fulltext/ED629465.pdf.

Toohey, Michael J. 2020. "Irritability Characteristics and Parameters in an International Sample." *Journal of Affective Disorders* 263: 558–67. https://doi.org/10.1016/j.jad.2019.11.021.

Tseng, Li-Ping, Mao-Te Chuang, and Yung-Ching Liu. 2022. "Effects of Noise and Music on Situation Awareness, Anxiety, and the Mental Workload of Nurses during Operations." *Applied Ergonomics* 99: 103633. https://doi.org/10.1016/j.apergo.2021.103633.

Valent, Paul, and Juan Moisés De La Serna. 2021. *Estrés Y Trauma En Tiempos De Pandemia*. 1st ed. Tektime. https://books.google.com.mx/books?id=ffUhEAAAQBAJ&printsec=frontcover&hl=ro&source=gbs_ge_summary_r&cad=0#v=onepage&q&f=false.

van As, Sven, Debby G.J. Beckers, Harm Veling, Wendela Hooftman, Michiel A.J. Kompier, and Sabine A.E. Geurts. 2022. "Sedentary Work and Participation in Leisure–Time Physical Activity." *International Archives of Occupational and Environmental Health* 95 (2): 509–25. https://doi.org/10.1007/S00420-021-01750-7.

Van Cutsem, Jeroen, Peter Van Schuerbeek, Nathalie Pattyn, Hubert Raeymaekers, Johan De Mey, Romain Meeusen, and Bart Roelands. 2022. "A Drop in Cognitive Performance, Whodunit? Subjective Mental Fatigue, Brain Deactivation or Increased Parasympathetic Activity? It's Complicated!" *Cortex* 155: 30–45. https://doi.org/10.1016/j.cortex.2022.06.006.

Vauhkonen, Anneli, Terhi Saaranen, Maria Cassar, Michelle Camilleri, Leandra Martín-Delgado, Elaine Haycock-Stuart, Andrea Solgajová, et al. 2024. "Professional Competence, Personal Occupational Well-Being, and Mental Workload of Nurse Educators – A Cross-Sectional Study in Four European Countries." *Nurse Education Today* 133 (February): 106069. https://doi.org/10.1016/J.NEDT.2023.106069.

Virtanen, Kai, Heikki Mansikka, Helmiina Kontio, and Don Harris. 2022. "Weight Watchers: NASA-TLX Weights Revisited." *Theoretical Issues in Ergonomics Science* 23 (6): 725–48. https://doi.org/10.1080/1463922X.2021.2000667.

Wick, Sara. 2021. "Subjectivity in Performance Evaluations: A Review of the Literature." *Accounting Perspectives* 20 (4): 653–85. https://doi.org/10.1111/1911-3838.12273.

Wolpe, Noham, Richard Holton, and Paul C. Fletcher. 2024. "What Is Mental Effort: A Clinical Perspective." *Biological Psychiatry* 95 (11): 1030–37. https://doi.org/10.1016/J.BIOPSYCH.2024.01.022.

Zagkas, Dimitrios G., George P. Chrousos, Flora Bacopoulou, Christina Kanaka-Gantenbein, Dimitrios Vlachakis, Ioanna Tzelepi, and Christina Darviri. 2023. "Stress and Well-Being of Greek Primary School Educators: A Cross-Sectional Study." *International Journal of Environmental Research and Public Health* 20 (7): 5390. https://doi.org/10.3390/IJERPH20075390.

Zhang, Han, Jingying Wang, Xinyi Geng, Chuantao Li, and Shouyan Wang. 2021. "Objective Assessments of Mental Fatigue During a Continuous Long-Term Stress Condition." *Frontiers in Human Neuroscience* 15: 733426. https://doi.org/10.3389/fnhum.2021.733426.

Zheng, Mao, Muhammad Asif, Muhammad Shahid Tufail, Saira Naseer, Shahid Ghafoor Khokhar, Xiding Chen, and Rana Tahir Naveed. 2022. "COVID Academic Pandemic: Techno Stress Faced by Teaching Staff for Online Academic Activities." *Frontiers in Psychology* 13 (July): 895371. https://doi.org/10.3389/FPSYG.2022.895371.

15 Ergonomic Approach to Restore Organizational Climate and Teamwork Potential in a Restructuring Context

Denis I. Chavez

15.1 INTRODUCTION AND CONTEXT

This chapter describes a step-by-step ergonomic intervention in a service area, which was implemented due to problems encountered during a period of transition and institutional restructuring. The organization is an Ecuadorian University, serving approximately 15,000 students and employing about 2,300 people, including 700 administrative staff and 1,500 professors and researchers, both full-time and lecturers. The main objective of the intervention was to take a comprehensive approach to addressing the work situation and human and production issues in the administrative department, which was recently created and has around 40 administrative employees. Additionally, the working environment of this department was influenced by the uncertainty associated with returning to in-person work after the pandemic and subsequent reorganization processes. This intervention implemented an integrated human factors and ergonomics approach to diagnose and propose improvements to the working conditions. This included identifying individual, organizational, and environmental factors that determine their activities and its sociotechnical impacts on the area's productivity. Particular consideration was given to the changing dynamics of the work team, the organization of the department's activity, and the formal and informal leadership structures.

This chapter is structured as follows. The first section describes the organizational context and the factors that triggered the intervention. It also presents the intervention's proposed logic, which is based on a preliminary diagnosis that reformulated the demand. The second section presents the methods and tools used throughout the intervention, including the ergonomics and human factors resources proposed to collaboratively build activity-based solutions. The third section summarizes the main findings of the intervention. The fourth section presents the intervention's overall recommendations for the institution. Finally, the fifth section discusses the advantages and limitations encountered and provides critical insights for developing the discipline and future interventions.

15.1.1 Intervention Context and Problem Setting

The department in question was created at the end of 2020 to optimize the institution's administrative services by centralizing functions that were previously dispersed among different academic programs (32 faculties and schools). Due to mobility restrictions caused by the pandemic, the department's first months of operation were conducted online. The department mainly used a series of technological platforms that the institution implemented less than a year before the pandemic. These platforms were mainly used for managing operational and administrative–financial processes, although the use of external, non-institutional tools, such as Zoom and WhatsApp, was also evident, primarily for communication. As pandemic-related constraints ceased, the department transitioned to a physical space, prompting the need to reorganize the physical space to

DOI: 10.1201/9781003582014-19

accommodate the department. To this end, the institution decided to completely refurbish an old cafeteria, spanning 400 m², located on the ground floor of one of the campus buildings. The space was fitted out so that the new department could function properly. This was done prior to the full return to in-person work by the employees. These processes of change, adaptation, and restructuring, first to a hybrid work model and later to an in-person model, triggered a series of human, social, and technological problems. After a year of operation, these issues had not been properly addressed, and the performance of the new department expected by the senior management was not achieved.

The intervention took place between February and March of 2022. It was a response to issues that the senior management expressed as urgent at the strategic planning meeting held at the end of 2021. The issues highlighted at the meeting included teamwork problems, leadership issues within the department, high staff turnover, and the resulting loss of skills and knowledge. These issues directly affected the department's service to both external customers (students) and internal customers (professors and administrative staff from other departments). These issues were evident in a number of formal and informal messages and comments.

The intervention began with a pre-diagnosis to identify and understand the department's main organizational demands of activity and enable a dialogue to renegotiate and clearly articulate the institutional demand with an adapted ergonomic intervention proposal. It is important to note that in our approach to intervention, we work from the idea of "activity" as the set of physical, psychological, emotional, and organizational commitments that the worker makes available to perform the task and which are continuously adjusted and fed back between the subject and the sociotechnical system, based on the contributions of Leplat (1993) and Rasmussen (1993).

15.1.2 Pre-Diagnosis, Problems, and Needs:

For the pre-diagnosis, we conducted a series of interviews with key stakeholders. These interviews were accompanied by a systematization of the problems and needs that reconfigured the initial demand and informed the intervention proposal.

- *Interviews with strategic actors*: In order to obtain a comprehensive view of the context, exploratory, in-depth interviews were conducted with six key figures involved in the management and functioning of the department. The interviewees included the Vice Chancellor (V), the Human Resources General Director (HR-CEO), the General Secretary (SG), the Human Resources Administration Coordinator (HR-C), a Human Resources analyst assigned to the department, and a former employee who quit the organization due to team-related issues. The interviews were recorded adding up a total corpus of 8 hours and 25 minutes. The analysis of this material presents diverse and complementary perspectives on the department's internal dynamics, personnel management challenges, and improvement expectations that could be leveraged by the intervention.
- *Systematization of problems*: Based on the information gathered during the interviews, a detailed analysis was conducted to categorize and prioritize existing problems. This exercise not only highlighted critical areas requiring immediate attention but also helped identify common patterns reflecting the tensions inherent in organizational reconfiguration and leadership changes. The pre-diagnosis laid the groundwork for defining a possible intervention roadmap aligned with the department's specific needs and the institution's strategic objectives. Combining a participatory approach allowed us to develop a more comprehensive understanding of the issues, which is essential for designing tailored and effective solutions. Below, we present the quantification of the problem and details of the systematization carried out in this pre-diagnosis (Table 15.1).

TABLE 15.1
Incidents for Category of Work

Category of Work	Incidents	Context
Information flow	4	Problems were detected in information management regarding "who has what." The hierarchical structure hinders easy access to information. There is no evidence of the logical flow of information necessary for developing *the activity*. The relationships between the analyzed department and other departments within the institution are unclear. There are no mechanisms for tracing student needs (inputs). Workers are unaware that available software tools allow for such traceability.
Organization of the activity	4	The department's management needs to group related activities or processes to seek their specialization and fluidity. Additionally, conflicts in the management of procedures have been identified, for which there may be multiple solutions. A logic of distributed group work has been identified, but it is not entirely clear to the work team.
Processes	3	The need for the department to be recognized and for its role in the institution's production process to be acknowledged is identified. It is necessary to clarify the direct relationships between the processes' inputs and outputs. Although there are several conceptually separate processes, they are connected during activity development. The logic of the department's synchronous and asynchronous activities must be determined and demonstrated.
Functions, evaluation of the activity, and psychosocial aspects	5	The functions and scope of action at the different functional levels of the department need to be formalized and clearly communicated. There is a discrepancy between the workers' and the institution's perception of the results of the department's activities (expectations vs. reality). It is necessary to identify how long it takes to fulfill requirements (inputs).
Leader competencies/tools	2	In the interviews, it is stated that there is an incongruence between the coordination activity and the competences brought into play during the organization of the team's activity. The need for a macro vision of the department's functioning is made explicit, and it is therefore requested to identify the needs to be included in the profile of a possible new coordinator. In addition, the objective is to identify the limitations of the IT tools available to the team for the management of the needs with three types of input: Counter, mail, or digital platform.

- *Systematization of needs*: The interviews revealed certain needs that should be incorporated into the intervention. The HR-CEO must articulate the "seedbed" program to establish a gateway in the initial personnel selection. This program is based on the logic of internal training that facilitates integration into the institution's productive dynamics and functioning, specifically in the academic and administrative–financial spheres. The top management requests working on work climate conflicts to improve internal and external customer service and relations between collaborators. The direct supervision line expressed the need to improve general productivity, emphasizing the resolution of student issues, including reducing errors and managing backlogs from months ago (over 2,000 unanswered requests). Another requirement was to introduce the new leader to the department's team. Finally, the HR-CEO expressed the need to highlight the potential of ergonomic and psychosocial actions as a new model for improving the workplace and the institution. The systematization of needs included information from an analysis of "critical nodes," which revealed project forecasts not addressed by other departments responsible for sustainability processes. The HR-CEO, the SG, and the HR-C participated in this analysis.

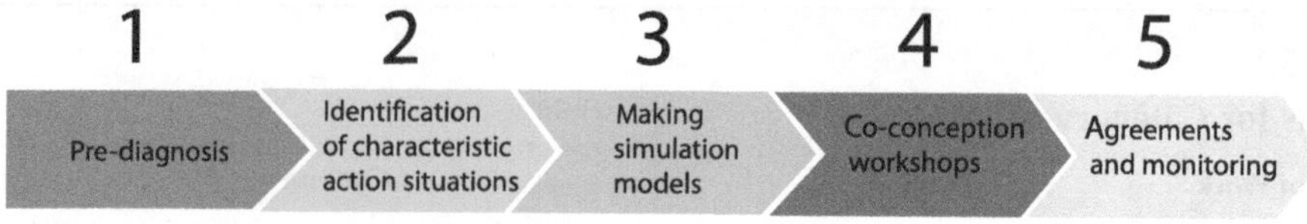

FIGURE 15.1 Intervention proposal.

Important points included the following: (a) The configuration of the physical space was not fully implemented as planned. The infrastructure area reported a delay due to the return to in-person work, which made it impossible to recover. (b) Limited use of the digital tools and systems available in the institution. The technology and information department presented statistics showing that certain tools are only marginally used by staff. These tools could generate better service controls.

15.1.3 Demand Reframing and Proposal for Ergonomic Intervention

According to the preliminary diagnosis, an ergonomic and human factor intervention is recommended. This intervention combines an approach based on activity-centered ergonomic theory ergonomics (Belleghem y Bourgeois 2003; Bourgeois y Van Belleghem 2003; Falzon 2004; Hubault 2017; Montmollin y Amalberti 1995) with psychodynamic theory of work (Caroly y Clot 2004; Clot 2015; Clot y Leplat 2005; Clot y Simonet 2015; Cuvelier et al. 2017; Dejours y Gernet 2016). This intervention aims to foster the collective aspects of work activity through a participatory approach, empowering workers to take action and devise customized solutions for their own activities (Beaujouan, Aubert, y Coutarel 2015; Folcher y Bationo-Tillon 2018; Pelayo et al. 2010; Rouat y Sarnin 2018). The intervention was proposed in three phases: identification of characteristic action situations, elaboration of supports and action scenarios, and workshops for co-construction of the activity. These phases are linked to the pre-diagnosis, *a priori*, and to the improvement monitoring activities, *a posteriori* (Figure 15.1).

15.2 METHODS AND TOOLS

A series of different methodologies and tools were proposed for each of the phases of the intervention.

15.2.1 Identification of Characteristic Action Situations

We use exploratory and systematic observation techniques to identify workers' activities and the discrepancies between reality and the task prescribed. These techniques were accompanied by exploratory and explicit interviews based on the activity analysis (Faïta 2001; Leplat 2008; Rabardel 1995; L. Van Belleghem y Bourgeois 2013). The observations and interviews were recorded using mobile phones and a notepad. The work was carried out by a team of three people: a chief ergonomist and two junior ergonomists.

This phase included 2.5 days of systematic observation of 20 workers and their workstations, accompanied by exploratory and confrontational interviews. This involved approximately 20 hours of corpus analysis and 20 hours of data processing and analysis. To obtain relevant information covering the dynamics of the area throughout the daily and weekly working day, this work was organized at different time points over the course of a week.

15.2.2 Development of Support and Performance Scenarios

Based on the data gathered, we generate a model of the department's activity and a set of characteristic actions. We used conception and design techniques (Lancry 2009; Mefteh 2018; Salembier y Pavard 2004; L. V. Van Belleghem 2018) to create supports that would allow the real activity to be expressed and visible in a simulated situation. Once the action scenarios clarified, DIY and layout techniques were employed to create the work situation in the simulation aids. We used foam board, fomix, thread, tacks, paint, markers, cardboard, and glue for this purpose.

15.2.3 Co-Construction Workshops

We developed two types of workshops:

- *Workshop on co-designing solutions*: Five of these workshops were conducted, each lasting 3 hours. Simulation techniques were employed to facilitate the co-creation of solutions (Daniellou 2007; Mefteh 2018; L. V. Van Belleghem 2018). These techniques enable participants to discuss their own activities using the designed resources, action scenarios, and previously defined characteristic situations as triggers. These workshops were recorded, both in audio and video. The technical setup used a webcam, a sound recorder, and two tripods for convenience. Three people facilitated the workshops: a lead ergonomist facilitator and two junior ergonomist workshop assistants. The assistants oversaw the material and supports, photographic recording with a Canon Rebel T7 SLR camera, video framing during the workshop, and taking written notes.
- *Workshop to share results, solutions, and define agreements*: This workshop was held at the institution's sports complex over a weekend, with no holiday charge for staff, in order to encourage commitment to attendance and a positive attitude during the process, and consisted of two 2-hour sessions: introduction and teamwork dynamics and presentation of results and agreements.

15.2.4 Analysis of Physical and Environmental Factors and Their Effect on Activity

Although the original intervention proposal did not include this analysis, the first findings and workshops suggested that an analysis of physical–environmental factors and their effects on the activity was necessary. This analysis was performed by taking samples at strategic workplace locations over a period of 7 days, divided into specific time blocks. We collected data and measured factors such as environmental noise, temperature, and air flow. The measurements were taken using a calibrated Sper Scientific 850069 Environmental Quality Meter, and the data were processed and recorded using a Toshiba Tecra A10-SP5802 laptop PC with a 9-pin serial port.

15.3 RESULTS

This section analyzes and presents the results of the different phases of the ergonomic intervention. First, we present a summary of the identification and exploration phase including its respective findings and contrast them with the initial findings of the pre-diagnosis phase. The systematization of the problems, needs, solutions, and opportunities generated during the co-construction workshops is included, as well as the logic behind the agreements reached during the socialization workshop. The methodological development and the results of the analysis of physical and environmental factors and their effects on activity are also included.

15.3.1 Identification and Exploration of Characteristic Situations

- *Activity modeling*: Figure 15.2a shows the model of activity organization constructed and validated based on interviews conducted during the pre-diagnosis phase. This model illustrates how the institution and senior management understand and attempt to manage the organization of the department's activities. Figure 15.2b, on the other hand, shows the actual organization of the activity, constructed from observations and interviews with department workers regarding their own work organization. When the two models are compared, cognitive dissonances are evident, particularly with regard to the department's functioning in terms of work organization and activity. While both models provide a service to academic areas (including deans, the academic council, students, and teachers), the differences in activity organization within the department demonstrate different perceptions of complexity. For the senior management, the institutional vision is that the activity is organized by hierarchical service logic between operators 1 and 2 and decision-makers, who should respond to input with a variable level of complexity in the requests generated by service users, mostly students. However, the workers organize the activity differently, according to the type of activity (archiving and call center), the use of specific technological tools for the service (virtual service platform), the configuration of the physical workspace (counter), and specific processes (degree certification). Finally, the senior management perceives the relationship with internal clients (administrative areas) as being more complex or involving more interaction than it actually does, as expressed by the department's employees. These differences represent the first line of work that was developed during the co-construction workshops to generate adapted solutions.
- *Modeling the quantification of problems and their effects on the service*: Figure 15.3a illustrates the senior management's vision of the problem areas of the activity and their quantification, as well as their impact on the department's overall service. This vision is based on senior managements and the person in charge of the area's attempts to manage the various issues raised prior to the intervention. The senior management validated this model during

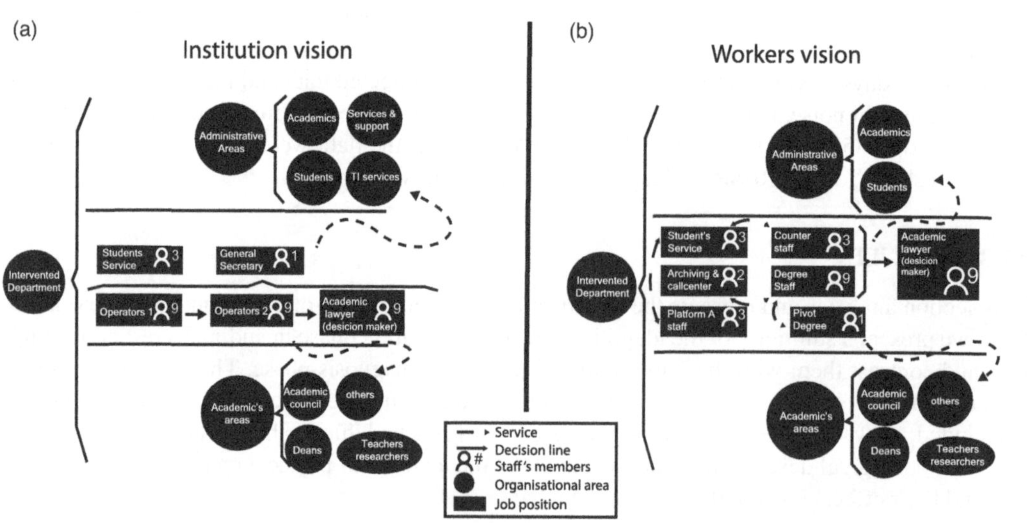

FIGURE 15.2 Activity organization: Comparison between the prescribed organizational model and the real activity. (a) institution's perspective; (b) workers' perspective.

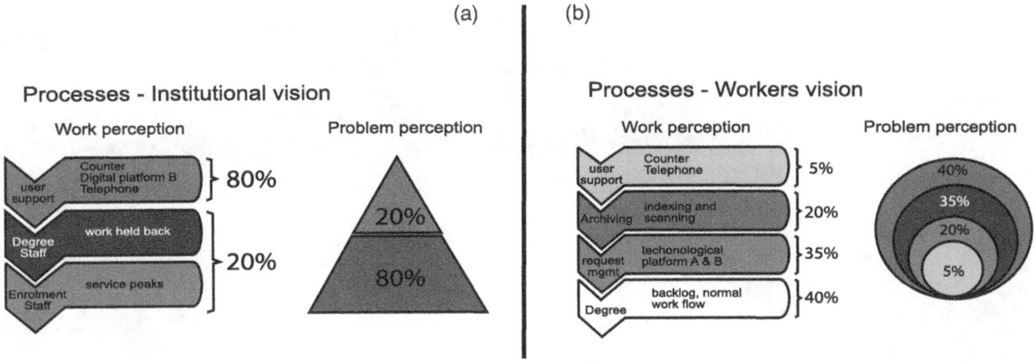

FIGURE 15.3 Work and problem perception: Comparison of the work-problem relationship: (a) institutional perspective; (b) workers' perspective.

the pre-diagnosis phase. Figure 15.3b, on the other hand, shows the model built from the perceptions of the department's workers. This model was validated with the workers during the first phase of the intervention. Comparing the two models reveals different approaches to quantifying the problems and their impact on the department's overall service.

15.3.2 Simulation Models and Validation of Action Scenarios

Based on the initial analysis and models of departmental activity in the intervention, we aimed to incorporate the various activity determinants into simulation models. These models would enable the team to engage in dialogue about their work and its relationship with the different aspects of their work activities. For this purpose, the point of convergence was defined as the intersection of the determinants of the work team, work processes, and physical workspace (Figure 15.4). Two simulation models were then proposed to support this dialogue on activity.

Once the materiality, use, and usability of the simulation models had been validated and it had been verified that the models enabled action scenarios and characteristic action situations to be activated, they were manufactured.

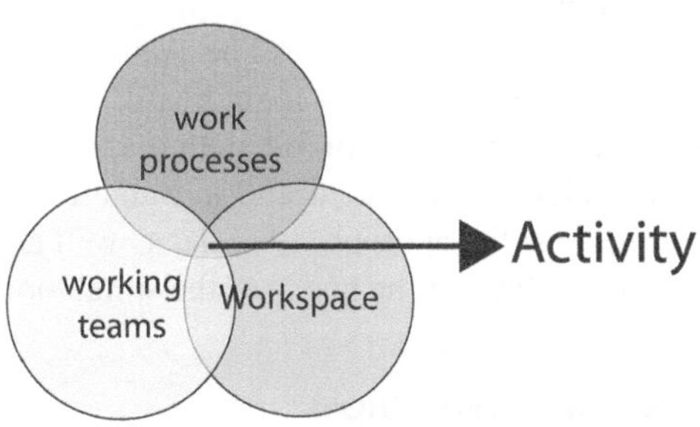

FIGURE 15.4 Cognitive model for simulation support generation.

FIGURE 15.5 Simulation Model 1, photograph.

- *Simulation Model 1: activities and processes:* Simulation Model 1 was built on a 1-m² flex foam sheet. It had spaces for external relationship departments (Faculties, DGE, DGA, and DI) on yellow glossy paper, spaces for internal processes (Degree, Enrollment, and Certification) on blue glossy paper, and spaces for inputs (Call Center, Counter, and Digital Platform) on pink glossy paper. Additionally, several action scenarios were determined and activated by an application-type trigger generated from a sample of 52 cases from the total number of held-back applications. Finally, a series of colored tacks served as avatars, representing different actors in the process, including department staff at various hierarchical levels and external department actors. Additionally, a group of generic avatars and colored threads were used to analyze the traceability of procedures, as required by the co-construction and thread (Figure 15.5).
- *Simulation Model 2: activity and workplace:* The Activity and Workplace simulation model was built on a 1.20 m × 0.90 m sheet of foam board. This support recreated the current layout of the intervention department in 3D, with corresponding characterizations of movable and fixed walls, columns, and fixed and movable furniture. Additionally, there was a set of avatars of people at the appropriate scale, as well as post-its and materials to build walls or furniture, depending on the needs of the simulation (Figure 15.6).

15.3.3 Solution Co-Conception Workshops

The workshops were divided into two parts. The first part included a dynamic presentation by the participants and facilitators of the workshop. It also included an explanation of the co-conception logic that would be used during the second part of the workshop. Additionally, there was a

FIGURE 15.6 Simulation Model 2, photograph.

presentation and discussion about the activity model (Figure 15.2) and the quantification of problems (Figure 15.3) that were raised in the previous phase. To ensure active participation, each workshop had ten participants. Forty people from the department under analysis were organized into five groups of eight people, and each group also included two guests. These guests came from other departments that interact with relevant processes, and the aim was to include interdepartmental coordination of activities in the workshop. Among the guests, there were two from human resources, two from infrastructure, two from academic management, and two from IT and technology. The second part of the workshop involved the participants' actions on the two simulation models created previously.

- *Activity on Simulation Model 1 (activity and processes)*: Figure 15.7 shows several photographs of the model in use. It is important to note that the model included colored threads to highlight collapsed system spaces and collective activity problems.
- *Activity on Simulation Model 2 (activity and workplace)*: Figure 15.8 shows several photographs of the support in use. It is important to note that this model helped reconfigure the workspace based on the activity and workers, especially the infrastructure team, who were not accustomed to this dynamic beforehand.

15.3.4 Analysis of Environmental Factors in the Activity

- *Noise:*
  a. *Sample*: For this analysis, samples were taken over 4 days at various locations within the workspace. A total of 186,146 records were collected between 8:00 a.m. and 5:00 p.m., distributed as follows: Call Center Area (56,085), Lawyer Secretary's Office

FIGURE 15.7 Simulation Model 1 in use.

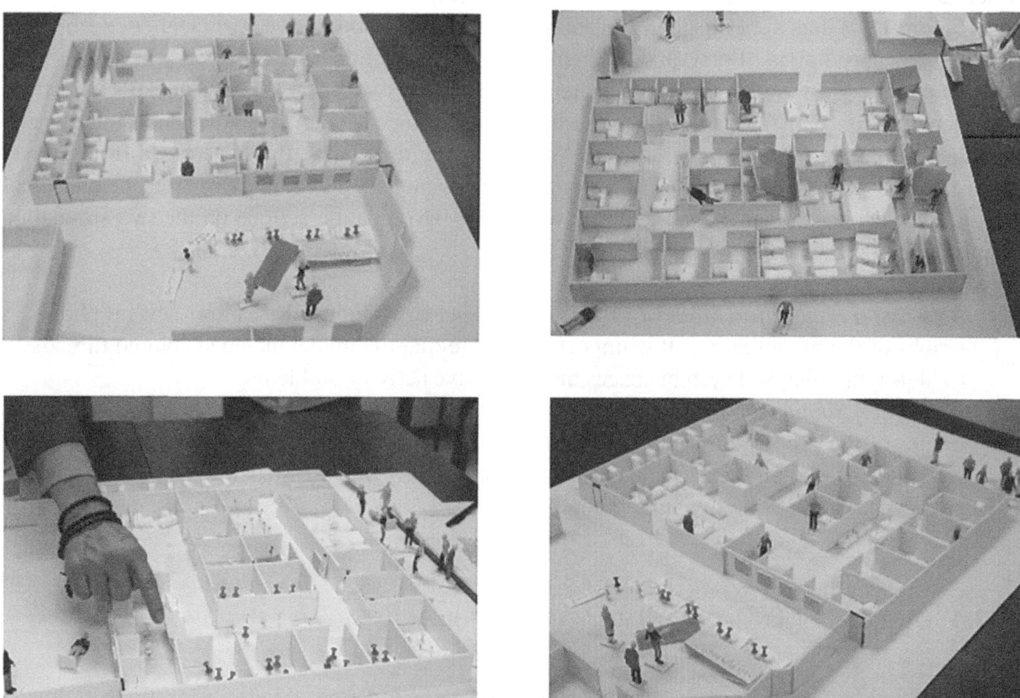

FIGURE 15.8 Simulation Model 2 in use.

(14,945), Customer Service Hall (60,334), and Counter Area (54,782). The sample data ensure a normal distribution with a 95% confidence interval.

b. *Conceptual framework of the analysis*: The sound pressure level is the measure considered, and it is expressed in type A decibels (dB). According to Ecuadorian and international technical standards, the daily noise dose limit in terms of risk is 85 dBA and 70 dBA for concentration work (Presidencia de la República del Ecuador, 1986, p. 29). While the noise dose does not pose an auditory risk, it is important to note that the noise factor can interfere with cognitive processes, affecting workers' performance. Each worker's regulatory processes are engaged in order to perform the requested activity in environments that do not ensure ideal conditions (Leplat 2006). Employees expend extra effort to cope with the activity, which reduces the quality of their work and directly affects their performance. The French standard NF S31–199, "Acoustics: Acoustic Performance in Open Office Spaces (AFNOR 2016)," was used as a reference.

Figure 15.9 shows the average noise levels during normal operating hours (8:00 a.m. to 4:00 p.m.) in different activity areas (Call Center, SA Offices, Customer Service Hall, and Counter). We also include references for harmful noise (85dB) and acceptable references for activities involving attention to the public (55 dB), telephone attention (52 dB), collaborative work (50 dB), and individual work (45 dB). A high degree of conversational interference (ambient noise) was identified in the different spaces, hindering the quality of customer service activities, whether collective or individual. The collected data confirm that noise levels in the different areas remain above 55 dB. The only time of day when noise levels ensure an acceptable level for proper activity development is at 1 pm (lunchtime), evidently due to the slowdown in service during this time.

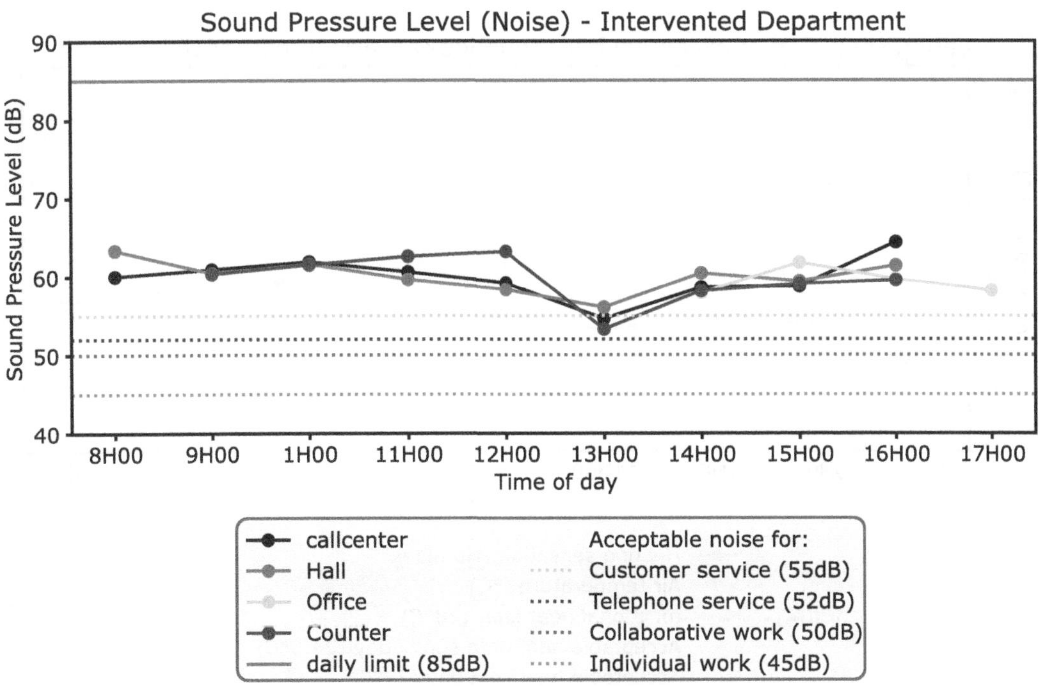

FIGURE 15.9 Sound pressure analysis (4-point sampling).

- *Heat sensation (excessive heat in the eastern part of the department):*
  a. *Sample*: To analyze this factor, samples were collected for 1 day at the midpoint of the offices on the eastern side of the department. A total of 57,753 records were collected between 8:00 a.m. and 5:00 p.m. The sample data ensured a normal distribution with a 95% confidence interval.
  b. *Conceptual framework of the analysis*: For the analysis of heat stress, the Humidex heat index (Masterton y Richardson 1979) is used to measure the level of discomfort a person may experience due to a combination of ambient temperature (°C) and relative humidity (% RH). Although there is no imminent danger of heat stress due to working conditions, it is important to note that heat sensation may affect cognitive response processes during activity, resulting in poor performance, errors, or unusual exhaustion during the workday. Workers' internal and external temperature regulation processes, as well as their capillary capacity, are engaged in order to perform the requested activity under such conditions. Long-term consequences may result in a reduction of the employee's working capacity. Studies in the Canadian standard CSA Z412-17, "Office Ergonomics: Application Standard for Office Workstation Ergonomics," address and analyze this issue by establishing optimal temperature parameters for summer (24.5°C) and winter (22°C).

     Figure 15.10 shows the average temperature sensation (Humidex) during normal operating hours (8 a.m. to 4 p.m.) in this area of the secretariat. We have also included the functional limit references for activity in terms of the perceived heat by employees (28°C), as well as the acceptable references for sunny (24°C) and cold (22°C) days. Our analysis revealed a high degree of discomfort caused by the perceived heat in the eastern part of the workspace. This could lead to reduced attentional performance, irritability, and drowsiness, among other issues. This is true for both individual and collective activities. The data confirm that the thermal sensation in this area exceeds

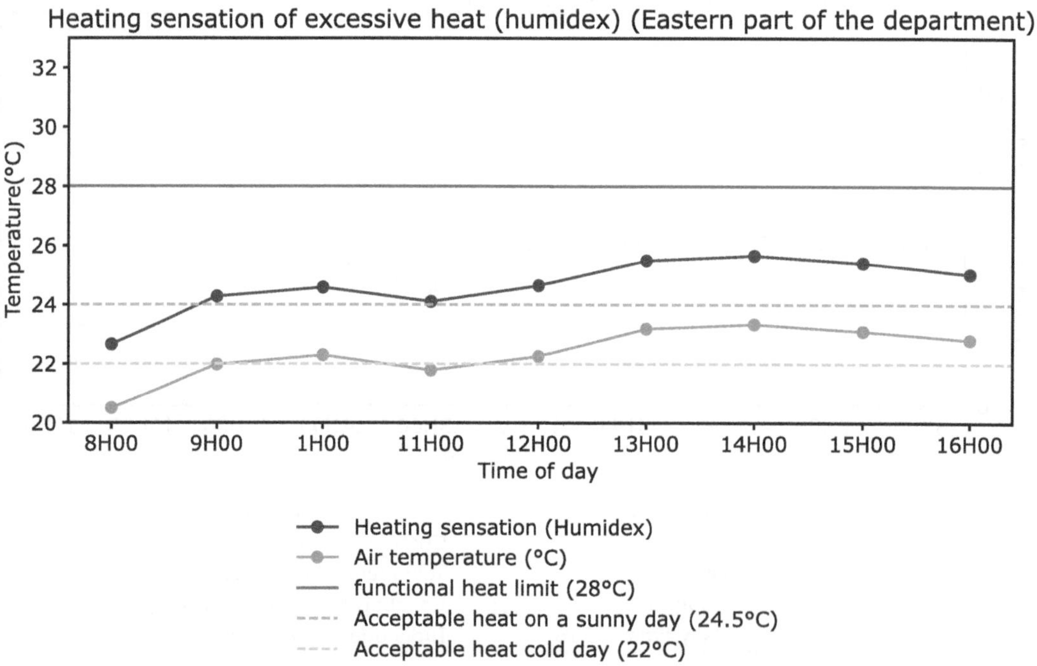

FIGURE 15.10 Sound pressure analysis (4-point sampling).

the acceptable reference from 9:00 a.m. onward and intensifies after midday. This is true for both hot and cold days.

- *A cold thermal sensation of excessive cold (counter zone):*
  a. *Sample*: To analyze this factor, samples were collected over 2 days at the two most frequently used customer service counters. A total of 42,101 records were collected between 8:00 a.m. and 4:00 p.m., ensuring a normal distribution with a 95% confidence interval.
  b. *Conceptual framework of the analysis*: To analyze wind chill, we used the Wind Chill Index (WCI) (Osczevski y Bluestein 2005) to measure the level of discomfort or cold sensation a person may experience due to the combination of ambient temperature (°C) and sudden cooling from air speed (km/h). Although there is no imminent danger of thermal stress from extreme cold, it is important to note that wind chill can affect cognitive response processes during activity, resulting in poor performance, errors, or a slower work rate during the workday. Long-term consequences may include constant respiratory illnesses, tension, irritability, and altered emotional states. This issue is addressed in the Canadian standard, CSA Z412–17, "Office Ergonomics — Application Standard for Office Workstation Ergonomics," which establishes optimal temperature parameters for summer (24.5°C) and winter (22°C).

Figure 15.11 shows the average WCI during the window's usual operating hours (8:00 a.m. to 4:00 p.m.). We also include cognitive functional limit references for activities related to perceived cold (4°C) and acceptable references for sunny (24°C) and cold (22°C) days. Our analysis of the activity revealed a high degree of discomfort caused by the perceived cold at the window, as well as obvious strategies employed by the employees to mitigate this issue. Our analysis also included the percentage of cold drafts during the different hours of the workday. This situation could lead to reduced

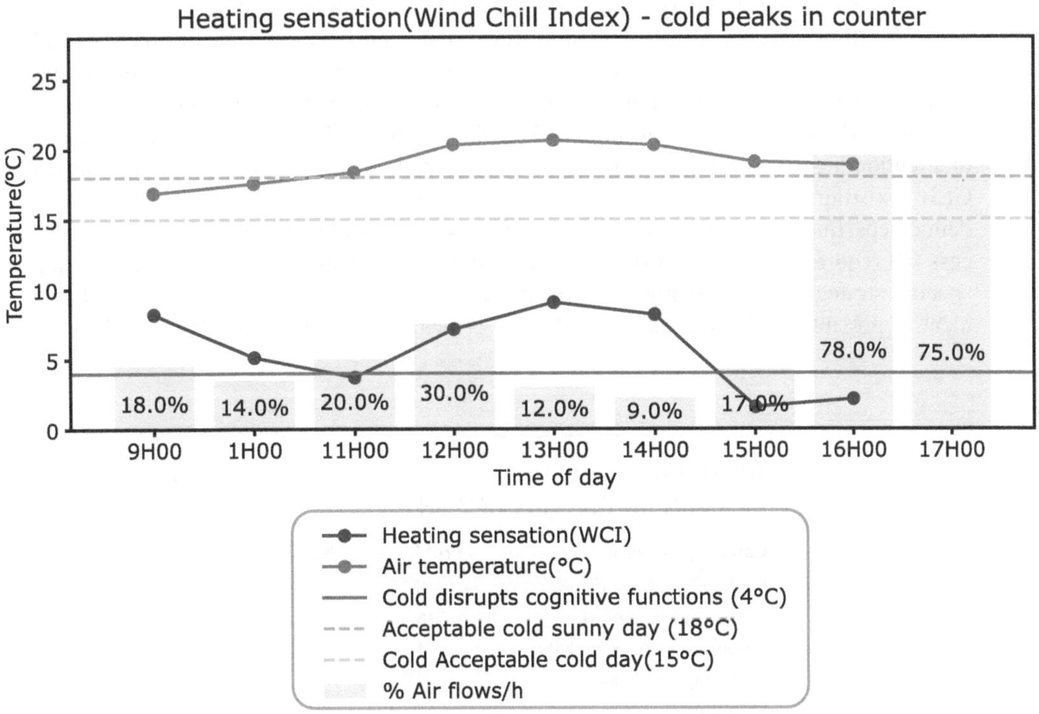

FIGURE 15.11 Cold thermal sensation, peaks in SUSG window area.

attentional performance, irritability, and lethargy. This is true for both individual and collective activities.

The data confirm that the thermal sensation in this area is far from comfort benchmarks and show that the sensation of cold remains below acceptable levels throughout the workday. It is important to emphasize that these data refer to the average peak of cold perceived during each hour of the workday. However, when we analyze these data in light of the percentage of drafts identified during the workday, we conclude that this situation requires specific attention because of the particularities of Ecuadorian geography (2,800 meters high in the middle of the Andes), where air currents are generally freezing and the ambient temperature is below 25°C. This atmospheric phenomenon does not allow for rapid compensation of workers' bio-thermodynamics, especially in the afternoon when the percentage of drafts exceeds 70% and the ambient temperature does not exceed 22°C.

15.4 RECOMMENDATIONS BASED ON THE INTERVENTION

Data processing was carried out after each workshop. During this process, the annotations and relevant points of the simulation workshop were collected. The video and audio corpus were reviewed, and the different problems, needs, and solutions were noted on an Excel spreadsheet, with each item on a different line. After this information was recorded, it was categorized in a different column. Once the information from the five workshops had been systematized, the entire corpus of findings was reviewed and re-categorized, revealing the following.

Six main lines of action were established regarding the articulation of solutions, as evidenced in the workshops and articulated from the global vision of the work situation and the gap between real and prescribed activity (Table 15.2).

- *Articulation of teams*: Four key activities to plan and implement were identified: (a) Clear definition of homogeneous working groups that take advantage of the knowledge and skills developed by the most experienced employees. (b) Define group strategies (small teams) and collective strategies (all department staff) for the temporary coverage of functions and key points in the routine. (c) Recruitment of a new coordinator. (d) Organize a functional restructuring by teams (reception and attention service and graduation service).
- *Workflow*: Four key activities to plan and implement were identified: (a) Establish levels of attention (instances) in the service balcony with the following logic: (1) Information, (2) GLB[1] (existing student), and (3) FD[2] (for a non-existing student. (b) Clarify with the academic departments the submission of all requirements to avoid disrupting the service process (all requirements are reported using the computer system). (c) Identify and socialize specific strategies on a weekly basis (in the case of postgraduate medicine students) to identify possible preventive actions and avoid critical knots in the workflow. (d) Create a

TABLE 15.2
Incidents for Category of Work

Category of Work	Problems	Needs
Health and safety at work	8	-
Technological platforms	6	4
Organization of the activity	4	2
Archiving problems	4	1
Workflow	4	12
Team coordination	3	1

defined and communicated strategy with all professors so that non-enrolled students cannot receive the courses.

- *Organization of the activity*: There were three activities to plan and implement: (a) Establish clear communication with professors defined as strategic actors. (b) The people who authorize the increase or opening of places for the academic programs must be part of the enrollment team (during the enrollment period). (c) Design collective strategies to work on the backlog of tickets, information packets and pending requests inherited, and the proper functioning of the archive.
- *Platforms*: Four activities were identified: (a) Think about implementing a sorting or pre-classification interface for the technological platforms (e.g., student ID). (b) Parameterize responses at night and on weekends, in addition to working hours. (c) Ask the Academic Director to coordinate so that a module on the use of the technological service platforms is included in the syllabus of the Information and Communication Technologies (ICT) course, which is mandatory for all students. (d) Implement the automatic generation of certificates (in accordance with regulations).
- *Psychosocial aspects*: Three activities were identified: (a) Taking environmental samples to adapt the workspace at critical points. (b) Redesigning an internal cafeteria space that can also be a space for decompression after tense attention and establishing strategies for support among colleagues. (c) Restructuring working hours and workspaces for peaks in attention.
- *Archive*: Two activities were identified: (a) Request the finalization of the proposal for the physical adequacy of the archive space by the infrastructure department, considering the limitations evidenced by the simulation. (b) Manage a temporary hiring of staff to support the process of indexing the faculties archive before being transferred to the new archive.

15.4.1 Workshop: Socialization of Solutions and Agreements

This workshop was organized as part of a larger event, coordinated with the senior management and the institution's Human Resources team. The workshop consisted of three parts. The first part consisted of an icebreaker activity and an outdoor team-building activity, which were coordinated and implemented by the institution's Human Resources team. The second part was our results and agreements workshop, which generated material for the afternoon. This material was used in a strategic planning workshop led by the director of the area and the new coordinator of the department.

During the workshop, we presented the systematization of the problems and needs that the workers discussed during the different workshops. The common points and the different solutions that the workers proposed to address individual and collective activities were presented, as well as the human, technological, and social resources of the work group. This enables to establish formal communication channels and a series of agreements to strengthen team cohesion and collective activity synergy.

15.4.2 Agreements and Consensus:

- *Articulation of teams*: The workshop defined a new team structure organized as follows: a) Administrative/legal manager and their Management Team (two persons per team): For the particular case of Medicine, one extra person per team is necessary due to the size and particularities of this academic program. Additionally, the Postgraduate Medicine Team will remain with the current work collective in the faculty due to the particularities of the postgraduate customer service process. b) The Service Balcony Team (Call

Center, Counter, and Facilitators—six people): It was necessary to coordinate an IT-driven improvement as the current interface does not allow one to consult the status of a procedure (Service Transaction Unit). c) The Archive Team (two people): The need for a temporary intervention team was confirmed to regularize the current situation of the student file archives and ensure operation within 2 months. d) The Service Balcony Team (Call Center, Counter, and Self-service) is considered a high-rotation team because it is also connected to the Seedbed Project (HR-CEO), where the career plan for the institution's administrative staff will begin.

- *Workflow*: (a) All active student applications will enter through technological platform A, and passive student applications will enter through technological platform B. (b) The counter and call center channels will focus on redirecting application entries through digital channels for tracking and control purposes. (c). An attention algorithm must be defined to streamline the workflow and manage the department. (d) The degree process will be managed with a single matrix for all academic programs by the degree team.
- *Organization of activity*: (a) A specific strategy will be defined for the enrollment period. This strategy will consist of temporarily installing several workstations and waiting rooms in the building's hall during the 3 weeks of the process. This is necessary since the redesigned workspace is not functional for this purpose. (b) Strategies will be established for the integration of the different degree teams of the academic programs. This will allow for a planned transition to the department's workspace. (c) During the enrollment period, the participation of the degree coordinators of the programs will be included and programmed into the strategies. This will speed up the processes.

15.4.3 Recommendations Regarding the Environmental Factors Analyzed

- *Noise*: It is recommended that the following particulars be considered regarding noise and the possibility of isolating certain critical areas to properly develop the different activities. Activities should be grouped according to the following criteria: call center, counter, customer service, collective work, and individual work. If workspaces cannot be adapted or isolated, it is recommended that collective strategies be implemented to reduce environmental noise. Examples include designing schedules for each activity to ensure adequate acoustic comfort, implementing volume reduction policies in certain areas, and other strategies that the group deems necessary and feasible for organizing the joint activity.
- *Noise hotspots*: In the counter space in particular, the noise level reduces the ability to understand messages during attention by at least 20%. In addition, there is a glass barrier with an audio system in this area, which limits communication independently by at least another 20%. By the entropy criterion in any communication process, there is a loss of at least 5% of the information, which means that the sum of factors means that the employees in this workplace are forced to work in a condition where at least 45% of the information is likely to arrive incomplete or incomprehensible, which does not allow the activity to be carried out properly or suggests an overexertion of the worker to carry out the activity. If to this condition, we add two extra factors: masks of the interlocutor or hearing impairment in collaborators, it is likely that the performance of the customer service activity at this point will decrease abruptly, and the degree of errors as well as the attention times may increase. It is recommended to consider these aspects for the organization of the activity and the adaptation of workspaces in this post.
- *Heat sensation*: Due to Ecuador's unique geography, controlling the environment's humidity is crucial. Humidity alters the perception of heat or cold, increasing the thermal sensation and causing discomfort. It is also important to ensure that workers are well-hydrated to compensate for the decrease in their capillary capacity for thermal regulation. Collective strategies should be developed for organizing work and using spaces to reduce the effects

of this factor on the performance of activities in the area. In terms of engineering and reconditioning the physical space, increasing curtain-type barriers or heat-absorbing sheets to reduce solar radiation and using dehumidifiers are possible solutions.

- *Cold*: A series of physical blocks (e.g., plants and furniture) should be installed in the entrance hall to the counter to prevent direct air current impact on workers. Redesigning the counter is advisable to prevent airflow from hitting employees directly in the upper trunk area (chest, neck, and head). While a permanent solution is being implemented, ensure that the control hatches through which the drafts enter are blocked. Additionally, it is recommended that collective strategies be developed for organizing work and using spaces to reduce the effects of this factor on the performance of different activities.

15.4.4 Workspace Reorganization Proposal

The infrastructure and general services team of the institution attended one of the five co-construction workshops to define and discuss points that were not originally considered in the adaptation of the workspace. The focus of this workshop was to find solutions to adapt and reorganize the department's workspace according to requirements generated by the construction of the collective activity.

- *Proposed use of space:*
  - *Applications management area*: Responsible for articulating the dynamics of requirements management in the department (19 people).
  - *Degree area*: This area is for confidentially articulating degree processes. Handling Kardex and organizing documents are essential tasks that require a high level of concentration (eight people).
  - *Information area—counter*: This space is designed for in-person articulation of information, as well as ensuring that requirements are entered through the service balcony or digitally (two people per shift).
  - *Call center information area*: This area is designed for telephone communication and to ensure that requirements are entered through the service balcony or digitally (two people).
  - *Service balcony area*: This space is intended for entering student requirements through digital service modules that connect the service platforms (A and B). It will be located outside the counter and must accommodate a mobile workstation.
  - *Archive management area*: This is the space for indexing and managing the student records archive (two people).
  - *Archive area*: This area stores student records in two sections: passive and active archives. Both areas must ensure confidentiality and adequate access control, allowing access only to authorized personnel. Additionally, physical and environmental security measures should be implemented to safeguard the documentary assets. This space should facilitate documentary activities. It is suggested that it maintain a work surface at least 1 m wide and 70 cm long. This space should be adjustable to within 10% of the average height of the equipment and allow for standing work. The artificial lighting in this space must be adjustable and provide a minimum of 300 lux. There must be at least one free electrical outlet throughout the workspace.
  - *Digitalization area*: This space is intended for digitizing documents for indexing and organizing the department's digital archive. This space must have a workstation suitable for the digitization process.
  - *Cafeteria and meeting room/decompression area*: This space is intended for social and team processes within the department and incorporates group activities through work breaks or collective workspaces. It must accommodate at least ten people for short periods of time and have electrical, network, and water connections.

- *All work modules must*: (a) Allow installation of a CPU, 19-inch monitor, and respective electrical and network connections. (b) Have separators that provide at least 50% privacy for individual activities and no more than 75% for group activities. (c) Provide secure storage for temporary filing of documents and personal items. (d) Ensure work surfaces are 72 cm high with an adjustable height between 0 and 7 cm. (e) Ensure at least one free electrical outlet for every two work modules. (f) Control the noise level, if possible, for the required activity in each area of continuous use (50 dB, 45 dB, 55 dB, 52 dB, and 45 dB of sound pressure type A). (g) Considering that the space has no major access to natural light, ensure that the artificial lighting of the work modules provides a luminance level of at least 300 lux on the work surface and does not produce reflections on the data display screens.

15.4.5 Proposal for Reorganizing the Working Team

The institution's Human Resources team attended one of the five co-construction workshops to define and discuss points not originally contemplated in the organizational restructuring. The workshop focused on finding solutions to adapt and reorganize the department's work team, as well as defining the characteristics and specifications to incorporate new people into the work team. This was done according to the requirements generated from the construction of the current and future collective activity foreseen by the senior management. Three technical criteria were identified for the incorporation of new personnel and internal movements of the work team: (a) An intervention team for archiving (temporary). (b) Information teams (counter and call center) with a 13/6 shift schedule (8:00 a.m. to 8:00 p.m., Monday to Saturday). (c) Facilitation teams (critical schedules and seasons). (d) Incorporating new personnel to complete existing work teams, as agreed upon with the senior management.

15.4.6 Final Recommendations for Institution

It was important to recommend and commit the department leader to implementing the change, develop the work team, and foster internal dynamics after the intervention. This ensures the work team's autonomy in the process. The new department coordinator will monitor the change process and develop continuous improvement strategies for the collective activity and department services.

The infrastructure must urgently resolve the security aspects of handling and storing confidential archives and files, as well as ensure the prompt functioning of both physical and digital archives. The new department coordinator must develop strategies and methods for updating and verifying digital data in the archive, which directly affects the department's service processes.

The new department coordinator must promote the development of collective and individual strategies to ensure customer service from 8:00 a.m. to 7:00 p.m., Monday through Saturday. The department should rely on experienced staff to develop effective knowledge transfer strategies that quickly and efficiently integrate new workers.

Finally, we recommended that a follow-up and progress meeting should be established with the HR-CEO at 6 and 12 months to adjust strategies for change or specific needs that allow the department to function properly.

15.5 LESSONS LEARNED AND LIMITATIONS

This intervention highlights the potential of the ergonomics and human factors approach as an integral, methodological, and conceptual framework to address and build solutions across an organization's complex sociotechnical system. The case presented in this chapter is based on a deep comprehension and dialogue with the actual activities of workers. The intervention uses ergonomics, human factors, and the psychodynamics of work to characterize individual and collective activity.

Based on these approaches, we built a discussion model that allows for the transformation of work, ensuring communication and the collaborative efforts of all stakeholders to create a more humane, efficient, and healthy work environment. What have we learned from this process?

- Proper recording of interviews and observations throughout the intervention was essential for systematizing information, generating explanatory and usage models, and demonstrating the concept of activity to both the institution's managers and the workers themselves.
- The intervention promotes using the concept of model as a dynamic vector of dialogue for ergonomists. From initial activity models to simulation or systematization models, these models are established as objects that allow each phase of the project to advance through dialogue. In the case of simulation models, they facilitate activity dynamics and systematization of data.
- The process and intervention demonstrated the importance of an interdisciplinary approach to finding adapted solutions. Articulating knowledge and discourses from the many areas involved in the activity is essential for generating adapted solutions.
- The intervention demonstrates the recurring problem of conception: a diffuse beginning, development without a roadmap, and unknown results until the end. However, the participatory component and the problem-setting exercise, supported by an ergonomic perspective, enable the intervention to progress effectively.
- The intervention demonstrates the potential of workers' voices and spaces for expressing them as valuable resources for constructing and improving the working environment, even in adverse situations, such as those described in the intervention.
- Finally, the intervention and its results demonstrate the importance of the collective construction of sociotechnical work conditions.

The most relevant limitation encountered in the process is considered to be the lack of knowledge about the scope of ergonomics and human factors in the organizational sphere. Managers are reluctant to implement this type of intervention because they believe that ergonomics is only associated with adapting chairs and measuring biomechanical factors in the workplace. However, once the space for dialogue and intervention is established, the results are recognized and take on new value within the organizational ecosystem. A second limitation is the limited availability of professionals in this field. Throughout this intervention, training and educational sessions were incorporated so that attendees could contribute more actively to the process. Finally, this type of intervention, based on real activity, can generate tension by taking stakeholders accustomed to vertical, hierarchical logic out of their comfort zone.

ACKNOWLEDGMENTS

This work would not have been possible without the support and commitment of the Vice Chancellor, the Human Resources General Director, and the General Secretary of Pontificia Universidad Católica del Ecuador.

NOTES

1 GoLiveBanner, virtual platform integrated to Academic RP.
2 FreshDesk, virtual platform for service and control of requests.

REFERENCES

AFNOR. 2016. "NF S31-199, Acoustique - Performances acoustiques des espaces ouverts de bureaux". https://www.boutique.afnor.org/fr-fr/norme/nf-s31199/acoustique-performances-acoustiques-des-espaces-ouverts-de-bureaux-/fa158972/1549.

Beaujouan, Joffrey, Sophie Aubert, and Fabien Coutarel. 2015. "Construction de l'intervention ergonomique. D'une préoccupation de montée en cadence à la décision d'investir pour transformer le travail : embûches et stratégies". *Perspectives interdisciplinaires sur le travail et la santé.* doi:10.4000/pistes.4599.

Belleghem, Laurent Van, and Fabrice Bourgeois. 2003. "La part de l'analyse de l'activité de travail dans l'intervention ergonomique: questionnements mutuels sur la pratique entre un jeune ergonome et un aîne". XXXVIIIème Congrès de la SELF, 493–502. https://www.ergonomie-self.org/wp-content/uploads/2019/12/acte-18-self-2003.pdf

Bourgeois, Fabrice, y Laurent Van Belleghem. 2003. "Avec l'approche travail dans l'évaluation des risques professionnels (décret 5/11/01), enfin du nouveau en prévention". *Actes du séminaire Paris* 1. 1–12.

Caroly, Sandrine, y Yves Clot. 2004. "Du travail collectif au collectif de travail : développer des stratégies d'expérience". *Formation Emploi* 88 (1): 43–55. doi:10.3406/forem.2004.1737.

Clot, Yves. 2015. *Le travail à coeur: pour en finir avec les risques psychosociaux.* Paris: La Découverte.

Clot, Yves, y J. Leplat. 2005. "La méthode clinique en ergonomie et en psychologie du travail". *Le travail humain* 68 (4): 289. doi:10.3917/th.684.0289.

Clot, Yves, y Pascal Simonet. 2015. "Pouvoirs d'agir et marges de manœuvre". *Le travail humain* 78 (1): 31. doi:10.3917/th.781.0031.

Cuvelier, Lucie., Falzon Pierre., Granry Jean-Claude Granry., y Orliaguet, Gilles. 2017. "Développement des collectifs de travail et développement de la sécurité : une étude sur les décisions à risque en anesthésie". *Psychologie du Travail et des Organisations* 23 (3): 255–72. doi:10.1016/j.pto.2017.05.004.

Daniellou, François. 2007. "Des fonctions de la simulation des situations de travail en ergonomie". *Activités* 04 (2). doi:10.4000/activites.1696.

Dejours, Christophe, y Isabelle Gernet. 2016. *Psychopathologie du travail.* Issy-les-Moulineaux: Elsevier Masson.

Faïta, Daniel. 2001. "L'analyse du travail et le statut de " l'activité " chez Bakhtine". *Travailler* 6 (2): 13. doi:10.3917/trav.006.0013.

Falzon, Pierre. 2004. *Ergonomie.* Paris: Presses Universitaires de France. https://www.cairn.info/ergonomie--9782130514046.htm.

Folcher, Viviane, y Anne Bationo-Tillon. 2018. "Approche instrumentale dans l'intervention ergonomique". *53ème Congrès International Société d'Ergonomie de Langue Française. 16–21.* https://www.ergonomie-self.org/wp-content/uploads/2019/03/acte-43-self-2018.pdf

Hubault, François. 2017. "A quoi sert l'analyse de l'activité en ergonomie?", 14.

Lancry, Alain. 2009. *L'ergonomie, Que sais-je?* Paris: Presses universitaires de France.

Leplat, Jacques. 1993. *L'analyse du travail en psychologie ergonomique: recueil de textes.* Tome 2 Tome 2. Toulouse, France: Octarès Editions.

Leplat, Jacques. 2006. "La notion de régulation dans l'analyse de l'activité". *Perspectives interdisciplinaires sur le travail et la santé*, 8–1. doi:10.4000/pistes.3101.

Leplat, Jacques. 2008. *Repères pour l'analyse de l'activité en ergonomie. Presses Universitaires de France.* doi:10.3917/puf.lepla.2008.01.

Masterton, Joan Mary., y Richardson Frank A.. 1979. *Humidex ; A Method of Quantifying Human Discomfort Due to Excessive Heat and Humidity.* Downsview, Ont.: Environment Canada, Atmospheric Environment.

Mefteh, Wafa. 2018. "Simulation-Based Design: Overview about Related Works". Mathematics and Computers in Simulation 152 (octubre): 81–97. doi:10.1016/j.matcom.2018.03.012.

Montmollin, Maurice de, y René Amalberti. 1995. *Vocabulaire de l'ergonomie.* Toulouse, Haute-Garonne: OCTARES Editions.

Osczevski, Randall, y Maurice Bluestein. 2005. "The New Wind Chill Equivalent Temperature Chart." B*ulletin of the American Meteorological Society* 86 (10): 1453–58. doi:10.1175/BAMS-86-10-1453.

Pelayo, Sylvia, Anthony Loiselet, Marie Catherine Beuscart-Zéphir, Janine Rogalski, y Françoise Anceaux. 2010. "Proposition méthodologique pour l'Analyse de la coopération dans une planification distribuée des actions:" *Le travail humain* Vol. 73 (4): 361–83. doi:10.3917/th.734.0361.

Presidencia de la Republica del Ecuador. 1986. "Decreto Ejecutivo 2393".

Rabardel, Pierre. 1995. *Les hommes et les technologies; approche cognitive des instruments contemporains.* Armand Colin. https://hal.archives-ouvertes.fr/hal-01017462/document.

Rasmussen, Jens. 1993. "Analysis of tasks, activities and work in the field and in laboratories". *Le travail humain* 56 (2/3): 133–55.

Rouat, Sabrina, y Philippe Sarnin. 2018. "L'intervention en santé au travail : l'intérêt de l'étude de cas d'intervention". *Le travail humain* 81 (4): 331–63. doi:10.3917/th.814.0331.

Salembier, Pascal, y Bernard Pavard. 2004. "Analyse et modélisation des activités coopératives situées: Évolutions d'un questionnement et apports à la conception". *Activites* 1 (1). doi:10.4000/activites.1170.

Van Belleghem, Laurent, y Fabrice Bourgeois. 2013. "La part de l'analyse de l'activité de travail dans l'intervention ergonomique : questionnements mutuels sur la pratique entre un jeune ergonome et un ainé". In *congresso self rester plus longtemps, dans quell travail* (vol. 38).

Van Belleghem, Laurent Van. 2018. "La simulation de l'activité en conception ergonomique : acquis et perspectives". *Activites* 15 (1). doi:10.4000/activites.3129.

16 A Descriptive Study of Companies' Management Perspective on the Critical Success Factors of Ergonomics Programs and Their Benefits in the Manufacturing Industry in Mexico

Julio César Ramos Rodríguez and
Aide Aracely Maldonado-Macías

16.1 INTRODUCTION

Ergonomics is the scientific study of human interactions with various systems, including workspace design and product development. It aims to improve efficiency and user satisfaction while minimizing injury risk (International Ergonomics Association 2000). Ergonomics seeks to optimize the physical environment, enhance work performance, and improve overall well-being (Wee, Kuo, and Ngu 2020).

Generally, implementing ergonomic practices has been approached in several ways, with programs being one of the most utilized methods. Ergonomics-based workplace programs identify problems and encourage changes to improve the job and working conditions, such as redesigning jobs and machines or changing processes and work environments to improve the quality of work life (Afroz and Haque 2021). However, several factors can inhibit the successful implementation of ergonomics programs. These factors may include changes in the administration, management, or personnel responsible for the program, lack of time and motivation, insufficient financial resources, task prioritization issues, inadequate training, resistance to change, poorly defined scopes of work, and a mismatch between job demands and employee capabilities, along with inappropriate equipment and work methods, among others (Villalobos 2018). Additional factors can be found in workplace design, such as the lack of focus on solutions to existing problems and employees' awareness of ergonomics (Yazdani and Wells 2018).

In Mexico, several standards have recently regulated the application of ergonomics. They define organizations' responsibilities and outline employees' shared duties in managing occupational health and safety. These standards specify the requirements for a safe work environment through ergonomic compliance. The primary goal is to control workplace risk factors to prevent employee health issues. One of the current regulations is the Mexican Official Standard NOM-036-1-STPS-2018, titled "Ergonomic Risk Factors at Work." This standard outlines the responsibilities of an organization's leader regarding ergonomic practices. Key elements include analyzing ergonomic risk

DOI: 10.1201/9781003582014-20

factors, developing and implementing prevention measures, monitoring employees' health, informing employees about potential health changes, providing training, and conducting medical examinations. Employees must engage with these elements to ensure shared responsibility, which leads to positive outcomes in the work environment and fosters a culture of prevention. Establishing a safety- and health-oriented culture at work not only enhances the quality of life at work for employees but is also associated with the timely prevention of accidents and occupational diseases. Accordingly, significant economic savings for organizations can be achieved by minimizing organizational costs and reducing occupational health and human resource management costs (Makhbul, Shukor, and Muhamed 2022).

For this study, the ergonomic program model proposed by Bradley S. (2003) and the research by Alferez-Padrón et al. (2021) were utilized to design a questionnaire for gathering data, which includes four stages: planning stage, start-up process, work improvement cycle, and long-term program development. This model has supported organizations in effectively implementing ergonomics, with benefits categorized into three main groups based on the literature review. Health and safety benefits are focused on health and safety in the workplace, significantly decreasing the number of injuries, the prevalence of occupational diseases, and workplace accidents. This proactive approach protects employees and fosters a safety culture that enhances productivity (Virmani and Ravindra 2021). The economic benefits include reduced medical expenses from fewer workplace injuries, lower absenteeism rates as employees remain healthier and more engaged, and an overall enhancement in product quality due to a more focused and motivated workforce (Abinaya and Rajkumar 2021). Finally, the benefits of quality of life at work improve employees' overall well-being by encompassing their job satisfaction, motivation, and well-being levels. A positive workplace environment contributes to higher morale, effective teamwork, and increased employee retention, leading to a more harmonious and productive organizational culture (Ishak et al. 2018).

Overall, these benefits contribute to a healthier, more productive workplace, considering that ergonomics positively affects employee effectiveness and efficiency (Abinaya and Rajkumar 2021; Heidarimoghadam et al. 2022; Rostami et al. 2021). On the other hand, to obtain such benefits, critical success factors are the key elements for a successful implementation within a project or organization that will help it achieve its objectives, and it is crucial to determine them properly (Vega, Bautista, and Martinez 2020). Buniya et al. (2021) found that factors such as employee involvement, the prevention system, safety control, and management commitment significantly influence the implementation of an ergonomics program that results in health and safety benefits. Employee participation in ergonomics management is significant as it is an effective behavior modification technique. This approach is beneficial because employees are closest to the risks, are actively involved in work activities, and are ideally qualified to suggest improvements in the workplace (Hong, Ramayah, and Subramaniam 2018). Therefore, understanding those key factors in managing ergonomics programs has become relevant for companies. Recognizing these factors enables manufacturing industries to appreciate the direct link between meticulous planning and optimal resource allocation, leading to advantages. When effectively implemented, these elements help ergonomics programs meet objectives and enhance workplace efficiency and employee well-being. This chapter presents a descriptive analysis of the critical success factors of implementing ergonomics programs and benefits from the company's perspective in Ciudad Juárez, Chihuahua, Mexico. A company's perspective plays a crucial role in the organization from the beginning of ergonomics programs as management commitments significantly promote occupational health and safety by minimizing risks and preventing workplace accidents (Berhan 2020). In 2020, the Mexican Social Security Institute reported a total of 597,713 cases of occupational accidents and illnesses in Mexico. Chihuahua, where the study area is located, ranked 8th with 21,436 incidents. Ciudad Juárez is the second-largest city in Mexico in terms of the number of manufacturing companies it hosts. With a robust industrial landscape, the city has 320 diverse industries, employing 318,566 people (Asociación de Maquiladoras, A.C. 2024).

16.2 METHODOLOGY

The methodology was based on a survey to identify the companies' activities crucial to the ergonomics program's success. After the surveys were completed, the data were collected and analyzed, and conclusions were drawn. The study was conducted in several phases, described below. The methodology used in this research is illustrated in Figure 16.1.

16.2.1 First Phase: Survey Implementation

A questionnaire developed by Maldonado-Macías and Alferez-Padrón was used to identify the critical success factors and the benefits of implementing ergonomics in the manufacturing industry in Ciudad Juárez (Alferez-Padrón et al. 2018). After conducting a prior validation using a database that considered missing, atypical, and out-of-range values, along with exploratory and confirmatory factor analyses to reduce the number of variables, the instrument consists of 105 items. These items utilize a five-point Likert scale to measure how frequently various factors are applied during the four stages of implementing an ergonomics program, as illustrated in Figure 16.2.

16.2.2 Second Phase: Creation of Database

For the creation of the database, the information was entered into an Excel® spreadsheet, and corrections were conducted using IBM SPSS® software version 26 by following these steps:

Missing values: Missing data were addressed through data imputation, replacing the missing values with statistically derived values. The median was used for ordinal data, and the mean was used for scalar data.

Out-of-range values: These values usually result from human errors during data entry or other factors and fall outside the specified ranges indicated in the questionnaire based on the Likert scale. Descriptive statistics were used to verify these outliers, including maximum and minimum values.

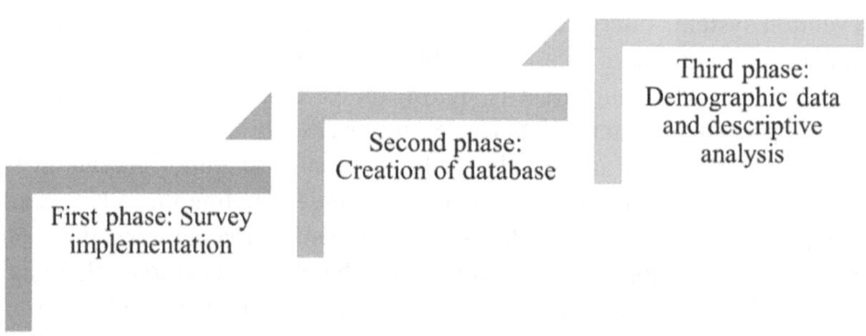

FIGURE 16.1 Methodology.

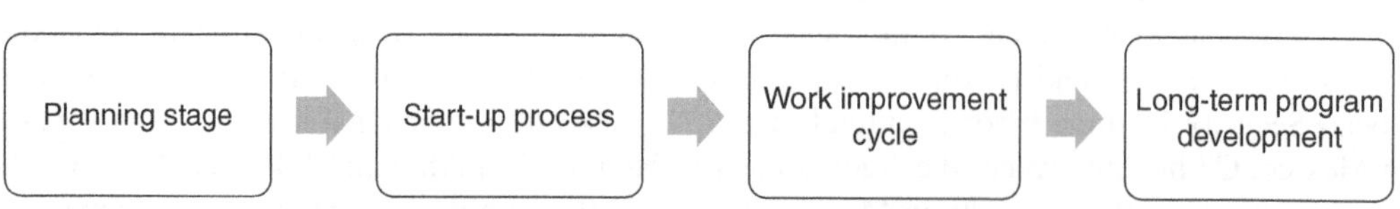

FIGURE 16.2 Ergonomic program stage implementation. (Proposal by the author Alferez-Padrón et al., 2018.)

Outlier values: A box plot was employed to analyze atypical values. This graph highlights data points that fall outside designated limits, considered outliers. According to Kwak and Kim (2017), these outliers should be excluded from the analysis.

16.2.3 Third Phase: Demographic Data and Descriptive Analysis

The subsequent analysis phase involved a comprehensive descriptive evaluation of the collected data. For the demographic data, a summary was developed for gender, job position, seniority, and manufacturing business line. Regarding the descriptive analysis, given that our dataset comprised numeric values, we utilized the median to assess the central tendency. A higher median value signified that a particular activity played a crucial role in the success of ergonomics programs. In contrast, lower median values suggested that the activities were deemed less significant.

In addition to the median, we computed the first and third quartiles, providing insights into the data distribution. The interquartile range, defined as the difference between the first and third quartiles, effectively represented the middle 50% of the dataset and encompassed the median at the second quartile. Notably, elevated interquartile range values indicated a pronounced lack of consensus among respondents regarding the perceived importance of the various activities. Conversely, lower interquartile range values pointed to a tighter clustering of responses, signaling a greater consensus among the participants about the significance of these activities (García Alcaraz, Alvarado Iniesta, and Maldonado-Mazías 2014).

16.3 RESULTS

16.3.1 Results of the First Phase: Survey Implementation

The fieldwork involved administering a questionnaire digitally using the Google Forms tool called "Critical Success Factors and Benefits of Implementing Ergonomics Programs with Lean Manufacturing Tools in the Manufacturing Industry in Ciudad Juárez" (Alferez-Padrón et al., 2018) with corresponding companies' permits. Participants signed a written informed consent letter, and the Autonomous University of Ciudad Juárez Research Ethics Committee approved the research protocol. The study requested confidentiality and used data solely for academic purposes. A nonprobabilistic sample of 248 employees from manufacturing companies in Ciudad Juárez, Chihuahua, was surveyed in 2022 and 2023, focusing on those in middle and senior management roles.

16.3.2 Results of the Second Phase: Creation of the Database

At this stage, no missing or out-of-range values were detected. However, four outliers were identified. According to Kwak and Kim (2017), outliers are observations that significantly differ from the other data points in a dataset. These values can impact the results of statistical analyses, so they were removed from the primary database.

16.3.3 Results of the Third Phase: Demographic Data and Descriptive Analysis

The descriptive analysis was divided into sections according to the five stages of ergonomics program implementation: Planning, start-up process, work improvement cycle, long-term program development, and benefits of ergonomic implementation.

Demographic data: 248 participants, representing various technical and management-level professionals, participated in the study. These individuals came from diverse manufacturing sectors in Ciudad Juárez, Chihuahua, Mexico, containing six distinct business lines. Regarding the respondents' gender, the male gender predominates with 73%. Figure 16.3 illustrates the gender distribution of the respondents.

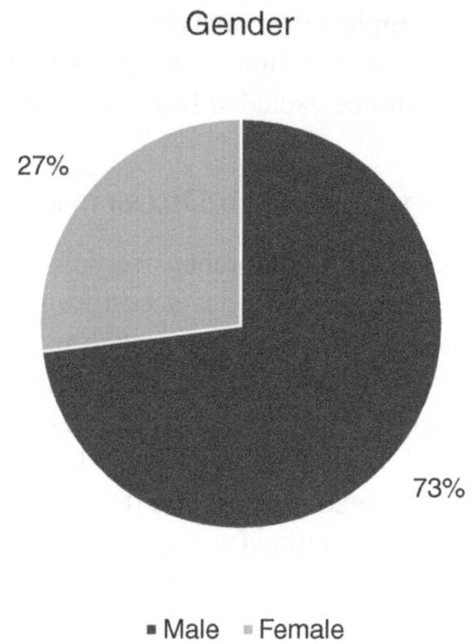

FIGURE 16.3 Gender of the respondents.

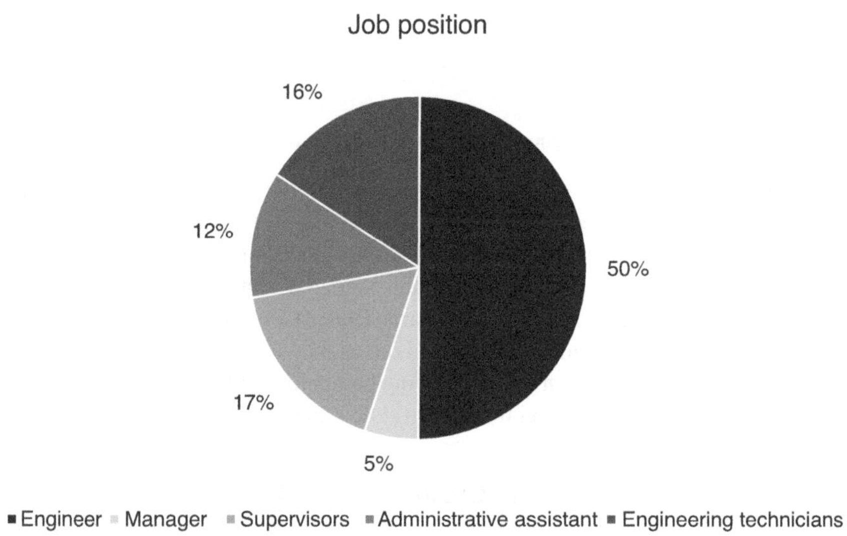

FIGURE 16.4 Distribution of respondents job positions.

The respondents' job positions primarily include engineers at 50% and supervisors at 17%. Figure 16.4 illustrates the distribution of respondents' job positions.

Regarding job seniority, the distribution of employees based on their experience is as follows: 21% have less than 2 years of experience, 25% have between 2 and 5 years of experience, 20% possess between 6 and 10 years of experience, and 34% have more than 10 years of experience in their current position. This stage details the respondents' manufacturing business lines. Figure 16.5 shows that the majority, 67%, are in the automotive sector. This is followed by the electrical–electronic industry, at 29%, with smaller contributions from the medical, transportation, plastics, and industrial machining sectors, each representing 1%.

Descriptive analysis: Dimensions conducted a descriptive analysis based on the questionnaire, which constitutes the planning stage. This stage encompasses planning costs, prevention activities, work environment, and management commitment. The second dimension is the start-up process, which comprises organization, training, capacity building, and project start-up costs. The third

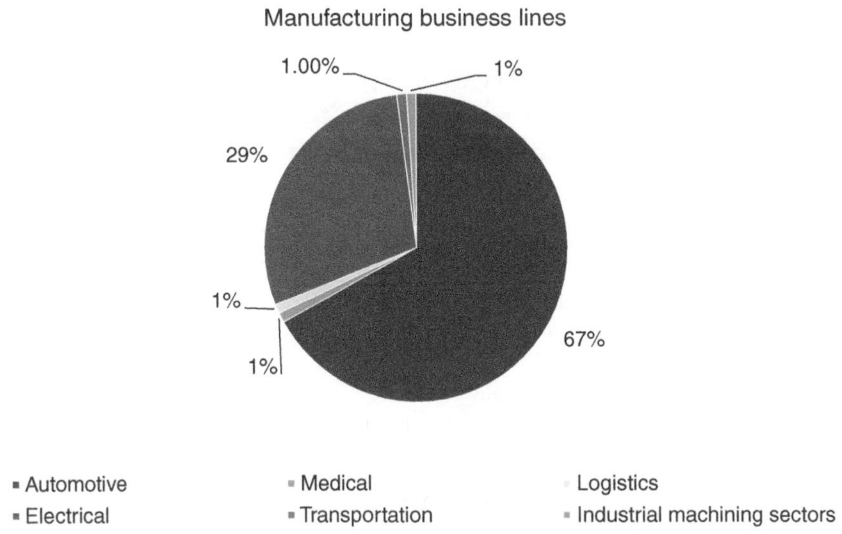

FIGURE 16.5 Manufacturing business lines.

dimension, the work improvement cycle, is confirmed by identifying priority work, conducting risk analysis, developing solutions, and implementing them. The fourth stage is called long-term program development and comprises health and safety monitoring and feedback. Finally, the benefits dimension includes the quality of life at work, economic benefits, and health and safety benefits.

Table 16.1 shows the list of 19 activities of the dimension called the planning stage. In addition, all the tables show the value of the median, 25th quartile, and 75th quartile, as well as the interquartile range. Based on the highest median values for this phase, it can be determined that, for the success of an ergonomics program within the manufacturing industries, from the perspective of the company, within the planning costs, organizations must consider disabilities, accidents, injuries, or work-related illnesses in employees. Organizations must integrate employees into a site, area, or workstation for prevention activities according to their capabilities and limitations. On the other hand, within the work environment, companies must identify ergonomic risk factors. Finally, regarding management commitment, companies must comply with regulations and the Law in Mexico based on ergonomics. The lower median values indicate those activities with the lowest impact on the success of ergonomics programs from the management perspective. Companies should be interested in ergonomics implementation regardless of the fines from the Instituto Mexicano del Seguro Social (IMSS) and the Ministry of Labor and Social Security because of the high incidence of workplace accidents and risks. In addition, other activities such as planning resting times according to the physical or mental demands of work, identifying and reducing the information load of tasks, and promoting relationships between employees and managers were less relevant for the success of ergonomics programs.

On the other hand, several activities have a low interquartile range, indicating a consensus between the survey responses and the activities' importance. Activities exhibiting the highest interquartile range suggest a lack of consensus among respondents, which reduces the perceived impact of these activities relative to their frequency of application among organizations. This means that the activities with the lowest interquartile range value are those on which organizations should invest in each of the stages of implementing the ergonomics program, based on the participants' recommendations. It is essential to highlight that the activities with a low median value have a high interquartile range value, which confirms that they are not key activities for the success of ergonomics programs.

TABLE 16.1
Descriptive Analysis of the Planning Stage

Planning Stage	Median	Percentiles 25	Percentiles 75	IR
Planning Costs				
Disabilities, accidents, injuries, or occupational illnesses of employees.	4.45**	3.47	N/A	N/A
Employees are absent and incapacitated due to work-related causes.	4.21	3.01	4.97	1.96
Medical care (internal and external: includes specialized care) for work-related injuries or illnesses.	4.33	3.24	N/A	N/A
Fines by IMSS, Ministry of Labor and Social Security.	3.63*	2.04	4.67	2.63
Train new employees to replace those injured and/or disabled due to work-related accidents or illnesses.	3.95	2.71	4.76	2.05
Prevention Activities				
Employees are integrated into a site, area, or workstation according to their capabilities and limitations.	4.00**	3.09	4.73	1.64
Periodic medical evaluation of company employees.	3.79	2.64	4.67	2.03
Resting periods are planned according to the physical or mental costs of the activities conducted by employees.	3.26*	2.15	4.33	2.18
Work Environment				
Decrease in complaints related to working conditions.	3.59	2.72	4.40	1.68
Improve plant layout.	3.88	3.07	4.63	1.56
Identification of ergonomic risk factors in the work environment.	3.94**	3.09	4.68	1.59
Analysis of the implementation of automation for operations that are risky for humans.	3.84	3.01	4.65	1.64
Reduction of human errors in tasks.	3.69	2.98	4.46	1.48
Identify and reduce the information load of tasks.	3.52*	2.67	4.32	1.65
Management Commitment				
There is a commitment from the management.	4.01	3.15	4.73	1.58
Foster relationships between employees and managers.	3.46*	2.44	4.41	1.97
Allocate resources to prevent disabilities, accidents, or work injuries.	3.85	2.88	4.66	1.78
Comply with regulations and laws in Mexico based on ergonomics.	4.37**	3.52	N/A	N/A
Strengthening competition in the market.	4.14	3.26	4.81	1.55

Note: The item with the lowest median value is indicated with an asterisk (*), and the item with the highest median value by dimensions is marked with a double asterisk (**).

Table 16.2 lists 11 activities of the dimension called the start-up process. Based on the highest median values, it can be determined that, for the success of an ergonomics program within the manufacturing industries from the company's perspective, companies must carry out activities to ensure communication between the ergonomics committee and management. Organizations must provide real case studies that demonstrate the effectiveness of training with the ergonomics person or group. Finally, within the project start-up cost, companies must measure the time invested in meetings. The lowest median values indicate those less relevant activities with the lowest impact on the success of ergonomics programs according to their frequency of application among organizations. This means that companies perceive the implementation of a work plan for the person or group (committee) on ergonomics and the use of computer programs specialized in ergonomics, training, and teaching of ergonomics topics through conferences, specialization, master's degrees, and doctorates as less relevant.

TABLE 16.2
Descriptive Analysis of the Start-Up Process

Start-Up Process	Median	Percentiles 25	Percentiles 75	IR
Organization				
Implement a work plan for an ergonomics specialist or group (committee).	3.55*	2.49	4.46	1.97
Activities are carried out to ensure communication between the ergonomics specialist or group (Ergonomics committee) and management.	3.49**	2.35	4.41	2.06
Activities are conducted to ensure communication between the ergonomics specialist or group (Ergonomics committee) and the employees.	3.46	2.39	4.38	1.99
Training				
Only topics (without explaining real cases).	3.35	2.38	4.25	1.87
Real case studies with the ergonomics person or group.	3.38**	2.33	4.31	1.98
Ergonomics manuals or guides.	3.01	2.05	3.94	1.89
Use of specialized computer programs in Ergonomics.	2.76*	1.63	3.91	2.28
Post-ergonomics training reassessments.	2.97	1.93	3.99	An increase in
Projects Start-Up Costs				
Promoting the ergonomics program within the company.	3.31	2.16	4.28	2.12
Time spent in meetings.	3.34**	2.20	4.34	2.14
Training and learning ergonomics topics (conferences, specialization, master's degrees, doctorates, etc.).	2.71*	1.66	3.72	2.06

Note: The item with the lowest median value is indicated with an asterisk (*), and the item with the highest median value by dimensions is marked with a double asterisk (**).

Various activities have a low interquartile range, which indicates a consensus between the respondents' responses and their importance. Those activities with a higher interquartile range show a poor consensus between the respondents' responses, which reduces the activity's impact. It is essential to highlight that the activities with a low median value have a high interquartile range value, which confirms that they are not key activities for the success of ergonomics programs.

Table 16.3 lists 17 activities of the dimension called the work improvement cycle. Based on the highest median values, it can be determined that, for the success of an ergonomics program within the manufacturing industries from the perspective of the company and within the identification of priority work, companies must involve employees. Regarding risk analysis, organizations must conduct walkthroughs. Regarding the development of solutions, the organization should solicit the opinions of employees from the ergonomics group. Finally, within the implementation costs, organizations must designate resources to purchase the tools and equipment necessary to execute the ergonomics program analyses.

The lower median values reflect activities with minimal impact on the success of ergonomics programs based on their application frequency across organizations. This means that companies perceive the medical service's recommendations, the use of specialized ergonomics software for risk investigation, and the development of solutions as less relevant. Several activities have a low interquartile range, which indicates a consensus between the respondents' responses and the activities' importance. Those activities with a higher interquartile range show a poor consensus between the respondents' responses, which reduces the activity's impact. It is essential to highlight that the

TABLE 16.3
Descriptive Analysis of the Work Improvement Cycle

Work Improvement Cycle	Median	Percentiles 25	Percentiles 75	IR
Identification of Priority Work				
Employees are subjected to repetitive movements, non-neutral postures, extreme temperatures, and vibration.	3.75	2.81	4.61	1.80
It has a higher rate of accidents, injuries, and/or occupational illnesses.	3.54	2.49	4.53	2.04
Based on the recommendations of the medical service.	3.55*	2.65	4.46	1.81
They have new processes, materials, equipment, tools, and workstations.	3.75	2.91	4.57	1.66
Employees are involved.	3.77**	2.84	4.61	1.77
Risk Analysis				
Tours.	4.02**	3.11	4.76	1.65
Use of questionnaires.	3.11	2.16	4.07	1.91
The analysis uses the ergonomic method (REBA, RULA, OWAS, or others).	2.93	1.87	4.01	2.14
Use of specialized ergonomics software for risk investigation.	2.45*	1.37	3.63	2.26
Development of Solutions				
Ergonomics manager or group based on employee feedback.	3.46**	2.32	4.42	2.10
The ergonomics manager or group is determined based on the area manager's opinion.	3.39	2.28	4.36	2.08
Use of specialized ergonomics software for the development of solutions.	2.52*	1.48	3.69	2.21
Research (scientific articles, internet archives, etc.).	2.79	1.76	3.82	2.06
Experts from outside the company (consultants, ergonomists, or university experts).	2.81	1.78	3.93	2.15
Gathering ideas and information by comparing aspects within or outside the company (benchmarking).	2.92	2.00	3.88	1.88
Implementation Costs				
For the purchase of tools (equipment and software).	3.74**	2.76	4.60	1.84
By an expert (consultant, ergonomist, or university).	2.96*	2.02	3.98	1.96
This is for stopping the line or station when a find is detected.	3.68	2.54	4.59	2.05

Note: The item with the lowest median value is indicated with an asterisk (*), and the item with the highest median value by dimensions is marked with a double asterisk (**).

activities with a low median value have a high interquartile range value, which confirms that they are not key activities for the success of ergonomics programs.

Table 16.4 lists nine activities of the dimension called long-term program development. Based on the highest median values, it can be determined that, for the success of an ergonomics program within the Mexican manufacturing industries from the company's perspective, companies must keep records and consider the costs associated with internal and external medical care after the program's implementation. Regarding health and safety monitoring, organizations must follow-up on ergonomic risks. Finally, with the feedback, organizations must ask middle and senior management to support solving the problem. Regarding the lower median values, they indicate those activities with the lowest importance according to the frequency with which they are applied for the success of ergonomics programs. This means that companies perceive the overall comparison of the total investment with the benefits obtained from the program or the comparison of the rate of accidents, injuries, or occupational diseases before and after implementing the program as less relevant.

TABLE 16.4
Descriptive Analysis of Long-Term Program Development

		Percentiles		
Long-term development	**Median**	**25**	**75**	**IR**
Feedback is provided to the management so that they can continue investing in the ergonomics program.	3.28	2.19	4.23	2.04
The total investment is compared with the benefits obtained from the program.	3.18*	2.21	4.09	1.88
Records are kept, and costs associated with completing the entire project are considered.	3.42	2.34	4.34	2.00
Records are kept, and internal and external medical care costs are considered after implementing the program.	3.44**	2.34	4.37	2.03
Health and Safety Monitoring				
Monitor ergonomic risks.	3.76**	2.76	4.62	1.86
Compare the rate of accidents, injuries, or occupational illnesses before and after implementing the program.	3.68*	2.68	4.55	1.87
Feedback				
The operational staff.	3.72	2.69	4.59	1.90
Those in charge of supporting and solving the problem.	3.77**	2.66	4.66	2.00
The manager of the department is affected by ergonomic risks.	3.69*	2.62	4.60	1.98

Note: The item with the lowest median value is indicated with an asterisk (*), and the item with the highest median value by dimensions is marked with a double asterisk (**).

Different activities show varying degrees of consensus among respondents, as indicated by their interquartile ranges. Activities with a low interquartile range suggest that respondents largely agree on their importance. In contrast, activities with a higher interquartile range indicate a lack of consensus, which diminishes their overall impact. It is important to note that activities with low median values and a high interquartile range further confirm that they are not critical for the success of ergonomics programs.

Table 16.5 lists 41 benefits of implementing ergonomics programs. Based on the highest median values, it can be determined that the company perceives the benefits of an ergonomics program in the quality of life at work, the increase in the skills to perform the work of employees, as well as the increase in the level of competitiveness of employees. Regarding the economic benefits, the image and reputation of the company can be improved.

Finally, regarding the benefits in health and safety, the reduction of employees' accidents, injuries, or work-related illnesses is perceived. Those benefits with the lowest impact on employees are indicated to have the lowest median values. Companies may perceive to a lesser extent, but not completely rule out, benefits such as increased chances of being promoted, decreased monotony of employees' work, and reduced or eliminated physical and mental stress at work for employees. Various benefits have a low interquartile range, which indicates a consensus between respondents' responses and their perception of the benefits. Benefits characterized by a higher interquartile range illustrate a lack of consensus among respondents, thereby reducing the perceived value of these benefits among the respondents.

16.4 CONCLUSIONS

While ergonomics programs are specifically designed to enhance efficiency, productivity, and user satisfaction while minimizing the risk of injury and discomfort, they ultimately are strongly based on people. That is why critical success factors are fundamental in ergonomics implementation.

TABLE 16.5
Descriptive Analysis of Ergonomics Benefits

Benefits of Implementing Ergonomics Programs	Median	Percentiles 25	Percentiles 75	RI
Benefits in the Quality of Life at Work				
Increment employees' morale.	3.58	2.68	4.41	1.73
Reduction of discomfort in the employees' work environment.	3.55	2.61	4.42	1.81
An increase in employee engagement across the company.	3.55	2.57	4.44	1.87
Increment employee motivation.	3.39	2.41	4.31	1.90
Increase positive employee attitude.	3.57	2.59	4.44	1.85
Increment employee self-esteem.	3.41	2.48	4.28	1.80
Increase employees' job performance skills.	3.71**	2.9	4.49	1.59
Increase employee satisfaction with their jobs.	3.51	2.57	4.36	1.79
Increase communication between the supervisor and the employee.	3.58	2.66	4.40	1.74
Increase teamwork.	3.67	2.78	4.49	1.71
Increase the chance of employee promotion.	3.30*	2.33	4.23	1.90
It helps prevent employees' compensation claims against the company.	3.56	2.57	4.45	1.88
Improvement of the performance of older or sick staff.	3.58	2.62	4.43	1.81
Enhance the competitiveness of employees.	3.71**	2.84	4.53	1.69
Enhance employees' problem-solving knowledge.	3.66	2.75	4.51	1.76
Economic Benefits				
Improve product quality.	3.82	3.07	4.58	1.51
Reduce employee absenteeism related to work-related accidents, injuries, or illnesses.	3.70	2.8	4.54	1.74
Reduce errors in product production.	3.75	2.92	4.55	1.63
Reduce days lost due to accidents, injuries, or work-related illnesses.	3.71	2.87	4.53	1.66
Increase productivity.	3.83	3.08	4.60	1.52
Increase the efficient use of working time.	3.81	3.07	4.56	1.49
Reduce waste (scrap).	3.58	2.68	4.43	1.75
Reduces the cost of medical care that employees may require.	3.71	2.85	4.53	1.68
Reduce material transport within work areas.	3.60	2.71	4.43	1.72
Reduce the monotony of employees' work.	3.47*	2.47	4.34	1.87
Decrease delays, i.e., on-time delivery of production requirements.	3.71	2.82	4.54	1.72
Help prevent damage/risks to company facilities.	3.79	3.01	4.58	1.57
Increase the quality of equipment maintenance.	3.60	2.73	4.43	1.70
Improve the company's image/reputation.	3.91**	3.01	4.67	1.66
Prevent/reduce sanctions from authorities.	3.78	2.84	4.61	1.77
Reduce the rate of employee compensation.	3.64	2.75	4.47	1.72
Save costs on cleaning your company and machinery.	3.54	2.65	4.38	1.73
Contribute to the efficient consumption of electrical energy and other inputs in the company.	3.79	2.94	4.59	1.65
Health and Safety Benefits				
Reduction of employee accidents, injuries, or occupational illnesses.	4.18**	3.29	4.86	1.57
Reduce fatigue during employees' workday.	3.75	2.94	4.53	1.59
Improve hygiene at work (excessive noise, dust, toxicity, etc.)	3.81	3.01	4.61	1.60
Reduce/eliminate physical/mental stress at work for employees.	3.69*	2.83	4.50	1.67
Increase the comfort level of environmental conditions (temperature, lighting, air quality, humidity, etc.)	3.75	2.95	4.54	1.59
Reduce the likelihood of environmental incidents.	3.88	3.04	4.64	1.60
Reduce the number of environmental complaints.	3.75	2.84	4.56	1.72

Note: The item with the lowest median value is indicated with an asterisk (*), and the item with the highest median value by dimensions is marked with a double asterisk (**).

For successful implementation and sustainability of ergonomics programs within Mexican manufacturing industries, companies must address several key components throughout different stages of the program. In the planning phase, organizations must consider the inclusion of employees with disabilities, accidents, injuries, or work-related illnesses by adapting workstations and environments to match their capabilities and limitations. Additionally, companies are responsible for identifying ergonomic risk factors in the workplace and ensuring compliance with Mexican regulations and laws related to ergonomics, demonstrating clear management commitment. During the start-up process, success depends on establishing a clear and consistent communication between the ergonomics committee and management. Training efforts should be strengthened through the presentation of real case studies, and the time invested in meetings should be systematically measured and accounted for as part of the program's initial costs. In the work improvement cycle, employee involvement is critical. Companies must prioritize engaging workers in identifying key areas for improvement, conducting thorough risk analyses through walkthroughs, and actively incorporating employee feedback into the development of ergonomic solutions. Furthermore, organizations need to allocate sufficient resources to procure the necessary tools and equipment required for implementing ergonomic interventions. For the long-term development of the program, companies should maintain detailed records and consider ongoing costs related to internal and external medical care. Continuous monitoring of ergonomic risks is essential, along with soliciting support from both middle and senior management to address any challenges. Activities with lower median values suggest less frequent application and perceived importance, highlighting areas where further awareness and emphasis may be needed.

Altogether, these findings underscore the importance of a holistic, integrated, and participatory approach, supported by strong management commitment and strategic resource allocation, to ensure the long-term success of ergonomics programs in the manufacturing sector.

REFERENCES

ABinaya, Gautam K., and Rajkumar, Dinesh. 2021. "Analysis of Ergonomic Risk Factors in the Construction Industry." *Materials Today: Proceedings, International Conference on Newer Trends and Innovation in Mechanical Engineering: Materials Science*, vol. 37, pp. 2415–18. doi:10.1016/j.matpr.2020.08.269.

Afroz, Saman, and Mohammad Israrul Haque. 2021. "Ergonomics in the Workplace for a Better Quality of Work Life." In *Ergonomics for Improved Productivity*, edited by Mohammad Muzammil, Abid Ali Khan, and Faisal Hasan, pp. 503–11. Singapore: Springer. https://doi.org/10.1007/978-981-15-9054-2_57.

Alferez-Padrón, César Roberto, Aide Aracely Maldonado-Macías, Manuel Alejandro Barajas-Bustillos, Oziely Daniela Armenta-Hernández, Arturo Realyvásquez Vargas y César Omar Balderrama-Armendáriz. 2021. "Ergonomics Implementation in Manufacturing Industries: Management Commitment for Financial Benefits." In *New Perspectives on Applied Industrial Ergonomics*, edited by Arturo Realyvásquez Vargas, Jorge Luis García-Alcaraz, and Emigdio Z-Flores, pp. 125–56. Cham: Springer International Publishing. doi:10.1007/978-3-030-73468-8_7.

Alferez-Padrón, César Roberto, Aide Aracely Maldonado-Macías, Jorge Luis García-Alcaraz y Liliana Avelar-Sosa. 2018. "Knowledge Management and Ergonomics Implementation in Manufacturing Systems: Development and Validation of a Questionnaire for Critical Success Factors." In *Knowledge Integration Strategies for Entrepreneurship and Sustainability*, pp. 188–213. IGI Global.

Asociación de Maquiladoras, A.C. 2024. *Información Estadística de las Empresas IMMEX*. 281. https://indexjuarez.com/estadisticas/infograma.

Berhan, Eshetie. 2020. "Management Commitment and Its Impact on Occupational Health and Safety Improvement: A Case of Iron, Steel and Metal Manufacturing Industries." *International Journal of Workplace Health Management* 13 (4): 427–44. https://www.emerald.com/ijwhm/article/13/4/427/160619/Management-commitment-and-its-impact-on

Bradley S, Joseph. 2003. "Corporate Ergonomics Programme at Ford Motor Company." *Applied Ergonomics* 34 (1): 23–28. https://www.sciencedirect.com/science/article/abs/pii/S0003687002000807?via%3Dihub

Buniya, Mohanad Kamil, Idris Othman, Riza Yosia Sunindijo, Ghanim Kashwani, Serdar Durdyev, Syuhaida Ismail, Maxwell Fordjour Antwi-Afari, and Heng Li. 2021. "Critical Success Factors of Safety Program Implementation in Construction Projects in Iraq." *International Journal of Environmental Research and Public Health* 18 (16): 8469. doi:10.3390/ijerph18168469.

García Alcaraz, Jorge Luis, Alejandro Alvarado Iniesta y Aide Aracely Maldonado-Mazías. 2014. *Human Factors Related to the Success of Total Productive Maintenance*, edited by Shrawan Kumar, Anil Mital, Arunkumar Pennathur. CRC Press Taylor & Francis Group. https://www.routledge.com/Human-Work-Productivity-A-Global-Perspective/Kumar-Mital-Pennathur/p/book/9780367379698.

Heidarimoghadam, Rashid, Iraj Mohammadfam, Mohammad Babamiri, Ali Reza Soltanian, Hassan Khotanlou, and Mohammad Sadegh Sohrabi. 2022. "What Do the Different Ergonomic Interventions Accomplish in the Workplace? A Systematic Review." *International Journal of Occupational Safety and Ergonomics* 28 (1): 600–624.

Hong, Chee Chong, T. Ramayah, and Chandrakantan Subramaniam. 2018. "The Relationship between Critical Success Factors, Internal Control and Safety Performance in the Malaysian Manufacturing Sector." *Safety Science* 104 (April): 179–88. doi:10.1016/j.ssci.2018.01.002.

International Ergonomics Association. 2000. "What Is Ergonomics?" https://iea.cc/what-is-ergonomics/.

Ishak, Siti Intan Diyana, Nordin Abd Razak, H. Hussin, Nur Suriaty Fhiri @ Daud, and Aida Shakila Ishak. 2018. "A Literature Review on Quality Teachers' Working Life." *MATEC Web of Conferences*, vol. 150, p. 05094. EDP Sciences. doi:10.1051/matecconf/201815005094.

Kwak, Sang Kyu, and Jong Hae Kim. 2017. "Statistical Data Preparation: Management of Missing Values and Outliers." *Korean Journal of Anesthesiology* 70 (4): 407–11. doi:10.4097/kjae.2017.70.4.407.

Makhbul, Zafir Khan Mohamed, Md Shafiin Shukor, and Ariff Azly Muhamed. 2022. "Ergonomics Workstation Environment toward Organisational Competitiveness." *International Journal of Public Health Science* 11 (1): 157–69.

Rostami, Matin, Alireza Choobineh, Mahnaz Shakerian, Majid Faraji, and Hossein Modarresifar. 2021. "Assessing the Effectiveness of an Ergonomics Intervention Program with a Participatory Approach: Ergonomics Settlement in an Iranian Steel Industry." *International Archives of Occupational and Environmental Health*, November. doi:10.1007/s00420-021-01811-x.

Vega, Javier Eduardo, María del Carmen Bautista y María del Carmen Martínez. 2020. "Orientación emprendedora en la innovación de las pequeñas y medianas empresas en México/Entrepreneurial orientation in the innovation of small and medium-sized companies in Mexico." *Revista de Ciencias Sociales* 26 (4). doi:https://doi.org/10.31876/rcs.v26i4.34651.

Villalobos, José. 2018. *Efectividad del Programa de Ergonomía para la Reducción de Molestias Musculoesqueléticas y Sobrecarga Postural en Trabajadores de Oficina que Utilizan Computadoras en una Empresa Bancaria. Lima-2018.* Lima: Universidad Nacional Mayor de San Marcos. https://cybertesis.unmsm.edu.pe/handle/20.500.12672/10392.

Virmani, Naveen, and Urmi Ravindra. 2021. "Assessment of Key Barriers for Incorporating Ergonomics Inventions and Suppress Work-Related Musculoskeletal Disorders." *Materials Today: Proceedings, International Conference & Exposition on Mechanical, Material and Manufacturing Technology (ICE3MT)*, 38 (January): 2601–6. doi:10.1016/j.matpr.2020.08.160.

Wee, Ian Jun Yan, Li-Jen Kuo, and James Chi-Yong Ngu. 2020. "A Systematic Review of the True Benefit of Robotic Surgery: Ergonomics." *The International Journal of Medical Robotics and Computer Assisted Surgery* 16 (4): e2113. doi:10.1002/rcs.2113.

Yazdani, Amin, and Richard Wells. 2018. "Barriers for Implementation of Successful Change to Prevent Musculoskeletal Disorders and How to Address Them Systematically - ScienceDirect." https://www.sciencedirect.com/science/article/pii/S0003687018301212.

17 Psychosocial Risk of Working Time Organization, Organizational Environment, and Lack of Control over Work during the COVID-19 Pandemic and Post-Pandemic among University Teachers

A Case Study

Mónica Gabriela Gutiérrez-Hernández,
Aide Aracely Maldonado-Macías,
Bertha Ivonne Sánchez Luján,
and Cesar Omar Balderrama-Armendariz

17.1 INTRODUCTION

The COVID-19 pandemic has created and/or exacerbated substantial psychosocial risks for university teachers, affecting their mental health, work–life balance, working time organization, and job satisfaction. During this period, numerous studies examined the effects of work on university teachers in these unprecedented conditions. These studies have suggested that the new nature of work has led to organizational changes, and that restructuring may adversely affect employee well-being. Teachers had to adapt to new working conditions caused by remote working, experiencing prolonged stress situations that, combined with psychosocial factors, detonated chronic work stress and psychosomatic health problems (Gómez-Domínguez et al. 2022). Working from home, transferring to new tasks, and team reorganizations increased due to the COVID-19 pandemic. However, it remains unclear how they may impact perceptions of psychosocial factors, work environment, and employee well-being (Ervasti et al., s/f). This work modality and working conditions prevail worldwide and have become relevant to a deeper understanding of the effects during post-pandemic periods.

Accordingly, several studies have found diverse effects of psychosocial factors during and post-pandemic among teachers. Evidence of these effects is related to the following aspects:

1. *Burnout and stress*: During the pandemic, university teachers experienced increased stress due to workload, role conflicts, and job insecurity. These factors led to burnout and psychosomatic health issues (Gómez-Domínguez et al. 2022).

DOI: 10.1201/9781003582014-21

2. *Work–life balance*: A study found that remote teaching extended working hours for 78.7% of teachers and disrupted work–life balance for 86%. Accordingly, these facts significantly lower the mental well-being of university teachers (Lizana and Vega-Fernadez 2021).
3. *Mental health decline*: College faculty reported moderate physical and mental health risks, including musculoskeletal issues and emotional strain from telework. These risks were more pronounced for instructors with extended online engagement (Ramirez 2023).
4. *Digital stressors*: The rapid shift to online teaching caused digital fatigue and stress, especially for teachers lacking adequate training in educational technologies (Navarro-Espinosa et al. 2021).
5. *Emerging psychosocial factors*: Post-pandemic evaluations highlighted continued stress linked to work overload and adjustments to evolving work demands, necessitating ongoing monitoring and intervention strategies (Panesso Patiño, Calderón Puerta, and Claribel Núñez, 2023).

In Mexico, academic activities at all levels during the COVID-19 pandemic were conducted from home for over 2 years, and working conditions demanded a reorganization and a new understanding of social interactions and dynamics (Quezada Díaz, Hindrichs, and Castellanos Simons 2024). Furthermore, during confinement, many workers began conducting their activities from home, necessitating various strategies for family organization (Quezada, Becquer, and Hindrichs 2022). These authors revealed emerging psychosocial factors that can promote or hinder working from home during confinement.

Other factors, such as comfort, children's adaptation to working from home, and new learning, may also encourage working from home in the post-pandemic period. However, psychosocial discomfort risk factors, such as long schedules and multitasking, are perceived as obstacles. Psychosocial risks during and post-pandemic periods may lead to health issues, reduced motivation, and decreased work effectiveness, ultimately impacting teaching quality (Prado-Gascó et al. 2020). Additionally, in Mexico, university teachers experienced high resilience during the pandemic, with perceived mild categories of stress and anxiety (Delgado-Gallegos et al. 2021). These studies suggest that work–life imbalance during and after the pandemic has been a significant psychosocial factor for university teachers in Mexico, contributing to work overload, stress, burnout, and mental health issues. In contrast, resilience and social support can mitigate these effects. Based on the above, this research has two main objectives. The first objective is to assess the psychosocial risk level resulting from the lack of control over the work, organizational environment, and work-time organization to which university professors were exposed during and post-pandemic. The second objective is to determine the type of relationship between the domain and the categories studied.

17.1.2 Literature Review

17.1.2.1 Work-Time Organization and Work–Life Imbalance as Psychosocial Risk Factors during and Post-Pandemic among University Teachers

Evidence suggests the importance of psychosocial risk factors (PRFs) among university teachers, particularly work-time organization and work–life balance, during and post-pandemic. This section summarizes studies on both elements, including PRFs among university teachers.

A study explored how the relationship between remote work, work stress, and work life developed during the pandemic in a Latin American context (Sandoval-Reyes, Idrovo-Carlier, and Duque-Oliva 2021). These authors found using structural modeling that remote work during pandemic times increased perceived stress (β=0.269; p<0.01), reduced work–life balance (β=−0.225; p<0.01) and work satisfaction (β=−0.190; p<0.01), and increased productivity (β=0.120; p<0.01) and engagement (β=0.120; p<0.01), and the perceived stress affects men's productivity more acutely than women's productivity. Another study evaluated the association between work hours, work–family balance, and quality of life (QoL) among teachers during the health emergency caused by the COVID-19 pandemic in Chile and found that teachers reported having increased their work hours during the COVID-19 pandemic due to teleworking, and 86% indicated adverse effects on their work–family balance (Lizana and Vega-Fernadez 2021).

Using a quantitative and correlational approach, differences were identified across age, gender, and university affiliation at three public universities in Ecuador. Female faculty members and those over the age of 50 reported more significant emotional distress. The lack of preparation and the unexpected shift to remote work during COVID-19 were key contributing factors (Cevallos Zambrano et al. 2023).

In addition, a study highlights the importance of work–life balance and the vitality of job autonomy on teachers' performance and overall well-being, where the role of organizations is to maintain a work environment where university teachers' well-being must be prioritized; elements such as training and support may help stabilize the work–life balance (Khawand & Zargar, 2022). Some other authors have revealed that perceived stress, burnout, job satisfaction, physical and mental health, psychosomatic problems, and quality of relations are antecedents of participants' work–life imbalance and concluded that teachers have altered their role during and post-pandemic by maintaining moderate levels of work–life imbalance and burnout. However, a robust, comprehensive framework is needed, considering the multiple and multilevel predictors of work–life imbalance (Bakracheva, Sofronieva, and Tsenov 2024).

Another study compared the perceptions among students and teachers during and post-pandemic and found that work overload correlated negatively with the emerging factor of "Making better use of time" and positively with "Work, school, and/or domestic activities overload." These authors found that students perceived more work overload than teachers (Quezada, Becquer, and Hindrichs, 2022).

Additionally, in the study conducted by Díaz Patiño, Anaya Velasco, and Santoyo Telles (2022), 156 participants completed an online survey, responding to Reference Guide III and Reference Guide V of the Mexican Official Standard 035 issued by the Secretariat of Labor and Social Welfare, as well as the Short Form-36 (SF-36) questionnaire on health-related quality of life (HRQoL). A total of 57.7% of the sample reported moderate to very high levels of PRFs, with vitality being the most affected dimension ($M = 65.3$). Work–family interference accounted for 19% of the variability in vitality, increasing to 23% when workplace violence was included as a variable. The most affected individuals were those aged 40 to 49, divorced, those holding a master's degree, full-time faculty members, and those with permanent contracts.

These studies suggest that university teachers' work–life imbalance, environment, and time management during and after the pandemic have been influenced by perceived stress, burnout, autonomy at work, organizational and family support, and workload, resulting in a moderate work–life imbalance and well-being. However, significant challenges in mental health and job satisfaction have also been identified.

17.1.2.2 Challenges to Attending to Psychosocial Risk in the Post-Pandemic Period for University Teachers

In post-pandemic settings, psychosocial risks remained significant, with intra-work demands and stressors such as organizational justice and resources continuing to affect university teachers (Panesso Patiño, Calderón Puerta, and Claribel Núñez 2023).

The authors have highlighted relevant challenges for the post-pandemic period, which include the following:

1. *Country and institutional variations*: Teachers in countries like Spain and Mexico faced resource inequity and higher job demands, amplifying psychosocial risks during lockdowns (Prado-Gascó et al. 2020).

2. *Protective factors*: Access to resources, organizational support, and proper communication reduced some psychosocial risks, preventing long-term mental health deterioration (Romeo et al. 2021).
3. *Mental health challenges*: University teachers experienced heightened stress, anxiety, and depression due to changes in work conditions, increased workloads, and teleworking demands during the pandemic (Koren et al. 2023).
4. *Digital training in new learning technologies*: During the pandemic and beyond, university faculty have faced an intensification of psychosocial risks, including work overload, emotional demands, work–family conflicts, conflicting interpersonal relationships, and lack of organizational support. These factors are associated with significant increases in stress, burnout, and symptoms of anxiety and depression, as well as decreases in job satisfaction and performance (Souto et al. 2018; Wischlitzki et al. 2020; Sanchis-Giménez et al. 2024; Pimenta et al. 2023). The studies suggest that it is necessary to implement organizational strategies that promote faculty members' psychological well-being and digital training, particularly in public universities and among older faculty, to mitigate the adverse effects of prolonged remote work, reduce the digital divide, and alleviate the associated stress associated with new digital learning technologies and environments (Antón-Sancho et al. 2023; Fernandes et al. 2023; Fernández-Arias, et al. 2024).

17.2 METHODOLOGY

17.2.1 Participants

The instrument was distributed among 350 teachers of the Tecnológico Nacional de México/Instituto Tecnológico de Ciudad Juárez on two occasions via Google Forms. The first period was during the pandemic, from January to June 2021, when classes were still taught remotely ($n = 202$ teachers). The second was a post-pandemic period from January to December 2024 (119 teachers) On both occasions, the samples were selected by convenience rather than being chosen probabilistically due to pandemic restrictions and the limited time available for data collection. Additionally, the instrument was completed voluntarily and anonymously, thereby expediting the research by allowing participants who wish to participate in the study to do so (Ortega 2018). Informed consent was obtained and approved by the Institutional Ethics Committee on November 16, 2019 (DEPI/ITCJ/035/19). Although the population under study was the same in both data collection occasions, the sample of individuals who participated during the pandemic period differed from the post-pandemic one, due to the same restrictions mentioned earlier.

17.2.2 Materials and Methods

17.2.2.1 Instrument

The Reference Guide III, proposed by NOM-035-STPS-2018, was used as an instrument for data collection. This instrument comprises 74 items, organized into five general categories, each divided into 10 domains. For this study, only the items corresponding to the category's organization of work time and organizational environment, as well as those that make up the domain's lack of control over the work, were used.

On the one hand, the category of working time organization is related to the length of the working day and the interference between the work–family relationship. The first domain only presents two items that analyze the amount of extra time that should be dedicated to work. The second domain evaluates the frequency with which work and family matters must be attended simultaneously (Table 17.1). The six items that comprise this category are Likert-type items with a five-point scale, where 4 = Always, 3 = Almost always, 2 = Sometimes, 1 = Almost never, and 0 = Never.

TABLE 17.1
Structure of the Working Time Organization Category

Category	Domain	Item	Description	Point Value
Organization of working time	Working hours	17	I work more than three times a week	4–0
		18	My job requires me to work on days off, holidays, and weekends.	
	Interference in the work–family relationship	19	I consider my work hours to be too long, which negatively impacts my family and personal life	
		20	I must attend to work matters when I am at home.	
		21	I think about family or personal activities when I am at work.	
		22	My family responsibilities affect my work.	

TABLE 17.2
Structure of the Organizational Environment Category

Category	Domain	Item	Description	Point Value
Organizational environment	Performance recognition	47	They inform me about what I do well in my work	0–4
		48	How my work is evaluated in my workplace helps me improve my performance.	
		49	At my workplace, I get paid my salary on time.	
		50	The payment I receive is what I deserve for the work I do	
		51	I am rewarded and recognized when I achieve the expected results in my work.	
		52	People who excel in their jobs can advance professionally.	
	Insufficient sense of belonging and instability	53	I consider my work to be stable	
		54[a]	In my work, there is continuous staff turnover.	
		55	I am proud to work in this workplace.	
		56	I feel committed to my work.	

[a] This item has an inverse scale.

On the other hand, the organizational environment category comprises 10 items, grouped into two domains: insufficient performance recognition (six items) and performance recognition (four items) (Table 17.2). This category assesses whether workers receive feedback on their performance, whether they receive their salary promptly, and their perception of job stability. Of the 10 items, nine are Likert-type items with a five-point scale, where: 0 = Always, 1 = Almost always, 2 = Sometimes, 3 = Almost never, and 4 = Never. Item 54 is inverse; that is, it ranges from 4 (always) to 0 (never).

Likewise, the 10 items that comprise the domain's lack of control over work were used (Table 17.3). They analyze autonomy, job growth possibilities, access to training, and the possibility

TABLE 17.3
Domain Structure: Lack of Control over the Work

Domain	Item	Description	Point Value
Lack of control over the work	23	My job allows me to develop new skills	0–4
	24	In my work, I can aspire to a better position.	
	25	During my workday, I can take breaks as needed.	
	26	I can decide how much work I do during the working day.	
	27	I can set the pace at which I perform my job tasks.	
	28	I can adjust the order in which I perform tasks at my job.	
	29[a]	The changes in my work make it more challenging.	
	30	When changes occur in my work, my ideas or contributions are considered.	
	35	They allow me to attend job-related training.	
	36	I receive helpful training to perform my job effectively.	

[a] This item has the inverse scale.

of participating in decision-making, where 0 = Always, 1 = Almost always, 2 = Sometimes, 3 = Almost never, and 4 = Never. Item 29 used an inverse scale, that is, it ranges from 4 (always) to 0 (never).

17.2.3 Methods

17.2.3.1 Descriptive Analysis

Once the instrument was applied, the database was created in SPSS® version 26. Subsequently, the responses were scored according to the values established by NOM-035-STPS-2018. It was corroborated that the database was free of outliers and missing values. In the event of missing values, these must be replaced with the median value as the variables are ordinal (IBM Corporation, 2019). Once the database was ready, the level of psychosocial risk to which the teachers were exposed was determined according to the established cut-off points (Table 17.4). Likewise, a graph compares the prevalence of risk during and post-COVID-19 pandemic.

17.2.3.1.1 Difference of Two-Proportion Test

In this stage, the risk values obtained during and after the pandemic were compared using the two-proportion difference test for independent samples, with a significance level of $\alpha = 0.05$.

17.2.3.1.2 Spearman Correlation Analysis

Spearman's Rho is a nonparametric statistical method used in small samples or when ordinal scales are present, as they do not meet the normality assumption (Croux and Dehon, 2010). This statistic measures the relationship between two variables in terms of their strength and direction (Easily 2023). In other words, correlation analysis allows us to determine the type of relationship between the variables under study.

TABLE 17.4
Cut-Off Points for Categories and Domains given by the NOM-035-STPS-2018

		Category: working time organization		
Null	**Low**	**Middle**	**High**	**Very high**
$C_{Time} < 5$	$5 \le C_{Time} < 7$	$7 \le C_{Time} < 10$	$10 \le C_{Time} < 13$	$C_{Time} \ge 13$
		Category: Organizational Environment		
Null	**Low**	**Middle**	**High**	**Very high**
$C_{Ent} < 10$	$10 \le C_{Ent} < 14$	$14 \le C_{Ent} < 18$	$18 \le C_{Ent} < 23$	$C_{Ent} \ge 23$
		Domain: Lack of control over the work		
Null	**Low**	**Middle**	**High**	**Very high**
$D_{ctrl} < 11$	$11 \le D_{ctrl} < 16$	$16 \le D_{ctrl} < 21$	$21 \le D_{ctrl} < 25$	$D_{ctrl} \ge 25$

TABLE 17.5
Characteristics of the Sample

Moment	Sample	Sex	Age	Marital Status	Contract
During the pandemic	202	Men: 102	20–29 years: 20	Single: 51	Base: 91
		Women: 100	30–39 years: 52	Married: 133	Temporary: 87
			40–49 years: 49	Divorced: 14	Mixed: 24
			50–59 years: 58	Widowed: 4	
			$_{60}$ Years: 23		
After the pandemic	119	Men: 61	20–29 years: 4	Single: 21	Base: 62
		Women: 58	30–39 years: 29	Married: 95	Temporary: 47
			40–49 years: 23	Divorced: 0	Mixed: 10
			50–59 years: 37	Widowed: 3	
			$_{60}$ Age: 26		

17.3 RESULTS

Information was collected during and after the pandemic. During the pandemic, the average response rate was 57.71%. On the other hand, a smaller sample was obtained with 34% of the population. Table 17.5 shows the sociodemographic data of the participants. It is observed that on both occasions, a greater response was obtained from men. Likewise, the age range of 50–59 years was the most common. On the other hand, most are married, and a higher percentage of those with a permanent contract are shown (base).

17.3.1 Descriptive Analysis Results

Once it was confirmed that the database contained no outliers or missing values, the level of psychosocial risk was determined using the proposed criteria in NOM 035 and the corresponding cut-off points. First, the domain's lack of control over work was analyzed. Figure 17.1 shows that the samples' behavior at both times was very similar. More than half of the participants had a low or null risk level. In other words, in higher education, the characteristics of academic freedom have been maintained, allowing teachers to decide the order and form of the class.

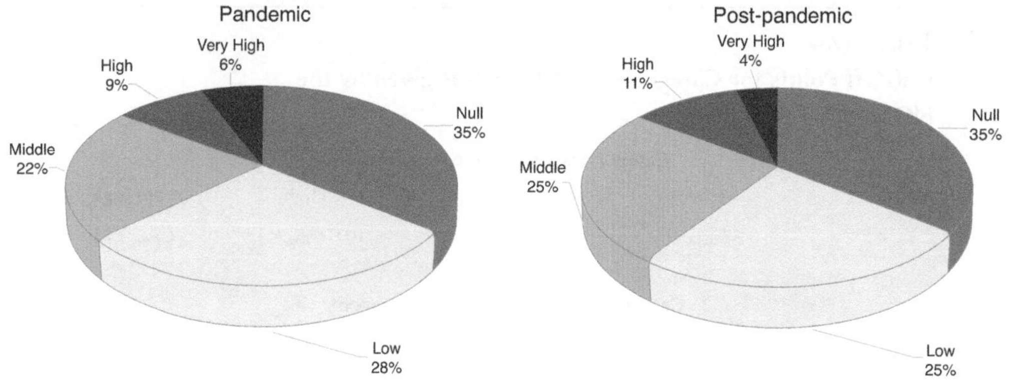

FIGURE 17.1 Prevalence of lack of control over work domain.

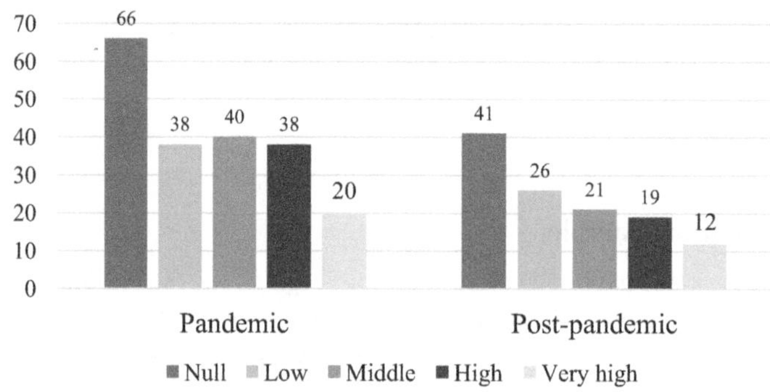

FIGURE 17.2 Organizational environment category.

On the other hand, when analyzing the organizational environment category, more than half of the participants reported a low or null risk level of 51.49% during the pandemic (Figure 17.2). After the pandemic, it was 56.30%.

On the other hand, when analyzing the working time organization category, it is evident that during the pandemic, most teachers reported a high level of risk, with 62.38% of them. On the contrary, this percentage decreased by 37.82% when returning to face-to-face classes (Figure 17.3). It is evident that, during the pandemic, beyond the technological challenge of abruptly shifting to telework, factors such as adhering to a work schedule and achieving a balance between family and work proved impossible to maintain.

17.3.2 The Two-Proportion Difference Test

The authors stated that although psychosocial risk levels increased during the pandemic, it is essential to determine whether these risks persist or decrease afterward. Thus, the differences between the two proportions were tested for the level of psychosocial risk in each category, and the domain was identified. The samples analyzed were 202 teachers during the pandemic and 119 in the post-pandemic period. To do this, the proportions by risk level concerning the moment were considered $P \neq P$.

Table 17.6 shows that the level of psychosocial risk derived from the lack of control over the work domain and the organizational environment category does not differ significantly. On the contrary,

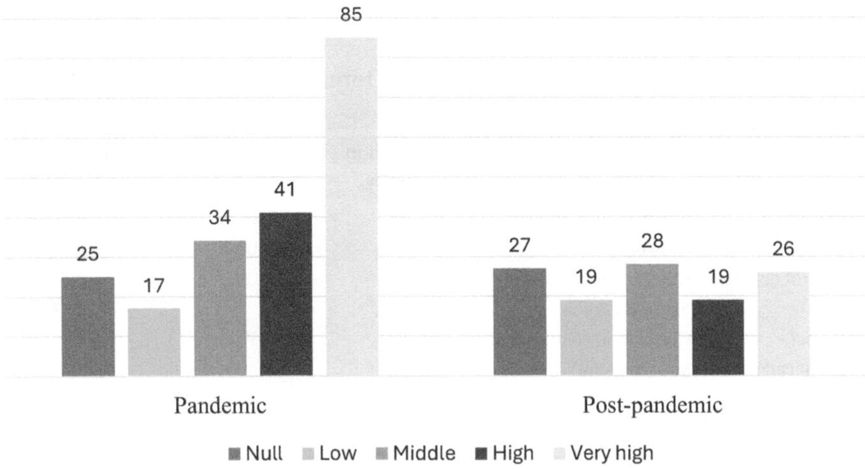

FIGURE 17.3 Category of working time organization.

TABLE 17.6
Differences between Two Proportions by Risk Level during and after the Pandemic

Domain/Category	Risk level	Events		p-Value
		Pandemic	Post-Pandemic	
Dominance lack of control over the work	Null	75	42	0.741
	Low	59	29	0.340
	Middle	46	30	0.623
	High	19	13	0.666
	Very high	13	5	0.376
Category: organizational environment	Null	66	41	0.745
	Low	38	26	0.516
	Middle	40	21	0.631
	High	38	19	0.512
	Very high	20	12	0.958
Category: working time organization	Null	25	27	0.021[a]
	Low	17	19	0.052
	Middle	34	28	0.154
	High	41	19	0.324
	Very high	85	26	0.000[a]

[a] These tests were significant with a proportional alpha $\alpha = 05$.

evidence of significant differences was presented regarding the levels of null and very high risk in the organization of working time category. Likewise, the low-risk level is marginal (p value = 0.05).

Subsequently, tests of unilateral proportions were performed on the significant risk levels with a significance level $\alpha = 0.05$. The results demonstrate sufficient statistical evidence to support the assertion that the prevalence of teachers with a very high level of psychosocial risk in work organizations was higher during the pandemic (p-value=0.000). In contrast, at the null risk level, the proportion of teachers who reported it during the pandemic was lower (p-value=0.011).

TABLE 17.7
Bivariate Correlation between Categories and Domain

		Time Organization Category	Lack of Control Domain	Organizational Environment Category
Time organization category	Correlation coefficient	1.000	0.217[a]	0.304[a]
	Sig. (bilateral)		0.000	0.000
Lack of control domain	Correlation coefficient	0.217[a]	1.000	0.576[a]
	Sig. (bilateral)	0.000		0.000
Organizational environment category	Correlation coefficient	0.304[a]	0.576[a]	1.000
	Sig. (bilateral)	0.000	0.000	

[a] The correlation is significant at the 0.01 level (two-tailed).

17.3.3 Spearman Correlation Analysis

At this stage, the bivariate correlation was calculated using Spearman's Rho because the instrument comprises ordinal variables that do not meet the normality assumption (Table 17.7). For this analysis, the two samples were combined: the pandemic sample (n=202) and the post-pandemic sample (n=119). Thus, the complete sample of 321 teachers was used for this analysis. The correlation between the categories and the domain studied was significant, with confidence levels of 95% and 99%, respectively.

17.4 DISCUSSION

The findings of this study are aligned with those of Delgado-Gallegos et al. (2021) and Sandoval-Reyes, Idrovo-Carlier, and Duque-Oliva (2021), which report a mild level of psychosocial risk during the pandemic. Specifically, the lack of control over the work domain was reported to have increased slightly. This may be since, so far, university professors continue to enjoy free academic (Creagh et al. 2025). This domain examines the possibility of determining the order and speed at which work is performed, among other factors. Similarly, the results for the organizational climate category indicate a low or no level of risk, with percentages exceeding 50% among the teachers. These are consistent with the results reported by Tapia, Otacoma, and Cerda (2024) and Tellechea-Acuña, Carpio-Camacho, and Valdés-Florat (2025).

On the other hand, when analyzing the category of organization of work time, a high level of risk was reported during the pandemic, which decreased when face-to-face teaching was resumed. These results are consistent with those of Cotonieto-Martínez (2021) and Gutiérrez-Hernández, Maldonado-Macías, and Sibaja Terán (2024). This is attributed to the need to dedicate additional time during the day to training and incorporating new information technologies into their teaching. Additionally, work and personal matters often had to be attended to simultaneously. In addition, the difference in the two-proportion test provides sufficient statistical evidence to indicate that the prevalence of the zero-risk and very high-risk levels in the working time organization category differed significantly. Through these results, the first objective of this research was achieved, : to determine the level of psychosocial risk reported by teachers during and post-pandemic. Specifically, the NOM-036 categories of lack of control over the work environment and the organization of work time were analyzed.

On the other hand, when examining the relationship between the domain of lack of control over work and each of the categories studied, significant positive correlation coefficients were obtained (Martínez Ortega et al. 2009). The correlation coefficient between this domain and the organizational environment category is strong and positive ($r=0.576$; $p<0.000$). Therefore, the greater the lack of control over the work, the greater the organizational environment. It should be noted that the organizational environment category of this instrument addresses the absence of performance recognition, the lack of a sense of belonging, and job instability. The correlation between this domain and the organization of working time category is weakly positive (r=0.217; p<0.001). This means that, despite a significant correlation, the organization of working time (long hours and interference with the work–family relationship) does not depend on the worker's control over the work. These results coincide with those reported by Cotonieto-Martínez (2021). With this, the second objective of the research was achieved by determining the type of relationship among the constructs studied.

17.5 CONCLUSION

The pandemic can intensify psychosocial risks for university teachers, contributing to mental health issues and work–life challenges. This study has demonstrated that the COVID-19 pandemic significantly affected psychosocial factors among university faculty, particularly in work-time organization, organizational environment, and work–life balance. A comparative analysis of the pandemic and post-pandemic periods shows that while some risk levels have decreased with the return to face-to-face teaching, others persist, continuing to affect faculty well-being.

The findings indicate that work-time organization had the highest risk level during the pandemic, with a 62.38% prevalence of high-risk levels, compared to 37.82% post-pandemic. This result highlights the difficulties that faculty members face in setting clear boundaries between their work and personal lives during remote education, increasing their workload and negatively affecting their well-being. Although this risk level decreased in the post-pandemic period, the persistence of demanding working conditions suggests the need for measures that promote better time management and more equitable working conditions.

Regarding the organizational environment, the comparison between the two periods shows that more than half of the faculty members reported null or low-risk levels in both periods: during the pandemic (51.49%) and after it (56.30%). However, the proportion analysis did not reveal significant differences in this category, suggesting that the pandemic did not substantially alter structural conditions related to recognition and job stability. Nonetheless, this remains a critical area that requires further reinforcement, particularly regarding performance recognition and contractual stability.

The lack of control over the work domain also exhibited similar patterns across both periods, with most faculty members reporting null risk levels. This result may reflect the relative autonomy characteristic of university teaching, allowing professors greater flexibility in managing certain aspects of their work. Nevertheless, the correlations found suggest that this domain remains interrelated with other psychosocial factors, highlighting the importance of a comprehensive evaluation of the work environment. From a methodological perspective, the proportion difference tests indicated that the very high risk level in work-time organization was significantly higher during the pandemic (p=0.000) than in the post-pandemic period. Conversely, the null risk level was lower during the pandemic (p=0.011), reinforcing the conclusion that the health crisis exacerbated adverse conditions in this area.

In terms of practical implications, these results underscore the need to develop organizational strategies that mitigate psychosocial risks in higher education, such as implementing faculty well-being programs, enhancing job recognition mechanisms, and promoting policies that strike a proper balance between professional and personal life. Furthermore, the evidence suggests that it is crucial to continue monitoring these factors in the post pandemic context, as the transition to new work dynamics may pose ongoing challenges for university faculty.

In conclusion, while the post-pandemic period has seen a reduction in some PRFs, the health crisis's impact on work organization and work–life balance remains a pressing concern. Therefore, educational institutions should consider these findings when developing policies to address the pandemic's immediate effects and foster healthier, more sustainable work environments in the long term. Interventions like workload management, improved resources, and psychosocial support can mitigate these impacts and enhance resilience in educational settings. Addressing these risks through organizational support, workload management, and targeted interventions is crucial to fostering teachers' well-being.

REFERENCES

Antón-Sancho, Álvaro, Vergara, Diego, Sánchez-Calvo, María y Fernández-Arias, Pablo. 2023. "On the Influence of the University Tenure on the Digital Pandemic Stress in Higher Education Faculty". *Behavioral Sciences* 13 (4): 335. doi:10.3390/bs13040335.

Bakracheva, Margarita, Sofronieva, Ekaterina y Tsenov, Martin. 2024. "Work-Life Balance of University Teachers after Two Years of Telework during the COVID-19 Pandemic". *Human Systems Management* 43 (5): 693–705. doi:10.3233/HSM-230089.

Cevallos Zambrano, Doris Patricia, Moreira-Choez, Jenniffer Sobeida, Lamus de Rodríguez, Tibisay Milene, y Sánchez-Urdaneta, Adriana Beatriz. 2023. "Psychosocial effects of COVID-19 in Ecuadorian university professors" XXIX (2): 15–31. https://produccioncientificaluz.org/index.php/rcs/index.

Cotonieto-Martínez, Ernesto. 2021. "Identificación y análisis de factores de riesgo psicosocial según la NOM-035-STPS-2018 en una universidad mexicana". *Journal of Negative and No Positive Results* 6 *(3). Asociación para el Progreso de la Biomedicina:* 499–523. doi:10.19230/jonnpr.3836.

Creagh, Sue, Thompson, Greg, Mockler, Nicole, Stacey, Meghan y Hogan, Anna. 2025. "Workload, Work Intensification and Time Poverty for Teachers and School Leaders: A Systematic Research Synthesis". *Educational Review* 77 (2). Routledge: 661–80. doi:10.1080/00131911.2023.2196607.

Croux, Christophe, y Dehon, Catherine. 2010. "Influence Functions of the Spearman and Kendall Correlation Measures". *Statistical Methods & Applications* 19 (4): 497–515. doi:10.1007/s10260-010-0142-z.

Delgado-Gallegos, Juan, Padilla-Rivas, Gerardo R., Zuñiga-Violante, Erika, Avilés Rodríguez, Gener, Arellanos Soto, Daniel, Villareal, Hector, Cosío León, María, Romo-Cardenas, Gerardo y Islas, Jose. 2021. "Teaching Anxiety, Stress and Resilience During the COVID-19 Pandemic: Evaluating the Vulnerability of Academic Professionals in Mexico Through the Adapted COVID-19 Stress Scales". *Frontiers in Public Health* 9 (mayo). doi:10.3389/fpubh.2021.669057.

Díaz Patiño, Diana Gisela, Anaya Velasco, Ana y Santoyo Telles, Felipe. 2022. "Factores de riesgo psicosocial y calidad de vida durante el confinamiento por covid-19 en universidades". *RIDE Revista Iberoamericana para la Investigación y el Desarrollo Educativo* 12 (24). doi:10.23913/ride.v12i24.1168.

Easily, Learn Statistics. 2023. "Comprensión de la correlación de Spearman en el análisis de datos". *LEARN STATISTICS EASILY.* mayo 19. https://es.statisticseasily.com/correlaci%C3%B3n-de-lancero/.

Ervasti, Jenni, Aalto, Ville, Pentti, Jaana, Oksanen, Tuula, Kivimäki, Mika, & Vahtera, Jussi. 2022.
"The Association of Changes in Work Due to Covid-19 Pandemic with Psychosocial Work Environment and Employee Health: A Cohort Study of 24, 299 Finnish Public Sector Employees." *Journal of Occupational Health* 64(1): e12310. https://doi.org/10.1136/oemed-2021-107745

Fernandes, Maria Neyrian de Fatima, Stabnow Santos, Floriacy, Pereira de Jesus Costa, Ana Cristina, Santos Neto, Marcelino, de Araújo e Silva, Roberta y Hunaldo dos Santos, Leonardo. 2023. "Stress during the Covid-19 Pandemic: Unusual Challenges for University Professors". *Medicina (Ribeirão Preto)* 56 (4): e–208086. doi:10.11606/issn.2176-7262.rmrp.2023.208086.

Fernández-Arias, Pablo, Antón-Sancho, Álvaro, Sánchez-Calvo, María y Vergara, Diego. 2024. "Teaching Experience as a Key Factor in Dealing with Digital Teaching Stress" 14. doi:10.3390/educsci14080809.

Gómez-Domínguez, Valentina, Gómez-Domínguez, Teresa, Navarro-Mateu, Diego y Giménez-Espert, María del Carmen. 2022. "The Influence of COVID-19 and Psychosocial Risks on Burnout and Psychosomatic Health Problems in Non-University Teachers in Spain during the Peak of the Pandemic Regressions vs. fsQCA". *Sustainability*. doi:10.3390/su142013030.

Gutiérrez-Hernández, Mónica Gabriela, Maldonado-Macías, Aidé Aracely y Sibaja Terán, Beatriz. 2024. "Organización del tiempo durante el teletrabajo como riesgo psicosocial en docentes". *Cultura Científica y Tecnológica* 21 (1): E2–8. doi:10.20983/culcyt.2024.1.2e.1.

IBM Corporation, Estados Unidos. 2019. "IBM SPSS - Valores perdidos 26". https://www.ibm.com/docs/en/SSLVMB_26.0.0/pdf/es/IBM_SPSS_Missing_Values.pdf.

Koren, Helena, Milaković, Marina, Bubaš, Marija, Bekavac, Petra, Bekavac, Barbara, Bucić, Lovro, Čvrljak, Jelena, Capak, Magdalena, & Jelicic, Pavle. 2023. "Psychosocial Risks Emerged from COVID-19 Pandemic and Workers' Mental Health." *Frontiers in Psychology* 14: 1148634. https://doi.org/10.3389/fpsyg.2023.1148634

Lizana, Paula, & Vega-Fernandez, Gustavo. 2021. "Teacher Teleworking during the COVID-19 Pandemic: Association between Work Hours, Work–Family Balance and Quality of Life." *International Journal of Environmental Research and Public Health* 18(14): 7566. https://doi.org/10.3390/ijerph18147566.

Martínez Ortega, Rosa María, Tuya Pendás, Leonel C., Martínez Ortega, Mercedes, Pérez Abreu, Alberto y Cánovas, Ana María. *2009. "EL COEFICIENTE DE CORRELACION DE LOS RANGOS DE SPEARMAN CARACTERIZACION". Revista Habanera de Ciencias Médicas* 8 (2). https://www.researchgate.net/publication/262544052_El_coeficiente_de_correlacion_de_los_rangos_de_Spearman_caracterizacion

Navarro-Espinosa, José Antonio, Vaquero-Abellán, María, Perea-Moreno, Alberto-Jesús, Pedrós-Pérez, Gema, Aparicio-Martínez, Patricia, & Martínez-Jiménez, María Pilar. 2021. "The Influence of Technology on Mental Well-Being of STEM Teachers at University Level: COVID-19 as a Stressor". *International Journal of Environmental Research and Public Health* 18. doi:10.3390/ijerph18189605.

Ortega, Cristina. 2018. "Muestreo no probabilístico: definición, tipos y ejemplos". *QuestionPro*. junio 14. https://www.questionpro.com/blog/es/muestreo-no-probabilistico/.

Panesso Patiño, Vanessa, Calderón Puerta, Sandra Marcela, y Claribel Núñez, Nadia. 2023. "Impact of Psychosocial Risk Factors in Post-Pandemic Times in a Colombian University Institution. | EBSCOhost". julio 1. doi:10.33423/jop.v23i4.6737.

Pimenta, Ana, Ramos, Delfina, Santos, Gilberto, Rodrigues, Matilde A. y Doiro, Manuel. 2023. "Psychosocial Risks in Teachers from Portugal and England on the Way to Society 5.0". *International Journal of Environmental Research and Public Health* 20 (14): 6347. doi:10.3390/ijerph20146347.

Prado-Gascó, Víctor, Gómez-Domínguez, María, Soto-Rubio, Ana, Díaz-Rodríguez, Laura, & Navarro-Mateu, Diego. 2020. "Stay at Home and Teach: A Comparative Study of Psychosocial Risks Between Spain and Mexico During the Pandemic". *Frontiers in Psychology* 11. doi:10.3389/fpsyg.2020.566900.

Quezada Díaz, Belem, Hindrichs, Imke y Castellanos Simons, Doris. 2024. "Emerging Psychosocial Factors and Work Overload Perceptions of Mexican University Teachers and Students Working and Studying from Home during the COVID-19 Pandemic". *Frontiers in Psychology* 15 (septiembre). Frontiers. doi:10.3389/fpsyg.2024.1349458.

Quezada, Belem, De la Hoz Becquer, Aime y Hindrichs, Imke. 2022. "Factores psicosociales emergentes y la diferencia percibida entre mujeres y hombres que trabajan desde casa por la COVID-19 en México". *Castalia - Revista de Psicología de la Academia, núm.* 39: 163–83. doi:10.25074/07198051.39.2231.

Ramirez, Sheryl H. 2023. "Perceived Health Risks among Teachers Working Remotely in the Era of Covid-19 Pandemic". *Journal of Healthcare and Biomedical Science*. doi:10.31098/jhbs.v1i1.1147.

Romeo, María, Yepes-Baldó, Marta, Soria, Miguel, & Marcos, Jayme. "Impact of the COVID-19 Pandemic on Higher Education: Characterizing the Psychosocial Context of the Positive and Negative Affective States Using Classification and Regression Trees". *Frontiers in Psychology* 12. doi:10.3389/fpsyg.2021.714397.

Sanchis-Giménez, Lucía, Tamarit, Alicia, Prado-Gascó, Vicente Javier, Sánchez-Pujalte, Laura y Díaz-Rodríguez, Luis. 2024. "Psychosocial Risks in Non-University Teachers: A Comparative Study between Spain and Mexico on Their Occupational Health". *Sustainability* 16 (16): 6814. doi:10.3390/su16166814.

Sandoval-Reyes, Juan, Idrovo-Carlier, Sandra y Duque-Oliva, Edison Jair. 2021. "Remote Work, Work Stress, and Work–Life during Pandemic Times: A Latin America Situation". *International Journal of Environmental Research and Public Health* 18 (13): 7069. doi:10.3390/ijerph18137069.

Souto, Isabel, Pereira, Anabela, Brito, Elisabeth, Sancho, Luís y Jardim, Jacinto. 2018. "Psychosocial Risk Factors And Distress In Higher Education Teachers". *European Proceedings of Social and Behavioural Sciences Health and Health Psychology-icH&Hpsy 2018 (noviembre)*. Future Academy. doi:10.15405/epsbs.2018.11.14.

Herrera Tapia, Mariela Alexandra, Toapanta Otacoma, Gladys Marlene y Calapucha Cerda, Lineth Rosenda. 2024. "Factores determinantes del clima organizacional y la satisfacción laboral en docentes universitarios". *Polo del Conocimiento* 9 (7): 939–61.

Tellechea-Acuña, Francis Yanett, Carpio-Camacho, Adilen y Valdés-Florat, Mirlandia Odilia. 2025. "Clima organizacional y la satisfacción laboral en docentes universitarios en Cumaná, Venezuela". *Noesis. Revista Electrónica de Investigación* 7 (13): 118–30. doi:10.35381/noesisin.v7i13.268.

Wischlitzki, Elisabeth, Amler, Nadja, Hiller, Julia y Drexler, Hans. 2020. "Psychosocial Risk Management in the Teaching Profession: A Systematic Review". *Safety and Health at Work* 11 (4): 385–96. doi:10.1016/j.shaw.2020.09.007.

18 Is Burnout Leading to Quiet Quitting, Counterproductive Performance, and Ultimately Turnover Intention?

Study in the Mexican Export Manufacturing Industry

Aurora Irma Máynez-Guaderrama
and Judith Cavazos-Arroyo

18.1 INTRODUCTION

Work activities and responsibilities can generate chronic stress, which can lead to burnout. Worldwide, burnout is recognized as a common public health problem that causes physical and mental discomfort, reduces work efficiency, and hinders social function and quality of life (Liu et al. 2024). This construct was officially recognized by the World Health Organization as an occupational phenomenon resulting from unmanaged chronic work-related stress (Demerouti 2024), and some experts consider it a 21st-century disease (Hildenbrand et al. 2018) that affects employees and employers as a result of psychological detachment and the potential loss of human talent (Liu et al. 2023; Malik and Singh 2020). Employees with high levels of burnout are more likely to leave their jobs, resulting in suboptimal work performance (Soelton et al. 2020).

Interest in studying burnout has prevailed for more than five decades, but the topic remains pertinent because workers in different sectors are affected, and organizations must take more preventive measures (Demerouti 2024). While many of the effects of burnout have already been studied, it is still necessary to investigate the effect of burnout on several variables, including quiet quitting and counterproductive performance (Demerouti 2024; Hamouche et al. 2023; Liu-Lastres et al. 2024).

Studies on burnout are conducted in the service sector, but conducting similar analyses in the manufacturing industry is essential as employees are also exposed to burnout, given their daily work in assembly lines, with high-intensity, monotonous, and mechanical work (Liu et al. 2023). Recent work has highlighted the need to improve manufacturing working conditions (Roman Bonilla 2023). Therefore, this chapter aims to evaluate the effect of burnout on employees' quiet quitting, counterproductive performance, and turnover intention in the export manufacturing industry in Ciudad Juárez, Chihuahua, Mexico. As a specific objective, this study analyzed the mediating role of quiet quitting and counterproductive performance in the relationship between burnout and turnover intention.

18.1.1 Literature Review

Burnout is a critical problem in several professional sectors (Bakker et al. 2023). It is a psychological response to chronic job stress that manifests itself in three dimensions: emotional exhaustion,

DOI: 10.1201/9781003582014-22

cynicism toward work, and decreased self-fulfillment (Liu et al. 2023). Employees experiencing burnout may desire to leave their company. However, some respondents consider that leaving is not an option and opt to quit quietly (Karrani et al. 2024). Quiet quitting is a coping mechanism practiced to avoid fatigue resulting from excessive and prolonged work overloads (Thu Trang and Thi Thu Trang 2024). This variable involves the adoption of poor work behavior (Galanis et al. 2023a) because employees only exert minimal effort to fulfill their work essential requirements and neglect to perform extra duties (Klotz and Bolino 2022; Scheyett 2023). Moreover, this quiet quitting entails a psychological disengagement related to a perception of material and moral devaluation (Atalay and Dağıstan 2024).

Burnout can affect work performance. Work performance has three dimensions: task, contextual, and counterproductive performance (Lousã et al. 2024). In this study, we focus on the third dimension, defined as intentional, deviant, and voluntary behavior that harms the organization (Zhang et al. 2024). Previous literature indicates that counterproductive performance involves purposeful behaviors that affect well-being, productivity, and firm performance (Chen et al. 2021; Galanis et al. 2024). Nonskilled or toxic employees may be considered counterproductive because their behavior affects the team and the organization's performance (Abedini et al. 2024). In practical terms, this dysfunctional behavior entails that a collaborator's repeated intentional actions (e.g., being late, lack of concentration at work, and reduced effort in work routines) harm other colleagues or damage the organization (Fernández-del-Río et al. 2021). Usually, it stems from pressure in the workplace, abusive supervision, internal conflict, or perceived unfair treatment (Ferdiansah et al. 2025).

Counterproductive employee behavior has a negative economic impact on organizations (Ahmed et al. 2024) due to issues such as work environment, rework, and excessive workloads. Counterproductive performance undermines significant organizational norms and triggers behavioral dysregulation in the workplace, resulting in a decline in operational activities and employee performance (Jawahar et al. 2018). As a result, this type of performance often encourages staff turnover (Valensiana and Sawitri 2024). Turnover intention is a vital issue for HR departments (Tutan and Kökalan 2024), and it is understood as the employee's intentional and deliberate choice to end their employment with the organization (Mercan Baspinar et al. 2024).

Like findings in other contexts, it is feasible that in the Mexican export manufacturing industry, burnout directly triggers quiet quitting, counterproductive performance, and quitting intention. Work conducted with service sector employees in Vietnam and Portugal revealed that burnout triggered quiet quitting (Prentice et al. 2024; Thu Trang and Thi Thu Trang 2024). Moreover, higher levels of burnout among nurses are associated with lower performance quality, affecting patient care and service levels (Agustin 2022). Finally, according to studies conducted with luxury hotel employees in Korea (Jung et al. 2023) and with nursing staff (Ohue et al. 2021; Turunç et al. 2024), burnout generates turnover intention. According to previous research, quiet quitting may result in counterproductive performance. Previous studies suggest that this may lead employees to improve their performance to the detriment of their colleagues, which would trigger counterproductive performance (Dutta 2024). Burnout and increased turnover intentions are associated with factors such as workload and intensity, low wages, and working conditions (Turunç et al. 2024). Previous research has identified that both quiet quitting and counterproductive behavior mediate in the organizational context (Rizky Parsasi and Prawitowati 2025; Sambung 2019; Varsney 2024). Therefore, the following hypotheses are proposed in the export manufacturing industry in Ciudad Juárez, Chihuahua:

- H1: Burnout positively influences quiet quitting
- H2: Burnout has a positive effect on counterproductive performance
- H3: Burnout exerts a positive effect on turnover intention
- H4: Quiet quitting positively influences counterproductive performance
- H5: Counterproductive performance positively influences turnover intention
- H6: Quiet quitting mediates burnout and counterproductive performance
- H7: Counterproductive performance mediates the relationship between burnout and turnover intention

18.2 METHODS

Study design. A quantitative, explanatory, and cross-sectional study was conducted with employees of the Ciudad Juárez, Mexico, export manufacturing industry. This study was approved by the ethics committee, following the guidelines of the Universidad Popular Autónoma del Estado de Puebla, registered as INV17–2024. The ethical principles for studies with human participants, including informed consent and voluntary participation, were also followed.

Population and sample: Nonprobabilistic convenience sampling was used. Data collection was conducted in October 2024. A total of 437 responses were received, of which only 421 were considered valid. In this sample, 52% of the respondents were men, 45% were women, and 2% did not reveal their sex. Regarding the educational background, 42.8% declared an unfinished bachelor's degree, and 35.5% stated that they had finished an undergraduate education. The largest proportion worked in automotive components (50.3%) and electronics (20.4%) companies. Concerning the job position, 19.9% of the respondents worked as engineers, 17.6% as clerical workers, and 17.2% as technicians. A total of 35.2% of the participants reported that they had worked in that company for between 1 and 3 years, and 28.4% reported that they had worked in that company for less than 1 year.

Instrument: Data were collected through an electronic survey that was shared with the participants. The instrument included latent variables assessed using scales adapted to the study's context. Professional inefficacy and cynicism were assessed using the Bang and Reio scale (2017). The former was measured with six items, such as the following: 'I have achieved many worthwhile things in this job', and the responses received were reverse-coded. The latter was scored with five items, such as 'I have become less interested in my job as time has continued'. Emotional exhaustion was assessed with eight items from the Maslach and Jackson (1981) scale, such as the following: 'I feel frustrated with my work'. Quiet quitting was rated with nine items by Galanis et al. (2023b), which includes the following: 'In my work, I do the basics or minimum without going any further'. Counterproductive performance was evaluated with five items from Ramos-Villagrasa et al. (2019), such as the following: 'In my job, I sometimes complain about minor work problems'. The turnover intention was assessed using four items, three from Máynez-Guaderrama and Cavazos-Arroyo (2021) and one from Bufquin (2020), such as the following: 'I frequently think of quitting my job'. The variance inflation factor (VIF) values of the latent variables were examined considering the possibility of a common method bias, and the results showed values lower than 3.3, indicating a low probability of collinearity problems (Lim 2024).

Statistical technique: Partial Least Squares Structural Equation Modeling (PLS-SEM) has become popular in a wide range of organizational and business research contexts, and numerous studies have shown its applicability in academic areas such as accounting, entrepreneurship, family business, higher education, human resource management, knowledge management, information systems, and marketing (Legate et al. 2023). A PLS-SEM was used to test the hypotheses using SmartPLSV41.1.4 ® software.

In the first stage of the SEM, the measurement model is tested to ascertain that it is reliable and valid. Factor loadings, reliability, convergent validity, and discriminant validity must be assessed. After confirming the reliability and validity, the results of the structural model should be reviewed. Figure 18.1 depicts the method steps followed.

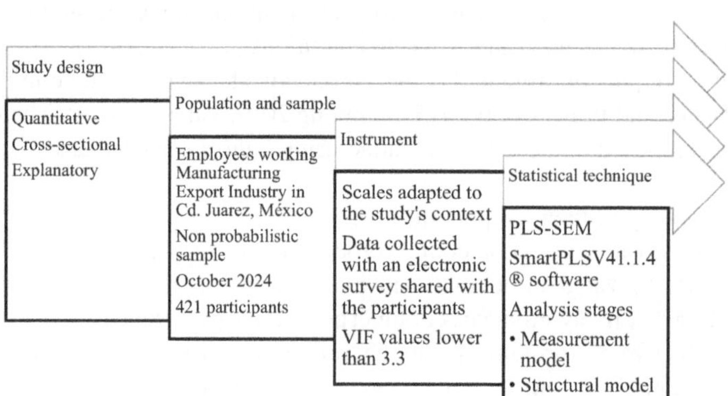

FIGURE 18.1 Method stages.

TABLE 18.1
Reliability and Convergent Validity

	Cronbach's Alpha	Composite Reliability (ρ_a)	Composite Reliability (ρ_c)	Average Variance Extracted (AVE)
EE	0.945	0.946	0.955	0.751
CY	0.932	0.934	0.949	0.787
CP	0.920	0.921	0.940	0.758
PI	0.881	0.893	0.909	0.625
TI	0.948	0.949	0.963	0.866
QQ	0.935	0.937	0.948	0.722

TABLE 18.2
Heterotrait–Monotrait Ratios

	EE	CY	CP	PI	IL
CY	0.760				
CP	0.625	0.761			
PI	0.187	0.270	0.179		
TI	0.704	0.834	0.759	0.255	
QQ	0.657	0.874	0.793	0.241	0.749

18.3 RESULTS

Measurement model: The initial measurement model consisted of 37 items, and the final one included 34 items as three of the original ones were eliminated because they did not meet the minimum threshold or caused multicollinearity problems within their construct. In the final model, the item factor loadings are statistically significant and greater than 0.708. The reliability and convergent validity were tested using Cronbach's alpha, ρ_a, ρ_c, and average variance extracted (AVE). Table 18.1 shows the values that meet the recommendation of being between 0.700 and 0.950 (initial three indicators) or exceeding the minimum cut-off point of 0.500 (fourth indicator) (Hair Jr. et al. 2020).

Subsequently, discriminant validity was assessed. The most recommended criterion is the heterotrait–monotrait ratio. Maximum values of 0.850 (conceptually different constructs) or 0.900 (conceptually similar constructs) are recommended (Ringle et al. 2020). Excluding the quiet quitting–cynicism ratio, for which a cut-off point of 0.90 was used, the cut-off point for all other ratios was 0.85. The ratios are lower than the suggested cut-off points, demonstrating the model's discriminant validity (Table 18.2).

Structural model: We found that Burnout had a positive and significant impact on quiet quitting (H1: β=0.767, t=35.790), counterproductive performance (H2: β=0.293, t=5.736), and turnover intention (H3: β=0.556, t=13.184). It was also confirmed that 'quiet quitting' triggers counterproductive performance (H4: β=0.512, t=9.767), which in turn leads to turnover intention (H5: β=0.328, t=7.310). Table 18.3 and Figure 18.2 present the results.

Subsequently, the R^2, f^2, and Q^2 results were evaluated. Regarding R^2 (Figure 18.2), the model can moderately explain not only quiet quitting (R^2=0.588) but also counterproductive work behavior (R^2=0.578) and intention to turnover (R^2=0.667). In parallel, significant effects were found between the variables: a small one in burnout → counterproductive performance (f^2=0.084), two medium effects in counterproductive performance → intention to turnover (f^2=0.171) and quiet quitting → counterproductive performance (f^2=0.256), and two large effects in burnout → intention

TABLE 18.3
Path Analysis of Outcomes from the Structural Model

Hypothesis	Path Relationship	*β* Coefficient	*t*-Value	Decision
H1	Burnout → Quiet Quitting	0.767	35.790***	Confirmed
H2	Burnout → Counterproductive Performance	0.293	5.736***	Confirmed
H3	Burnout → Turnover Intention	0.556	13.184***	Confirmed
H4	Quiet Quitting → Counterproductive Performance	0.512	9.767***	Confirmed
H5	Counterproductive Performance → Turnover Intention	0.328	7.310***	Confirmed

*** Statistical Difference at 95%, 99% and 99.9% level of significance.

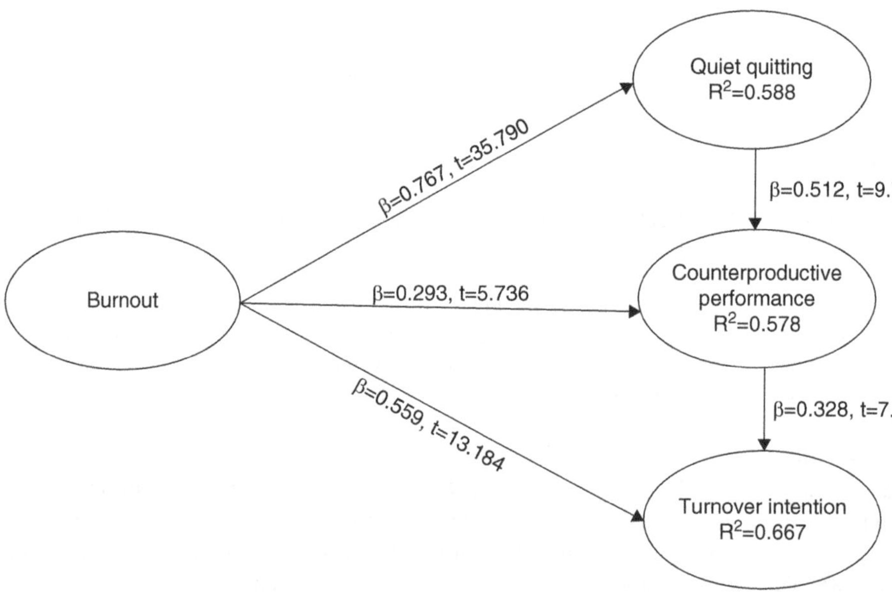

FIGURE 18.2 Structural model.

to turnover (f^2=0.493) and burnout → quiet quitting (f^2=1.425). Finally, regarding the predictive power of the model within the sample, the Q^2 indicator revealed a moderate predictive capacity for counterproductive performance (0.465) and a high predictive capacity for turnover intention (0.608) and quiet quitting (0.584).

Indirect effects were examined to assess potential mediating effects. First, the results revealed that quiet quitting plays a complementary mediating role between burnout and counterproductive performance; therefore, H6 was confirmed (β=0.393, t=9.298). Quiet quitting and counterproductive performance have complementary mediating effects on the relationship between burnout and turnover intention; therefore, H7 was confirmed (β=0.225, t=6.772). Table 18.4 presents this information.

TABLE 18.4
Mediation

Mediation	β	t-Value
Burnout → Quiet quitting → Counterproductive performance	0.393	9.298
Burnout → Quiet quitting → Counterproductive performance → Turnover intention	0.225	6.772
Burnout → Counterproductive performance → Turnover intention		

18.4 DISCUSSION AND CONCLUSIONS

This study assessed the effect of burnout on quiet quitting, counterproductive performance, and turnover intention within the export manufacturing industry. Additionally, the effect of quiet quitting on counterproductive performance and the influence of counterproductive performance on turnover intention were analyzed.

The findings show that burnout has a positive and significant influence on quiet quitting, counterproductive behavior, and turnover intention. The strongest influence was on quiet quitting, followed by turnover intention and counterproductive performance. This confirms H1, H2, and H3. These findings support those of previous studies. Burnout causes quiet quitting (Prentice et al. 2024; Thu Trang and Thi Thu Trang 2024), counterproductive performance (Agustin 2022), and turnover intention (Jung et al. 2023; Ohue et al. 2021; Turunç et al. 2024). Exhaustion was found to be the most relevant dimension of burnout, and the strongest manifestation of this is that employees feel tired because of their work. Regarding quiet quitting, the most important item refers to the employee's avoidance of giving opinions or ideas due to the fear of their boss increasing their workload. In counterproductive work behavior, the indicator that stands out is the one that assesses whether the employee pays more attention to the negative aspects rather than the positive ones. In the case of turnover intention, the most relevant items refer to the employee's desire to leave the company and how often they consider doing so.

Moreover, H4 and H5 were supported. The results demonstrate that quiet quitting triggers counterproductive performance, thus confirming Dutta's (2024) findings. Counterproductive performance provokes turnover intention within the study context, similar to other studies (Turunç et al. 2024). In addition, the mediating role of quiet quitting and counterproductive performance in the relationship between burnout and turnover intention was assessed. Both variables exert a complementary mediation between burnout and turnover intention in both cases.

In conclusion, burnout has a highly negative effect on organizations. This study revealed that burnout triggers quiet quitting, counterproductive behavior, and turnover intention. Consequently, all organizations must be aware of the consequences of burnout and strive to prevent its occurrence. Moreover, quiet quitting as a phenomenon is not new but has increased sharply after the COVID-19 pandemic, probably because of the reassessment of the work–life balance. Through this research, it became clear that burnout is a precursor to quiet quitting, and its consequences are counterproductive behavior and turnover intention. Therefore, organizations must design and implement strategies that avoid the psychological disengagement of employees who are 'quietly quitting' by, for example, continuously recognizing their work and effort, creating conditions to help with work–life balance, avoiding excessive workloads, and providing fair and equitable remuneration. Reducing burnout and 'quiet quitting' prevents counterproductive performance and, ultimately, turnover.

This study also has some limitations, given that a nonprobabilistic sample was collected for convenience; therefore, the results cannot be extrapolated to the population. Furthermore, less tech-savvy employees may have been excluded from the sample because the survey was conducted electronically. Future studies might include antecedents of quiet quitting, such as leadership style and breach or violation of the psychological contract. It is also advisable to explore the effects of quiet quitting on variables such as task performance and organizational citizenship behaviors.

REFERENCES

Abedini, Mahsa, Schreurs, Bert, Jawahar, I.M. and Hamstra, Melvyn R.W. 2024. "Does worrying about money motivate counterproductive work behavior? A time-lagged study." *Journal of Managerial Psychology* 39(4): 469–82. https://doi.org/10.1108/JMP-04-2023-0223

Agustin, Leni. 2022. "Impact of burnout on nurse performance in the inpatient ward: A literature review." *D'Nursing and Health Journal (DNHJ)* 3(1): 10–5. https://doi.org/10.36835/DNURSING.V3I1.380

Ahmed, Mohamed Abdelkhalek Omar, Zhang, Junguang, Fouad, Ahmed Sabry, Mousa, Kawther and Nour, Hamdy Mohamed. 2024. "The dark side of leadership: How toxic leadership fuels counterproductive work behaviors through organizational cynicism and injustice." *Sustainability* 17(1): 105. https://doi.org/10.3390/su17010105

Atalay, Murat, and Dağıstan, Umut. 2024. "Quiet quitting: A new wine in an old bottle?." *Personnel Review* 53(4): 1059–74. https://doi.org/10.1108/PR-02-2023-0122

Bakker, Arnold B., Xanthopoulou, Despoina and Demerouti, Evangelia. 2023. "How does chronic burnout affect dealing with weekly job demands? A test of central propositions in JD-R and COR theories." *Applied Psychology* 72(1): 389–410. https://doi.org/10.1111/apps.12382

Bang, Hyejin, and Reio Thomas G.2017. "Examining the role of cynicism in the relationships between burnout and employee behavior." *Journal of Occupational Psychology and Organizations* 33(3): 217–27. https://doi.org/10.1016/j.rpto.2017.07.002

Bufquin, Diego. 2020. "Coworkers, supervisors, and frontline restaurant employees: Social judgments and the mediating effects of exhaustion and cynicism." *Journal of Hospitality and Tourism Insights* 3(3): 353–69. https://doi.org/10.1108/JHTI-11-2019-0123

Chen, Chang, Zhang, Zhe and Jia, Ming. 2021. "Stretch goals and unethical behavior: Role of ambivalent identification and competitive psychological climate." *Management Decision* 59(8): 2005–23. https://doi.org/10.1108/MD-04-2019-0433

Demerouti, Evangelia. 2024. "Burnout: A comprehensive review." *Zeitschrift* für *Arbeitswissenschaft* 78(4): 492–504. https://doi.org/10.1007/s41449-024-00452-3

Dutta, Sumedha, Thomas, Asha, Shiva, Atul, Papa, Armando and Cuomo, Maria Teresa. 2024. "The hustle behind knowledge: Role of workplace ostracism and knowledge hiding towards quiet quitting in knowledge-intensive organisations." *Journal of Knowledge Management* 28(10): 2911–39. https://doi.org/10.1108/JKM-01-2024-0035

Ferdiansah, Muhammad Irdam, Chong, Vincent K., Wang, Isabel Z. and Woodliff, David R. 2025. "The effect of stretch goals and moral justification on counterproductive work behavior." *Journal of Accounting & Organizational Change*. Vol. ahead-of-print (No. ahead-of-print). https://doi.org/10.1108/JAOC-06-2024-0196

Fernández-del-Río, Elena, Barrada, Juan Ramón and Ramos-Villagrasa, Pedro J. 2021. "Bad behaviors at work: Spanish adaptation of the Workplace Deviance Scale." *Current Psychology* 40(3): 1660–71. https://doi.org/10.1007/s12144-018-0087-1

Galanis, Petros, Katsiroumpa, Aglaia, Moisoglou, Ioannis, Derizioti, Konstantina, Gallos, Parisis, Kalogeropoulou, Maria and Papanikolaou, Vasiliki. 2024. "Emotional intelligence as critical competence in nurses' work performance: A cross-sectional study." *Healthcare* 12(19): 1936. https://doi.org/10.3390/healthcare12191936

Galanis, Petros, Katsiroumpa, Aglaia, Vraka, Irene, Siskou, Olga, Konstantakopoulou, Olympia, Moisoglou, Ioannis, Gallos, Parisis and Kaitelidou, Daphne. 2023b. "The quiet quitting scale: Development and initial validation." *AIMS Public Health* 10(4): 828–848. https://doi.org/10.3934/publichealth.2023055

Galanis, Petros, Katsiroumpa, Aglaia, Vraka, Irene, Siskou, Olga, Konstantakopoulou, Olympia, Katsoulas, Theodoros, Moisoglou, Ioannis, Gallos, Parisis and Kaitelidou, Daphne. 2023a. "The influence of job burnout on quiet quitting among nurses: The mediating effect of job satisfaction." *Research Square* PREPRINT (Version 1). https://doi.org/10.21203/rs.3.rs-3128881/v1

Hair Jr., Joe F., Howard, Matt C. and Nitzl, Christian. 2020. "Assessing measurement model quality in PLS-SEM using confirmatory composite analysis." *Journal of Business Research* 109: 101–110. https://doi.org/10.1016/j.jbusres.2019.11.069

Hamouche, Salima, Koritos, Christos and Papastathopoulos, Avraam. 2023. "Quiet quitting: relationship with other concepts and implications for tourism and hospitality." *International Journal of Contemporary Hospitality Management* 35(12): 4297–312. https://doi.org/10.1108/IJCHM-11-2022-1362

Hildenbrand, Kristin, Sacramento, Claudia A. and Binnewies, Carmen. 2018. "Transformational leadership and burnout: The role of thriving and followers' openness to experience." *Journal of Occupational Health Psychology* 23(1): 31. https://doi.org/10.1037/ocp0000051

Jawahar, I. M., Schreurs, Bert and Mohammed, Shawn J. 2018. "How and when LMX quality relates to counterproductive performance: A mediated moderation model." *Career Development International* 23(6/7): 557–75. https://doi.org/10.1108/CDI-05-2018-0134

Jung, Hyo-Sun, Jung, Yoon-Sik and Yoon, Hye-Hyun. 2023. "The effects of workaholism on employee burnout and turnover intent at deluxe hotels during the COVID-19 pandemic: Evidence across generations." *Sustainability* 15(6): 5227. https://doi.org/10.3390/su15065227

Karrani, Mariam Arif, Bani-Melhem, Shaker and Mohd-Shamsudin, Faridahwati. 2024. "Employee quiet quitting behaviours: Conceptualization, measure development, and validation." *The Service Industries Journal* 44(3–4): 218–36. https://doi.org/10.1080/02642069.2023.2286604

Klotz, Anthony C., and Bolino, Mark C. 2022. "When quiet quitting is worse than the real thing." *Harvard Business Review,* 1: 1–5.

Legate, Amanda E., Hair, Joe F. Jr, Lambert Chretien, Janice and Risher, Jeffrey J. 2023. "PLS-SEM: Prediction-oriented solutions for HRD researchers." *Human Resource Development Quarterly* 34(1): 91–109. https://doi.org/10.1002/hrdq.21466

Lim, Weng Marc. 2024. "Validity typology: Content, face, convergent, discriminant, nomological, and predictive validity." *Journal of Trade Science* 12(3): 155–79. https://doi.org/10.1108/JTS-03-2024-0016

Liu, Ruihong, Zhang, Hanzhong, Feng, Chunyuan, Wu, Xueyi, Pan, Zhenyu, Li, Wanyu and Jia, Liping. 2024. "The impact of telecom industry employees' stress perception on job burnout: Moderated mediation model." *BMC Public Health* 24(1):1623. https://doi.org/10.1186/s12889-024-18704-6

Liu, Xiaoman, Wang, Jin, Zhang, Qiaoyun, Chen, Huiquing, Yuan, Fang, Lou, Jianlin, Zhao, Rong, Li, Jue, Jia, Xiaodong, Liu, Jing and Li, Shuang 2023. "Job burnout and influencing factors among Chinese manufacturing employees" *Journal of Environmental and Occupational Medicine* 40(4): 396–404. https://doi.org/10.11836/JEOM22367

Liu-Lastres, Bingjie, Karatepe, Osman M. and Okumus, Fevzi. 2024. "Combating quiet quitting: Implications for future research and practices for talent management." *International Journal of Contemporary Hospitality Management* 36 (1): 13–24. https://doi.org/10.1108/IJCHM-08-2023-1317

Lousã, Eva Petiz, Pereira Alves, Marta and Koopmans, Linda. 2024. "Adaptation and validation of the individual work performance questionnaire into a Portuguese version." *Administrative Sciences* 14(7): 150. https://doi.org/10.3390/admsci14070150

Malik, Amina Raza, and Singh, Parbudyal. 2020. "The role of employee attributions in burnout of "talented" employees." *Personnel Review* 49(1): 19–42. https://doi.org/10.1108/PR-02-2018-0064

Mann, Sara L., Budworth, Marie-Hélene and Ismaila, Afisi S. 2012. "Ratings of counterproductive performance: The effect of source and rater behavior." *International Journal of Productivity and Performance Management* 61(2): 142–156. https://doi.org/10.1108/17410401211194653

Maslach, Christina, and Jackson, Susa E. 1981. "The measurement of experienced burnout." *Journal of Organizational Behavior* 2: 99–113. https://doi.org/10.1002/job.4030020205

Máynez-Guaderrama, Aurora Irma, and Cavazos-Arroyo, Judith. 2021. "Consecuencias del agotamiento laboral en tiendas de conveniencia." *Investigación administrativa* 50 (128): 1–18. https://doi.org/10.35426/iav50n128.03

Mercan Baspinar, Melike, Sevgi Demiray, Zerafet Birer, Basar, Hakan and Tolga Adas, Gokhan. 2024. "A hospital-based nursing study on intention to leave job and nurse-physician collaboration." *Nursing Forum* 1: 9503988. https://doi.org/10.1155/2024/9503988

Ohue, Takashi, Aryamuang, Supaporn, Bourdeanu, Laura, Church, Jean N., Hassan, Hamidah, Kownaklai, Jaruwan, Pericak, Arlene and Suwannimitr, Amorn. 2021. "Cross-national comparison of factors related to stressors, burnout and turnover among nurses in developed and developing countries" *Nursing Open* 8: 2439–51. https://doi.org/10.1002/nop2.1002

Prentice, Catherine, Dominique-Ferreira, Sergio, Wang, Xuequn, Tuominen, Jesse, Duarte, Márcia and Rocha, Helena. 2024. "Work-life imbalance, burning out, feeling down, I will quit, but quietly – The case of hospitality employees." *Journal of Hospitality Marketing & Management* 34(1): 24–45. https://doi.org/10.1080/19368623.2024.2389074

Ramos-Villagrasa, Pedro J., Barrada, Juan R., Fernández-del-Río, Elena and Koopmans, Linda. 2019. "Assessing job performance using brief self-report scales: The case of the individual work performance questionnaire." *Journal of Work and Organizational Psychology* 35(3): 195–205. https://doi.org/10.5093/jwop2019a21

Ringle, Christian M., Sarstedt, Marko, Mitchell, Rebecca and Gudergan, Siegfried P.. 2020. "Partial least squares structural equation modeling in HRM research." *The International Journal of Human Resource Management* 31(12): 1617–43. https://doi.org/10.1080/09585192.2017.1416655

Rizky Parsasi, Wiannisa and Prawitowati, Tjahjani. 2025. "The influence of Employeee engagement on Turnover intention mediated by job satisfaction and quiet quitting of Generation Z employee." *Jurnal Sistem informasi, Managemen, dan Akuntansi (SIMAK)* 23(1): 31–44. https://doi.org/10.35129/simak.v23i01.621

Roman Bonilla, Enrique. 2023. "Achieving well-being for employees with work-related stress in manufacturing firms: A qualitative exploratory case study." *Doctoral dissertation,* University of Phoenix. https://www.proquest.com/openview/155c0d1a71af3e6053ebc44c2c0e8083/1?pq-origsite=gscholar&cbl=18750&diss=y

Sambung, Roby. 2019. "Job satisfaction on employee performance; counterproductive work behavior and organizational citizenship behavior as mediations." *International Journal of Economic Behavior and Organization* 7(3): 50–6. https://doi.org/10.11648/j.ijebo.20190703.12

Scheyett, A. 2023. "Quiet quitting." *Social Work* 68(1): 5–7. https://doi.org/10.1093/sw/swac051

Soelton, Mochamad, Hardianti, Devy, Kuncoro, Suryo and Jumadi, Jumadi . 2020. "Factors affecting burnout in manufacturing industries". In *4th International Conference on Management, Economics and Business* (ICMEB 2019) (pp. 46–52). Atlantis Press. https://doi.org/10.2991/aebmr.k.200205.010

Thu Trang, Pham, and Thi Thu Trang, Nguyen. 2024. "Job burnout and quiet quitting in Vietnamese banking sector: The moderation effect of optimism." *Cogent Business & Management* 11(1): 2371549. https://doi.org/10.1080/23311975.2024.2371549

Turunç, Ömer, Çalışkan, Abdullah, Akkoç, Irfan, Köroğlu, Özlem, Gürsel, Güney, Demirci, Ayhan, Hazir, Köksal and Özcanarslan, Neslihan. 2024. "The impact of intensive care unit nurses' burnout levels on turnover intention and the mediating role of psychological resilience" *Behavioral Sciences* 14(9): 782. https://doi.org/10.3390/bs14090782

Tutan, Abdulhanit and Kökalan, Özgür. 2024. "The mediation role of work–family conflict in the effect of workplace violence on job satisfaction and intention to leave: A study on health care workers in Turkey." *Frontiers in Psychology* 15: 1322503. https://doi.org/10.3389/fpsyg.2024.1322503

Valensiana, Brigita Febi, and Sawitri, Dewi Khrisna. 2024. "Investigating turnover intention: The role of counterproductive work behavior and job stress." *Dinasti International Journal of Economics, Finance & Accounting (DIJEFA)* 5(4): 3316–26. https://doi.org/10.38035/dijefa.v5i4.3136

Varsney, Deepanjana. 2024. "Employee quiet quitting behavior and performance: A moderated mediation model." *Proceedings of the 8th International Conference on Advanced Research in Management, Business and Finance*. https://www.dpublication.com/proceeding/8th-icmbf/

Zhang, Wen, Xia, Bohang, Derks, Daantje, Pletzer, Jan Luca, Breevaart, Kimberley and Zhang, Xichao 2024. "Perceived overqualification, counterproductive work behaviors, and withdrawal: A moderated mediation model." *Journal of Managerial Psychology* 39(5): 539–54. https://doi.org/10.1108/JMP-06-2022-0310

Section 5

Macroergonomics and Policy

19 Applying Conjoint Analysis to Assess Ergonomists' Preferences for Range of Motion Categories and Associated Ergonomic Risk Levels in Posture Evaluation

Patricia Eugenia Sortillón González,
Aide Aracely Maldonado-Macías,
David Sáenz-Zamarrón, Juan Luis Hernández Arellano,
and Enrique Javier De la Vega Bustillos

19.1 INTRODUCTION

Posture evaluation is a critical component of ergonomics as it plays a central role in identifying and mitigating musculoskeletal risks associated with various tasks and work environments (Trovato et al., 2022). Poor posture, especially when sustained over extended periods, can lead to excessive strain on the musculoskeletal system, increasing the likelihood of discomfort, fatigue, and development of musculoskeletal disorders (MSDs) (Fan et al., 2024). By systematically analyzing posture, ergonomists can identify awkward or extreme joint angles (JAs), repetitive movements, and static positions that contribute to these risks. Posture evaluation not only helps assess the biomechanical demands placed on the body but also serves as a basis for designing interventions, such as ergonomic workstation adjustments, task modifications, or training programs, to promote safer and more efficient work practices (Hoe et al., 2018). As workplaces increasingly prioritize employee well-being, accurate assessment of posture is essential to prevent injuries, improve productivity, and promote sustainable occupational health outcomes (Juba, 2024). JAs are a fundamental measure in the analysis of human movement and posture, providing critical insights into the alignment and orientation of body segments during various tasks (Faisal et al., 2019). By quantifying the angular relationship between adjacent body segments, such as the wrist, elbow, or trunk, JAs allow researchers and practitioners to evaluate the biomechanical efficiency and ergonomic safety of postures (Baklouti et al., 2024). These data are essential for identifying awkward or extreme postures that may place undue stress on the musculoskeletal system, potentially leading to discomfort, fatigue, or injury over time. JA has been a cornerstone of traditional ergonomic assessment methods, serving as a key indicator for assessing the posture and identifying potential ergonomic risks (Lowe, Dempsey and Jones, 2019). Methods such as the Rapid Upper Limb Assessment (RULA) (McAtamney and Corlett, 1993) and the Rapid Entire Body Assessment (REBA) (Hignett and McAtamney, 2000) rely heavily on JA measurements to determine the degree of deviation from neutral or optimal postures. These tools classify postures into risk categories based on the angular relationships of body segments such as the neck, trunk, and upper limbs, helping practitioners identify tasks or

DOI: 10.1201/9781003582014-24

movements that may lead to MSDs. In RULA, JAs are scored according to their range of motion (ROM), with higher scores indicating more extreme or awkward postures that pose greater risks. Similarly, REBA evaluates whole-body postures, combining JAs with task load factors to produce a comprehensive risk score. In ergonomic assessments, JA measurements are often compared against established ROM standards to classify postures into risk levels (Nath, Akhavian, and Behzadan, 2017), enabling targeted interventions to minimize ergonomic hazards. ROM categories and ergonomic risk levels are essential in posture analysis because they provide a structured framework for evaluating the biomechanical demands placed on the body during various tasks (Wei, McQuade, and Smidt, 1993). ROM defines the extent of movement available to a joint, and deviations from safe ROM thresholds often indicate postures that may strain muscles, tendons, and ligaments (Cejudo et al., 2020). By categorizing the ROM into specific ranges and associating these with ergonomic risk levels, practitioners can systematically identify postures that pose minimal, moderate, or high risks of MSDs (McAtamney and Corlett, 1993; Hignett and McAtamney, 2000). These categories are especially valuable for prioritizing interventions and tailoring ergonomic solutions. They enable the assessment of postural deviations in relation to task demands, repetition, and force application. Incorporating ROM-based risk levels into posture analysis not only enhances the accuracy of risk identification but also provides a quantifiable basis for improving workplace design and promoting health and safety in diverse settings.

Conjoint analysis has been widely applied in diverse fields, demonstrating its versatility in understanding user preferences and decision-making processes. It has been used in sports marketing to assess fan membership programs' perceived value (Dias and Monteiro, 2020), in workstation design to optimize setups for online learning (Gumasing, Ong, and Bare, 2022), and in environmental management to evaluate the economic value of conservation benefits (Ortiz De Orue, 2021). Additionally, it has helped analyze students' online learning preferences during the COVID-19 pandemic (Consing, 2023), segment users based on privacy concerns (Kuzmanovic and Savic, 2020), assess consumer preferences for yogurt packaging (Wang, Wang, and Hyung, 2022), and determine key attributes influencing hybrid electric vehicle adoption (Silaen and Windasari, 2022). Unlike these studies, where conjoint analysis has primarily been used to evaluate consumer and user preferences, in this research, it has been applied to define ROM categories and ergonomic risk levels. This distinct approach focuses on classifying biomechanical factors rather than consumer behavior, making it a novel contribution to ergonomic risk assessment methodologies.

Previous research has extensively explored the role of JA and ROM thresholds in posture evaluation. Singh et al. (2016) and Peppoloni et al. (2016) examined the relationship between joint kinematics and musculoskeletal risk, while Nath, Akhavian, and Behzadan (2017) focused on establishing ROM cut-off points for categorizing postural risks. Similarly, Lowe, Dempsey, and Jones (2019) and Baklouti et al. (2024) highlighted the importance of JA analysis in evaluating ergonomic safety, and methods such as RULA and REBA operationalize these thresholds into standardized risk scores. Collectively, these studies provide a strong foundation for understanding the biomechanical basis of posture analysis. However, despite this progress, little research has examined how ergonomists themselves interpret and prioritize ROM categories and risk levels during real-world evaluations.

This gap is significant as ergonomists' preferences can influence the consistency and effectiveness of posture assessments and subsequent interventions. Previous studies have highlighted that professional judgment and interpretation play a central role in ergonomic evaluations, often affecting the reliability and consistency of outcomes (Kee and Karwowski, 2007; Spielholz et al., 2001). Additionally, the lack of insights into these preferences limits the ability to refine existing tools and develop more intuitive, user-centered methodologies (van der Beek and Frings-Dresen, 1998). Addressing this research gap is essential for ensuring that ROM and risk-level classifications are both scientifically robust and practically applicable, ultimately enhancing the accuracy and impact of ergonomic evaluations in diverse settings.

Besides, there is lack of understanding of how ergonomists' preferences for ROM categories and ergonomic risk levels influence the development of assessment tools and methodologies.

While current tools and frameworks often rely on predefined thresholds and risk classifications (David, 2005; Takala et al., 2022), they may not fully align with the practical judgment and priorities of practitioners in real-world scenarios. This disconnect can result in tools that are either overly complex or insufficiently tailored to meet the needs of users, potentially reducing their effectiveness in identifying and mitigating musculoskeletal risks (David, 2005; Takala et al., 2022). Additionally, without a clear understanding of how preferences shape the use and interpretation of these tools, opportunities to enhance their usability and consistency are missed. Investigating these preferences is essential for creating tools and methods that are not only scientifically sound but also aligned with the intuitive decision-making processes of ergonomists, ensuring more effective and widely adopted posture evaluation practices.

The study focuses on understanding how different ROM thresholds are perceived by professionals in the field and how these perceptions can be quantified to inform the design of more effective and user-friendly ergonomic assessment tools. By exploring these preferences, the research aims to bridge the gap between theoretical ergonomic standards and their practical application in real-world settings. The findings of this study will provide valuable insights into improving posture evaluation methodologies. This will ultimately contribute to the development of more accurate, standardized, and accessible ergonomic practices. These practices will enhance workplace safety, reduce injury risks, and promote long-term health outcomes for workers.

19.2 METHODOLOGY

19.2.1 Study Design

This study adopts a quantitative research approach. It systematically analyzes ergonomists' preferences for ROM categories and associated ergonomic risk levels in posture evaluation. Conjoint analysis was selected as the primary methodology due to its ability to capture and quantify individual preferences and decision-making processes in a structured and rigorous manner. Unlike traditional survey methods, conjoint analysis enables the decomposition of complex choices into their underlying attributes, providing deeper insights into the relative importance of each factor. This makes it particularly well-suited for this study, where understanding the interplay between ROM categories and ergonomic risk levels is essential for improving posture evaluation practices. By employing this approach, the study aims to generate actionable data that can inform the development of standardized tools and methodologies for ergonomic risk assessment, ultimately contributing to enhanced workplace safety and health outcomes.

19.2.2 Participants

The study included a diverse group of ergonomics professionals with expertise in posture evaluation and ergonomic risk assessment. Participants were recruited through professional ergonomics associations, academic networks, and industry contacts to ensure a representative sample. The study enrolled 16 ergonomists in conducting ergonomic assessments and familiarity with ROM categories and ergonomic risk classification methods. Demographic data such as age, gender, educational background, and areas of specialization within ergonomics were collected to provide a context for interpreting the results and are as follows: participants had a mean average age of 38.5 years (standard deviation=7.2), with 55% identifying as male and 45% as female. Regarding educational background, 62.50% held a master's degree, 27.5% had a doctoral degree, and 10% had a bachelor's degree as their highest level of education. In terms of areas of specialization within ergonomics, 56.25% focused on occupational health, 25% specialized in MSDs, 12.5% worked in biomechanics, and 6.25% were engaged in safety. This diverse sample allowed the study to capture a wide range of preferences and decision-making criteria relevant to the development of effective posture evaluation tools. To ensure the study's relevance and reliability, specific inclusion and exclusion criteria were established for participant selection. Inclusion criteria required participants to be ergonomists or professionals with at least 3 years of experience in ergonomics, specifically in posture evaluation and ergonomic risk assessment. Participants were also required to have familiarity with ROM categories and ergonomic risk classification methodologies. Those who had published research or conducted practical assessments in workplace ergonomics were prioritized to ensure a high level of expertise. Exclusion criteria excluded individuals with no formal ergonomics training, limited or no experience in posture evaluation, or who were unable to commit to the study's requirements. This approach ensured that the participant group consisted of qualified, experienced professionals capable of providing informed, meaningful input to support the research objectives.

19.3 ETHICAL CONSIDERATIONS

To protect participant anonymity and ensure informed consent, several measures were implemented throughout the study. Before participation, all individuals were provided with a detailed informed consent form that outlined the purpose of the research, the nature of their involvement, and their right to withdraw at any time without consequence. Participants were assured that their personal information would remain confidential and that all data collected would be anonymized, with no identifying details associated with their responses. Unique participant codes were used to track responses, eliminating any connection to personal identities. Additionally, all data were securely stored in encrypted databases and were only accessible to the research team. These measures ensured that participant privacy was safeguarded and that they were fully informed about their rights and the study procedures before providing consent to participate.

19.4 CONJOINT ANALYSIS DESIGN

The conjoint analysis was designed to systematically assess ergonomists' preferences by evaluating specific attributes related to posture evaluation. Conjoint analysis was performed using SPSS ver. 25.0 software (IBM, Armonk, NY, United States of America), and a significance level of 0.05 was used. Two primary attributes were identified for the conjoint analysis: ROM categories for each posture considered and the corresponding ergonomic risk levels (1 to 5) associated with those ROM categories. ROM categories were defined as angular intervals specific to each posture (e.g., wrist pronation, wrist supination, and wrist radial/ulnar deviation). Each interval was then classified into five ergonomic risk levels, ranging from low (Level 1) to high (Level 5), based on established ergonomic standards and commonly applied thresholds for joint movement.

These attributes and their levels were selected to align with common practices in ergonomic assessments and to represent realistic scenarios encountered in workplace evaluations. In this study, the term "scenarios" refers to the unique combinations of ROM categories and ergonomic risk levels assigned to each body posture. Each scenario, therefore, represents a distinct evaluation case, simulating realistic conditions that ergonomists may face during workplace assessments. Table 19.1 represents the body segment postures, their corresponding ROM categories (Options 1–3), and the associated risk levels (1 to 5), creating posture evaluation scenarios for participants to rank. This design enabled the quantification of the relative importance of each attribute and provided insights into how ergonomists prioritize factors when assessing posture-related risks.

The attributes selected for the conjoint analysis—ROM categories and ergonomic risk levels—were chosen based on their critical role in posture evaluation and their direct connection to musculoskeletal health. ROM categories reflect the extent of joint movement and are fundamental to determining whether a posture falls within safe, biomechanically efficient limits. Ergonomic risk levels indicate the likelihood of developing MSDs or discomfort associated with specific postures. They are essential for assessing workplace safety.

These attributes were identified through a review of ergonomic standards, such as those outlined by Panero and Zelnik (2012), and risk assessment tools like RULA and REBA, which emphasize

TABLE 19.1
Body Segment Postures, Ranges of Movement Categories (Options), and Associated Risk Levels

		Ergonomic Risk Level by Range of Motion				
Body Segment and Posture	**Range of Motion Categories by Body Posture (Options)**	**1**	**2**	**3**	**4**	**5**
Wrist pronation	Option 1:	0°< Θ≤18°	19°<Θ ≤37°	38°<Θ≤56°	57°<Θ ≤75°	76°<Θ≤90°
	Option 2:		0°< Θ ≤23°	24°<Θ≤47°	48°<Θ ≤71°	72°<Θ≤90°
	Option 3:			0°<Θ≤30°	31°<Θ ≤60°	61°<Θ≤90°
Wrist supination	Option 1:	0°<Θ≤18°	19°<Θ≤37°	38°<Θ≤56°	57°<Θ ≤75°	76°<Θ≤90°
	Option 2:		0°<Θ≤23°	24°<Θ≤47°	48°<Θ ≤71°	72°<Θ≤90°
	Option 3:			0°<Θ≤30°	31°<Θ ≤60°	61°<Θ≤90°
Wrist radial deviation	Option 1:	0°< Θ≤5°	6°<Θ≤11°	12°<Θ≤17°	18°<Θ ≤23°	24°<Θ≤30°
	Option 2:		0°<Θ≤8°	9°<Θ≤17°	18°<Θ ≤26°	27°<Θ ≤30°
	Option 3:			0°<Θ≤10°	11°<Θ ≤21°	22°<Θ ≤30°
Lower arm internal rotation	Option 1:	0°< Θ≤18°	19°< Θ≤37°	38°<Θ≤56°	57°<Θ ≤75°	76°<Θ≤90°
	Option 2:		0°<Θ≤23°	24°<Θ≤47°	48°<Θ ≤71°	72°<Θ≤90°
	Option 3:			0°<Θ≤30°	31°<Θ≤60°	61°<Θ≤90°
Lower arm external rotation	Option 1:	0°<Θ≤9°	10°< Θ≤19°	20°<Θ≤29°	30°<Θ≤39°	40°<Θ≤45°
	Option 2:		0°<Θ≤11°	12°<Θ≤23°	24°< Θ≤35°	36°<Θ≤45°
	Option 3:			0°<Θ≤15°	16°< Θ≤30°	31°< Θ ≤45 °
Lower arm abduction	Option 1:	0°<Θ≤18°	19°< Θ≤37°	38°<Θ≤56°	57°< Θ≤75°	76°<Θ≤90°
	Option 2:		0°< Θ≤23°	24°<Θ≤47°	48°<Θ≤71°	72°<Θ≤90°
	Option 3:			0°<Θ≤30°	31°<Θ≤60°	61°<Θ≤90°
Lower arm adduction	Option 1:	0°<Θ≤18°	19°<Θ≤37°	38°<Θ≤56°	57°<Θ≤75°	76°<Θ≤90°
	Option 2:		0°<Θ≤23°	24°<Θ≤47°	48°<Θ≤71°	72°<Θ≤90°
	Option 3:			0°<Θ≤30°	31°<Θ≤60°	61°<Θ≤90°
Upper arm extension	Option 1:	0°< Θ≤9°	10°< Θ ≤19°	20°<Θ≤29°	30°<Θ≤39°	40°<Θ≤45°
	Option 2:		0°< Θ ≤11°	12°<Θ≤23°	24°<Θ≤35°	36°<Θ≤45°
	Option 3:			0°<Θ≤15°	16°<Θ≤30°	31°< Θ≤45°
Upper arm internal rotation	Option 1:	0°<Θ≤13°	14°<Θ≤27°	28°<Θ≤41°	42°<Θ ≤55°	56°< Θ≤70°
	Option 2:		0°<Θ≤18°	19°<Θ≤37°	38°<Θ≤56°	57°< Θ≤70°
	Option 3:			0°<Θ≤23°	24°< Θ≤47°	48°< Θ≤70°
Upper arm external rotation	Option 1:	0°< Θ≤18°	19°< Θ ≤37°	38°<Θ≤56°	57°< Θ≤75°	76°< Θ≤90°
	Option 2:		0°< Θ ≤23°	24°<Θ≤47°	48°< Θ≤71°	72°< Θ≤90°
	Option 3:			0°< Θ≤30°	31°< Θ≤60°	61°< Θ≤90°
Upper arm abduction	Option 1:	0°< Θ≤5°	6°< Θ ≤11°	12°< Θ≤17°	18°< Θ≤23°	24°< Θ ≤30°
	Option 2:		0°< Θ ≤23°	24°< Θ≤47°	48°< Θ ≤71°	72°< Θ≤90°
	Option 3:			0°< Θ≤30°	31°<Θ≤60°	61°<Θ≤90°
Upper arm adduction	Option 1:	0°<Θ≤5°	6°< Θ≤11°	12°<Θ≤17°	18°<Θ≤23°	24°<Θ ≤30°
	Option 2:		0°<Θ≤8°	9°<Θ≤17°	18°<Θ≤26°	27°<Θ ≤30°
	Option 3:			0°<Θ≤10°	11°<Θ≤21°	22°<Θ ≤30°
Neck rotation	Option 1:	0°<Θ≤20°	21°<Θ≤30°	31°<Θ≤40°	41°<Θ≤50°	51°<Θ≤55°
	Option 2:		0°<Θ≤14°	15°<Θ≤29°	30°<Θ≤44°	45°<Θ≤55°
	Option 3:			0°<Θ≤20°	21°<Θ ≤40°	41°<Θ≤55°

ROM categories and ergonomic risk levels as key factors in evaluating posture. Including these attributes ensures that the study addresses the core elements that ergonomists consider during posture evaluation, enhancing the relevance and applicability of the findings to real-world ergonomic practices. Combinations of the selected attributes, ROM categories, and ergonomic risk levels were systematically constructed into scenarios using an experimental design framework (SPSS ver. 25.0 software (IBM, Armonk, NY, United States of America)). Each option represented a unique posture evaluation task, in which participants assessed three cases that combined different ROM ranges with varying ergonomic risk levels (1, 2, 3, 4, 5).

To ensure statistical rigor and minimize bias, a fractional factorial design was employed, generating a balanced set of combinations that captured all possible interactions between the attributes while reducing the total number of tasks to a manageable size. These options were presented to participants in a randomized order to counteract potential ordering effects. By structuring the tasks in this way, the study facilitated the collection of reliable data on ergonomists' preferences and decision-making processes when evaluating posture and associated ergonomic risks.

19.5 DATA COLLECTION

Data collection involved presenting the constructed options to participants through an online survey platform, which allowed for efficient and standardized data gathering. Each scenario was displayed individually, providing a clear description of the associated ROM categories and ergonomic risk levels. Participants were instructed to evaluate each scenario based on their professional judgment and to indicate their preferences or rank the scenarios according to perceived importance or relevance to posture evaluation.

The survey interface was designed to be user-friendly, ensuring that participants could focus on the content without technical distractions. Instructions and examples were provided at the beginning of the survey to familiarize participants with the task format. Responses were automatically recorded and securely stored in a database, ensuring data accuracy and participant confidentiality throughout the process. The conjoint analysis data collection was administered using an online survey form. This form supported the creation and randomization of options, ensuring that each participant received a unique sequence of attribute combinations to minimize order effects. Interactive features, such as drag-and-drop ranking, were utilized to enhance participant engagement and ease of use. The tool also included built-in validation mechanisms to ensure responses were complete and consistent before submission. Data encryption and secure storage protocols were implemented to safeguard participant confidentiality and comply with ethical research standards. This platform provided robust analytical outputs, streamlining the transition from data collection to analysis and ensuring high-quality data for the study.

19.6 DATA ANALYSIS

19.6.1 Conjoint Analysis

To analyze the data, conjoint analysis was employed to derive insights from participants' preferences. A conjoint analysis was made for each body segment posture. First, part-worth utility values were estimated to quantify the relative importance of each option (ROM categories and ergonomic risk levels) by assigning numerical values to the levels within each option. These part-worth utility values showed participants' preferences, with higher values indicating more favorable attitudes toward specific attribute combinations. Importance weights were then calculated to determine the relative contribution of each attribute to participants' overall preferences. This step allowed for an understanding of how much ROM categories and ergonomic risk levels influenced decision-making. This technique yielded a comprehensive understanding of underlying preferences and generated actionable insights to improve ergonomic assessment practices.

19.7 MODEL VALIDATION

To ensure the robustness of the conjoint analysis, a validation process involving holdout tests was implemented. This approach divided the data set into two subsets: one used for model estimation and the other reserved as a holdout sample for model validation. The holdout sample, which was not used in the initial model training, provided an independent data set to assess the generalizability of the results.

Two statistical measures were applied in the validation stage: Pearson's correlation coefficient and Kendall's Tau. Pearson's r was used to assess the strength of the linear relationship between the model's predicted utilities and the observed preferences in the holdout sample. Kendall's Tau, in turn, assessed the consistency of the rank ordering between predicted and actual responses, indicating how well the model preserved preference orderings.

By comparing the predicted preferences from the model with the actual preferences observed in the holdout data, the accuracy and reliability of the part-worth values and importance weights were evaluated. This process identified potential overfitting or biases in the model, ensuring that the findings would hold up when applied to different participant groups.

19.8 RESULTS AND DISCUSSION

19.8.1 Conjoint Analysis Results

The conjoint analysis revealed the relative importance of the studied attributes (ROM categories and ergonomic risk levels) in posture evaluation. Table 19.2 summarizes the utility scores and importance values obtained. Ergonomic risk level emerged as the most influential attribute across body segments, while variations in ROM categories also significantly affected risk perception.

TABLE 19.2
Part-Worth Utilities and Importance for Range of Movement and Ergonomic Risk-Level Options

Posture	ROM and Ergonomic Risk-Level Options	Part-Worth Utility Estimation	Std. Deviation	Importance
Wrist pronation	Option 1	0.180	0.335	41.2644
	Option 2	0.359	0.669	
	Option 3	0.539	1.004	
Wrist supination	Option 1	−0.359	0.335	25.963
	Option 2	−0.719	0.669	
	Option 3	−1.078	1.004	
Wrist radial deviation	Option 1	0.160	0.335	32.772
	Option 2	0.320	0.669	
	Option 3	0.480	1.004	
Lower arm internal rotation	Option 1	2.261	−0.261	31.198
	Option 2	−0.523	−0.523	
	Option 3	−0.784	−0.784	
Lower arm external rotation	Option 1	0.111	0.433	30.696
	Option 2	0.222	0.866	
	Option 3	0.333	1.300	

(Continued)

TABLE 19.2 (*Continued*)
Part-Worth Utilities and Importance for Range of Movement and Ergonomic Risk-Level Options

Posture	ROM and Ergonomic Risk-Level Options	Part-Worth Utility Estimation	Std. Deviation	Importance
Lower arm abduction	Option 1	−1.778	0.433	19.429
	Option 2	−1.556	0.866	
	Option 3	−2.333	1.300	
Lower arm adduction	Option 1	0.092	0.433	18.677
	Option 2	0.183	0.866	
	Option 3	0.275	1.300	
Upper arm extension	Option 1	1.667	1.723	18.565
	Option 2	3.333	3.446	
	Option 3	5.000	5.169	
Upper arm internal rotation	Option 1	3.278	1.723	25.298
	Option 2	0.556	0.446	
	Option 3	0.833	0.169	
Upper arm external rotation	Option 1	1.894	1.723	22.902
	Option 2	0.889	0.446	
	Option 3	0.833	0.169	
Upper arm abduction	Option 1	−0.778	0.433	17.529
	Option 2	1.556	0.866	
	Option 3	2.333	1.300	
Upper arm adduction	Option 1	−2.833	1.723	15.706
	Option 2	−5.667	3.446	
	Option 3	8.500	5.169	
Neck rotation	Option 1	0.382	0.486	100.000
	Option 2	0.765	0.971	
	Option 3	1.147	1.457	

Ergonomists were presented with three options involving different combinations of ROM ranges and associated risk levels for various body segment postures (wrist pronation, wrist supination, wrist radial deviation, lower arm internal rotation, lower arm external rotation, lower arm abduction, lower arm adduction, upper arm rotation, upper arm internal rotation, upper arm external rotation, upper arm abduction, upper arm adduction, and neck rotation).

By asking participants to rank or choose between these options, the researchers could quantify preferences. The significance level of 0.05 ensured that the findings were statistically robust, minimizing the likelihood of random variation influencing the results. Part-worth utility scores provided a quantitative measure of preference. A positive part-worth utility value indicates that the option is viewed favorably, while a negative part-worth utility value suggests that the option is less desirable. Based on the conjoint analysis (Table 19.2), ergonomists demonstrated a preference for wider ROM categories for body segments with higher mobility, such as the upper arm external rotation (part-worth utility value: +1.8), likely because this segment is involved in a wide range of dynamic tasks (e.g., lifting, reaching, and rotating).

A broader ROM allows for greater flexibility in posture evaluation, accommodating natural movement variability without overestimating risk. Additionally, like the upper arm, the lower arm also showed a preference for wider ROM categories, though slightly less pronounced. This underscores the importance of wrist and elbow mobility for tasks such as gripping, typing, and operating tools. There was less preference for narrower ROM categories in segments with limited mobility, such as the neck (utility score: −0.3). This indicates that ergonomists prioritize flexibility in defining ROM for highly mobile body segments, as these segments are more likely to be involved in dynamic and varied postures in the workplace. Besides, ergonomists preferred narrower ROM categories for the neck, indicating a higher perceived risk when this segment moves outside a limited range. This is likely due to the neck's susceptibility to strain and injury, especially in static or awkward postures (e.g., prolonged head tilt or excessive rotation).

Additionally, the analysis revealed that ergonomists associate broader ROM categories with lower ergonomic risk levels for highly mobile segments, suggesting that they perceive greater movement capacity as less risky when properly managed. Conversely, for less mobile segments like the neck, narrower ROM categories were linked to higher perceived risk, likely due to the potential for strain or injury when these segments are forced into extreme or unnatural positions.

These findings could lead to the development of updated ergonomic guidelines that emphasize tailored ROM thresholds for different body segments. Guidelines might recommend wider acceptable ROM ranges for the upper and lower arms during tasks such as lifting or reaching, while advocating stricter limits on neck movements to reduce the risk of MSDs.

This targeted approach could improve workplace safety by aligning ergonomic interventions with expert preferences and risk assessments. Regarding risk perception and ROM associations, ergonomists associated wider ROM categories with lower risk for highly mobile segments, as these segments are naturally designed for greater movement. For example, the wide ROM of the upper arm might be considered safe during overhead tasks, provided the movement is controlled and within functional limits.

Conversely, narrower ROM categories were linked to a higher risk for less mobile segments. For instance, a narrow ROM for the neck might reflect concerns about cervical spine strain when the neck is extended or rotated beyond a comfortable range. Besides, the analysis revealed that ergonomists are willing to tolerate slightly higher ROM variability in highly mobile segments to avoid overly restrictive guidelines that could hinder task performance. However, for less mobile segments, they prioritize stricter limits to prevent injury.

The findings suggest that ergonomic guidelines should not adopt a "one-size-fits-all" approach to ROM categories. Instead, they should differentiate between body segments based on their natural mobility and risk profiles. For example, ergonomic guidelines for the upper arm might recommend a ROM of 0–180° for flexion/extension, acknowledging the need for wide range of motion. Neck guidelines might restrict the ROM to 0–45° for rotation and 0–30° for flexion/extension to minimize strain. Ergonomists could use these findings to refine posture assessment tools (RULA and REBA) by incorporating segment-specific ROM thresholds. This would make risk evaluations more accurate and aligned with expert preferences.

This research could inform the design of workstations and equipment to accommodate the natural ROM of highly mobile segments while protecting less mobile ones. For instance, adjustable chairs and monitors could reduce neck strain by promoting neutral postures. Tools with ergonomic handles could allow for greater wrist and arm mobility without increasing risk. Training programs could educate workers on maintaining safe ROMs for different body segments during tasks. While conjoint analysis provides valuable insights, future studies could validate these findings in real-world workplace scenarios.

Observational studies could assess whether the preferred ROM categories effectively reduce injury rates. Preferences for ROM categories might vary across cultures or industries. For example, manual workers might have different ROM needs compared to office workers. Expanding the research into diverse populations could enhance the generalizability of the findings. Combining conjoint analysis with biomechanical measurements (electromyography and motion capture) could provide a more comprehensive understanding of how ROM preferences align with physiological limits and injury risks. This research highlights the importance of segment-specific ROM categories in ergonomic evaluations. By quantifying ergonomists' preferences, the study provides an evidence-based foundation for developing more nuanced and effective ergonomic guidelines.

These insights could lead to safer workplace designs, reduced MSDs, and improved alignment between expert evaluations and practical interventions.

19.8.2 Model Validation: Holdout Test

To further assess the robustness of the conjoint analysis model, a holdout test was conducted to validate its predictive performance. Two key statistical measures were employed for this purpose: Pearson's correlation coefficient and Kendall's Tau. Pearson's *r* quantifies the strength of the linear relationship between predicted and observed values, with a high r value (accompanied by a low p-value) indicating that the model effectively captures the magnitude of consumer preferences. In parallel, Kendall's Tau assesses the consistency of the rank ordering between predicted and actual responses, providing insights into whether the model correctly prioritizes profiles, as preferred by ergonomists. Together, these metrics offer a comprehensive view of the model's accuracy and reliability, confirming that the conjoint analysis robustly reflects both the intensity and the ranking of ergonomists' preferences. Table 19.3 summarizes Pearson's r and Kendall's Tau, along with their corresponding p-values, for each body segment, thereby offering a multifaceted evaluation of the conjoint analysis model's predictive performance. In this context, Pearson's *r* measures the linear correlation between predicted utilities and actual responses, with higher values indicating that increases in predicted utility are strongly associated with increases in observed consumer preferences. For instance, a Pearson's *r* value near 0.6 or above, when coupled with a statistically significant *p*-value (typically below 0.05), suggests that the model reliably quantifies the magnitude of consumer preference across the evaluated segments. Concurrently, Kendall's Tau assesses the consistency of the rank ordering between predicted and actual outcomes, reflecting how well the model preserves the order of preferences. A robust Kendall's Tau value indicates that the profiles are ranked similarly in both the model predictions and the observed data, which is critical for validating the model's practical applicability, especially in contexts where the relative importance of ergonomic risk levels is at stake. Together, these metrics not only affirm the strength of the linear relationship but also confirm the consistency in ranking, thereby reinforcing the overall reliability of the conjoint analysis model in capturing true ergonomist choice behavior across different body segments.

The holdout test results for the conjoint analysis reveal varying levels of predictive accuracy across different body segments. For the wrist, Pearson's *r* is 0.621, and Kendall's Tau is 0.461, both with *p*-values of 0.000, indicating a strong and statistically significant correlation between the predicted utilities and actual responses.

This suggests that the conjoint model performs particularly well for the wrist, reliably capturing both the magnitude and ranking of ergonomists' preferences. In contrast, the lower arm segment shows a Pearson's *r* of 0.273 and Kendall's Tau of 0.231, with *p*-values of 0.008 and 0.004, respectively. Although these correlations are statistically significant, they are notably weaker, implying that

TABLE 19.3
Holdout Test Results by Body Segment, Pearson's r, Kendall's Tau, and p-Values

Body Segment	Pearson's r	p-Value	Kendall's Tau	p-Value
Wrist	0.621	0.000	0.461	0.000
Lower arm	0.273	0.008	0.231	0.004
Upper arm	0.486	0.005	0.309	0.012
Neck	0.458	0.004	0.398	0.026

the model's predictions for the lower arm are less robust. The upper arm and neck segments display moderate performance, with Pearson's r values of 0.486 ($p=0.005$) and 0.458 ($p=0.004$) and Kendall's Tau values of 0.309 ($p=0.012$) and 0.398 ($p=0.026$), respectively.

19.9 CONCLUSIONS

This research underscores the critical role of ROM categories in ergonomic posture evaluation and risk assessment. By applying conjoint analysis, the study quantified ergonomists' preferences for ROM thresholds, revealing a distinction between highly mobile body segments (upper and lower arms) and less mobile ones (neck). Ergonomists demonstrated a strong preference for wider ROM categories in segments with greater mobility, reflecting the need for flexibility in dynamic tasks, while favoring narrower ROM categories in less mobile segments to mitigate the risk of strain and injury. These findings have profound implications for ergonomic practice and workplace safety. They advocate for a tailored approach to ROM guidelines, moving away from generic standards toward segment-specific recommendations that align with the natural movement capacities and risk profiles of different body parts. By integrating these insights into ergonomic tools, workplace designs, and training programs, organizations can better address posture-related risks, reduce MSDs, and enhance worker well-being. Ultimately, this research bridges the gap between expert preferences and practical applications, offering a data-driven foundation for more effective and targeted ergonomic interventions. It highlights the importance of collaboration between researchers and ergonomists to create safer, healthier, and more productive work environments. This study paves the way for a future where ergonomic guidelines are not only scientifically robust but also practically actionable, ensuring that workplaces are designed with the human body's capabilities and limitations in mind. Besides, the results highlight the importance of conducting holdout tests and segment-specific analyses when using conjoint analysis. The variation in predictive accuracy across body segments not only validated the model for some areas but also signaled the need for targeted improvements in others. Future research could focus on exploring the underlying reasons for these discrepancies, such as potential measurement errors, attribute misspecification, or differences in ergonomists' understanding, and on refining the model accordingly. By addressing these issues, researchers and practitioners can enhance the robustness of the conjoint analysis model, ensuring more reliable and actionable insights across all segments. Moreover, the application of conjoint analysis across various disciplines highlights its effectiveness in capturing user preferences and decision-making factors, from marketing and product design to environmental management and consumer behavior. While previous studies have leveraged this method to understand subjective preferences, this research takes a different approach by applying conjoint analysis to define ROM categories and ergonomic risk levels. This methodological shift extends the utility of conjoint analysis beyond consumer-based evaluations, demonstrating its potential for classifying biomechanical factors and enhancing ergonomic risk assessment. This innovative use case contributes to the field of ergonomics by providing a data-driven approach to evaluating physical strain and optimizing workplace safety. Finally, a review of the current literature reveals that no similar studies have employed conjoint analysis to define ROM categories and their associated ergonomic risk levels. This innovative approach offers a novel method for delineating risk categories in a more nuanced, user-centered manner. The absence of a precedent in this specific application underscores the study's originality and highlights its potential to fill an important gap in ergonomic research. By integrating conjoint analysis into the evaluation of ROM and ergonomic risks, this work paves the way for new methodological perspectives and encourages further investigation into its applicability across diverse occupational settings.

REFERENCES

Baklouti, Souha, Chaker, Abdelbadia, Rezgui, Taysir, Sahbani, Anis, Bennour, Sami and Laribi, Med Amine. 2024. "A Novel IMU-Based System for Work-Related Musculoskeletal Disorders Risk Assessment." *Sensors* 24 (11): 3419. doi:10.3390/s24113419.

Cejudo, Antonio, Sainz de Baranda, Pilar, Ayala, Francisco, De Ste Croix, Mark and Santonja-Medina, Fernando. 2020. "Assessment of the Range of Movement of the Lower Limb in Sport: Advantages of the ROM-Sport I Battery." *International Journal of Environmental Research and Public Health* 17 (20): 7606. doi:10.3390/ijerph17207606.

Consing, Jennifer Pasang. 2023. "Preference Analysis on Online Learning Attributes among Senior High School Students during the COVID-19 Pandemic: A Conjoint Analysis Approach." *World Journal of Advanced Research and Reviews* 18 (3): 1552–55. doi:10.30574/wjarr.2023.18.3.1248.

David, G. C. 2005. "Ergonomic Methods for Assessing Exposure to Risk Factors for Work-Related Musculoskeletal Disorders." *Occupational Medicine* 55 (3): 190–99. doi:10.1093/occmed/kqi082.

Dias, Pâmela, and Monteiro, Plínio. 2020. "Sports Marketing and Perceived Value: An Application of the Conjoint Analysis in the Fan Membership Programs of Football." *Brazilian Business Review* 17 (3): 253–74. doi:10.15728/bbr.2020.17.3.1.

Faisal, Abu Ilius, Majumder, Sumit, Mondal, Tapas, Cowan, David, Naseh, Sasan and Deen, M. Jamal. 2019. "Monitoring Methods of Human Body Joints: State-of-the-Art and Research Challenges." *Sensors* 19 (11): 2629. doi:10.3390/s19112629.

Fan, Jialin, Tan, Xiaotong, Smith, Andrew Paul and Wang, Jing. 2024. "Work-Related Musculoskeletal Disorders, Fatigue and Stress among Gas Station Workers in China: A Cross-Sectional Study." *BMJ Open* 14 (7). doi:10.1136/bmjopen-2023-081853.

Gumasing, Ma. Janice, Kester Ong, Ardvin and Bare, Maria Angelica. 2022. "User Preference Analysis of a Sustainable Workstation Design for Online Classes: A Conjoint Analysis Approach." *Sustainability* 14 (19): 12346. doi:10.3390/su141912346.

Hignett, Sue, and McAtamney, Lynn. 2000. "Rapid Entire Body Assessment (Reba)." *Applied Ergonomics* 31 (2): 201–5. doi:10.1016/s0003-6870(99)00039-3.

Hoe, Victor CW, Urquhart, Donna M., Kelsall, Helen L., Zamri, Eva N and Sim, Malcolm R. 2018. "Ergonomic Interventions for Preventing Work-Related Musculoskeletal Disorders of the Upper Limb and Neck among Office Workers." *Cochrane Database of Systematic Reviews* 2018 (10). doi:10.1002/14651858.cd008570.pub3.

Juba, Omolara Oluseun. 2024. "Impact of Workplace Safety, Health, and Wellness Programs on Employee Engagement and Productivity." *International Journal of Health, Medicine and Nursing Practice* 6 (4): 12–27. doi:10.47941/ijhmnp.1819.

Kee, Dohyung, and Karwowski, Waldemar. 2007. "A Comparison of Three Observational Techniques for Assessing Postural Loads in Industry." *International Journal of Occupational Safety and Ergonomics* 13 (1): 3–14. doi:10.1080/10803548.2007.11076704.

Kuzmanovic, Marija, and Savic, Gordana. 2020. "Avoiding the Privacy Paradox Using Preference-Based Segmentation: A Conjoint Analysis Approach." *Electronics* 9 (9): 1382. doi:10.3390/electronics9091382.

Lowe, Brian D., Dempsey, Patrick G., and Jones, Evan M. 2019. "Ergonomics Assessment Methods Used by Ergonomics Professionals." *Applied Ergonomics* 81 (November): 102882. doi:10.1016/j.apergo.2019.102882.

McAtamney, Lynn, and Corlett, E. Nigel. 1993. "Rula: A Survey Method for the Investigation of Work-Related Upper Limb Disorders." *Applied Ergonomics* 24 (2): 91–99. doi:10.1016/0003-6870(93)90080-s.

Nath, Nipun D., Reza Akhavian, and Behzadan, Amir H. 2017. "Ergonomic Analysis of Construction Worker's Body Postures Using Wearable Mobile Sensors." *Applied Ergonomics* 62 (July): 107–17. doi:10.1016/j.apergo.2017.02.007.

Ortiz De Orue, L. Haydee. 2021. "Conjoint Analysis: A Preference Based Approach for the Accounting of Conservation and Tourism Benefits in Choquequirao Management." *Journal of Ecology & Natural Resources* 5 (3). doi:10.23880/jenr-16000253.

Panero, Julius, and Zelnik, Martin. 2012. *Las Dimensiones Humanas En Los Espacios Interiores*. Barcelona: Gustavo Gili.

Peppoloni, Luca, Filippeschi, Alessandro, Ruffaldi, Emanuele, & Avizzano, Carlo Alberto. 2016. "A Novel Wearable System for the Online Assessment of Risk for Biomechanical Load in Repetitive Efforts." *International Journal of Industrial Ergonomics* 52 (March): 1–11. doi:10.1016/j.ergon.2015.07.002.

Silaen, Randy Vinstein, and Windasari, Nila Armelia. 2022. "Customer Preference Analysis on Attributes of Hybrid Electric Vehicle: A Choice-Based Conjoint Approach." *International Journal of Current Science Research and Review* 05 (12). doi:10.47191/ijcsrr/v5-i12-30.

Singh, Ruchira, Carranza Leon, Daniel A., Morrow, Melissa M., Vos-Draper, Tamara L., Mc Gree, Michaela E., Weaver, Amy L., Woolley, Sandra M., Hallbeck, Susan and Gebhart, John B. 2016. "Effect of Chair Types on Work-Related Musculoskeletal Discomfort during Vaginal Surgery." *American Journal of Obstetrics and Gynecology* 215 (5). doi:10.1016/j.ajog.2016.06.016.

Spielholz, Peregrin, Silverstein, Barbara, Morgan, Michael, Checkoway, Harvey and Kaufman, Joel. 2001. "Comparison of Self-Report, Video Observation and Direct Measurement Methods for Upper Extremity Musculoskeletal Disorder Physical Risk Factors." *Ergonomics* 44 (6): 588–613. doi:10.1080/00140130110038035.

Takala, Esa-Pekka, Pehkonen, Irmeli, Forsman, Mikael, Hansson, Gert-Åke, Mathiassen, Svend Erik, Neumann, W. Patrick, Sjøgaard, Kaj Bo, Westgaard, Rolf H., & Winkel, Jørgen 2022. *Systematic Evaluation of Observational Methods Assessing Biomechanical Exposures at Work*, October. doi:10.32920/ryerson.14639964.

Trovato, Bruno, Roggio, Federico, Sortino, Martina, Zanghì, Marta, Petrigna, Luca, Giuffrida, Rosario, & Musumeci, Giuseppe 2022. "Postural Evaluation in Young Healthy Adults through a Digital and Reproducible Method." *Journal of Functional Morphology and Kinesiology* 7 (4): 98. doi:10.3390/jfmk7040098.

van der Beek, A J, and Frings-Dresen, M H. 1998. "Assessment of Mechanical Exposure in Ergonomic Epidemiology." *Occupational and Environmental Medicine* 55 (5): 291–99. doi:10.1136/oem.55.5.291.

Wang, Fa, Wang, Haifeng, and Cho, Joung Hyung. 2022. "Consumer Preference for Yogurt Packaging Design Using Conjoint Analysis." *Sustainability* 14 (6): 3463. doi:10.3390/su14063463.

Wei, Shun-Hwa, McQuade, Kevin J. and Smidt, Gary L. 1993. "Three-Dimensional Joint Range of Motion Measurements from Skeletal Coordinate Data." *Journal of Orthopaedic & Sports Physical Therapy* 18 (6): 687–91. doi:10.2519/jospt.1993.18.6.687.

20 Ergonomic Intervention Experiences for the Strengthening of Forest Fire Coordination and Control Centers

Maureira C. Fabiola, Meyer C. Felipe, Espinoza B. Jorge, Chávez M. Francisca, and Farías A. Marcelo

20.1 GENERAL INTRODUCTION

This chapter compiles four ergonomic investigations in the field of forest fire management, given that this is a complex issue with a high impact in our region and in many countries of the world. This phenomenon has economic, social, environmental and safety implications for those who are directly or indirectly affected by these events.

The management and control of forest fires in Chile represent a shared responsibility between the State and private forestry companies (CONAF 2022). In the Latin American context, ergonomic research on forest fires has historically focused its attention on the work of the brigades who fight forest fires (Apud, Meyer and Maureira 2002). Their recommendations constitute a reference material that guides the practices and work organization of those working on the front line in forest firefighting, both nationally and internationally. Subsequently, further progress was made in understanding the work of another key link in forest fire management, control and coordination centers. These units play a strategic role in the institutional value chain since they manage strategic and logistical processes to ensure an effective response to emergencies. Paradoxically, in our continent, there are no publications that systematically address ergonomic working conditions, human factors or decision-making processes in these environments. Given the above, the purpose of this chapter is to provide academic research evidence and contribute to the framework of undergraduate and graduate theses developed in the Ergonomics Department of the Universidad de Concepción in Chile (Maureira 2015; Chávez 2023; Farías 2024). On the other hand, this unit of analysis constitutes an emerging and innovative field, with high potential to contribute to the design of more efficient, safe and resilient systems. The following discussion not only aims to make visible the work of this key link in the crisis management system but also to illustrate the role of the ergonomist as a transforming agent of the systems in which this person intervenes.

The cases that will be approached make organizational problems visible, where the professionals have a key role in the creation of knowledge with a participatory, user-centered approach, aimed at improving the efficiency of work systems, safeguarding safety, integral health and development potential of the people who participate in them.

 DOI: 10.1201/9781003582014-25

20.2 CHARACTERIZATION OF THE FOREST FIRE CONTROL AND COORDINATION CENTERS IN CHILE

Forest fire control and coordination centers correspond to sociotechnical systems that operate in complex dynamic environments in which situations are under partial control of the operator and can evolve without the operator's intervention (Hoc, Amalberti, Cellier, and Grosjean 2004). To cope with these scenarios, their members establish compromises and articulate adaptive, personal, and collective regulation strategies to keep the situation within acceptable limits. These systems depend on the concerted participation of committed individuals to regulate disturbances to the ongoing situation (Weill-Fassina and Benchekroun 2000). At the same time, they operate under the model of distributed social cognition since people are in a co-dependent relationship and must interact to pool their information sources and negotiate their differences (Darses, Falzon, and Munduteguy 2004). The main task is to dispatch resources to fight a forest fire as quickly as possible, causing minimal damage, and is broken down into a set of subtasks. These correspond to the detection and registration of outbreaks, the analysis of the situation, the evaluation of resource dispatch alternatives, and the evaluation and monitoring until confirmation of fire extinction and brigade deployment actions (Maureira 2015; Chávez 2023; Farías 2024).

Regarding the nature of the work, it is a sedentary job with significant cognitive and emotional demands. That is, those who work in this occupation perform analytical and strategic work, including understanding the situation, evaluating action alternatives, anticipating scenarios, making decisions, evaluating the impact of actions, and monitoring the evolution of the process, under constant pressure. Operationally, they exchange information to coordinate actions, using technical and codified language, supported by technological tools and equipment to coordinate actions with other actors in the field.

The above description contextualizes the complexity of the work, helping better understand the problems that the following research will address.

20.3 ERGONOMIC INTERVENTIONS DEVELOPED IN CHILE ON FOREST FIRE COORDINATION AND CONTROL CENTERS

The four cases of ergonomic research and interventions are presented below, each of which will be organized according to the problem addressed, research objectives, methodological framework, results, discussion, and limitations.

20.3.1 Organizational Conditions for the Development of Competencies of an Ephemeral Team

The following description corresponds to an excerpt from a PhD thesis in Ergonomics (Maureira 2015), along with subsequent communications by the author, motivated by the understanding of the work and activities of the members of a center, whose configuration and operation were intriguing in terms of system reliability.

20.3.2 Problems

The problem focused on the system's fragility because the team's configuration was composed of temporary employees due to the institution's budgetary constraints. The research problem, which gave rise to the objective, was formulated in terms of understanding the mechanisms through which an ephemeral collective turns into a learning collective in a complex, dynamic and risky work environment.

20.3.3 Theoretical Positioning

The theoretical basis in which this research was based on was collective work (Caroly Clot 2004; Cellier, Eyrolle and Marine 1997; De la Garza Weill-Fassina 2000; Weill-Fassina Benchekroun 2000), the management of risky complex dynamic environments (Hoc et al. 2004), the organizational factors that accelerate training and learning (Paquay 2005), the development of competencies (Rogalski and Marquié 2004; Vidal-Gomel and Rogalski 2007), and the emergence of learning collectives (Flageul-Caroly 2001).

20.4 METHODOLOGY

This was a qualitative research with an ethno-methodological approach where the analysis unit is focused on human practices, taking advantage of the story lived by the actors themselves in both their individual and collective dimensions (Urbano Gil 2007). At the same time, it is based on Hutchins' theory of distributed social cognition (1995) since the focus is on the mental processes necessary to coordinate actions and make decisions in interaction with people, mediated by artifacts in a specific context.

The center was composed of five experienced operators, three novice operators and two coordinators. It should be noted that only one of the members, the chief coordinator, had a permanent employment contract.

The data collection techniques were as follows: individual semi-structured interviews, review of documentary material and both open and systematic observations in the field during the initial training period and later during the operational working period. Observables, in the latter case, were defined as individual actions and verbal interactions between team members oriented to training. There were also typified situations that triggered an observable who intervened, their content and the modality of the interaction for training purposes. The data collection and content analysis of the verbal interactions was supported by audio and video recordings of the activities in progress.

20.5 RESULTS

In organizational terms, at the internal level, it was possible to visualize a set of informal regulation strategies to favor the development of competencies from the beginning of the fire season during the training period and during the forest fire season:

- Informal self-training actions and discretionary social interactions of a constructive type.
- Use of theoretical–practical didactic resources, with emphasis on case analysis, role-playing and simulations, together with decision-making and problem-solving exercises.
- Leadership style of the chief favorable to the articulation of informal regulation strategies, transmitted orally and oriented to the transmission of rules of action that stimulate the recognition of differential competences, self-training, collective work, training and mutual help (Maureira 2015; Maureira, Benchekroun and Falzon 2019).
- Establishment of an operative mode or collective strategy defined as follows: "epistemo-vigilance" (Maureira 2015), through which each of the operators is attentive to the verbal interactions of his co-worker, while performing his work, to assist him in case of need and availability.
- Arrangement of a common and shared workspace, with a spatial distribution favorable to mutual listening and visibility.
- Use of informal strategies of constructive regulation, defined by the author as follows: "organizational accelerators of training, learning and reinforcement of knowledge"

TABLE 20.1
Evolution of Accelerators Used by Members of the Center Team during the Season, Biobio Region Chile, 2010–2011 (Source: Maureira 2015; Maureira, Benchekroun, Falzon and Apud 2019).

Team Member	Start of Season	Mid-season	End of Season	Total Frequency
Experienced	26	34	6	66
Novice	32	59	9	100
Team	0	3	3	6
Coordinator	39	17	9	65
Total	97	113	27	237

(AOFAR). Seven types of AOFAR were identified in the field: (a) animation of training situations, (b) retrospective analysis of the activity, (c) assistance to assistance, (d) direct assistance focused on learning, (e) self-teaching, (f) cooperation for learning purposes and (h) evaluation and reinforcement of knowledge. In this regard, the most used strategy was direct assistance focused on learning, activated predominantly by a novice to receive support and guidance in their actions and decisions in office management or by one of the experienced ones, who were attentive to constructively guide the novices in their work. In contrast, the least mobilized strategy was assistance to assistance, which occurred when the coordinator interacted with an experienced person with a training objective so that the experienced person could in turn assist a novice.

These accelerators were recorded throughout the season, with variations in their frequency of use depending on the period and the person triggering the situations of action or interaction with a constructive purpose. As shown in Table 20.1, the highest number of actions and interactions with a constructive purpose was mobilized at the initiative of novice dispatchers, while both experienced dispatchers and coordinators contributed significantly, almost equally.

It is important to note that the frequency of constructive actions and interactions increases toward the middle of the season because operators are expected to have greater mastery of operations. However, workload changes and the operators must deal with an increased frequency of complex forest fires due to adverse weather conditions, requiring the management of more forest fire-fighting resources. By the end of the season, the situations of interaction with constructive purposes are considerably reduced, but they persist

20.5.1 Discussion

This research reveals an organization that seeks to address its reliability problem by fostering an organizational culture favorable to learning and the development of action competencies. At the same time, the system is aligned with a type of management that promotes the use of formal and informal strategies of action and interaction, oriented to exploit productive activity for constructive purposes, typical of opportunistic learning (Falzon and Pascualetti 2000). In fact, they take advantage of moments of greater work demand and complex events to exploit them as situations for constructive interaction. This finding constitutes an unusual practice, since, in general, when periods of high demand occur, construction activity is postponed to prioritize productive work. In any case, toward the end of the season, the proportion of interactions for construction purposes is significantly lower. This background could reveal the acquisition of trade skills, although it is not expected that this type of constructive interaction dynamics will disappear, since, in such a complex dynamic environment, new situations will always emerge that can be intentionally exploited as resources for the development of constructive activity. Finally, the research contributes novel notions and concepts for our discipline, such as AOFARs, situations of interaction for learning purposes (SIA), and "epistemo-vigilance", a notion that could be proposed as a specific collective competence that complements the theory of ecological safety described by Amalberti (2012).

20.5.2 Research Limitations

This study has limitations: it is a unique case of forest fire management within a state institution, with results mainly applicable to similar dynamic sociotechnical systems. Its validity is influenced by the reliance on experienced leaders and dispatchers as transmitters of knowledge, thereby limiting generalizability to other contexts. Although the Biobío region is representative due to the high occurrence of fires, the findings need to be contrasted in different areas, institutions, and working conditions. Nevertheless, they provide valuable elements of informal learning and work design that may inspire future ergonomic research.

20.6 ERGONOMIC DESIGN PROPOSAL FOR A FOREST FIRE CONTROL AND COORDINATION CENTER

In the following section, an ergonomic intervention is described in which the professional challenge of the ergonomist was oriented toward the technical assistance service in the field of design and prototyping of a work environment.

20.6.1 Problems

The aim of the intervention was to develop a functional design prototype for a new forest fire coordination center. In this case, the ergonomist participated in an intermediate stage of the design process, gathering the experiences and knowledge of those who carry the identity of the trade to empathize with their needs and expectations, identify reference situations, devise, co-construct solutions, and prototype future scenarios of action contextualized to their reality and requirements.

20.7 THEORETICAL POSITIONING

The intervention was supported as a frame of reference by the research of Maureira (2015), where ergonomic design criteria, conditions, or requirements adapted to the context and nature of the work team configuration were defined. Additionally, spatial and functional design considerations and restrictions were identified in accordance with national regulations (Ministry of Health 1999) and international standards, such as the ISO 11064-1:2000 series on the ergonomic design of control centers and ISO 9241-210:2019, which establishes ergonomic requirements for interactive systems and office workstations. In addition, ergonomic intervention involves a systematic sequence of stages, from the formulation of the demand, the identification of reference situations, the diagnosis, and the simulation of future scenarios (Guérin et al. 2007). Moreover, design ergonomics constitutes a constructive activity that also demands specific skills (Barcellini, Van Belleghem, and Daniellou 2013) since it involves iterative stages, defining the roles of the different actors involved in this creative and innovative process, and managing social skills. Particularly, it is a dynamic of social construction, with negotiations and tensions (Benchekroun 2016), and the margins of maneuver and degrees of freedom within which the ergonomist or other participants can negotiate in this process.

20.8 METHODOLOGY

Using the action intervention methodology and Design Thinking (Brown 2020; Kelly 2016), seven iterative work sessions were conducted in accordance with the design stages. In total, 13 people participated, including representatives of the operational unit with positions of responsibility, experienced former dispatchers working in other fire management-related areas, representatives of the institution, and technical project managers.

During the pre-diagnostic stage, several sessions were conducted to identify needs and expectations regarding a future design. Two ethnographic techniques were used. The first one was used to identify their expectations and needs based on the question "what would an ideal working day at the plant be like?" (Fresno García and Hernández-Echegaray 2019). The second one, the AEIOU technique created by Robinson, Prokopoff, Cain, and Pokorny in 1991 (Martin and Hanington 2012), is oriented toward characterizing the work, from the point of view of the person–task–environment relationship, and the simulation technique with mock-up (Brown 2009).

20.8.1 Results

From a sociotechnical perspective, the use of ethnographical techniques allows us to systematize a set of requirements, expectations, and design considerations, focused on user experience and considering existing restrictions. In this regard, their main concerns were related to the spatial configuration of the premises and the type of furniture. With respect to the physical environment, they raised requirements for acoustic insulation from external agents, an air-conditioning system to uniformly regulate the area, and an efficient lighting system, with generators for autonomous operation. Other requirements included more robust, up-to-date technological tools and support systems to ensure an uninterrupted flow of communications. They also mentioned clothing needs and the need for an active break during the working day.

Afterward, in the session focused on characterizing both the interaction among the person, task, and environment and the reference situations, users identified two critical or key tasks and prioritized a functional design to focus on them. These are as follows: decision-making regarding resource management that combats forest fires and the on-the-job informal training process. In addition, they worked together on the anticipation of future scenarios, defined the critical positions of the unit, and agreed on criteria regarding restrictions on access to the premises by persons outside these critical positions.

Subsequently, another work session was held to model to scale the layout of workstations, circulation areas, furniture, tools, and support equipment at the new worksite. Using the mock-up simulation technique, users recreated different work scenarios, with materials designed to scale, in accordance with pre-established ergonomic design policies and regulatory constraints.

Finally, after several iterations with the users, a prototype of the plant design was agreed upon (see Figure 20.1). The intervention and its results were presented to the project team, together with a report on ergonomic specifications and recommendations to ensure comfort, balance, and efficiency from the perspectives of exposure to agents in the physical work environment, clothing, and furniture design.

This prototype was validated by the team responsible for executing the works and was built in accordance with the study's recommendations.

20.9 DISCUSSION

This intervention illustrates the work of the ergonomist in processes of social construction, in facilitating interaction and problem-solving, in producing ideas, and in co-constructing solutions. In this sense, the emerging methodologies of Gaming and Design Thinking constitute an

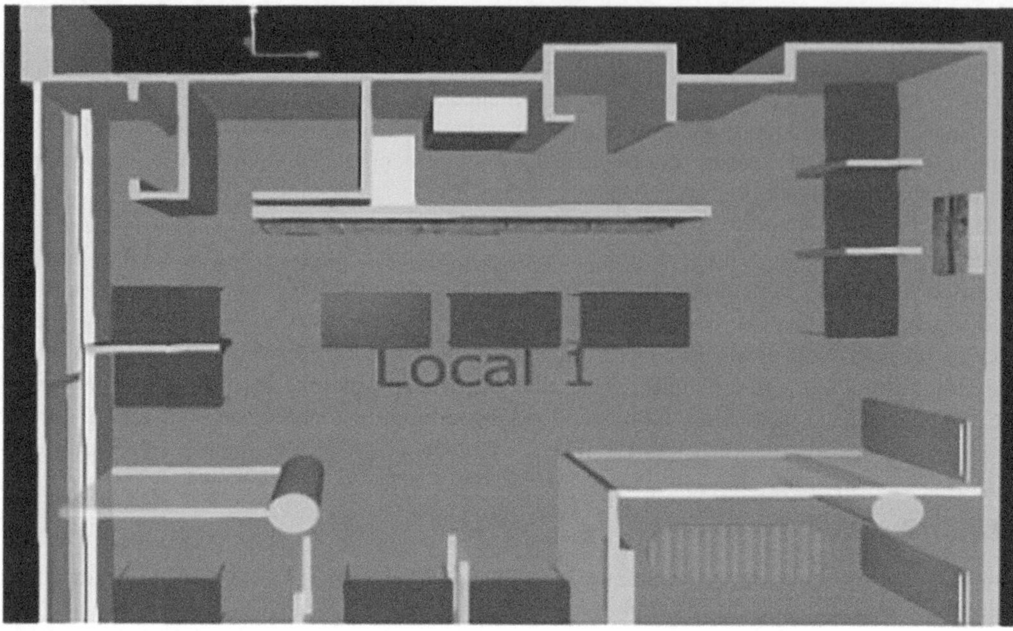

FIGURE 20.1 Modalization prototype built and agreed upon by the users in Dialux.

interesting complement to the conduct and intent of each iterative phase of ergonomic design. The interactive dynamics articulated facilitate exploration, divergent thinking, dialogue, and metacognition regarding their own actions and their impact on both their performance and internal state.

Regarding the final product that was co-constructed, it was possible to model, in a mock-up, a workspace that complies with the pre-established ergonomic specifications and restrictions, responds to the critical tasks, and adjusts to the context and the persisting problematic issues related to the temporality of the work team. The prototyped design favors proxemic spaces of cooperation, productive and constructive activity, and the development of a learning collective, i.e., the collective competence of episteme-vigilance and a work scenario free of irruptions and invasions of people alien to the critical tasks of the trade.

20.9.1 Limitations of the Intervention

The team of ergonomists did not have the opportunity to participate in the subsequent stages of confirming the specifications defined in the prototype, which prevents greater control over the impact of this design experience on users. Finally, despite this being a mock-up proposal for a specific operator context, it could serve as a reference for eventual redesign and adaptation projects of this type of sociotechnical system that can manage complex, dynamic environments.

20.9.2 Workload and Functional Configuration for Work Teams of the Forest Fire Control and Coordination Center (CENCOR)

The following description corresponds to an extract of an undergraduate thesis for the degree of Industrial Civil Engineer, developed by Chavez (2023) under the supervision of the Ergonomics department.

20.10 PROBLEM

The problem at the base persists, stemming from the ephemeral and heterogeneous configuration of the teams, resulting from the lack of contractual continuity for most members and the consequent loss of human capital (Maureira 2015). On this occasion, the scenario was even more uncertain, given that all the CENCOR dispatchers were inexperienced and only one of the three coordinators had a permanent contract. The questions were posed around identifying the strategies and operating modes that CENCOR members need to articulate and the commitments they incur to meet performance standards, personnel rotation, and internal status in the face of the mental and temporal demands imposed by the fire season. Thus, the research was proposed with the general objective of analyzing the distribution of workload at CENCOR and its impact on the performance and mental health of the operators (Chávez 2023).

20.10.1 Theoretical Positioning

The research is part of the line of extension of Maureira's doctoral thesis (2015), the activity analysis model (Leplat 1985) for the understanding of work, analysis of regulation strategies and commitments to respond to the objectives set by the organization. This research focuses on the analysis of the mental workload perceived by each member and on precursor factors that affect the system, such as collective work and learning collectives (Caroly and Clot 2004; Rogalski et al. 2004).

20.10.2 Methodology

This was a descriptive mixed-type research study carried out on the premises of the plant designed in 2018. We worked with 63% of the CENCOR members over a 4-month period. The measurement techniques used were open and systematic observations, individual interviews, the application of the NASA-TLX (Hart and Staveland, 1988), and a documentary review. Data processing was based on content analysis of the interviews. As for the activity analysis, observations were made of the interactions between CENCOR members. The observables were based on the interactions between members, typified according to their purpose, whether functional or constructive (Maureira 2015), their frequency, and their directional flow, according to position.

20.10.3 Results

The institution has a defined and formalized hierarchical structure, with positions of responsibility and general job descriptions, training programs for each position, operational criteria for decision-making regarding the dispatch of resources, indicators of management or regional success, and performance evaluation scales. This internal organization is complemented by dynamic, informal regulatory strategies, defined and promoted by management to carry out daily tasks (Maureira 2015). Among these, we note that management implemented another informal regulation strategy by assigning the role of "dispatcher supervisor" to the only dispatcher who had joined toward the end of the previous season.

As shown in Table 20.2, in the analysis of the activity, the predominance of top–down vertical communications was observed between headquarters and dispatchers.

The latter evidences the mobilization of both individual and collective regulatory strategies. On the one hand, management establishes an important personal commitment to guide the dispatchers' activity in the room. But, in addition, collective regulatory modes of assistance and mutual support persist, both for interactions with a functional and constructive purpose, although the latter occur proportionally less.

TABLE 20.2
Frequency, Flow and Type of Communicational Interactions Related to the Dispatch Decision-Making Task

Type of Interaction	Relationship	Total Frequency (C/F)
Vertical downward	Foreman → Coordinator	4 (0/4)
	Foreman → Dispatcher	6 (0/6)
	Coordinator → Dispatcher	22 (8/14)
Vertical ascending	Coordinator → Dispatcher	4 (0/4)
	Dispatcher → Coordinator	6 (0/6)
Horizontal	Coordinator ↔ Coordinator	5 (0/5)
	Dispatcher ↔ Dispatcher	12 (4/8)
Total		59 (12/47)

When analyzing the impact of work on the perceived workload of the different members of CENCOR, a significant variability was found in the perception of the workload of the group members:

- Coordinators feel overwhelmed by the responsibilities associated with their role in operational and formative terms. On the one hand, they must be vigilant in guiding and correcting the actions of novice dispatchers, while also performing administrative functions and coordinating with other actors outside the institution in cases of more complex fires.
- Variations were found among the dispatchers in charge of radio traffic with respect to perceived mental load, with the most overloaded being those in charge of areas with the highest occurrence and complexity of forest fires. Among the floating or flying dispatchers, there was significant variability in the types of tasks assigned. On the one hand, some of them had a balanced mental workload, while others had a working mental underload because they said they only performed routine administrative tasks. There were also those who experimented with working under mental overload for the commitment to deal with the fulfillment of their own responsibilities and the support they should provide to others.

20.11 DISCUSSION

The findings provide a deeper understanding of a system that operates under budget constraints for hiring permanent staff, using informal regulatory strategies to sustain day-to-day work. However, to the extent that the system relies on inexperienced people, team leaders must make significant compromises in physical and mental effort to meet performance standards. Among the internal debilitating precursors, previous experience was identified, although, paradoxically, it was found that those with greater experience had a perceived mental working overload since they had to constantly monitor the decisions of the novice dispatchers, support them, and transmit knowledge to ensure adequate control of the process. This additional responsibility also negatively impacted their ability to fulfill their administrative tasks. As for the external debilitating precursor factors identified, those are the type of position and role attributed formally and informally, the variety of tasks assigned, the historical complexity of the area they guard and incidents, the early rotation of dispatchers who leave the post in the middle of the season and, therefore, the late incorporation of replacement dispatchers who deserted after the initial training period.

In addition, it was possible to systematize a set of protective precursor factors through which the sociotechnical system dynamically adapts to the limitations and restrictions imposed on it. These correspond to the mobilization of informal, individual, or collective regulation strategies, such as vigilance regarding the actions and verbal interactions of co-workers (see epistemo-vigilance in Maureira 2015), which trigger behaviors of mutual support and assistance in the workplace. In any case, this organizational configuration maintains a state of reliability and causes physical and mental exhaustion among its members.

Thus, the main ergonomic proposals insist on ensuring minimum staffing with indefinite contracts for the critical positions identified, reviewing and updating job descriptions based on an ergonomic analysis of the work, updating and deepening a design plan for a distribution of workloads and incentives according to real roles and responsibilities, and reducing the observed imbalances. In this way, moving toward a more stable and balanced configuration would not only allow it to meet performance standards but also to protect the physical and mental health of its members.

20.11.1 Research Limitations

One of the main limitations of this research was associated with the time available for its development. Since this was an undergraduate thesis, the study had to be adjusted to a maximum period of 6 months, which restricted the possibility of prolonged follow-up. Consequently, data from the previous season were used to evaluate performance, which allowed us to obtain relevant information but limited our ability to analyze in depth the performance of the same team studied throughout the season. Undoubtedly, a longitudinal analysis would have significantly enriched the results.

20.12 ESTABLISHING THE BASIS FOR THE DEVELOPMENT OF AN OPERATIONAL DECISION-MAKING MODEL FOR THE PLANT

The following is an excerpt from an undergraduate thesis for the degree of Civil Industrial Engineer, developed by Farías (2024) under the supervision of the Ergonomics Department of the Universidad de Concepci6n in Chile.

20.12.1 Problem

This research was conducted at a forest fire coordination and control center in the Biobío region. This private organization was interested in evaluating its performance to optimize its operational processes. As a general objective, the research was oriented toward establishing theoretical and methodological bases for developing an operational decision-making model for the team's fire management.

20.12.2 Theoretical Positioning

The research is part of the extension of previous ergonomic studies on coordination centers (Maureira 2015), the management of complex dynamic environments (Samurçay and Rogalski 1992), and distributed social cognition (Hollan, Hutchins, and Kirsh 2000) to understand how collective decision-making processes are affected by the interaction between actors, roles, and artifacts. To understand and characterize the factors affecting management and decision-making processes, the analysis is complemented by the sociotechnical model (Trist and Emery 1951; Carayon and Smith 2000), which highlights the interdependence between organization, technology, and human labor.

20.12.3 Methodology

This was a descriptive mixed-type research. The universe consisted of all the members of the center: two supervisors, an operations manager, five dispatchers, and a camera manager. The following techniques were used to collect information: documentary review of the institution and of standards and indicators of efficiency prescribed in forest fire management, application of surveys and semi-structured individual interviews, open and systematic observations, and critical incident techniques. The treatment of the information considered the content analysis of the accounts extracted in each of the techniques applied.

20.12.4 Results

A set of factors that affected the operations of its members and their decision-making processes was identified. Among the main findings, the following precursor factors were identified as weakening factors:

- Absence of formal, structural, and operational prescriptive elements. The institution lacked an organizational chart, written protocols, or systematized performance indicators. This implies a predominance of an informal system of operation and division of labor.
- Unstable working conditions related to temporary employment and high staff turnover. Only the managers have an indefinite contract.
- Lack of technical competency development plans related to the use of technological support tools, technical language, and fire behavior.
- Perception of work overload among dispatchers with less experience is associated with the effort required to meet high time pressure for timely and pertinent decision-making, as well as the pressures arising from the complexity and interdependence of coordinating actions with other actors at a distance.
- Absence of a systematic, formal, and validated protocol for the identification of critical factors that affect emergency assessment and decision-making in forest fire management.

In contrast, the following were identified as protective precursor factors:

- The dispatch team is made up of experienced dispatchers who return each season or who have worked in other centers with similar characteristics.
- Clarity of roles on the part of the dispatchers who have experience in the trade.
- Use of collective regulation strategies, such as coordination and mutual support.
- An open, shared spatial arrangement of workstations, which facilitates communication and supervision between operators.

20.12.5 Discussion

Although the organization's managers were interested in identifying and weighing the critical factors to optimize their dispatchers' decision-making processes, when analyzing the work and its operational modes, other underlying, debilitating elements in the system emerged. From the sociotechnical approach to work proposed by Carayon and Smith (2015), the analysis and diagnosis of the system identify weakening factors of the socio-organizational context that interfere in the functioning of the internal work system where the activity takes place. Those results reinforce what Maureira (2015) pointed out about the fragility of the system, to the extent that, for example, personnel training relies on oral transmission and informal transmission from the headquarters, without a structured process to ensure the continuity of organizational learning.

Thus, a proposal to improve the optimization of dispatchers' decision-making criteria may not significantly impact system efficiency, unless the other sociotechnical components across the different levels of the system are part of an integrated, systemic improvement program.

From the point of view of the development of technical competencies in ergonomic intervention, this experience illustrates the challenges of the social construction of demand in the initial stages of an intervention: understanding the problem, identifying needs and expectations, negotiating, and reformulating the demand.

20.12.6 Limitations of the Study

The exploratory nature of the study and its thematic delimitation do not allow us to generalize the results, but rather allow us to offer an initial basis for designing future work scenarios, maintaining a dynamic of participatory action intervention with those who carry the identity in these highly complex and uncertain trades.

20.13 FINAL DISCUSSION

The research compiled in this chapter constitutes an innovative contribution to ergonomics in our continent as it shifts the focus of the discipline toward thematic areas less exploited in the scientific community, such as informal learning and the development of competencies during the operational phase, the acceleration of training and learning processes, the construction of a learning collective and the informal regulation strategies of sociotechnical systems in complex dynamic contexts. In this sense, we seek to contribute conceptual notions such as epistemo-vigilance and organizational accelerators of training proposed by Maureira (2015) as components that contribute to the organizational reliability and resilience of these sociotechnical systems.

Regarding the field of research, although there is abundant scientific production and empirical evidence on forest fires, this has been more oriented toward the ergonomic conditions of those who directly face them or toward providing technological support tools for decision-making in forest fire management. In this sense, the focus on understanding the workload conditions of forest fire coordination and control centers provides an unprecedented view of the dynamic equilibrium of sociotechnical systems under high-pressure conditions. The evidence presented invites us to reflect on the redesign of organizational structures and processes by identifying weakening precursor factors that undermine system reliability and inform the design of more robust, inclusive, and effective decision-making models.

The experience of participatory design for the habilitation of a new CENCOR working area reaffirms the need for a continuous process of training and enrichment of its technical and nontechnical competencies to facilitate and lead constructive, innovative processes through collective dialogue, producing creative ideas. The intended dialogue conducted by the ergonomist with the main actors involved in designing processes becomes a transforming action, a source of individual and collective learning that contributes to anticipating future scenarios that are user-centered. In that sense, participatory and co-construction intervention methodologies offer opportunities for organizational innovation through collective learning and cultural transformation.

From an academic point of view, the two undergraduate university theses described in this chapter seek to raise awareness of the challenges of our discipline, in the role of mentors or trainers of future industrial engineers, by teaching in the field. In this sense, ergonomics humanizes work, providing a theoretical and methodological framework that connects process design to real human needs, thereby enriching their future professional profile. In this sense, it not only provides technical competencies but also soft skills linked to observation, activity analysis, and co-design. In this way, it enables them to understand how human factors influence the reliability of highly complex systems and prepares them to lead projects with an inclusive, safe, and sustainable approach.

In terms of the limits described in each research, although they are all case studies, investigated in a local reality and in a specific thematic area of intervention, they can be useful as a source of inspiration for other colleagues in Chile and Latin America to broaden their horizons of intervention and challenge themselves to intervene in contexts of high social and technological complexity. There is a need to consolidate a reference framework for future research and practice in the region.

REFERENCES

Amalberti, R. 2012. *Construir la seguridad*. Madrid: Modus Laborandi, S. L.

Apud, Eduardo, Meyer, Felipe, & Maureira, Fabiola. 2002. Ergonomía en el combate de incendios forestales. Santiago, Chile: Valverde.

Barcellini, F., Van Belleghem, L. y Daniellou, F. 2013. "Les projets de conception comme opportunité de développement des activités." *En Ergonomie constructive, editado por P. Falzon*, 249. Paris: Presses Universitaires de France.

Benchekroun, T.-H. 2016. *L'ergonomie de conception participative: Méthodes et pratiques*. Toulouse: Octarès Éditions.

Brown, T. 2009 *Change by design: How design thinking creates new alternatives for business and society*. New York: Harper Business.

Brown, T. 2020. *Diseñar el cambio: Cómo el design thinking transforma organizaciones e inspira la innovación*. Barcelona: Ediciones Urano.

Carayon, P.,Hancock, P., Leveson, N.,Noy, I., Sznelwar,L. y van Hootegem, G. 2015. "Advancing a sociotechnical systems approach to workplace safety – Developing the conceptual framework." *Ergonomics* 58 (4): 548–564. https://doi.org/10.1080/00140139.2015.1015623.

Carayon, P., y Smith, M. J. 2000. "Work organization and ergonomics." *Applied Ergonomics* 31 (6): 649–662. doi:10.1016/s0003-6870(00)00040-5.

Caroly, S., y Clot, Y. 2004. "Du travail collectif au collectif de travail. Des conditions de développement des stratégies d'expérience." *Formation et Emploi* 88: 43–55.

Cellier, J. M., Eyrolle,H. y Marine, C. 1997. "Expertise in dynamic environments." *Ergonomics* 40 (1): 28–50. https://doi.org/10.1080/001401397188350.

Chávez, F. 2023. *Propuesta de configuración funcional de equipos de trabajo en la central de control y coordinación de incendios forestales*. Memoria de título, Universidad de Concepción.

Chile. Ministerio de Salud. N° 594: 1999. "Decreto Supremo N° 594: Reglamento sobre condiciones sanitarias y ambientales básicas en los lugares de trabajo." *Diario Oficial de la República de Chile*, 29 de abril de 2000.

CONAF. 2022. *Plan Estratégico 2022–2026: Programa de protección contra incendios forestales*. Santiago, Chile: Ministerio de Agricultura.

Darses, F., Falzon, P. y Munduteguy, C. 2004. "Paradigmes et modèles pour l'analyse cognitive des activités finalisées." En *Ergonomie*, editado por P. Falzon, 191–212. Paris: Presses Universitaires de France.

De la Garza, C., y Weill-Fassina, A. 2000. "Régulations horizontales et verticales du risque." En *Le travail collectif. Perspectives actuelles en ergonomie*, editado por T.-H. Benchekroun y A. Weill-Fassina, 217–234. Paris: Éditions Octarès..

Falzon, P., Pasqualetti, L., and In, T. H. 2000. *L'apprentissage opportuniste*. In *Le travail collectif: perspectives actuelles en ergonomie*, 121–133 Paris: Éditions Octarès.

Farías, M. 2024. Bases para el desarrollo de un modelo de toma de decisiones operacionales en la central de coordinación de incendios forestales de la empresa privada CAMBIUM S.A. Memoria de título, Universidad de Concepción.

Flageul-Caroly, S. 2001. *Régulations individuelles et collectives de situations critiques dans un secteur de service: Le guichet de la Poste*. Tesis doctoral, Ecole Pratique des Hautes Etudes, Paris.

Fresno García, M. D., and Hernández-Echegaray, A. 2019. *Técnicas de diagnóstico, intervención y evaluación social*. Madrid: UNED.

Guérin, F., Laville, A. Daniellou,F. Duraffourg,J. y Kerguelen, A.. 2007. "La construction de l'intervention." En *Comprendre le travail pour le transformer*, 136–157. Lyon: ANACT.

Hart, S. G., y Staveland, L. E.. 1988. "Development of NASA-TLX (Task Load Index): Results of empirical and theoretical research." En *Human Mental Workload*, editado por P. A. Hancock y N. Meshkati, 139–183. North-Holland. https://doi.org/10.1016/S0166-4115(08)62386-9.

Hoc, J. M.,Amalberti, R., Cellier, J. M., y Grosjean,M.2004. "Adaptation et gestion des risques en situation dynamique." En *Psychologie ergonomique: tendances actuelles*, editado por J. M. Hoc y F. Darses, 15–48. Paris: Presses Universitaires de France.

Hollan, J., Hutchins,E. y Kirsh, D.. 2000. "Distributed cognition: Toward a new foundation for human-computer interaction research." *ACM Transactions on Computer-Human Interaction* 7 (2): 174–196.

Hutchins, E. 1995. *Cognition in the Wild*. Cambridge, MA: MIT Press.

International Organization for Standardization. 2000. *ISO 11064-1:2000 Ergonomic design of control centres — Part 1: Principles for the design of control centres*. Geneva: ISO. Confirmed 2020.

International Organization for Standardization. 2019. *ISO 9241-210:2019 Ergonomics of human-system interaction — Part 210: Human-centred design for interactive systems*. Geneva: ISO.

Kelly, R. 2016. *Creative development: Transforming education through design thinking, innovation, and invention*. Brush Education Toronto, Canada: Brush Education.

Leplat, J. 1985. *L'analyse du travail en psychologie ergonomique*. Paris: PUF.

Martin, Bella, andHanington,Bruce . *Universal methods of design: 100 ways to research complex problems, develop innovative ideas, and design effective solutions*. Beverly, MA: Rockport Publishers, 2012.

Maureira, F. 2015. *Favoriser la construction d'un collectif apprenant: les conditions organisationnelles du développement des compétences d'un collectif* éphémère. Tesis doctoral, Conservatoire National des Arts et Métiers, Paris.

Maureira, F., Benchekroun,T.-H. y Falzon, P. 2019. "Worker's management and skills development in a temporary team." En *Advances in Intelligent Systems and Computing* 821, editado por W. Karwowski, T. Ahram, y R. Taiar, 405–411. Springer. https://doi.org/10.1007/978-3-319-96080-7_49.

Paquay, L. 2005. "Devenir des enseignants et formateurs professionnels dans une organisation apprenante? De l'utopie à la réalité!" *European Journal of Teacher Education* 28 (2): 111–128.

Rogalski, J., y Marquié, J. C.. 2004. "Evolution des compétences et des performances." En *Psychologie ergonomique: tendances actuelles*, editado por J. M. Hoc y F. Darses, 141–173. Paris: PUF.

Samurçay, R., y Rogalski, J. . 1992. "Cooperative work and decision making in emergency management." *Le Travail Humain* 55 (1): 53–77.

Trist, E. L., y Emery, F. E. . 1951. *The social engagement of social science: A tavistock anthology*. Vol. II: The Socio-technical Perspective london UK. Tavistock Institute.

Urbano Gil, H. 2007. "El enfoque etnometodológico en la investigación científica." *Liberabit* 13: 89–91. https://pepsic.bvsalud.org/scielo.php?script=sci_arttext&pid=S1729-48272007000100011&nrm=iso.

Vidal-Gomel, C., y Rogalski, J.. 2007. "La conceptualisation et la place des concepts pragmatiques dans l'activité professionnelle et le développement des compétences." *Revue @ctivités* 4 (1). https://journals.openedition.org/activites/1401

Weill-Fassina, Anne, & Benchekroun, Taoufik-Hassen. 2000. "Diversité des approches et objets d'analyse du travail collectif en ergonomie." En Le travail collectif. Perspectives actuelles en ergonomie, editado por Taoufik-Hassen Benchekroun & Anne Weill-Fassina, 1–15. [Ciudad]: [Editorial].

21 A Macroergonomic Approach to Calculating the Optimal Workforce

Igor Bello

21.1 INTRODUCTION

Often, life confronts us with the problem of finding the best way to do something. For example, a doctor may try to find the exact number of doses of a drug to cure the patient in the shortest time possible, but with minimal side effects; a manufacturer will want to get the maximum possible production of machinery at a point where this is not compromised by excessive use and can continue running for a long time. We commonly find situations that require choosing an optimal point at which the descriptor curve reaches its peak, and after which, it starts to decline. At that point, since the time of Leibniz[1] is known as the maximum value of the function, it is continuous and within the range of possible values.

From a macroergonomic perspective, such a diatribe is as follows: on the one hand, the company seeks the highest possible productivity based on more intensive use of all available resources; on the other hand, we have human beings as a key productive resource, with their abilities and needs, to which work must be adapted to provide adequate working conditions, enabling them to enjoy the best attainable state of physical, psychological, and social well-being possible. However, in the center of the variables that appear to move inversely (Productivity vs. Welfare), is the notion of sustainability. It is not possible to maintain production levels over time if the process capabilities are not maintained due to loss of productive resources, wear, fatigue, and stress on the human factor. Then, the problem is translated into a slider bar to find the right spot where a good rate of productivity converges with the ability to sustain it over time, through ergonomic strategies that allow, for example, a suitable recovery rate, frequency of uprisings safe for the back, or a level of mental burden that can be handled by the subjects with their own coping resources.

Another key concept to understanding this work is that of diminishing marginal returns. This economic law states that the addition of a new production factor, keeping all others constant (*ceteris paribus*[2]), will cause the performance that adds this new factor to be less than that of the above additions.

A clear example of this theory is the frequency of manual lifting of loads. If a production manager asks his workers to increase this frequency, probably for each lift/additional minute, you will get a higher marginal performance; but with increasing frequency, it reaches the point at which workers begin to need more recovery time (fatigue) or sick leave for low back pain, which ends up lowering the production rate, and therefore a marginal performance is observed for each lift/additional minute. The central idea of this research is to find a system for estimating the maximum marginal working capacity under an ergonomic perspective and labor productivity and then determine the optimal workforce to maintain the human factor sustainability, answering some fundamental questions:

- What is the maximum level of marginal productivity that can be maintained over time without affecting people's health and safety?
- What estimation models can be used to calculate the maximum workload capacity from a biomechanical, physiological, and psychosocial point of view?

DOI: 10.1201/9781003582014-26

- Could these models be used to determine the optimal workforce in a production process, such that people's health and well-being are not affected?

To achieve these answers, three different approaches were used. First, the approach methodologies' biomechanical, physiological, and psychological evaluation was sufficiently validated as a starting point to build an ergonomically viable model. Second, the historical data of a large group of companies grouped in the Chamber of Industries of the State of Carabobo (Venezuela) on productivity rates contrasted with working arrangements reported by the respective Industrial Health and Safety Services; these data allowed the construction of mathematical models of productivity/resources based on empirical evidence. Finally, it is based on the differential calculus to find the optimal points within mathematical constructs based on a production function approach. This model was validated using real work data, obtaining a simulation model of productive functions that maximizes performance, but within a range that ensures macroergonomic sustainability and, thus, a system for the estimation of maximum working capabilities that may lead to maintenance of optimal and sustainable marginal returns, and using this information to calculate the optimal workforce necessary to guarantee the desired production level as well as human sustainability in terms of health and safety.

In this work, the human factors engineering approach begins at the trailing edge of Taylorian time and motion studies.

21.2 THE METHOD

To answer the questions of this research, a theoretical–mathematical construction was generated that was empirically validated. This validation is based on the compilation of bibliographic sources collected longitudinally and by cohort, without an experimental component. Research has a projective character as it aims to design a system for estimating the optimal workforce for a process based on a set of defined variables related to human sustainability. Documentary research focused on the study of health and safety results, on the one hand, and the resulting business productivity, on the other hand. The data were based on the results reported to the Industrial Chamber of Carabobo State (Venezuela) for 159 companies in the automotive, chemical, metallurgical, food, and tire industries, with a combined workforce of about 45,000 workers; data were collected by the production volume, number of employees, working time, absenteeism, occupational diseases, accidents, and others. We studied data collected between 2008 and 2014. The respective matrix was constructed stating the data analysis of total production, average product, and marginal product, according to the model proposed in Table 21.1.

TABLE 21.1
Summary Table of Production Data (Proposed)

Control Date	Added Factor	TPP	Increased Factor	Increased Production	MgPP	APP
	x	y	Δx	Δy	$\frac{\Delta y}{\Delta x}$	$\frac{y}{x}$
(0)	(1)	(2)	(3)	(4)	(5)=(4)/(3)	(6)=(2)/(1)
Date on which the cutting data are taken	Contributed resources (man-hour, frequency, weight, and mental load)	Retrieved total physical product	Additional amount of factors contributed	Additional amount of production obtained	Marginal physical product	Average physical product

The diminishing marginal returns function, in its most complete form, involves an initial curve with increasing marginal returns, a turning point, and, finally, diminishing marginal returns.

The cubic production function (or third-degree polynomial) seems best suited to capturing these three stages in the behavior of the data since, by having an inflection point, it reflects the moment at which the marginal product reaches its maximum before beginning to decline. This behavior is crucial for short-run analysis, as described in the literature on production theory. Another crucial aspect is the model's ability to show negative marginal returns. A cubic function, unlike the Cobb–Douglas function, has a maximum point and then declines, meaning the marginal product becomes negative. This phenomenon, in the context of production, occurs when the addition of a variable input (such as more workers in a fixed space, for example) begins to hinder total production, a behavior that Cobb–Douglas models cannot replicate.

Marcos Portillo Vázquez (2015), in his work *The Cubic Production Function*, explores the suitability of this model for the analysis of inputs in agricultural production. George J. Stigler (1947), in his work *The Theory of Price*, uses the short-run production model in the form of an "S" curve (cubic function) to illustrate the three stages of production in the analysis of the total product, average product, and marginal product.

From a calculus perspective, the superiority of the cubic function is evident. Its second derivative, which measures the change in the rate of change of output, changes sign. Specifically, it changes from positive (positive concavity, increasing marginal returns) to negative (negative concavity, diminishing marginal returns) at the inflection point. This mathematical property makes it the ideal tool for modeling the dynamic behavior of production. Applying the generic formulation of the yield curve as diminishing marginal returns, the cubic equation theorized by Portillo Vázquez(2015) is considered a more convenient element of differential calculus to describe and study this behavior of production factors (for one variable, with a constant)[3]:

$$f(x) = a + bx + cx^2 - dx^3$$

This function can be represented graphically and generically by curves, as in Figure 21.1.

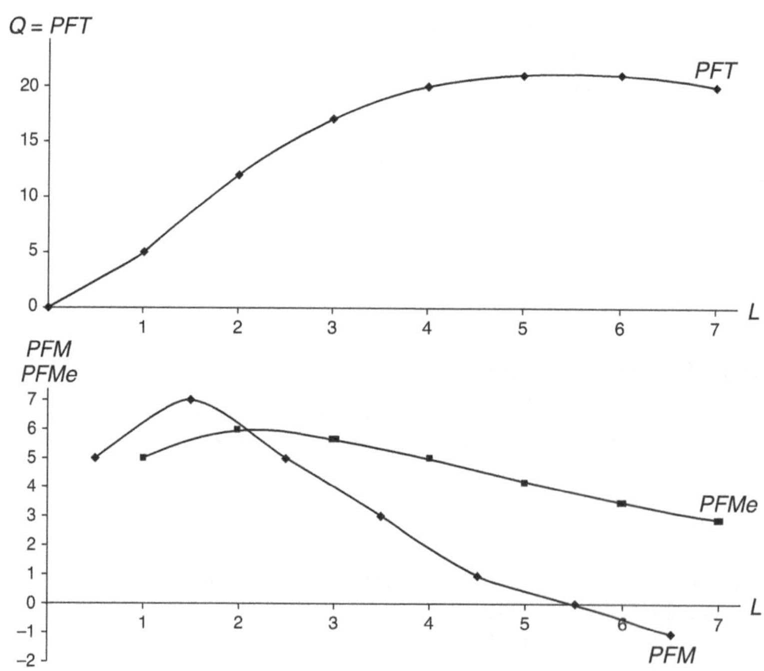

FIGURE 21.1 Curve descriptor of the production function.

So it can be approximated in the mathematical construction of several production function evaluation points:

- Maximum Point Total Physical Product (TPP)
- Maximum Point Marginal Physical Product (MgPP)
- Maximum Point Average Physical Product (APP)
- Maximum Point Human Factor (from the corresponding ergonomic assessment).

For inserting into the construction, ergonomic evaluation models were used to empirically validate the theoretical model:

- NIOSH lifting equation (1994)[4]
- OCRA Check List (Occupational Repetitive Actions)[5]
- JSI Method (Job Strain Index)[6]
- Method of Metabolic Consumption Calculation (ISO 8996)[7]
- Queuing theory

To facilitate complex calculations and simulations of analytical models, was used the web Ergonautas (application developed by Polytechnic University of Valencia) and Ergo/IBV v14.0 (application developed by the Institute of Biomechanics of Valencia). The basics of differential calculus were used to complement the assessment model: derivatives, maxima and minima, monotony, and concavity. It is appreciated that the data models provided were found in the form of discrete data; however, through the differential calculus, it was possible to approximate descriptive models and continuous ones, rather close to the observed phenomenon, useful for discrete results of practical value. Although this is not a research project on biomechanics or occupational physiology, some ergonomics models[8] were used to allow validation of the mathematical models with empirical information obtained from industrial data:

- *Manual lifting loads*: Builders in the tire rubber industry were evaluated.
- *Repeatability of movements*: The workstation operator packing snacks in the food industry was evaluated.
- *Cardiometabolic consumption*: The operator position in a vitrified isolating electrical parts manufacturing industry was evaluated.

The correlation of production data was analyzed in relation to variations in working arrangements to validate the model's consistency using historical data, since the researcher had no experimental control. Based on the information obtained from the analysis, an estimation model based on work schemes – an analytical approach and another graphical one – was constructed. These models are the starting points for developing a method to calculate the optimal workforce, ensuring that processes achieve both the desired productivity and the sustainability of human factors, combining operations research techniques, differential calculus, and queuing theory.

21.2.1 Literature Review. Time and Motion Studies: From Taylor to Artificial Intelligence (AI)

The idea of finding an optimal point that balances maximum performance with minimal resource wastage is not new, and this framework allows for increasing the comprehension of the new model.

The study of time and motion began in the 18th century with the studies carried out by Perronet (1755) on the manufacture of pins. However, this technique was not widely known until the end of the 19th century, with the proposals of Frederick Taylor. Also known as the "father of scientific administration", he began to study time and motion, timing the time that workers used to perform each task within a specific work cycle. The concept of "efficient task" was then developed, in which he proposed that the management should oversee work planning and that each job should have an optimal time standard based on the work of a highly qualified worker (Taylor, 1911).

Based on Taylor's studies, Frank and Lilian Gilbreth (1917) expanded this research work and developed the study of motion, dividing the work into 17 fundamental movements called *Therbligs*. Depending on their effect on productivity, these movements are classified as efficient or effective (physical or muscular in nature: reach, move, release, and preposition; or objective or concrete in nature: use, assemble, and disassemble) and inefficient or ineffective (mental or semi-mental: search, select, position, inspect, and plan; delays or procrastination: avoidable delay, unavoidable delay, rest, and hold). The Gilbreths applied the scientific method to consolidate their proposals. Their philosophy was based on increasing effectiveness by performing as few movements as possible in a technique or work stage.

Time and motion studies were initially applied in the manufacturing industry during the Industrial Revolution: Taylor and the Gilbreths implemented these techniques in factories to improve mass production. They focused on simplifying tasks, minimizing unnecessary movements, and structuring efficient work schedules. The Gilbreths also implemented their studies in the construction industry to optimize workflow and improve safety and efficiency. Today, time and motion studies are applied in a variety of industries, not only in manufacturing but also in customer service activities (retail stores and call centers to reduce waiting times and improve customer satisfaction); digital and technology sector (software development and project management); health (in hospitals and clinics to improve efficiency in patient management and medical processes); or logistics and transportation (from warehouses to delivery services, optimizing routes and processes to reduce costs and improve delivery times). While these methods provide advantages from a production point of view, minimizing the waste of resources and reducing operating costs, they have also had their drawbacks. Trade unions, for example, considered time study as a covert management tool to standardize and intensify the pace of production. Gilbreth, Cadbury, and Marshall harshly criticized Taylor for having a production-centered perspective that did not consider the needs and capabilities of workers. Other criticisms of the method stem from its overly mechanistic approach, which can ignore the human and creative aspects of work. The focus on maximizing efficiency can lead to ignoring the effects of fatigue and physical and mental stress on workers. Repetitive and standardized tasks can be physically exhausting and monotonous, affecting workers' health and well-being. Although optimal time is measured based on the time it takes for an efficient worker to complete a task, it is inappropriate to take this as a reference for the entire process, without a systemic view. First, it must be considered that the study takes a sample, generally with the worker aware that he or she is being studied, which can generate a bias in the measurement. The bias inherent to the transversality of the sample must also be taken into consideration as work rates can vary throughout the day, week, or month. Learning curves are also not considered: an expert worker tends to complete tasks more quickly than an apprentice. Differences in gender, age, anthropometry, and other individual characteristics that explain marginal productivity differences are also important. Therefore, it is very difficult to determine the time required for each process without considering the process's ergonomics. Often, rework and process delays are better explained by a suboptimal design of the workflow, workstations, machines, or tools, rather than by differences in workers' performance. So, an approach to optimizing work processes from the perspective of the most efficient times and movements is, at the very least, reductionist.

Times and movements can vary greatly from worker to worker, and this does not mean that someone is doing their job poorly, but rather that they have different abilities. A more anthropocentric view of this problem is required. The most modern ergonomic models do not allow us to identify the optimal times for each task, but they do allow us to understand which movements are harmful to the worker and the repetitions per unit of time that are safe from the musculoskeletal point of view and from the fatigue of the worker, to establish healthy and therefore humanly sustainable ranges for each task. With the help of artificial intelligence (AI), many organizations are generating real-time video analysis of tasks, analyzing times and automatically evaluating postures and repetitive tasks undertaken by workers, reducing up to 85% of the time needed to perform an ergonomic analysis, with the ability to predict potential ergonomic risks by analyzing large amounts of historical data and movement patterns of workers. This allows us to understand not only the marginal performance of each worker and the biomechanical or physiological risks to which they are exposed but also to understand the overall performance of the system from a macroergonomic perspective to carry out redesigns, simulations, and interventions more efficiently, while ensuring safe and healthy work for operators.

21.3 RESULTS: FINDING SUSTAINABLE WORK PACES

One of the most important factors reported was total monthly production, which was correlated with a group of variables mandated under Venezuelan law, including weight limits used, lift frequency, work schedules, and task duration. These data allow linkage of proportionality between the resources provided by the human factor and the achieved productive results. The analysis of the data showed a correlation coefficient between 0.72 and 0.91 regarding the variation in production in relation to the variation in application rates of the human factor, within a model that could be characterized mathematically, to achieve production function optimization based on total, marginal, and average product points. The analysis focused on production functions analogous to the *decreasing marginal returns* production function.

Due to extension considerations, this chapter shows the results of the NIOSH index analysis based on the manual lifting point guard in a tire manufacturer. Additional analyses are included in the full-extended research report, available by email request to the author.

This modeling was performed with data collected between 2009 and 2014 based on the analysis that the production management had to implement to find the optimal point of production for several workers who were being included again at the tire builder after sick leave due to low back pain. So, in this case, the company was setting in reverse: the regime of "normal" production was decreasing the level of frequency of lifting (the critical factor for this job) to obtain acceptable levels of fatigue in workers, as measured by the application of Kuorinka questionnaires. The task studied consists of Radial Builder Module 15 Phase II, where the level of high risk is due to the high load handling oscillating at about 17 kg, with a frequency variable lifting up to 18 times per hour (it has been tested in various frequency options), which is equal to 144 rubbers, each turn, charged by the operator. In addition to this, you must flex the trunk to take the sides of the container when it is located below 58.7 cm, but high above 32 cm height (measured from the ground) and flex the trunk to take the shells found in the lower level of the container hanger (96 cm height), to place the covers on the drum of the machine, and to place crude rubbers in the container so that the position requires repetitive movements of back-bending during working hours. According to this description of the task, it was selected for modeling the frequency parameter of uprisings.

An initial approach to finding the optimal frequency factor was an analytical assessment using the NIOSH Index Liftings. Ergo/IBV application was used for this task, with the Simple Manual Handling module applied. The first analysis was done using the final 2014 survey data, which aimed for a lifting frequency of up to 18 times per hour, which translates to 0.3 times per minute. This assessment yielded a recommended weight limit of 9.08 kg in contrast to the 17 kg that was manipulating effectively, leaving the worker exposed to a biomechanical risk index of 1.87, which represents a risk of back injury, requiring intervention.

Using the analysis capabilities of the tool, we proceeded to perform a computer simulation of the task, varying only the frequency of lifting and leaving the rest of the parameters unchanged, to force a frequency rate that would provide an acceptable rate increase. Thus, it was found that at around 0.2 liftings per minute (12 per hour), a biomechanical index of 1.02, which is located within the limits of worker safety, is obtained. In this vein, one might understand that if the work regime of 12 liftings per hour is low, workers with musculoskeletal involvement in the task are protected. However, this work regime may be inconvenient for production, in terms of productivity, due to factors such as job fatigue, high heat, the annoyance of noise, and the aspiration of particles and vapors that arise from the production process.

On the other hand, it was considered an empirical approach. The results obtained from a group of operators, varying the frequency of lifts per hour between 1 and 25 times, correlated with the average monthly production:

From the above table, it is shown that peak production occurs at a frequency of 20 lifts per hour, yielding a monthly output of 23,316 tires. However, the marginal increase in the frequency contribution is negligible, and the average throughput per lifting is also low (1,166), compared with 1,311 units produced at a rate of 15 lifts per hour. It can also be seen that, at a rate of 10 lifts per hour, the delivery rate increases, thereby improving productivity (Table 21.2).

TABLE 21.2
Table Summary Data – Reinforced Tire

Added factor	TPP	Increased Factor	Increased Production	MgPP	APP
x	y	Δx	Δy		
(1)	(2)	(3)	(4)	(5) = (4)/(3)	(6) = (2)/(1)
1	174	1	174	174	174
2	661	1	487	487	331
3	1,427	1	766	766	476
4	2,436	1	1,009	1,009	609
5	3,654	1	1,218	1,218	731
6	5,046	1	1,392	1,392	841
7	6,577	1	1,531	1,531	940
8	8,213	1	1,636	1,636	1,027
9	9,918	1	1,705	1,705	1,102
10	11,658	1	1,740	**1,740**	1,166
11	13,398	1	1,740	1,740	1,218
12	15,103	1	1,705	1,705	1,259
13	16,739	1	1,636	1,636	1,288
14	18,270	1	1,531	1,531	1,305
15	19,662	1	1,392	1,392	**1,311**
16	20,880	1	1,218	1,218	1,305
17	21,889	1	1,009	1,009	1,288
18	22,655	1	766	766	1,259
19	23,142	1	487	487	1,218
20	**23,316**	1	174	174	1,166
21	23,142	1	–174	–174	1,102
22	22,585	1	–557	–557	1,027
23	21,611	1	–974	–974	940
24	20,184	1	–1,427	–1,427	841
25	18,270	1	–1,914	–1,914	731

1. The factor provided represents the frequency of liftings per hour.
2. FTP is measured as monthly production.

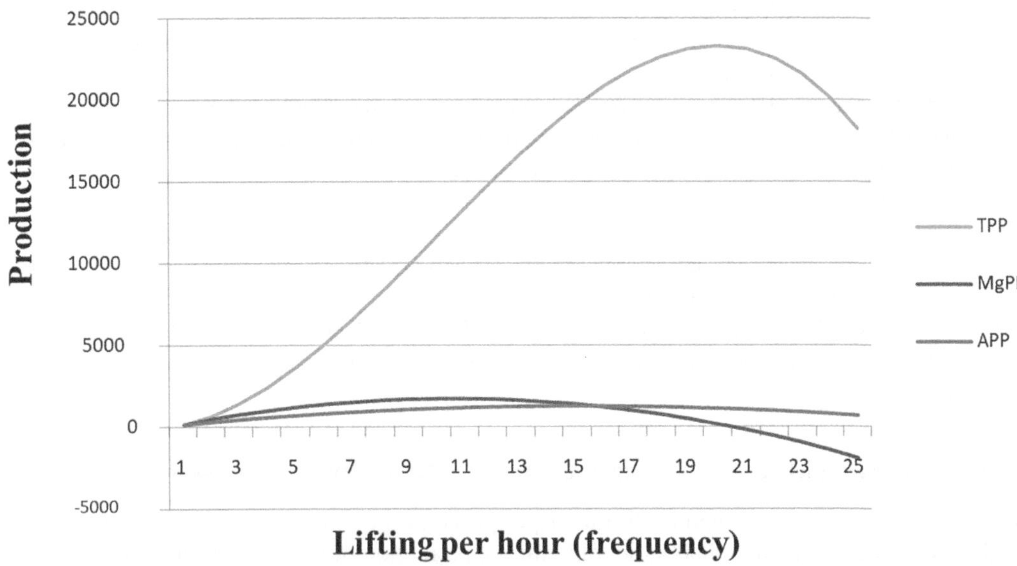

FIGURE 21.2 Graphical Analysis Case 1: tire production builder.

Another approach to estimating the optimal working regimes is graphical analysis. The graph shows special behavior at critical points. At the turning point (10 lift/h) of the total output curve, the marginal product is highest. Where the curves of the marginal product and average product converge, the average product is optimal. At the point at which the marginal product is zero, the total production reaches its peak. The curves can be seen in Figure 21.2.

Another approach is the use of modeling differential calculus. Suppose the cubic equation that is closest to the discrete values obtained in this case (by least squares modeling[9,10]):

$$f(x) = x + 30x^2 - x^3$$

We can get the peak of the production curve where the derivative is zero[11]:

$$f(x)' = 1 + 60x - 3x^2 = 0 \text{ for } x = 20$$

and obtain the peak of the curve's marginal product, where the second derivative is zero[12]:

$$f(x)'' = 60 - 6x = 0 \text{ for } x = 10$$

and analogously calculate the peak of the curve of the average product, where the first derivative of the curve $f(x)/x$ is zero or where the values of average product and marginal product are equal[13,14]:

$$\text{PFMe} = \frac{x + 30x^2 - x^3}{x} = 1 + 30x - x^2$$

$$\text{PFMe}' = 30 - 2x = 0 \text{ for } x = 15$$

In short, they can denote four areas of the curve, differentiated by the critical points:

- Zone 1 (0 to 10 lift/h), ranging from zero frequency to the tipping point where the marginal return is highest. In this area, marginal and average performance are increasing, but total production remains low.
- Zone 2 (10 to 15 lift/h), ranging from marginal performance peak to the peak of the average yield, in what may be called "biological optimum." This is an area where it is convenient to operate from a viewpoint of productivity; it also draws attention to the outcome of biomechanical analysis, which yielded a result in this area (12 lift/h).
- Zone 3 (15 to 20 lift/h), ranging from the peak of the average yield to the peak of the production curve. Note that in this area, although the production is maximum, average yields are decreasing, and the marginal return reaches zero.
- Zone 4 (over 20 lift/h): after the peak of the curve of total production, production begins to decrease despite the implementation of major efforts. Marginal returns are negative in this area.

21.3.1 Maximum Marginal Working Capacity: A New Model Development

Although from a theoretical approach, this paper provides the basis for the reconciliation of the production model with human labor, not only for the promotion of the workers' health but also for strengthening the productivity and sustainability of organizations; it constitutes an effort to deepen the mathematical modeling of productive situations and to harness the power of differential calculus, but taking account of biological variables involved in the work, and mainly fatigue, based on the premise that it is not enough to preserve illness-free workers but also at an optimum production level that can maintain throughout working life and allows to enjoy a full life quality.

Taking the example of the tire industry (manual lifting and NIOSH index), we can determine that under ideal conditions of rising for a long duration, a load of 5 kg can be lifted safely to the lumbar region of the back, even with a rate of 5.7 lifts/minute (a lifting every 10.52 seconds determined by the speed of the conveyor belt). However, if one considers only this argument to set the frequency of uprisings, one could be making a mistake, neither in the sense that this calculation does not reflect physiological and psychological variables or the biomechanics associated with the cumulative effect of the repetitions nor of the values of the environmental conditions such as temperature, humidity, lack of lighting, or vibration.

Then, an empirical approach can shed light on the problem through an organized collection of data on the productivity obtained in various schemes of work, from a low speed (two to three repetitions per minute), and collection of data in different regimes so that the total productivity begins to decline. This allows for a "map" of the overall system performance in hot conditions and thus monitors the evolution of marginal returns and averages.

Following the general methodology for determining the optimum for biologically marginal results, an average maximum efficiency operating range is described. To get the values associated with this range, the following methodology consisting of three phases can be used: (a) preparation of data; (b) estimation of the critical points, and (c) establishment of the optimum operating range.

21.3.1.1 PHASE 1: Preparation of Data

- Define the dependent and independent variables (e.g., task duration vs. productivity obtained or repetitions per minute vs. gross production).
- Determine a model valid for the type of analysis that aims to reach out (NIOSH, REBA, OCRA, OWAS, etc.) evaluation. Formulate the equation and equalize it with the acceptable upper limit; this establishes a "safe" point based on the ergonomic analysis within the curve.
- Build a data table with information from the experimental evidence of the production process. It is important to raise varied factors of dependent production units(repetitions, duration of the task, load weight, operating temperature, etc.). The data table shown in

Table 21.1 must contain the date of data collection, Factor Contributed (x), Physical Product Total (y), Increased Factor (Δx), Production Increase (Δy), Marginal Physical Product ($\Delta y/\Delta x$), and Physical Product Medium (y/x).

- Confirm that there is a high degree of correlation between variables (>0.8) for the analysis to be relevant (the variation of an essential consequence of the change in the other). The other independent factors must remain constant.

21.3.1.2 PHASE 2: Estimation of the Critical Points

At this point, you can choose between two methods: one graphic and one analytical.

Graphical Method

- Plot the data in the table as follows: Place the resource provided on the horizontal axis (x). In the vertical axis (y), the resulting total corresponding product (TPP) is expressed. You can take the same graph to locate on the vertical axis the result of Marginal Product ($\Delta y/\Delta x$) and the Middle Product (y/x), or you can use a parallel axis under the graph of TPP, changing the scale to the vertical axis, to better appreciate the MgPP and APP curves. An example of chart and graph construction is provided in the *Results* section.

- Set the three critical points on the graph. The first step is to locate the turning point of the total output curve where the peak of the maximum marginal product (concave and continuous curves, decreasing marginal returns function) is also located; this is the point "A". Where the curves of marginal product and average product are equal, the average product is optimal; this is the point "B". At the point at which the marginal product is zero, the total production reaches the peak; this is the "C" point.
- Define the areas bounded by the critical points. Zero point "A" will be zone 1 (zone of increasing marginal returns, but low production); zone 2 is defined from point "A" to point "B" (zone between the maximum marginal performance and maximum average yield); point "B" to point "C" will be zone 3 (the production is maximized, but the marginal contribution is very inefficient and tends to zero); and point "C" on will be zone 4 (production decreases, negative marginal contribution). The model construction is given in Figure 21.3.

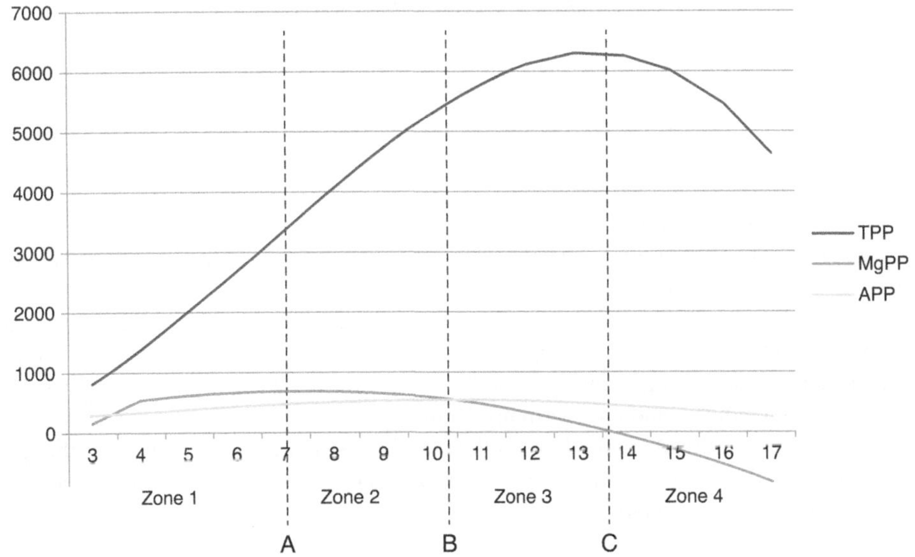

FIGURE 21.3 Construction of the graph to estimate optimum operating ranges.

Analytical Method

- Build an equation that emulates the behavior of the data, in the form of a function. Remember that the data obtained empirically are discrete, so basically, the mathematical model emulates the production function continuously. To do this, we use the emulation model of decreasing marginal returns and determine the method of logarithmic correlation indices, linearizing the equation using logarithms ten-based. The basic form of the function is defined by the following[15]:

$$f(x) = ax + bx^2 - cx^3$$

- We obtain the height of the marginal product curve where the second derivative is zero (and the point of change of concavity)[16]:

$$f(x)'' = 2b - 6cx = 0 >> x = \frac{b}{3c}(A)$$

- Similarly, calculate the height of the average product curve, where the first derivative of the curve *f*(*x*)/*x* is zero or where the values of the average product and marginal product are equal[17,18]:

$$\text{PFMe} = \frac{ax + bx^2 - cx^3}{x} + a + bx + cx^2$$

$$\text{PFMe}' = b - 2cx = 0 >> x = \frac{b}{2c}(B)$$

- Get the peak of the production curve where the derivative is zero[19,20]:

$$f(x)' = a + 2bx - 3cx^2 = 0$$

$$x = \frac{-b \pm \sqrt{b^2 + 3ac}}{a}(C)$$

21.3.1.3 PHASE 3: Setting the Optimal Operating Range

- Validate that the "safe point" determined by the ergonomic task analysis is between "A" and "B" of the equation.
- Take a production decision with this input. The biological optimum production is located between points "A" and "B". Placing it in "B" will yield the highest average of the human factor, but producing at a point between "A" and "safe point" will yield the best sustainability in the medium- and long-term production.
- By using the graphical method, we could easily estimate these points on the graph (Figure 21.2). By using the analytical method, we could obtain more accurate values, using the following generic equation[21]:

$$\frac{b}{2c} > x > \frac{b}{3c}c > 0$$

When this "safe point" is referred to a frequency resource (or number of repetitions of an action in function of a unit of time, e.g., lifts per minute), it is termed **F***.

21.3.2 Optimal Workforce for the Human Factor Sustainability: Model and Appliances

Although we use the Decreasing Marginal Returns model and ergonomic models to determine the optimal points for variables such as the frequency of lifts per minute, this optimal work frequency (**F***) can be used to calculate the number of workers necessary to maintain a stable and sustainable production from the perspective of human factors.

There are several ways of calculating the optimal number of workers for a production function:

- *Simple production functions*: These involve a single variable, where production remains stable and continuous during the workday.
- *Composite production functions*: These involve several variables, with rest time, different processes, and work rhythms during the workday.
- *Production functions with queues*: These involve waiting times by internal or external clients, where maximum waiting times must be taken care of.

21.4 SIMPLE PRODUCTION FUNCTIONS

When production is a given volume of a single product, and each item's production is related to a defined quantity of single tasks, the optimal workforce Q^* is determined using the following expression[22]:

$$Q^* \frac{P}{\mathbf{F}^* t}$$

where

Q^*=Optimal workforce (number of workers required)
F*=Safe frequency (repetitions per hour, each worker)
P=Production goal in units per day
t=Time (hours), per day

21.4.1 Application Example

Assuming a simple production function, with the daily production target being 1,000 tires, if the F* calculated in the tire production model studied for manual load handling of 12 repetitions per hour is followed, for 8-hour work shifts, there would be an optimal workforce defined by the following:

$$Q^* \frac{1{,}000}{12 * 8} = 10{,}41 \approx 11 \text{ workers}$$

21.5 COMPOSITE PRODUCTION FUNCTIONS

In many cases, a production process is not only composed of one task but also involves many activities, including tasks and subtasks. For the production analyst, it is often a challenge to calculate the number of workers required in each one since the required times vary considerably, and normally, this "balancing" of the workloads is done through an empirical trial-and-error model.

In this sense, the proposed model can provide valuable information to balance workloads, justifying the placement of more resources in tasks that take more time, while generating a more uniform production system with a human sustainability criterion. In this type of production situation, tasks may be divided by departments, which generate outputs that serve as a semi-finished product for another production line; in this case, the optimal workforce will be the sum of each of the optimal workforces of each task or subtask, that is[23]:

$$Q^* = Q_1 + Q_2 + Q3 + \ldots Q_n$$

where

Q^* = Total optimal workforce

Q_n = Optimal workforce of task n (each task Q has to be calculated by the simple production function)

Another possibility is the sequential combination of tasks, that is, the tasks are linked one after the other, and the tasks must be done in an ordered sequence to produce the final item, which is common in assembly line production systems. In this case, the optimal workforce will be the one determined for the work process that requires the largest Q^* since this will be the task with the greatest physical demand, complexity, or slowness. From this workforce, a process of balancing the workloads between the different tasks in the sequence can be started.

However, the relationship between production and labor costs must also be considered. Perhaps, a lower level of production per unit of time can be considered to better balance the sequence if very large differences in Q^* are found from one task to another. In this case, a workforce size that is due to the Q^* of another task could be used, but with the understanding that to maintain stable production from the point of view of human factors, the **F*** of each task must be respected, and therefore, a lower level of total production could be obtained.

21.5.1 Application Example

We assume a production function composed of three departments: (a) tire assembly, (b) labeling, and (c) packaging. The daily production goal is of 1,000 tires, if the F1* calculated in the tire production model studied for manual load handling of 12 repetitions per hour (assembly) is followed, F2* is 18 repetitions per hour (labeling), and F3* is 22 repetitions per hour (packaging); in 8-hour work shifts, the optimal workforce is defined by the following:

$$Q^* \frac{1,000}{12*8} + \frac{1,000}{18*8} + \frac{1,000}{22*8} \approx 11 + 7 + 6 = 24 \text{ workers}$$

21.6 PRODUCTION FUNCTIONS WITH QUEUES

In many cases, and especially in the service sector, work involves serving internal or external customers who wait in an orderly queue for their turn. This necessitates the worker serving one customer after another as soon as the service is finished. Examples of this type of operation include supermarket cashiers, store order dispatchers, and teleoperators. In this case, each customer's waiting time and the average operation time are taken into account to calculate the number of operators required for the system to function efficiently. In this case, Queue Theory can be applied.

To find the Q^* in this type of production scenario, we must also know the following:

- *Arrival rate* (λ): The rate at which customers arrive to the system.
- *Service rate* (μ): The rate at which servers can serve customers. This rate is defined by calculating **F*.**
- *Queue discipline*: The order in which customers are served (e.g., First In, First Out (FIFO) and Last In, First Out (LIFO)).

- *Capacity*: The maximum number of customers who can wait in line or be served at the same time.
- *Time*: The maximum period for which a customer can normally wait, on average. It is made up of the time spent in line and the time it takes to serve him.

In a FIFO system, the formula for the average queue time W_q is as follows[24]:

$$W_q \frac{L_q}{\lambda}$$

where L_q is the average number of people in queue and is calculated as follows[25]:

$$L_q \frac{(\lambda/\mu)^c \cdot \frac{\lambda\mu}{(c!\cdot(c\mu-\lambda)^2}}{P_0}$$

And P_0 is the probability that the system is empty[26]:

$$P_0 = \left[\sum_{n=0}^{c-1}\frac{(\lambda/\mu)^n}{n!} + \frac{(\lambda/\mu)^c}{c!\cdot\left(1-\frac{\lambda}{c\mu}\right)}\right]^{-1}$$

c is the number of servers in the system, and μ is the product of calculating the maximum work capacity **F***; the equation must be evaluated with different numbers of servers, in an iterative process, until finding ***c*** that satisfies the maximum established waiting time (W_q). This will then be the Q^* for this task.

21.7 FINAL CONSIDERATIONS

Finding an optimal point is a challenge for both those responsible for production and for those who care for the health and well-being of workers. The use of productive resources must be managed with sustainability criteria so that, just as machinery must not be forced beyond its limits to avoid breaking, people who work within the ergonomic limits relevant to their activity can do so in a safer and healthier way, for longer in their working life and with less probability of making mistakes.

However, it is important to consider that ergonomic work analysis models have important limitations, linked to the premises within which they are validated. It can be concluded that they do not represent an absolute and universal solution for decision-making on issues such as setting work regimes, repetition, and the maximum physical or psychological loads to which workers may be subjected. On the other hand, the traditional ergonomics evaluation models (NIOSH Lifting Index, OCRA, JSI, etc.) are focused on the prevention of musculoskeletal and/or psychosocial pathologies, but they do not have a direct link to the estimation of physical and mental fatigue, which constitutes a "grey" area where the appearance of physical or psychological disorders may not be observed in the short term, but which definitely affects the production, as well as a deterioration in health in the long term.

Note that, although at first glance it might be thought that, if work regimes increase, production should also increase proportionally. This, however, does not work linearly in this way because (taking for example, lifting loads) of the following:

- A higher lifting regime also causes a higher rest regime.

- A higher lifting regime results in a greater appearance of "short breaks", also called fatigue supplements, to go to the bathroom, hydrate, etc., which is difficult to quantify (personal allowance and delay allowance).
- A higher regime must be aligned with other vertically integrated elements (raw materials management, production, storage, and packaging) across departments, with consequences for the system's global performance.
- A higher work regime produces more fatigue, which has short-, medium-, and long-term effects on the maintenance of the work regime, through (for example) presenteeism, cardiometabolic imbalance, absenteeism, a chronic decrease in functional capacity, or the open manifestation of occupational pathologies.

According to the results obtained in this study, a general methodology can be described to identify the optimal point for biomechanical, physiological, and/or psychological demands relative to the model of increasing marginal returns and quantitative ergonomic analysis tools, and from that point, calculate the optimal staff size. But these results must be understood as a theoretical optimal point, whose uncertainty will be linked to some fundamental premises:

1. The correlation between the human component and total production in semi-automated processes. The contribution of the human factor can be very small in certain processes, to the point of being virtually uncorrelated with production. For example, if the amount of human load is incorporated when a robot arm is damaged or under maintenance, a decrease in the human load (the robot is working) implies an increase in productivity.
2. The phenomena describing the human factor involved (repetitiveness, physical or mental load, etc.) must be able to be evaluated quantitatively, through a validated method. That is, through an equation such as that of NIOSH, OCRA, JSI, and OWAS.
3. The production function must be continuous. The data must behave in a "normal" way (Gaussian Bell curve) following the distribution analogous to the decreasing marginal returns curve. Some processes may present total proportionality.

Finally, humanity must be considered. Each human being is different in their capacities and needs; at work, the worker tries to give the best of themselves by virtue of their skills and abilities, their physical and mental capacities, or their general state of health. Therefore, it is very risky to calculate precisely what the ideal work regime is for each person, which can vary depending on factors such as mood, nutrition, gender, and age, leading to significant variations between different workers and between different stages in the lifecycle of the same worker.

The main utility of this work lies in its ability to combine qualitative and multifactorial variables from microergonomic models (such as cognitive load or the risk of musculoskeletal disorders) within dynamic production functions to model the labor process, introducing an approach that allows companies to determine the number of workers that not only maximizes production but also avoids work overload and health deterioration, making the process more sustainable for humans.

The most significant innovation is the integration of diverse theories. The study combines queuing theory with production theory to go beyond purely descriptive time-and-motion studies. This dual approach allows not only the analysis of workflow efficiency but also the linking of it to human biomechanical, physiological, and cognitive limitations, which are used as inputs within a macroergonomic equation. The proposed model aligns with the macroergonomics approach developed by Hendrick (1991), who also examined the interactions among humans, the environment, work, and organization, considering the number of workers (an organizational factor) as a key variable. Although the production model is based on the law of diminishing marginal returns popularized by economists such as David Ricardo (1817) and Thomas Malthus (1798), its application to ergonomics is a significant advance. For example, authors such as B.H. Workie (2007) has studied worker performance in relation to factors such as fatigue, but this study takes the analysis to a higher level by combining it with queuing theory for predictive calculations. The citation of Portillo Vázquez (2015) and his use of the cubic production function suggest that the proposed model shares a similar mathematical basis for modeling the entire production curve, which aligns with and supports its conclusions.

However, a structured observation of production phenomena can provide relevant empirical evidence that, in conjunction with the analytical methods of macroergonomic work analysis, can be closer to determining the work regimes that are most convenient for the health and safety of workers, as well as convenient for the productivity of the organization, within a scheme of sustainability and labor bioavailability. By combining differential calculus with production models and queuing theory, a robust analytical tool is proposed that enables operations managers to make data-driven decisions to optimize the workforce so that productivity and worker health cease to be inversely proportional variables and become synergistic objectives.

NOTES

1 Gottfreid Leibniz (1646–1716) was a German polymath who co-invented calculus. His contributions span binary arithmetic, differential equations, and philosophy, influencing multiple fields.
2 Latin phrase meaning "all or other things being equal or held constant".
3 Equation Model for Decreasing Marginal Returns, using constant a value.
4 Equation developed by the National Institute for Occupational Safety and Health (USA, Rev.1994).
5 Method for Repetitive Actions Assessment. E. Occhipinti & D. Colombini (Italy).
6 Method to analyze workplaces for risk of upper extremity disorders. J. Steven Moore & Arun Garg (USA).
7 Physiologic method for metabolic rates. Standard ISO 8996. Include FRIMAT & CHAMOUX criteria (France).
8 For this chapter, only NIOSH index 1 is developed to illustrate the validation. Other results are available upon request.
9 The least squares method is a statistical and computational technique used to find the best "line of fit" to a dataset, whether a straight line or a curve. Its objective is to generate a mathematical function that represents the relationship between the data as accurately as possible.
10 Production function reinforced tire constructed from production data.
11 TPP optimization curve. The first derivative is zero.
12 MpPP optimization curve. The second derivative is zero.
13 Average Product function.
14 APP optimization curve. The first derivative is zero.
15 Equation Model for Decreasing Marginal Returns, without constant values.
16 Generic model of the second derivative of the equation (point "A").
17 Generic model of equation midpoint of the equation.
18 Generic model of the first derivative of the equation APP (point "B").
19 Generic model of the first derivative of the equation.
20 Generic model of quadratic equation. Solve the equation derived (point "C").
21 Optimal range determined by the analytical method.
22 Optimal workforce value applied to a simple production function.
23 Optimal workforce value applied to a composite production function.
24 Equation for the average queue time.
25 Equation for average number of people in queue.
26 Equation to calculate the probability that the system is empty.

BIBLIOGRAPHY

Ahsberg, Elizabeth. "Dimension of Fatigue in Different Working Populations." *Scandinavian Journal of Psychology* 41, no. 3 (2000): 231–41.

Bello, Igor., and Salmen,Acran.. "Ergonomics and Increasing Marginal Returns." In *Advances in Intelligent Systems and Computing*. Springer, 2016.

Cobb, C. W., and P. H. Douglas. "A Theory of Production." *American Economic Review* 18 (1928): 139–165.

Drinkaus, Philip., Bloswick, Donald. Sesek,Richard. Mann, Cla. and Bernard,Thomas. *The Strain Index: Using Task Level Outputs to Evaluate Job Risk*. Salt Lake City: National Occupational Research Agenda (NORA), 2003.

Estrada, Jairo. *Ergonomics*. 3rd ed. Medellín: Editorial Universidad de Antioquia, 2011.

Guelaud, F., Beauchesne,M. Gautrat,J. Roustang, G. and Barbieri, R.. *Para un análisis del trabajo obrero en la empresa* [*For an Analysis of Worker's Labor in the Company*]. Lima: IFEA Net. Laboratorio de Economía y Sociología del Trabajo, 1982.

Hendrick, Hal. W., and Kleiner,Brian. M.. *Macroergonomics: Theory, Methods, and Applications*. Laurence Erlbaum Associates Publishers, 2002.

Jansen, Nicole. W., and van Amelsvoort,Ludovic. G.. "Work Schedules and Fatigue: A Prospective Cohort Study." *Occupational and Environmental Medicine* 60 (2003): 47–53.

Janssen, Nicky. W.,Kant, Ijmert. J. and Swaen, G. M. H.. "Fatigue as a Predictor of Sickness Absence: Maastricht Cohort Study on Fatigue." *Occupational & Environmental Medicine* 60, suppl. 1 (2003): 71–76.

Koontz, Harold., and Weihrich, Heinz. *Administration: A Global Perspective*. 11th ed. México: Mc Graw-Hill Interamericana, 2004.

Kumashiro, Masaharu. "Practical Measurement of Psychopysiological Functions for Determining Workloads." In John R. Wilson and E. N. Corlett (Eds.), *E: Evaluation of Human Work*, 864–887. London: Taylor and Francis, 2001. doi:10.1201/9781420055948.ch21

Llaneza, Javier. *Ergonomics* and *Applied Psychosociology. Manual for the Specialist Training*. 15th ed. Valladolid: Lex Nova, 2009.

Manero, Alfert R.., Barreras, I. and González, M. . "Un estudio integral para una paletización exigente" [A Comprehensive Study for a Demanding Palletizing]. *MAPFRE Medicina* 11, no. 2 (2000): 126–135.

Márquez, M. *Ergonomía [Ergonomics]*. San Cristóbal, Venezuela: Fondo Editorial UNET, 2015.

Moore, J Steven., and. Garg,Arun. "The Strain Index: A Proposed Method to Analyze Jobs for Risk of Distal Upper Extremity Disorders." *American Industrial Hygiene Association Journal* 56 (1995): 443–458. https://doi.org/10.1080/15428119591016863

Morales, E. *Fatiga Industrial [Industrial Fatigue]*. México: Jefatura de Servicios de Medicina del Trabajo, IMSS, 1986.

Park, Jungsun., Kim,Yanghoo. Chung,Ho Keun. et al. "Long Working Hours and Subjective Fatigue Symptoms." *Industrial Health* 39, no. 3 (2006): 250–54.https://doi.org/10.2486/indhealth.39.250

Peaucelle, Jean-Louis. "Adam Smith's Pin Factory: The Original Source Identified." *Organization Science* 18, no. 4 (2007): 515–528.

Portillo Vázquez, Marcos., Pérez, Francisco. Figueroa,Esther. Godínez,Lucila. Pérez Soto,Maria Teresa. and . Barrios. Gerónimo."La función de producción cúbica y su aplicación" [The Cubic Production Function and its Application]. *Revista Mexicana de Agronegocios* 37 (2015): 11–24.

Purcell, Edwin., and Varberg,Dale.. *Calculus with Analytic Geometry*. 6th ed. México DF: Prentice Hall Hispanoamericana, 1993.

Santana, Dulce. "Estudio de la aparición de las lumbalgias en la Industria del Caucho" [Study of the Occurrence of Low Back Pain in the Rubber Industry]. Specialist Degree Thesis, Universidad de Carabobo, 2014.

Taylor, Frederick. Winslow. *The Principles of Scientific Management*. New York: Harper & Brothers, 1911.

Waters, T. R., Putz-Anderson, V., and Garg, A. . *Applications Manual for the Revised NIOSH Lifting Equation*. Cincinnaty: National Institute for Occupational Safety and Health, 1994.

22 Understanding and Managing Human Error in Manufacturing Environments

Yaniel Torres and Carlos Real

22.1 INTRODUCTION

Manufacturing plays a pivotal role in the economic development and industrialization of Latin America. Although its contribution to national gross domestic product (GDP) varies—from around 20% in Mexico to lower percentages in countries like Colombia and Chile—it remains a key driver of growth, employment, and foreign investment across the region (World Bank 2023). Moreover, the challenges faced by the sector are inherently transversal and require coordinated strategies to enhance resilience and competitiveness across borders. The global manufacturing landscape is marked by increasing complexity and intense competition. This dynamic has led manufacturing organizations to adopt innovative practices and the introduction of Industry 4.0 technologies (Nadeau and Landau 2018). However, despite growing levels of automation and digitization, manual assembly processes remain indispensable in contexts that require high adaptability to product variety and complexity. Manual assembly refers to the process of combining pre-manufactured components or subassemblies into a finished product or unit. This process relies heavily on the dexterity, skill, and decision-making abilities of human operators to complete necessary tasks (Swift and Booker 2013). Manual assembly plays a critical role in manufacturing sectors such as automotive, consumer electronics, aerospace, and the production of industrial machines and tools (Correia et al. 2018; Falck et al. 2010; Judt et al. 2020). However, it also presents challenges in maintaining consistent quality and product reliability.

Historically, the study of human error has been more prominent in safety-critical sectors such as nuclear energy, chemical processing, oil and gas, and aviation. A series of major accidents—including the Tenerife airport disaster (1977), the Three Mile Island nuclear incident (1979), the Challenger space shuttle explosion (1986), and the Chernobyl disaster (1986)—triggered a significant shift toward understanding the underlying mechanisms and contributing factors of human error. These events underscored the limitations of purely technical approaches and highlighted the critical role of organizational, cognitive, and systemic factors in accident causation (Reason 1990; Senders and Moray 1991).

In contrast, human error as a phenomenon grounded in Human Factors and Ergonomics (HFE) has not been sufficiently addressed within the manufacturing sector. Traditional quality management approaches, particularly those rooted in statistical quality control, have historically focused on data-driven methods to detect and reduce variability in processes (Montgomery 2020). While effective in many respects, these methods often overlook the complex human and organizational dimensions that contribute to error (Drury 2000; Falck and Rosenqvist 2014). This gap has limited the integration of HFE insights into manufacturing systems, especially in manual assembly environments where human performance plays a central role.

Human operator–induced assembly errors are a significant source of defects in manufacturing systems (Botti et al. 2017; Falck and Rosenqvist 2014; Su et al. 2010; Torres et al. 2021a). These errors

DOI: 10.1201/9781003582014-27

lead to major downstream consequences, such as defects, rework, delays, and increased operational costs. Within the lean manufacturing framework, they are categorized as waste (muda), emphasizing the importance of error reduction in streamlining operations (Arunagiri and Gnanavelbabu 2014; Gupta and Jain 2013).

Poorly designed manual assembly systems that disregard the principles of HFE can reduce manufacturing efficiency and increase production costs (Drury 2000; Lotter and Wiendahl 2009). Poor physical and cognitive ergonomics have been identified as contributing factors to higher error rates and diminished product quality in manufacturing (Falck and Rosenqvist 2014; Real and Torres 2024; Torres et al. 2021a). Integrating HFE has demonstrated tangible benefits, such as improved product quality and lower production costs (Zare et al. 2016).

The objective of this chapter is to analyze the origins and contributing factors of human error in manufacturing—particularly in manual assembly tasks—and to examine intervention strategies that integrate HFE into production and quality management systems. The chapter aims to provide practical insights to support continuous improvement, reduce human error, and enhance system performance in manufacturing environments.

22.2 METHODOLOGICAL APPROACH

This chapter is based on a narrative literature review complemented by the authors' professional experience in the manufacturing sector, particularly in the contexts of automotive and aerospace assembly. The review draws on both foundational works in the field of human error—such as the skill–rule–knowledge (SRK) model (Rasmussen 1983), the Generic Error Modeling System (Reason 1990), and concepts related to error-proofing and procedural tasks—as well as recent studies published in peer-reviewed journals and technical reports. Rather than following a systematic review protocol, this narrative approach allows for the integration of theoretical perspectives with practical insights derived from real-world industrial contexts. The selection of sources was guided by their relevance and recurrence in the literature, with particular emphasis on models and strategies applicable to manual assembly environments. Throughout the chapter, the reviewed literature is illustrated and contextualized using representative examples inspired by real scenarios from the authors' field experience, serving to bridge academic knowledge and industrial practice. The results of this narrative review are presented in a series of thematic sections that reflect both the structure of the reviewed literature and the authors' interpretation based on industry experience.

22.3 RESULTS

22.3.1 Cognitive Models and Classifications of Human Error

Various definitions of human error have been proposed in literature. Most of these definitions align with the fundamental aspects of error. For example, Swain and Guttmann (1983) defined human error as "any member of a set of human actions that exceed some limit of acceptability, i.e., an out-of-tolerance action, where the limits of tolerable performance are defined by the system." The definition adopted in this chapter is that of Reason (1990): "a generic term to encompass all those occasions in which a planned sequence of mental or physical activities fails to achieve its intended outcome, and when these failures cannot be attributed to the intervention of some chance agency."

Sharit (2012) outlined three key perspectives for studying human error: (a) the ergonomics perspective, which views error as resulting from a mismatch between task demands and the physical or mental capacities of the individual; (b) the cognitive perspective, which examines how humans acquire, represent, and use information to guide their actions; and (c) the sociotechnological perspective, which focuses on the influence of managerial policies and organizational culture in shaping the context in which humans operate. The study of human error often requires an integrated approach that considers all three perspectives.

Jens Rasmussen's SRK framework provides a foundational model for classifying errors based on the cognitive control mode employed during task execution (Rasmussen 1983). This model identifies three levels:

1. *Skill-based errors*: These occur in automatic and routine tasks governed by learned behaviors. Such errors often arise when actions deviate from intended routines.
2. *Rule-based errors*: These result from the misapplication of established rules or procedures. These errors are common in contexts where workers rely on procedural guidelines.
3. *Knowledge-based errors*: These arise in novel or uncertain situations due to inadequate understanding or incorrect problem-solving strategies.

In manufacturing, particularly in manual assembly, tasks are performed within structured and predictable environments, typically following a series of sequential operations designed to achieve specific objectives. Operators make decisions and carry out tasks based on previously acquired action patterns to produce the desired outcomes (Dezfouli et al. 2014). A key characteristic of manual assembly tasks is that they are often sequence-constrained procedural tasks. According to Bovo et al. (2020), a sequence-constrained procedural task consists of a series of steps that must be performed in a specific order. Errors associated with sequence-constrained procedural task include performing steps out of sequence, omitting steps, or unintentionally repeating steps. Ensuring a high-quality, defect-free product requires precise execution of various operations, including selecting, handling, and fitting parts; inspecting components; applying force; and retrieving and analyzing information. The work of authors such as Dezfouli et al. (2014) and Bovo et al. (2020) has helped contextualize human error research within the procedural and temporal dynamics of manual assembly tasks.

Most errors in manual assembly occur at the skill-based level of human behavior, as described in the SRK model. Common assembly errors include incorrect component selection, improper installation, loose connections, missing components, damage during assembly, and contamination with foreign debris. These errors can be classified as slips of action or lapses of memory under the Generic Error Modeling System proposed by Reason (1990). A slip occurs when an action is executed incorrectly, whereas a lapse involves failing to carry out an intended action.

Rule-based errors can occur in assembly tasks because these tasks rely on well-established procedures tailored to specific situations. While process standardization is common in assembly lines, variations in circumstances can be significant. For instance, in a mixed-assembly line, an operator may be accustomed to installing a specific trim piece in a fixed sequence for a particular car model. However, when a new variant or an entirely different model is introduced, the installation sequence may need to be adjusted. If the operator inadvertently follows the previously learned sequence—such as installing a fastener before aligning the corresponding part—it can result in a misaligned component. This underscores the need for adaptable procedures and comprehensive training or real-time assistance to accommodate variations in mixed-assembly environments. However, since there is minimal decision-making under uncertainty, the occurrence of knowledge-based errors is significantly reduced.

22.3.2 Challenges of Complex Manual Assembly

In manufacturing, complex manual assembly poses significant challenges as tasks require a high degree of specialized knowledge and skill to ensure successful execution. These assemblies often involve multiple components and numerous potential configurations, with the final product typically exhibiting several planes of geometric symmetry (Richardson et al. 2004). Such structural factors substantially affect cognitive workload and increase the likelihood of error.

In a study conducted by Torres et al. (2021b), the NASA-TLX method was used to assess the subjective perception of workload in the aerospace engine manufacturing context. The results

FIGURE 22.1 Example of an aircraft engine and its components, known as externals. (Free for use under the Pixabay Content License.)

demonstrated that operators rated the mental component of workload as higher than the physical component, highlighting the critical role of cognitive aspects in complex assembly.

Figure 22.1 illustrates an example of an aircraft engine, showing various components and parts assembled—particularly the so-called externals, which are manually installed during the final stages of the process. One example of such externals is brackets that support flexible elements within the aircraft engine. Errors may occur during bracket installation, including incorrect orientation, improper placement, or omission of installation altogether (Torres et al. 2021a). Additionally, an incorrect type of bracket may be selected. Fastening screws may be left loose, overtightened, or omitted entirely. Such assembly errors can compromise product quality, reduce productivity, and pose safety risks.

Figure 22.2 presents a schematic of the assembly instructions for installing three different brackets used to secure cables to the engine structure. The instructions are displayed on a computer monitor in a fixed two-dimensional (2D) format, which cannot be manipulated. Since the assembly parts are presented in a predetermined orientation, the operator must mentally construct a three-dimensional (3D) representation that aligns with the actual assembly process.

The automotive manufacturing sector is another industry where assembly errors significantly impact production quality. In high-volume manufacturing (mass production), even negligible errors can have a significant economic impact on production quality and safety. In the automotive industry, recalls of already sold vehicles are common due to assembly or design errors that may pose risks during use.

A study conducted at a Volvo plant in Sweden analyzed the quality outcomes of 47,061 cars and identified several assembly errors (Falck and Rosenqvist 2014). The findings showed that workstations with poor ergonomics had a higher incidence of errors. Similarly, Real and Torres (2024) reported multiple assembly errors at a Mexican car manufacturing plant, including missing safety

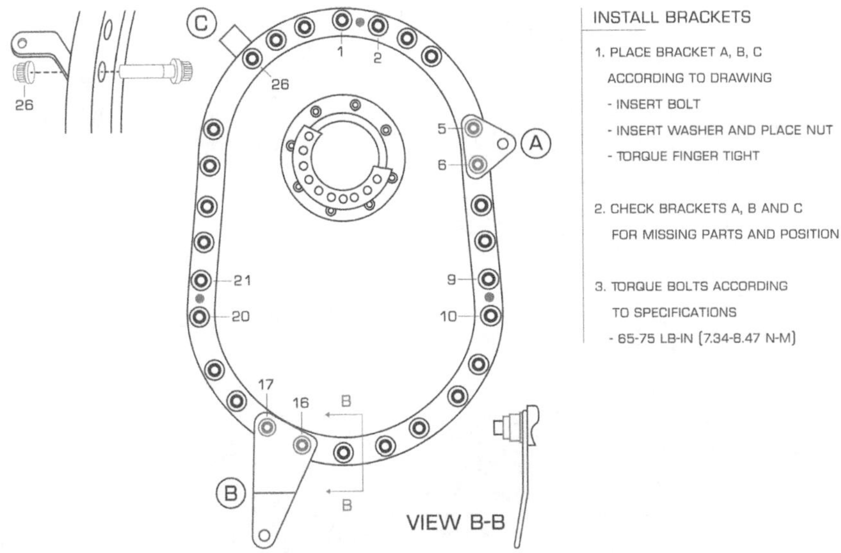

FIGURE 22.2 Example of a 2D drawing used in assembly instructions, indicating the locations where different brackets should be installed on the aircraft engine structure. (Torres et al. (2021a), reprinted under CC BY 4.0.)

FIGURE 22.3 Schematic representation of an assembly worker tightening the wiper arm bolt using a DC nut runner. Digital human models were created using IPS IMMA software (FleXstructures). 3D vehicle data were obtained from Caresoft Global. (Image provided by one of the authors.)

lock labels on doors, loose screws in windshield wipers and panoramic roofs, and the installation of an incorrect armrest model. Figures 22.3 and 22.4 illustrate the installation of the windshield wiper and armrest.

FIGURE 22.4 Schematic representation of an assembly worker installing the armrest inside the car. Digital human models were created using IPS IMMA software (FleXstructures). 3D vehicle data were obtained from Caresoft Global. (Images provided by one of the authors.)

Selection errors are among the most common assembly errors in car manufacturing, largely due to the complexity of managing the vast number and variety of components required for assembly, many of which share subtle similarities. Assembly workers must often select the correct component from multiple options that vary slightly based on the vehicle's version, trim level, accessories, or the specific region where it will be used. Additionally, assembly lines frequently produce different vehicle types, such as sedans and small SUVs, which share the same platforms or frames, further increasing the complexity of the process.

22.3.3 Factors Influencing Assembly Errors

Human errors do not occur in isolation; rather, they emerge from the complex interactions between human operators and their working environments. These interactions are influenced by various factors, commonly referred to as performance shaping factors (PSFs) or error-producing conditions, which either trigger errors or increase their likelihood. In practice, numerous factors can affect human performance. The number of PSFs documented in specialized studies on HFE ranges from dozens to hundreds (Kirwan 1994). However, a relatively small subset of these factors tends to gain prominence depending on the specific task and context. For instance, in a task requiring a worker to visually inspect small details of a product, workplace lighting becomes a critical environmental factor influencing performance.

In a literature review of 73 empirical studies, Kolus et al. (2018) identified 64 quality risk factors that significantly affect human performance in manufacturing and assembly. These factors were categorized into four key groups: products, processes, workstations, and individuals. This categorization provides a structured framework for understanding how different elements of the work system contribute to performance variability and error occurrence. This section focuses on factors particularly relevant to human error in manual assembly and manufacturing environments, using the four categories outlined by Kolus et al. (2018) as a reference framework.

A critical factor in manual assembly is complexity. A high level of complexity is associated with increased cognitive workload. Assembly complexity is a construct that encompasses multiple factors contributing to cognitive load. At its core, complexity relates to human cognition, particularly information processing and reasoning. The literature identifies two primary types of complexity: static complexity and dynamic complexity (Rodríguez-Toro et al. 2004). In the first case, complexity is associated with the product; in the second, it is linked to the process, following the categories defined by Kolus et al. (2018).

Static complexity is an inherent attribute of an assembly object, influenced by factors such as the number of components, subassemblies, interactions, selections, and geometric properties, including symmetry (Richardson et al. 2004). It arises from product design. However, design choices can simplify assembly. In contrast, dynamic complexity emerges from the interaction between the worker, instructions, and the assembly object, evolving throughout the assembly process. Workers must construct mental representations to manipulate spatial information (Richardson et al. 2004), and unclear or poorly designed instructions can increase cognitive load, reducing performance and raising error probabilities (Li et al. 2018; Torres et al. 2021a).

Other significant factors are time pressure and work cadence. Manufacturing environments often impose stringent time constraints to meet productivity targets. However, time pressure increases cognitive load and reduces the time available for careful decision-making and execution. Under these constraints, workers may rush tasks, compromising precision in selecting, fitting, or fastening components. This heightened urgency can impair fine motor skills and reduce attention to detail. Additionally, time pressure can shorten rest periods, elevate stress levels, and contribute to fatigue.

Environmental conditions—such as lighting, noise levels, and the thermal environment—also influence operator performance in manual assembly. For instance, inadequate lighting can lead to component misidentification, while excessive noise can disrupt concentration and increase the likelihood of errors. These factors fall under the workstation category proposed by Kolus et al. (2018), which also includes workspace design and the tools used by operators. The organization of components within the workspace is another critical factor affecting error rates. Poorly arranged, inadequately labeled, or disorganized components—especially in environments with multiple product variations—can complicate the selection process, leading to delays or incorrect assemblies.

Tools are key determinants of operator performance and error rates. Variability in psychomotor skills among operators can amplify the impact of tool design on task execution. For instance, using a standard wrench without a force-measuring mechanism creates a scenario where psychomotor variability remains uncontrolled. In such cases, the applied torque depends entirely on the operator's judgment, manual dexterity, and consistency, increasing the potential for error. Even when using a torque wrench, inconsistencies in torque application can arise if the tool lacks precise calibration or clear visual indicators.

Distractions and interruptions, although distinct, disrupt attention and workflow continuity, increasing the likelihood of errors. Distractions momentarily divert focus due to external factors such as noise or conversations but can also stem from internal preoccupations. Interruptions, on the other hand, actively break workflow, requiring reorientation upon resumption. These may arise from external disruptions (e.g., a supervisor's intervention) or process-related issues, such as an improperly sequenced assembly task. Both impair concentration and accuracy, particularly in complex environments where continuity and precision are critical. Research suggests that although initially considered separate phenomena, distractions and interruptions similarly degrade human reliability by depleting attentional resources (Williams and Bell 2015).

Abnormalities, a common form of interruption in automotive manufacturing, occur when deviations in process elements disrupt assembly continuity (Real and Onieva 2011). The case illustrated in Figure 22.5 shows how operators commonly request assistance for issues such as tool failures, component shortages, or installation difficulties, usually through an assistant call button.

FIGURE 22.5 Assistance call buttons at the workstation for operators to request support. The red button sends a signal and stops the line, the yellow button sends a signal while keeping the assembly line running, and the green button releases a pressed signal. (Image provided by one of the authors.)

Data from a mid-sized vehicle assembly plant with 320 workstations revealed that for every 1,000 vehicles produced, nearly 2,000 assistance calls were made—averaging two calls per operator per shift. Moreover, a strong correlation was found between abnormality frequency and human-related defects, highlighting their role in increasing the likelihood of errors (Real and Onieva 2011).

The sequence of operations in manual assembly significantly impacts error rates, especially when designed solely for productivity rather than cognitive efficiency. Workers' limited cognitive resources increase the likelihood of errors when tasks include functionally unrelated or isolated steps (Rasmussen 1982). A specific type of sequence-related error is the post-completion error, which occurs after the primary task goal has been achieved (Byrne and Davis 2006). Common examples include forgetting to retrieve a bank card from an ATM or failing to replace a gas cap after refueling.

In automotive assembly, Real and Torres (2024) documented a post-completion error during door trim panel installation, where a child lock label was omitted. Since this step was unrelated to the main task and positioned at the end of the cycle, its misalignment with the task goal significantly increased the likelihood of error.

Worker fatigue is another critical factor affecting manual assembly performance. It is typically associated with three primary sources: physical workload, mental demands, and chronobiological factors, such as shift work and irregular schedules. Fatigue can manifest in various forms, including sleepiness, mental exhaustion, and physical fatigue, depending on its underlying cause. Research, such as the study by Torres et al. (2021b), highlights the impact of fatigue in complex manual assembly settings involving shift work. Findings indicate that the key predictors of high fatigue levels include workload estimates, evening shifts, and baseline fatigue at the start of a shift. Among these, mental demand emerged as the most significant predictor, underscoring the role of cognitive strain in intensifying fatigue in the context of complex manual assembly tasks.

The impact of shift work and extended hours on human performance is particularly significant as they can greatly increase the risk of errors and reduce productivity. In environments requiring shift work or atypical schedules, careful attention should be given to chronobiological sources of fatigue, such as circadian and homeostatic processes, which vary based on shift duration, start time, and end time (Miller 2008).

Regarding mental fatigue, studies have shown that prolonged cognitive effort leads to glutamate accumulation in the brain, impairing decision-making processes (Wiehler et al. 2022). The same study found that mentally fatigued participants increasingly opted for low-effort choices, even when those choices offered lower rewards. This highlights how mental fatigue not only hinders cognitive performance but also influences behavior and decision-making, ultimately affecting productivity and efficiency.

Among the four categories of factors proposed by Kolus et al. (2018)—product, process, workstation, and individual—individual factors represent one important dimension to consider, particularly in relation to how they interact with the other categories. Two of the most prominent factors in this group are experience and age. Age, in particular, is a multifaceted factor as it encompasses changes in both psychomotor and cognitive abilities. Research indicates that these abilities decline more noticeably after age 60–65 (Harada et al. 2013), which has significant implications for tasks requiring high levels of precision, attention, or rapid decision-making. Employers must consider these age-related changes when designing work systems, assigning tasks, or implementing training programs.

Some individual factors are closely tied to health and fitness and are commonly referred to as "fit-for-duty" in industries such as aviation. Medical conditions or medications may impair a worker's ability to perform tasks safely and effectively. For example, certain medications can cause drowsiness, impair motor coordination, or reduce cognitive alertness, increasing the likelihood of errors (Do and Schnittker 2020).

Additionally, neurocognitive conditions, such as attention-deficit hyperactivity disorder (ADHD), contribute to performance variability. Although ADHD is often associated with childhood and is thought to diminish with age, studies indicate that up to 6.75% of adults experience persistent symptoms (Song et al. 2021). Adults with ADHD are more prone to impulsivity and distraction, which can have direct consequences in safety-critical work environments. Similarly, psychological conditions, including anxiety and depression, affect executive functions such as decision-making, attention, and problem-solving (Snyder 2013).

22.3.4 Strategies to Reduce Assembly Errors

An effective strategy for preventing assembly errors in manufacturing is reducing product complexity during the design phase. This is often achieved through design for assembly (DFA), an engineering methodology focused on minimizing the number of parts and assembly operations to lower costs and improve efficiency (Boothroyd 1994). By simplifying design, DFA not only reduces assembly time and costs but also enhances quality by decreasing potential failure points during assembly (Kent 2016).

The DFA methodology was developed primarily to simplify manual assembly processes. Its core principle is that products should incorporate the fewest possible number of parts while ensuring ease of assembly through features such as improved handling, positioning, and alignment

(Kent 2016; Mital et al. 2014). The methodology optimizes assembly by enhancing component accessibility and visibility, reducing part count through multifunctional designs, and simplifying processes by eliminating unnecessary elements. It prioritizes vertical insertion to leverage gravity, minimizes the need for rotations or adjustments, and replaces threaded fasteners with alternatives such as press-fits or rivets to improve efficiency.

Despite its advantages, DFA has limitations. Some products must prioritize robustness and reliability over assembly simplicity. As a result, fully implementing DFA principles is not always feasible or desirable in every manufacturing process. For example, aircraft engine design emphasizes maximizing reliability and durability rather than reducing assembly complexity (Dhillon 1999). Therefore, in complex manufacturing environments, sustainable error prevention strategies must extend beyond the design phase to address challenges throughout the product lifecycle.

The error-proofing technique, also known as poka-yoke, is an effective strategy for mitigating assembly errors (Blais 2011). It prevents or significantly reduces errors in manual assembly by making them physically or logically difficult to occur. Developed as part of lean manufacturing principles, poka-yoke focuses on designing processes, tools, or systems that either eliminate the possibility of human mistakes or detect and correct them immediately. Poka-yoke techniques often ensure that parts can only be assembled correctly. Examples include keyed components that fit in only one orientation or fixtures that guide a worker's actions.

Selecting the correct parts during assembly requires extensive information processing and decision-making. To address this challenge, kitting systems have been explored as a strategy to reduce the cognitive load on workers (Brolin et al. 2011). Kitting is a logistics technique in which groups of parts are organized and delivered together in containers known as kits (Caputo et al. 2015). These kits are typically assembled in a warehouse and then delivered to specific workstations on the assembly line in accordance with the production schedule. When properly prepared, kits ensure that parts are readily available, inspected, and pre-positioned for easy and rapid access (Christmansson et al. 2002).

In manual assembly, part-kitting systems are commonly implemented using plastic boxes with compartments, which facilitate organization and quick access to components. However, fabric bags with compartments may also be used, particularly when greater flexibility or portability is required for transporting parts. Figure 22.6 shows an example of a part-kitting system using fabric bags.

According to Brolin et al. (2017), kits can act as a "carrier of information" for assembly, embedding work instructions directly into the kit itself. Medbo (2003) further emphasized that well-structured kits not only support assemblers in their tasks but also facilitate skill development and learning. Additionally, Caputo et al. (2017) highlighted that kitting systems enable in-process quality control, allowing additional checks in the kitting room or at the workstation. This approach significantly reduces the risk of assembling incorrect or missing parts by providing direct feedback to the assembler. Research suggests that kitting systems enhance both quality and productivity compared to other part-feeding methods in manufacturing logistics.

The effectiveness of manual assembly processes heavily depends on the quality, presentation, and usability of work instructions (Backstrand et al. 2008; Brolin et al. 2017; Li et al. 2018; Mattsson et al. 2018). To fully leverage assemblers' cognitive abilities, work instructions must provide clear and precise guidance on which components to use and how to assemble them. By simplifying and structuring information effectively, work instructions can significantly reduce workers' cognitive demands, particularly by minimizing dynamic complexity in tasks (Frizelle and Woodcock 1995).

According to Mattsson et al. (2016), instructions need to be more perceptual, meaning they should provide richer and more immediate sensory inputs to assemblers. Using 3D models in work instructions presents the assembly process in a more realistic, accurate, and intuitive manner. Model-based instructions offer multiple views and easy-to-follow assembly procedures (Geng et al. 2015; Hagan et al. 2009). Enhancing work instructions involves incorporating visual elements

FIGURE 22.6 Example of a kit with components for door line assembly. In this case, the kit is a fabric bag with multiple compartments for organizing components. It is located to the left of the car door, as seen in the image. (Image provided by one of the authors.)

such as diagrams, sketches, and drawings, as well as dynamic formats such as 3D animations or videos. Optimizing interface design with clear, logical sequences (e.g., left to right and clockwise) improves comprehension while ensuring instructions remain easily accessible at assembly workstations. Additionally, the strategic use of colors enhances visual clarity and contrast, further supporting effective task execution.

Changing the sequence of assembly operations can be an effective intervention for decreasing error rates by reducing the operator's cognitive load. The operations within the cycle must be closely aligned with the task goal, and maintaining cycle continuity is crucial. For instance, requiring an assembler to pick up an assembly part from the workbench only to install it three steps later can disrupt the mental process. Similarly, split sequences or cycles that start at one workstation and are completed at another should be avoided. Research indicates that assembly sequences can be designed to be both time-efficient and cognitively intuitive for assemblers. This approach minimizes errors and enhances workflow without compromising productivity (Real and Torres 2024).

In addition to the strategies previously discussed, which may be specific to manual assembly and manufacturing, improving the workstation design, environment, and management using HFE principles enhances worker performance. Basic yet well-known principles—such as ensuring logically sorted, clearly identified, and easily accessible components—can significantly reduce cognitive load and improve efficiency. Additionally, for visually demanding tasks, consideration should be given to the lighting system, including localized supplementary lighting if necessary. Job rotation can help mitigate visual fatigue, and organizations must ensure that workers have adequate or corrected vision. Similarly, noise, distractions, and interruptions should be controlled at the source to minimize disturbances to workers.

Fatigue management is well-established in safety-critical industries such as aviation, where strict regulations govern work–rest cycles and are supported by automated time management systems. In manufacturing and manual assembly, similar approaches can be applied to shiftwork scheduling. One such tool is the fatigue and risk index (FRI), developed by the Health and Safety Executive in the United Kingdom (UK) (Spencer et al. 2006). The FRI is widely used in the UK rail industry and has proven effective in designing and comparing different shift configurations in other sectors, including manufacturing and manual assembly (Torres et al. 2021c). This tool enables organizations to evaluate shift structures based on accident risk and the likelihood of elevated fatigue levels. It considers factors such as commute time, shift duration, number and length of breaks, workload, attention levels, and the proposed sequence of shifts. By leveraging tools such as the FRI, organizations can optimize shift schedules to minimize fatigue, improve worker well-being, and enhance overall productivity.

In terms of work tools, it is essential to prioritize ergonomically designed tools that minimize physical strain and support intuitive operation. These tools significantly reduce worker fatigue and enhance cognitive performance over time. Investing in high-quality, user-centered tool design not only enhances operational efficiency but also reduces error rates, resulting in improved product quality and decreased rework.

For example, wireless-enabled tightening systems are becoming industry standards for improving the reliability of torque applications. These systems address common fastener-related errors, such as omissions, incorrect torque applications, and deviations from specifications. Equipped with decentralized intelligence, they perform checks and verifications to guide operators through the tightening process. Advanced features, such as integrated scanners, enable the system to identify components, select appropriate programs, and detect assembly errors—including missing parts or incorrect thread lengths—by analyzing torque and angle values (Brands 2018). Although implementation may require significant effort and investment, these systems have great potential for reducing manual assembly errors and supporting zero-defect manufacturing.

With the Industry 4.0 paradigm, several technologies are emerging to support humans in task execution, offering a unique opportunity to enhance worker performance and reduce assembly errors. According to Romero et al. (2016), these technologies assist workers in both the physical and cognitive aspects of their work. Examples include collaborative robots, virtual and augmented reality, digital twins, exoskeletons, and biometric tracking systems. The integration of these technologies within Industry 4.0 has led to the emergence of the Operator 4.0 concept (Romero et al. 2018). Operator 4.0 is defined as "a smart and skilled operator who performs not only cooperative work with robots but also work aided by machines as and if needed by means of human cyber-physical systems (...)" (Romero et al. 2016).

Numerous technologies can enhance the performance of assembly operators. Figure 22.7 illustrates some of the digital technologies designed to support workers in manufacturing contexts. However, the literature highlights several challenges, including investment costs, compatibility with work environments, impacts on physical and mental workload, reliability and safety risks, and concerns regarding the acceptability of these technologies—particularly in terms of data collection and handling sensitive information (Nadeau and Landau 2018). While digital technologies hold great potential for driving progress, their implementation requires careful planning to ensure that human considerations remain central to the process.

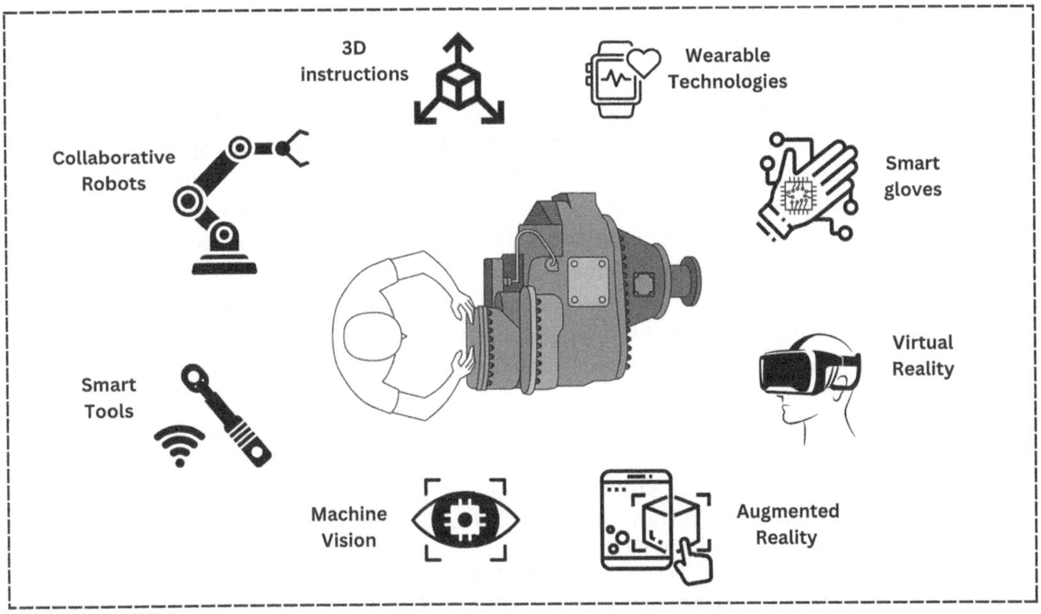

FIGURE 22.7 Examples of technologies associated with the concept of Operator 4.0 within the framework of the Industry 4.0 paradigm. (Figure created by the authors.)

22.4 LIMITATIONS AND FUTURE PERSPECTIVES

This chapter offers a narrative synthesis of foundational and recent literature on human error in manufacturing, supported by practical insights drawn from the authors' experience in the aerospace and automotive sectors. This approach allows for a meaningful integration of theory and practice, while also defining the scope and focus of the analysis. As a narrative review, the chapter does not follow a systematic protocol for source selection or appraisal. References were selected based on their conceptual relevance, recurrence in the field, and applicability to manual assembly environments. Likewise, the practical examples are inspired by real-world industrial contexts and are intended to illustrate key concepts, though they may not reflect the full diversity of manufacturing settings.

This chapter includes established intervention strategies and acknowledges recent developments related to Industry 4.0. However, emerging technologies—such as advanced cognitive support systems, digital twins, and industrial augmented reality—are not examined in depth. Given their growing relevance in intelligent manufacturing, future work could further explore how these tools align with HFE principles to enhance error prevention and operational performance. Despite its limitations, this chapter provides a solid foundation that bridges theory and practice, while also opening the door to further exploration of HFE integration into production and quality management within modern manufacturing systems.

22.5 CONCLUSIONS

Effectively managing assembly errors in manufacturing requires a comprehensive approach that considers the interaction between worker actions, task characteristics, and systemic factors. Errors rarely result solely from operator mistakes but often arise due to workstation design, task complexity, environmental conditions, and organizational processes. Addressing these root causes enables the development of more effective strategies to reduce errors and enhance system performance. Integrating HFE into manufacturing design improves production quality, leveraging both traditional methods—such as part kitting, DFA, and poka-yoke—alongside advanced technologies, including augmented reality, digital torque wrenches, and intelligent feedback systems.

Beyond technology, enhancing work conditions, managing physical and mental workload, and optimizing shift schedules are essential for minimizing errors. Structured work instructions and optimized task sequencing further support accuracy and efficiency. While Industry 4.0 technologies provide valuable tools for enhancing productivity, their integration must prioritize human-centered design. A balanced approach that combines systemic analysis, ergonomic principles, and both classic and modern methodologies fosters safer, more efficient, and higher-quality manual assembly processes, strengthening resilience in increasingly complex manufacturing environments.

REFERENCES

Arunagiri, P., and Gnanavelbabu, A. 2014. Identification of major lean production waste in automobile industries using weighted average method. *Procedia Engineering* 97: 2167–2175. https://doi.org/10.1016/j.proeng.2014.12.460

Backstrand, Gunnar, Thorvald, Peter, De Vin, Leo J., Hogberg, Dan, and Case, Keith. 2008. The impact of information presentation on work environment and product quality: A case study. *40th Annual Conference of the Nordic Ergonomics Society, NES 2008 Conference*, Reykjavik, Iceland.

Blais, Gilles. 2011. Méthodologie de sélection d'anti-erreur à l'assemblage par calcul de l'efficacité et du coût global d'une application. M.Sc. thesis, École de technologie supérieure de Montréal. https://espace.etsmtl.ca/id/eprint/1133/

Boothroyd, Geoffrey. 1994. Product design for manufacture and assembly. *Computer-Aided Design* 26(7): 505–520. https://doi.org/10.1016/0010-4485(94)90082-5

Botti, Lucia, Mora, Cristina, and Regattieri, Alberto. 2017. Integrating ergonomics and lean manufacturing principles in a hybrid assembly line. *Computers and Industrial Engineering* 111: 481–491. https://doi.org/10.1016/j.cie.2017.05.011

Bovo, Ricardo, Binetti, Nicola, Brumby, Dunca P., and Julier, Simon. 2020. Detecting errors in pick and place procedures: Detecting errors in multi-stage and sequence-constrained manual retrieve-assembly procedures. *Proceedings of the 25th International Conference on Intelligent User Interfaces*, Cagliari, Italy. https://doi.org/10.1145/3377325.3377497

Brands, J. 2018. *Tightening technology 4.0: Flexible and transparent production with decentralized intelligence*. Bosch Rexroth. March 5. www.boschrexroth.com/en/tr/blog/tightening-technology-flexible-and-transparent-production/

Brolin, Anna, Bäckstrand, Gunnar, Högberg, Dan, and Case, Keith. 2011. The use of kitting to ease assemblers' cognitive workload. *43rd Annual Conference of the Nordic Ergonomics Society, NES 2011*, Oulu, Finland.

Brolin, Anna, Thorvald, Peter, and Case, Keith. 2017. Experimental study of cognitive aspects affecting human performance in manual assembly. *Production and Manufacturing Research* 5(1): 141–163. https://doi.org/10.1080/21693277.2017.1374893

Byrne, Michael D., and Davis, Elizabeth M. 2006. Task structure and postcompletion error in the execution of a routine procedure. *Human Factors* 48(4): 627–638. https://doi.org/10.1518/001872006779166398

Caputo, Antonio C., Pelagagge, Pacifico M., and Salini, Paolo. (2017). Modelling human errors and quality issues in kitting processes for assembly lines feeding. *Computers and Industrial Engineering*, 111, 492–506. https://doi.org/10.1016/j.cie.2017.04.004

Caputo, Antonio C., Pelagagge, Pacifico M., and Salini, Paolo 2015. Modeling errors in kitting processes for assembly lines feeding. *IFAC-PapersOnLine* 48(3): 338–344. https://doi.org/10.1016/j.ifacol.2015.06.104

Christmansson, Marita, Medbo, L., Hansson, Gert Åke., et al. 2002. A case study of a principally new way of materials kitting—an evaluation of time consumption and physical workload. *International Journal of Industrial Ergonomics* 30(1): 49–65. https://doi.org/10.1016/S0169-8141(02)00077-X

Correia, Damáso, Silva, Francisco José Gomes, Gouveia, R. M., Pereira, Teresa, and Ferreira, Luis Pinto. 2018. Improving manual assembly lines devoted to complex electronic devices by applying Lean tools. *Procedia Manufacturing*, 17: 663–671. https://doi.org/10.1016/j.promfg.2018.10.115

Dezfouli, Amir, Lingawi, Nura W., and Balleine, Bernard W. 2014. Habits as action sequences: Hierarchical action control and changes in outcome value. *Philosophical Transactions of the Royal Society B: Biological Sciences* 369(1655): 20130482. https://doi.org/10.1098/rstb.2013.0482

Dhillon, B. S. 1999. *Engineering Maintainability: How to Design for Reliability and Easy Maintenance.* Gulf Professional Publishing.

Do, Duy, and Schnittker, Jason. 2020. Utilization of medications with cognitive impairment side effects and the implications for older adults' cognitive function. *Journal of Aging and Health* 32(9): 1165–1177. https://doi.org/10.1177/0898264319895842

Drury, Colin G. 2000. Global quality: Linking ergonomics and production. *International Journal of Production Research* 38(17): 4007–4018. https://doi.org/10.1080/00207540050204876

Falck, Ann.-Christine, and Rosenqvist, Mikae. 2014. A model for calculation of the costs of poor assembly ergonomics (part 1). *International Journal of Industrial Ergonomics* 44(1): 140–147. https://doi.org/10.1016/j.ergon.2013.11.013

Falck, Ann-Christine, Örtengren, Ronald, and Högberg, Dan. 2010. The impact of poor assembly ergonomics on product quality: A cost–benefit analysis in car manufacturing. *Human Factors and Ergonomics in Manufacturing and Service Industries* 20(1): 24–41. https://doi.org/10.1002/hfm.20172

Frizelle, Gerald, and Woodcock, Eric. 1995. Measuring complexity as an aid to developing operational strategy. *International Journal of Operations and Production Management* 15(5): 26–39. https://doi.org/10.1108/01443579510083640

Geng, Junhao, Zhang, Sumei, and Yang, Bin. 2015. A publishing method of lightweight three-dimensional assembly instruction for complex products. *Journal of Computing and Information Science in Engineering* 15(3): 031004. https://doi.org/10.1115/1.4029753

Gupta, Shaman, and Jain, Sanjiv Kumar. 2013. A literature review of lean manufacturing. *International Journal of Management Science and Engineering Management* 8(4): 241–249. https://doi.org/10.1080/17509653.2013.825074

Hagan, R., Lichtner, D., and Senesac, C. 2009. Three-dimensional model based manufacturing work instructions. *65th Annual Forum- American Helicopter Society,* TX, United States.

Harada, Caroline N., Natelson Love, Marissa C., and Triebel, Kristen L. 2013. Normal cognitive aging. *Clinics in Geriatric Medicine* 29(4): 737–752. https://doi.org/10.1016/j.cger.2013.07.002

Judt, David, Lawson, Craig, and Lockett, Helen. 2020. Experimental investigation into aircraft system manual assembly performance under varying structural component orientations. *Proceedings of the Institution of Mechanical Engineers, Part B: Journal of Engineering Manufacture* 234(4): 840–855. https://doi.org/10.1177/0954405419883047

Kent, Robin. 2016. Chapter 8- Design quality management. In *Quality Management in Plastics Processing*, edited by R. Kent. Elsevier. https://doi.org/10.1016/B978-0-08-102082-1.50008-3

Kirwan, Barry. 1994. *A Guide to Practical Human Reliability Assessment.* CRC Press.

Kolus, Ahmet, Wells, Richard, and Neumann, Patrick. 2018. Production quality and human factors engineering: A systematic review and theoretical framework. *Applied Ergonomics* 73: 55–89. https://doi.org/10.1016/j.apergo.2018.05.010

Li, Dan, Mattsson, Sandra, Salunkhe, Omkar, Fast-Berglund, Åsa, Skoogh, Anders, and Broberg, Jasper. 2018. Effects of information content in work instructions for operator performance. *Procedia Manufacturing* 25: 628–635. https://doi.org/10.1016/j.promfg.2018.06.092

Lotter, Bruno, and Wiendahl, Hans. 2009. Changeable and reconfigurable assembly systems. In *Changeable and Reconfigurable Manufacturing Systems*, edited by H. A. El Maraghy. Springer London.

Mattsson, S., Fast-Berglund, Å., and Li, D. 2016. Evaluation of guidelines for assembly instructions. *IFAC-Papers OnLine* 49(12): 209–214. https://doi.org/10.1016/j.ifacol.2016.07.598

Mattsson, Sandra, Li, D., and Fast-Berglund, Åsa. 2018. Application of design principles for assembly instructions – evaluation of practitioner use. *Procedia CIRP* 76: 42–47. https://doi.org/10.1016/j.procir.2018.02.011

Medbo, Lars. 2003. Assembly work execution and materials kit functionality in parallel flow assembly systems. *International Journal of Industrial Ergonomics* 31(4): 263–281. https://doi.org/10.1016/S0169-8141(02)00220-2

Miller, James C. 2008. Fundamentals of shiftwork scheduling. *Ergonomics in Design: The Quarterly of Human Factors Applications* 16(3): 13–17. https://doi.org/10.1518/106480408x333109

Mital, Anil, Desai, Anoop, Subramanian, Anand, and Mital, A. 2014. Designing for assembly and disassembly. In *Product Development (Second Edition)* edited by A. Mital, A. Desai, A. Subramanian, and A. Mital. Elsevier. https://doi.org/10.1016/B978-0-12-799945-6.00007-7

Montgomery, D. C. 2020. *Introduction to Statistical Quality Control*. John Wiley and Sons
Nadeau, Sylvie, and Landau, K. 2018. Utility, advantages and challenges of digital technologies in the manufacturing sector. *Ergonomics International Journal* 2(6): 1–15. https://doi.org/10.23880/eoij-16000188.
Rasmussen, Jens. 1982. Human errors. A taxonomy for describing human malfunction in industrial installations. *Journal of Occupational Accidents* 4(2): 311–333. https://doi.org/10.1016/0376-6349(82)90041-4
Rasmussen, Jens. 1983. Skills, rules, and knowledge; signals, signs, and symbols, and other distinctions in human performance models. *IEEE Transactions on Systems, Man, and Cybernetics* 13(3): 257–266. https://doi.org/10.1109/TSMC.1983.6313160
Real, Carlos, and Onieva, Luis. 2011. Mejora de la productividad a través de la reducción de la variabilidad de las entradas en procesos de montaje manual. *Dirección y Organización* 43: 42–50. https://doi.org/10.37610/dyo.v0i43.361
Real, Carlos, and Torres, Yaniel. 2024. Effect of changes in the sequence of assembly operations on error rates: A case study from the car manufacturing industry. *IEEE Access* 12: 34644–34655. https://doi.org/10.1109/ACCESS.2024.3371980
Reason, J. 1990. *Human Error*. Cambridge University Press.
Richardson, Miles, Jones, Gary, and Torrance, Mark. 2004. Identifying the task variables that influence perceived object assembly complexity. *Ergonomics* 47(9): 945–964. https://doi.org/10.1080/00140130410001686339
Rodríguez-Toro, C., Jared, G. E. M., and Swift, K. G. 2004. Product-development complexity metrics: A framework for proactive-DFA implementation. *8th International Design Conference - Design 2004*, Dubrovnik, Croatia.
Romero, D., Mattsson, S., Fast-Berglund, Å., Wuest, T., Gorecky, D., and Stahre, J. 2018. Digitalizing Occupational Health, Safety and Productivity for the Operator 4.0. *International Conference on Advances in Production Management Systems (APMS)*, Seoul, South Korea.
Romero, David, Stahre, Jahan, Wuest, Thorsten, et al. 2016. Towards an operator 4.0 typology: A human-centric perspective on the fourth industrial revolution technologies. *46th International Conference on Computers and Industrial Engineering 2016 (*CIE46*)*, Tianjin, China. https://doi.org/10.1007/978-3-319-99707-0_59
Rosenqvist, Mikael, Falck, Ann -Christin, Lars, Lindkvist, and Söderberg, Rikard. 2014. Geometrical robustness analysis considering manual assembly complexity. *Procedia CIRP* 23: 98–103. https://doi.org/10.1016/j.procir.2014.10.072
Senders, John. W., and Moray, Neville. P. 1991. *Human Error: Cause, Prediction, and Reduction*. CRC Press.
Sharit, Joseph. 2012. Human Error and human reliability analysis. In *Handbook of Human Factors and Ergonomics*. https://doi.org/doi:10.1002/9781118131350.ch26
Snyder, Hannah. R. 2013. Major depressive disorder is associated with broad impairments on neuropsychological measures of executive function: A meta-analysis and review. *Psychological Bulletin* 139(1): 81–132. https://doi.org/10.1037/a0028727
Song, Peige, Zha, Mingming, Yang, Qingwen, Zhang, Yan, Li, Xue, and Rudan, Igor. 2021. The prevalence of adult attention-deficit hyperactivity disorder: A global systematic review and meta-analysis. *Journal of Global Health* 11: 04009. https://doi.org/10.7189/jogh.11.04009
Spencer, M., Robertson, K., and Folkard, S. 2006. The development of a fatigue/risk index for shiftworkers. Research Report RR446. Health and Safety Executive, Sudbury. Available at: https://webarchive.nationalarchives.gov.uk/ukgwa/20241206191010/https://www.hse.gov.uk/research/rrhtm/rr446.htm
Su, Qiang, Liu, Lei, and Whitney, Daniel E. 2010. A systematic study of the prediction model for operator-induced assembly defects based on assembly complexity factors. *IEEE Transactions on Systems, Man, and Cybernetics - Part A: Systems and Humans* 40(1): 107–120. https://doi.org/10.1109/TSMCA.2009.2033030
Swain, A. D., and Guttmann, H. E. 1983. *Handbook of Human-Reliability Analysis with Emphasis on Nuclear Power Plant Applications*. Sandia National Labs. https://doi.org/10.2172/5752058
Swift, K. G., and Booker, Julian D. 2013. Assembly systems. In *Manufacturing Process Selection Handbook* edited by K. G. Swift and J. D. Booker. Butterworth-Heinemann. https://doi.org/10.1016/B978-0-08-099360-7.00010-0
Torres, Yaniel, Nadeau, Sylvie, and Landau, Kurt. 2021a. Classification and quantification of human error in manufacturing: A case study in complex manual assembly. *Applied Sciences* 11(2): 749. https://doi.org/10.3390/app11020749
Torres, Yaniel, Nadeau, Silvie, and Landau, Kurt. 2021c. Evaluation of shift work scheduling and overtime allocation practices using the HSE Fatigue Index: Application on complex manual assembly. *Kongress der Gesellschaft für Arbeitswissenschaft*, Bochum, Germany.

Wiehler, Antonius, Branzoli, Francesca, Adanyeguh, Isaac, Mochel, Fanny, and Pessiglione, Mathias. 2022. A neuro-metabolic account of why daylong cognitive work alters the control of economic decisions. *Current Biology* 32(16): 3564–3575. https://doi.org/10.1016/j.cub.2022.07.010

Williams, Jeremy C., and Bell, Julie L. 2015. Consolidation of the error producing conditions used in the human error assessment and reduction technique (heart). *Safety and Reliability* 35(3): 26–76. https://doi.org/10.1080/09617353.2015.11691047

BankWorld. 2023. *Manufacturing, Value Added (% of GDP).* World Bank Group, December 11. https://data.worldbank.org

Zare, Mohsen, Croq, Michel, Hossein-Arabi, Farhard, Brunet, Rene, and Roquelaure, Y. 2016. Does ergonomics improve product quality and reduce costs? A review article. *Human Factors and Ergonomics in Manufacturing and Service Industries* 26(2) 205–223. https://doi.org/10.1002/hfm.20623

23 Informal Work of Indigenous Women

Characterization of Street Vending in Quito

Caridad González-Maldonado

23.1 INTRODUCTION

23.1.1 Indigenous People and Agriculture in Ecuador

In Ecuador, according to data from the Secretariat for the Management and Development of Peoples and Nationalities, 7.7% of the total population self-identifies as indigenous, equivalent to 1,301,887 people in 2022 (SGDPN 2023). These data are contextualized within a national population of 16,938,986 inhabitants registered, for the same year, by the National Institute of Statistics and Census (INEC 2022). Such data reflect the demographic proportion of the indigenous population relative to the country's total population, highlighting its numerical significance and its presence within Ecuador's population structure. Of this indigenous population, women represent a slight majority at 51%, compared to 49% of men, indicating a relatively balanced gender distribution. This proportion is also part of a historical pattern of predominantly rural settlements, at 73.9% (SGDPN 2023), which reflects the need for access to land for production, linked to a collective subsistence economy based on agricultural activities.

This geographic and occupational distribution reflects an adaptation strategy related to traditions, customs, and ancestral practices that have characterized over time more than 18 indigenous peoples and 14 indigenous nationalities (CODENPE 2024), which are constitutionally recognized in the country (Constitution of the Republic of Ecuador 2008), and find in agriculture a mode of social, cultural, and economic survival (PNUD 2008). According to various authors (Houtart and Laforge 2016; Herrera 2021), indigenous agriculture falls within the category of Peasant Family Farming (AFC), a productive model characterized by high labor intensity and a focus on respect for natural cycles. The main objective of this system is to ensure family and community food security, prioritizing local and sustainable consumption. Based on data from the Food and Agriculture Organization of the United Nations (FAO 2023), the AFC contributes approximately 64% of Ecuador's agricultural production and supplies 60% of the food consumed locally. These indicators show the interdependence between indigenous ancestral practices and national food security, highlighting the value of these collectivities in Ecuador's economic productivity.

23.1.2 The Role of Women in the Informal Trade of Agricultural Products: Metropolitan District of Quito-DMQ

Ecuadorian indigenous women are key figures in agricultural production. Historically, they have maintained an intimate relationship with "Pachamama," an ancestral concept from the Andean worldview that understands the Earth as a sacred maternal entity, which guides their economic responsibilities in harmony with the preservation of their territory (Torres 2015). However, this

DOI: 10.1201/9781003582014-28

fundamental role is accompanied by a marked labor precariousness, characterized by scarce salary recognition. According to data from the Ministry of Agriculture and Livestock of Ecuador, of the women linked to the AFC, 523,789 do not receive any remuneration, while only 603 have a permanent salary (MAG 2023). This disparity reflects a significant gap in the economic recognition of women's work in the agricultural sector, which translates into high poverty rates at the individual level and in their households. These conditions are aggravated by the persistently low educational levels of these communities or, in many cases, the total absence of access to education (Macías et al. 2025), factors that limit their employment opportunities and perpetuate cycles of marginalization. In this way, the indigenous woman also assumes a participation in commercial roles, preserving the autonomy to set the prices of their products, an aspect that has traditionally been influenced by the participation of intermediaries in the processes of distribution and agricultural sales (Houtart 2014). This phenomenon is associated, then, with a process of indigenous migration from the periphery to the centrality, that is, from rural areas to urban areas, motivated by the search for better living conditions and proximity to spaces of mass consumption. In the Metropolitan District of Quito (DMQ), internal mobility has resulted in an indigenous population of 80,829 people in 2022, representing 4% of the total population living in the capital (Council for the Protection of Rights 2024).

In this context, street vending of agricultural products becomes a key strategy for income generation and family subsistence. This activity not only reflects their adaptation to urban environments but also their ability to maintain sustainable agricultural practices, valued for their alignment with healthy consumption preferences and in contrast to agro-industrial models based on the intensive use of chemical inputs with adverse environmental impacts and equally harmful to human health.

The significant increase in street vending within the capital's public space has established indigenous women as an archetype in the marketing of healthy and economically affordable products. However, this transition into the urban environment exposes them to persistent vulnerabilities. Among the most prominent are the lack of social protection and exposure to ergonomic risks associated with long work hours, manual handling of heavy loads, and repetitive movements (Ministerio del Trabajo 2024), conditions that are exacerbated within the context of labor precariousness and informality (International Labour Organization 2020).

Consequently, this research aims to characterize the labor activity of indigenous women involved in street vending of fruits and vegetables in Quito, based on their life experiences and with particular consideration of human factors and ergonomic risks. To this end, this study is guided by the following research questions: What ergonomic risks do you face at work? What are the main socio-technical elements, understood as a system with its own inherent logic, specific to both the activity and informality, where tools, spaces, and social practices are intertwined (International Labour Organization 2018), that are involved in their work dynamics within the urban context? What is the business logic they employ? What problematic nodes can be identified?

23.2 METHODOLOGY

23.2.1 MATERIALS AND METHODS

This research was framed within a qualitative design and employed an exploratory, descriptive methodological approach. It focused on the analysis of street vending of agricultural products by indigenous women in public spaces in the DMQ, Ecuador, in 2024. The research sample was intentionally selected and comprised 10 indigenous women who reported proficiency in Spanish, which facilitated data collection by minimizing language barriers. All participants were street vendors in the El Inca sector, located in the north of the city on an avenue of the same name. This sector was chosen for its relevance as a space for socioeconomic interaction, its high presence of indigenous people engaged in commerce, and its official designation as one of the 38 critical points of informal street vending in the city (El Comercio 2019). It is characterized as a linear commercial axis of approximately 1.0 km that connects the urban area with rural areas of the DMQ, generating constant vehicular and pedestrian traffic.

In this context, it was observed that indigenous women have predominantly engaged in the sale of agricultural products on public roads under informal conditions, lacking regulation and protection under established legal frameworks. It is crucial to note that in Ecuador, formality is mainly governed by a tax logic tied to the payment of taxes, which does not necessarily guarantee access to basic social protection, such as health care or retirement. This situation accentuates the structural precariousness in which these workers operate, despite their significant role in the local economy.

Verbal informed consent was obtained for the research. This method was deemed most appropriate given that several participants had low levels of education and were wary of signing documents. Each participant was clearly informed of the purpose of the study, the voluntary nature of their participation, their right to withdraw at any time, and the confidentiality of their data.

23.2.2 Data Collection and Triangulation Strategy

Data collection was based on a triangulation strategy using different data sources (Flick, 2018), through the combination of two complementary qualitative techniques that allowed for a comprehensive approach toward the research phenomenon: (a) systematic nonparticipant observation and (b) interviews based on micro life stories. Figure 23.1 shows the triangulation strategy and how the techniques complement each other.

Systematic nonparticipant observation was conducted using two recording instruments: field notes and documentary photography. This technique enabled systematic documentation of the social, spatial, and commercial dynamics of vendors in their work environment. The use of photography played a fundamental role in capturing nonverbal aspects: gestures, expressions, and spatial arrangement, which enrich the understanding of commercial practices that are often invisible (Bonetto 2016). This approach was particularly relevant given the communicative nature of the

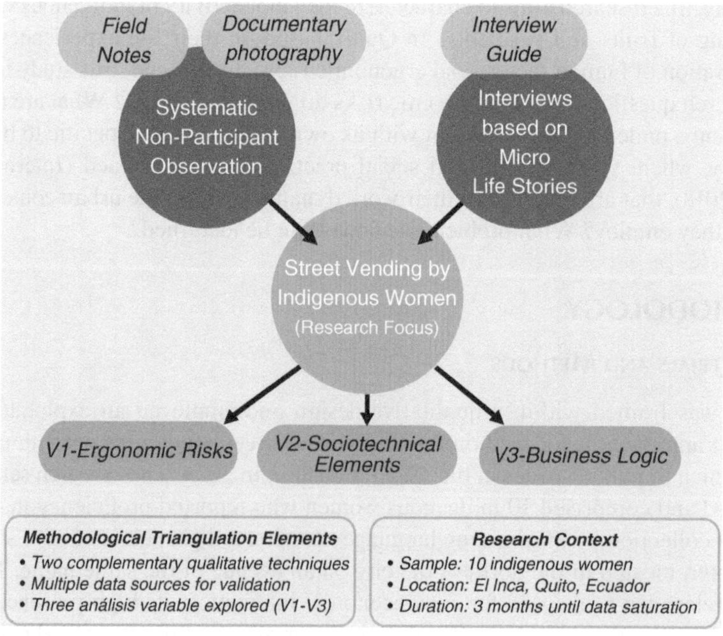

FIGURE 23.1 Research methodology design.

vendors, who predominantly use indigenous languages, making images a key resource for accessing nondiscursive dimensions during their work activity.

The interviews were based on the micro life stories approach (Moriña 2017), whose value lies in its ability to delve into significant fragments of everyday life through the participants' narratives. This technique provided deep insights into work dynamics and sociotechnical aspects usually overlooked by quantitative methods. The interviews were conducted through direct engagement in the work environment. Three visits were scheduled at different time points and on different days, including weekdays and holidays, to capture contextual variations. Each session was guided by a semi-structured interview guide designed to explore the central variables of the research: (a) ergonomic risks, (b) sociotechnical elements, and (c) business logic.

The data collection process was carried out over 3 months, until saturation of information was reached on the topics associated with the variables of analysis. To ensure the credibility and robustness of the findings, a strategy of methodological rigor was implemented through triangulation of data sources. This process was operationalized through a systematic comparison between interview narratives, contextual observation notes, and visual evidence from photographs, which allowed for validation of the consistency of the topics and reduced the risk of biased interpretations. Together, these strategies strengthened the validity of the research and enriched the depth of the analysis.

The following methodological limitations are acknowledged: The small sample size and intentional selection limit the statistical generalization of the findings, although they are adequate for achieving an in-depth analysis of the experiences of the activities under investigation. Likewise, caution is required when transferring the results to other urban contexts, given that they reflect particular conditions in the El Inca sector that may not be representative of other realities. Interpretative subjectivity was managed through methodological transparency strategies: reflective records of the researcher's personal impressions, emotions, and assumptions during the process were incorporated into the field notes. The systematic triangulation of sources allowed for the comparison of interpretations, the mitigation of biases, and the highlighting of the voices and experiences of the participants. Although these limitations are inherent to qualitative studies, it is considered that the strategies described, while not avoiding subjectivity, make it manageable, visible, and productive for analysis.

23.3 RESULTS

In terms of sociodemographic characteristics, the 10 indigenous women who participated in the research presented a profile that reflects structural vulnerability. The average age was 51, ranging from 42 to 60, indicating that commercial activity related to the sale of fruits and vegetables within the group and context studied is concentrated among middle-aged women. In terms of education, 80% were illiterate and 20% had only completed primary education, which limits their opportunities for access to the formal labor market.

Regarding marital status, 90% of the participants were married or in a stable *de facto* union. The remaining 10% were single women who formed an intergenerational household with their youngest child and their mother. All participants came from the provinces of Chimborazo and Imbabura and self-identified as indigenous, falling within the residual category of self-employed informal workers according to the Ecuadorian national classification (INEC 2015). Regarding job diversification, 60% were exclusively engaged in informal street vending, while 40% supplemented this activity by supporting their spouses in construction work, also informal. All of the participants were also responsible for unpaid domestic work, which created a situation of double work presence that required strategies to reconcile commercial work and daily family responsibilities. The results are grouped by analysis variables in Table 23.1 and are further detailed below.

TABLE 23.1
Data Triangulation by Research Variables

Variable	Research Question	Interviews (Micro Life Stories)	Nonparticipant Observation	Documentary Photography	Findings
Ergonomic Risks	What ergonomic risks do you face at work?	Narratives of: Physical pain Working hours Handling loads Risk normalization	Recording of: Body postures Working hours Uneven surfaces Environmental conditions	Images of: Heavy loads Work postures Footwear and clothing Conditions of public spaces	Working: 8–10 hours per day Loads up to 7.5 kg per hand Dorsolumbar conditions Inflammation in the lower limbs
Sociotechnical Elements	What are the main sociotechnical elements in their work dynamics?	Organizational urban and adaptation strategies Use of the body as a tool Work cycles	Light-synced work Trees Interaction Spatial mobility patterns	Improvised tools Use of boxes and buckets as furniture Use of public space. Mobile displays	Traffic integration Multifunctionality of tools Appropriation of space Social innovation
Business Logic	What is the business logic they employ?	Descriptions of sales strategies Familiar work patterns Differentiated roles Economic income	Mixed strategies (fixed–mobile) Participation of minors Family role assignment Strategic location	Documented product organization Own vehicles as points of sale Family involvement Packaging strategies	Uniform price of $1 USD Income of $390 USD per month Cross-generational work Survival strategies

23.3.1 About Ergonomic Risks and the Demands of the Activity

Commercial activity at the Inca takes place from Tuesday to Sunday between 9:30 a.m. and 6:30 p.m., with working hours ranging from 8 to 10 hours per day throughout the year. Mondays are reserved for product supply. Work intensity fluctuates depending on the achievement of the minimum daily sales target of USD 15, intensifying on weekends, with peaks between 1:00 p.m. and 4:00 p.m. According to the women, the average income amounts to $390 USD per month, which is 17% less than Ecuador's Unified Basic Salary (SBU) for 2025, set at $470 USD. During traditional festivities that coincide with national holidays such as Christmas, New Year's, Holy Week, Mother's Day, and All Souls' Day, the daily income reaches up to $30 USD. In terms of ergonomic risks, daily handling of sustained and repeated loads weighing up to 7.5 kg per hand was documented. This weight is distributed in 2.5 kg grips per finger, according to the load of fruit packed in plastic mesh bags, excluding the little finger and thumb. Simultaneous frontal elevation of the upper limbs was observed, with both arms at shoulder height and always with the hands in a supine position, increasing the risk of musculoskeletal disorders in saleswomen with height between 1.50 and 1.62 meters. In the case of one of the saleswomen, the aggravating factor of simultaneously carrying and moving a baby on her back during her working day was recorded. The health conditions reported included back, shoulder, wrist, and hand pain. In 60% of the participants who wore flat school-type shoes and anaco (traditional indigenous skirt), inflammation was observed in both feet. Of these women,

80% indicate that because they depend on daily work to secure their income and lack affiliation with the social security system due to their informal status, their ailments remain without timely medical attention, forcing them to prolong their suffering and resort to palliative solutions such as traditional practices related to the use of natural medicine. The remaining 20%, who are older, do visit free municipal health services when their ailments are acute and prevent them from working. The participants expressed normalization of precarious conditions through recurring phrases that reflected habituation to work demands, such as "I just ask God to bring me customers," "I just want to be given the opportunity to work," "I'm used to carrying things," "I can't let the pain get the better of me," and "Nothing happens, ceñito [a colloquial diminutive for *señora*, meaning 'lady']," the latter expression referring to the risk of selling on the street.

23.3.2 About the Sociotechnical Elements

Commercial traffic is recorded in areas of high vehicular traffic on the avenue and on unstable urban surfaces. Tree roots in the area cause cracks in the pavement and raised surfaces on the sidewalks, increasing the risk of accidents due to tripping. It has been found that work cycles are synchronized with traffic light changes, as indicated in Figure 23.2. Women walk between 30 and 40 m while displaying their products between cars, with their arms raised for greater visibility. When the light is red, they make their offers and collect payment, and while when the light is green, they return to their initial location where unsold products are stored and packed. This dynamic generates permanent exposure to vehicular traffic and combustion gases, without access to basic services.

Products are sold in plastic mesh bags containing between 4 and 25 units, depending on the type: avocados, corn cobs, pitahayas, and papayas in packs of four; apples in groups of six; strawberries by the kilogram; and naranjillas, tangerines, tomatoes, lemons, garlic, onions, and potatoes in dozens of 12 to 24. The low-price strategy sets a uniform cost of USD 1 for all packages, facilitating quick transactions. The body constitutes the primary work tool: arms, hands, and legs are engaged in mobility and load-carrying; the raised voice functions as a resource for direct sales; and traditional indigenous clothing operates as a symbolic marker of commercial trust. Urban trees play an essential role as environmental protection elements, providing shade against intense solar radiation and shelter from frequent rainfall. This function is particularly relevant in Quito, where an altitude of over 2,800 meters exposes the population to high levels of ultraviolet radiation and to abrupt climatic variations that, within a single day, may alternate between intense Sun, rain, and temperature drops. In the workspace, wooden boxes, plastic crates, and empty buckets perform three functions: they serve as product containers, as low tables for packing, and as improvised seating during breaks, as shown in Figure 23.3. According to the vendors, these implements are repurposed from

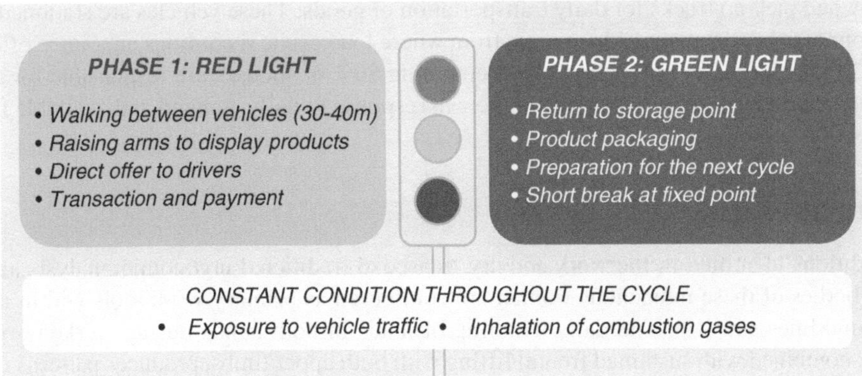

FIGURE 23.2 Work cycle synchronized with traffic lights.

FIGURE 23.3 Improvised tools for work.

the household, which prevents additional expenses. At the same time, their use facilitates spatial optimization in transport vehicles such as taxis and pick-up trucks for moving goods and reduces the number of objects required in the narrow sidewalks of the area, where limited space demands an efficient management of the sales post.

23.3.3 About Business Logic

The commercial strategy can be characterized as mixed, combining both fixed and mobile sales, with family members collaborating in most cases. Fixed selling points are strategically located along blind walls without doors or windows, thus avoiding direct conflict with formal, regulated businesses. This modality responds to a sedentary and proximity-based logic, mainly targeting local pedestrians. In contrast, street vending operates along the main road, oriented toward a mobile clientele such as drivers of cars, trucks, and motorcycles, thereby adapting to vehicular circulation flows. The activity reflects a clear intergenerational participation: children between 8 and 12 years of age are involved at least 4 out of 6 working days per week, with a more regular presence on weekends and holidays, when higher demand requires greater family support. Male household members generally assume roles different from those in direct sales. Around 20% act as drivers, using mainly family-owned pick-up trucks for daily transportation of goods. These vehicles are stationed on side streets connected to the main El Inca road, from where loud cyclic recordings announce offers such as "Get 20 mandarins for one dollar." The remaining 80% of spouses are responsible for continuous supply, purchasing from wholesale markets in response to daily demand and available financial resources.

23.4 DISCUSSION

The conditions identified in the work activity transcend traditional ergonomic analysis and show how the bodies of these indigenous women are configured as primary work tools within a system that commodifies survival. The daily handling of loads in both hands during workdays of up to 10 hours, combined with sustained frontal lifting with both upper limbs, produces patterns of physical overload that are intensified by the exclusion of these women from effective healthcare services. The records on work postures and load handling coincide with the research by Samiksha et al.

(2021) on street vendors in India, who identified ergonomic risks associated with frontal lifting of upper limbs and prolonged manual loads. Their results establish that these work postures generate musculoskeletal disorders that may be comparable regardless of the geographical context and suggest that universal patterns of risk in street vending transcend cultural particularities. Simultaneous exposure to multiple biomechanical, environmental, and psychosocial risk factors generates synergistic effects that amplify health inequalities in this population. The interaction of these multiple health determinants allows us to understand how the conditions identified in El Inca operate as part of a broader system of bodily vulnerabilities that disproportionately affects socially excluded populations.

The normalization of precariousness expressed in the phrases recorded from the interviews reveals psychosocial processes of adaptation that transcend simple resignation, shaping complex strategies that combine subordination and resistance. Subordination manifests itself in adaptation to precarious conditions and remains institutionally invisible in order not to lose work spaces. Resistance takes place in the collective appropriation of public spaces to secure economic income, where vendors create a socially legitimized territory that attracts customers due to the concentration of products located and segmented spatially in the sector, ensuring the sustainability of their activity, a situation that coincides with that reported by Martínez and Short (2022) in their analysis of informality in Cali, Colombia.

This dynamic coincides with patterns documented by Silva-Peñaherrera et al. (2022) in Peruvian informal workers, where 73% of the informal workforce develops cognitive normalization mechanisms that function as psychological buffers in the absence of government protection. López-Ruiz et al. (2017) add that in Spanish-speaking Central American countries, these processes operate differently by gender, with women developing more elaborate strategies to reconcile productive work and household tasks.

Synchronizing work with traffic light cycles represents a sophisticated form of urban knowledge that transcends individual adaptation to shape collective knowledge that enables street vendors to function effectively in the city. This practice demonstrates what Castillo and López (2023) identify as "social innovation" among informal traders in the Sinincay parish in Cuenca, Ecuador, in the form of creative solutions that emerge from the margins of the formal economic system to resolve structural problems of labor exclusion. The multifunctional use of reused wooden boxes, plastic crates, and empty buckets as product containers, improvised tables, and temporary seats demonstrates capabilities for spatial optimization and efficient management of limited resources. The strategy highlights a challenge to the limitations imposed by the lack of capital and the absence of formal infrastructure. The multifunctionality of these implements as containers and seats reflects what is conceptualized in this research as "resilience technology," recognizing that these women innovate and actively adapt to their circumstances, where constant need generates creative solutions that maximize the usefulness of minimal resources.

In terms of the appropriation of public space, this research reveals processes of territorialization that challenge traditional conceptions of the use and function of the urban environment. The points of sale located next to blank walls and the synchronization of work with traffic light changes are examples of how these women develop geographies of survival by mapping territory according to their needs, allowing them to operate in parallel to formal urban planning. The protective function of urban trees against intense solar radiation and frequent rainfall, particularly relevant in Quito, establishes how vendors integrate natural elements as basic work infrastructure. This adaptation reflects local environmental knowledge that compensates for the absence of urban-planning services.

The mixed business logic that combines fixed sales spaces with street vending and family participation reveals an adaptive commercial organization that maximizes market opportunities within deep structural constraints. Fixed points are strategically located next to blind walls without doors or windows, avoiding direct conflict with regulated formal businesses, while street vending is geared toward passing customers, mainly vehicle drivers.

Intergenerational participation in work activities among children aged 8 to 12, according to the group analyzed, although on specific dates and on days of high demand, reveals mechanisms of knowledge transmission that can be interpreted as family survival strategies but which perpetuate patterns of educational exclusion and reproduction of informality. Likewise, the observed sexual division of labor reveals how patriarchal structures operate in subtle but effective ways to maintain gender hierarchies even within family strategies for collective survival. While men control the links in the production chain with greater capital and security, involving transportation and wholesale supply at specific time points, women take on more vulnerable activities with direct exposure to risk in public spaces and on a sustained basis throughout the working day. This family-based organization follows patterns of subordination in which women's bodies become the main economic asset on display, assuming both the physical risks and the responsibility for the success or failure of daily sales. As UN Women (2022) points out, gender inequalities are structural, social, and institutional in nature, historically linked to the sexual division of labor and the stereotypes and biases that establish behaviors considered appropriate according to gender. The apparent complementarity of roles in informal work conceals a significant asymmetry in terms of formality and informality. Men maintain a certain degree of institutional protection by operating registered vehicles and conducting transactions in formal markets, while women face constant criminalization for their unregulated presence when carrying out their commercial activities in public spaces.

The uniform pricing strategy reveals a paradox of informal work: by simplifying transactions, these women sacrifice potential income for immediate survival. By setting all fruit and vegetable packages at $1 USD, these indigenous women consciously sacrifice incremental profitability in favor of agile transactions. It is understood that the immediate need for liquidity and daily survival forces them to forego more elaborate commercial strategies, keeping them in a cycle that prevents them from accumulating capital to diversify and improve their economic or commercial position. The fact that their income remains consistently below the minimum wage reflects how the Ecuadorian state externalizes the social costs of sustaining family and urban life to indigenous family economies, thus allowing these women to indirectly subsidize the cost of living in the DMQ by offering fresh food at prices below the real value. The higher income earned during the holidays represents temporary prosperity, accompanied by a seasonal demand that allows vendors to briefly earn a decent income, revealing the economic potential that the formal system systematically denies them.

The intersectional analysis of the conditions identified reveals how gender, ethnicity, social class, level of education, and age operate as a system of interconnected factors that produce specific forms of labor vulnerability. The convergence of these factors, identified through the methodological triangulation developed, allows us to understand the living conditions of these indigenous women as concrete expressions of inequalities and social exclusion that manifest themselves in the urban space of the El Inca sector.

This research concludes that the precarious working conditions among indigenous informal vendors constitute a form of systematic exclusion. This is reflected in the lack of access to labor rights, social security, and official recognition under municipal regulations, which perpetuates and exacerbates the structural vulnerabilities these women face on a daily basis in their work. It is imperative to enable regulatory frameworks that transcend the formal–informal dichotomy, recognizing the specificities of the survival work carried out by indigenous women. The documented capacities for social innovation demonstrate potentialities that can be harnessed by public policies aimed at facilitating transitions toward more dignified working conditions.

The ergonomic and sociotechnical findings also reveal important opportunities to design culturally appropriate solutions that improve working conditions without disregarding existing organizational logic. The development of adapted ergonomic tools, such as modular containers or displays that optimize weight distribution, load-bearing systems or supports that reduce the permanent forward lifting of upper limbs, or specialized footwear for uneven surfaces, could significantly mitigate the documented musculoskeletal risks. Likewise, the design of urban furniture that respects identified patterns of spatial appropriation or adapts to natural elements, together with clothing that integrates Sun and rain protection while maintaining the cultural identity, could represent feasible interventions. These proposals should emerge from participatory processes that recognize vendors as experts in their own work, ensuring that design solutions strengthen their existing social innovation processes and resilience strategies. This approach from the field of design and creative practices opens up important lines of applied research.

In relation to the identified lack of basic social protection, social security systems need to be developed to guarantee access to health care regardless of the employment status. This implies recognizing informal work as a legitimate form of economic activity that requires specific regulation aimed at protecting fundamental labor rights.

The limitations of the research relate to the intentional sample size and the contextual specificity of the El Inca sector, which implies caution in generalizing the results. However, the triangulation strategy implemented strengthens the internal validity of the findings and allows for the identification of consistent patterns that can guide future research in similar urban contexts and populations. The main contribution of this study lies in the systematic documentation of the specific working conditions of indigenous women in street vending, a population that has been historically overlooked in occupational health studies. The findings provide empirical evidence that can guide public policies on labor inclusion and social protection differentiated by gender and ethnicity, as well as fuel academic debates on the need for intersectional approaches in the analysis of informal urban work.

REFERENCES

Bonetto, María José. 2016. "El uso de la fotografía en la investigación social." *Relmis. Revista Latinoamericana de Metodología de la Investigación Social* 11 (6): 71–83. https://www.relmis.com.ar/ojs/index.php/relmis/article/view/81

Castillo, Yonimiler, and López, Bryam Stalin. 2023. "Boas práticas de inovação social em comerciantes informais da paróquia de Sinincay, cantão de Cuenca, Equador." *SciELO Preprints*. https://doi.org/10.1590/SciELOPreprints.6023

CODENPE (Consejo de Nacionalidades y Pueblos del Ecuador). 2024. *Pueblos y Nacionalidades Indígenas del Ecuador*. Quito: CODENPE. https://www.goraymi.com/es-ec/ecuador/culturas-nacionalidades/pueblos-nacionalidades-indigenas-ecuador-a0ubmq0jf

Consejo de Protección de Derechos (CPD). 2024. *Documento ayuda. Pueblos y nacionalidades indígenas en el Distrito Metropolitano de Quito*. Quito: CPD. https://protecciondederechosquito.gob.ec/wp-content/uploads/2024/03/Informacion-Estadistica-PIndigena-DMQ.pdf

Constitución de la República del Ecuador. 2008. Asamblea Constituyente. *Registro Oficial 449*. https://www.asambleanacional.gob.ec/es/constitucion

El Comercio. 2019. "En 38 puntos de Quito proliferan las ventas ambulantes en espacio público." *El Comercio*, 17 de diciembre de 2019. https://www.elcomercio.com/actualidad/quito/quito-ventas-ambulantes-controles-informalidad.html

FAO. 2023. *Ecuador en una mirada*. Quito: FAO. https://www.fao.org/ecuador/fao-en-ecuador/ecuador-en-una-mirada/es/

Flick, Uwe. 2018. *An Introduction to Qualitative Research*. 6th ed. Los Angeles: SAGE.

Herrera, Santiago. 2021. *La agricultura familiar campesina: Un modelo de agricultura con la potencia para cambiar la vida*. Quito: Instituto de Estudios Ecuatorianos. https://ocaru.org.ec/la-agricultura-familiar-campesina-un-modelo-de-agricultura-con-la-potencia-para-cambiar-la-vida/

Houtart, François, and Laforge, Marc. 2016. *Manifiesto para la agricultura familiar campesina e indígena en Ecuador*. Quito: Instituto de Altos Estudios Nacionales (IAEN).

Houtart, François. 2014. "El desafío de la agricultura familiar campesina en Ecuador." *Agropecuaria. Boletín Electrónico*, 2014. https://agropecuaria.org/2014/11/el-deafio-de-la-agricultura-familiar-campesina-en-ecuador/

INEC (Instituto Nacional de Estadística y Censos). 2015. *Actualización metodológica: Empleo en el sector informal y la clasificación de los ocupados según sectores*. Quito: INEC. https://www.ecuadorencifras.gob.ec/documentos/web-inec/EMPLEO/2015/Junio-2015/Metogologia_Informalidad/notatecnica.pdf

INEC (Instituto Nacional de Estadística y Censos). 2022. Censo Ecuador. Quito: INEC. https://www.censoecuador.gob.ec

International Labour Organization (ILO). 2018. *Women and Men in the Informal Economy: A Statistical Picture*. 3rd ed. Geneva: ILO.

López-Ruiz, Montserrat, Benavides, Fernando G., Vives, Alejandra and Artazcoz, Lucia. 2017. "Informal Employment, Unpaid Care Work, and Health Status in Spanish-Speaking Central American Countries: A Gender-Based Approach." *International Journal of Public Health* 62 (2): 209–18. https://doi.org/10.1007/s00038-016-0871-3

Macías, Juan, Pérez, María, Torres, Ana y Almeida, Carlos 2025. "Sector informal y pobreza 2019–2023: un análisis socioeconómico desde Ecuador." *Revista InveCom* 5 (1). https://doi.org/10.5281/zenodo.10892442

MAG (Ministerio de Agricultura y Ganadería). 2023. *Boletín Situacional. Agricultura Familiar Campesina.* Quito: MAG. https://sipa.agricultura.gob.ec

Martínez, Lina, and Short, John Rennie. 2022. "The Informal City: Exploring the Variety of the Street Vending Economy." *Sustainability* 14 (12): 7213. https://doi.org/10.3390/su14127213

Ministerio del Trabajo. 2024. *Norma Técnica en Seguridad e Higiene del Trabajo, Anexo 3.* Quito: República del Ecuador. https://www.trabajo.gob.ec/wp-content/uploads/2024/11/Anexo-3_Norma-Tecnica-de-Seguridad-e-Higiene-del-Trabajo-signed-signed-signed-signed.pdf

Moriña, Anabel. 2017. *Investigar con historias de vida: Metodología biográfico-narrativa.* Madrid: Narcea Ediciones.

OIT (Organización Internacional del Trabajo). 2020. *Informalidad laboral y género en América Latina.* Lima: OIT. https://www.ilo.org

ONU Mujeres. 2022. "El trabajo doméstico y de cuidados conforman un sector feminizado con alta informalidad y bajos salarios." Septiembre de 2022. https://lac.unwomen.org/es/stories/noticia/2022/09/el-trabajo-domestico-y-de-cuidados-conforman-un-sector-feminizado-con-alta-informalidad-y-bajos-salarios

PNUD (Programa de las Naciones Unidas para el Desarrollo). 2008. *Informe sobre desarrollo humano en Ecuador.* Quito: PNUD. https://www.undp.org

Samiksha, S., Singh, L. P., Garg, R. K. and Atri, M. 2021. "Ergonomic Evaluation of Street Vendors as Determined by Rapid Entire Body Assessment Method." *International Journal of Current Research and Review* 13 (8): 62–66. https://doi.org/10.31782/IJCRR.2021.13819

SGDPN (Secretaría de Gestión y Desarrollo de Pueblos y Nacionalidades). 2023. *Análisis preliminar Censo 2022 con enfoque en Pueblos y Nacionalidades.* Quito: SGDPN. https://www.secretariapueblosynacionalidades.gob.ec/wp-content/uploads/2023/12/Presentacion-CENSO-2022-Pueblos-y-Nacionalidades.pdf

Silva-Peñaherrera, María, Ayala-García, Andrés, Mayer, Edward A., Sabastizagal-Vela, Iván and Benavides, Fernando G. 2022. "Informal Employment, Working Conditions, and Self-Perceived Health in 3098 Peruvian Urban Workers." *International Journal of Environmental Research and Public Health* 19 (10): 6105. https://doi.org/10.3390/ijerph19106105

Torres, Sofía. 2015. "El rol de la mujer indígena en la comunidad Inti Waykupunku del cantón Otavalo – provincia e Imbabura." Tesis de grado, Pontificia Universidad Católica del Ecuador. https://repositorio.puce.edu.ec/handle/123456789/22196

24 Analysis of Laws and Regulations Related to Ergonomics and Human Factors in Latin American Countries

Aide Aracely Maldonado-Macías,
Mauricio Santos Morales, Andrés Arce Guadalupe,
Martín Rodríguez, and Daniel Lizama Álvarez

24.1 INTRODUCTION

Human factors and ergonomics (HFE) is the scientific discipline concerned with understanding the interactions among humans and other elements of a system, as well as the profession that applies theory, principles, data, and methods to design and optimize human well-being and overall system performance (International Ergonomics Association, 2000). This discipline focuses on enhancing performance and well-being by optimizing the system as an integrated whole and integrating humans into the system. This aligns with the environment, job requirements, and human characteristics (Dul et al., 2012).

Nowadays, HFE has been evolving to address multiple challenges arising from highly demanding, rapidly changing situations, leading to significant shifts in the types of work, workplaces, work environments, and work conditions worldwide. New perspectives and topics have emerged, making HFE a critical component of occupational health and safety. Among the most relevant issues is sustainable development, for which HFE can play a crucial role by contributing to industrial and product design, architecture, health and safety, and human–computer interaction (HCI) categories (Bolis et al., 2023; Radjiyev et al., 2015). Another relevant aspect is that due to the growing interest in the discipline worldwide and addressing these multiple challenges, there is a need for established formal manners of applying it by developing a legal framework constituted by laws, regulations, and decrees that may help the appropriate application of principles, standards, techniques, and methods for the benefit of humans and systems in a rapid changing world.

Accordingly, the International Association for Ergonomics (IEA) together with the International Labour Organization (ILO), recognizing the critical need for HFE principles and guidelines in the design and management of work systems, have published the principles and guidelines that are intended to serve as a technical basis for the ILO to develop an international labor standard on good workplace practices about human factors/ergonomics (*Principles and Guidelines for Human Factors/Ergonomics (HF/E) Design and Management of Work Systems*, s. f.). These guidelines highlight the unprecedented disparity of work situations, comparing the digital economy and technology with the informality of work, and expose the new challenges posed by information technology, robotics, artificial intelligence, and digitalization. Additionally, from the rise of a wide range of labor scenarios, standards, and guidelines that must address the safety, health, well-being, and

DOI: 10.1201/9781003582014-29

sustainability issues of workers from a variety of aspects, such as musculoskeletal disorders or injuries derived from physiological diseases and biomechanical, cognitive, psychological, and relational threats from old practices, the pertinence of these guidelines and the necessity of laws and regulations are emphasized even more. In addition, it must be recognized that the influence, prevalence, and adverse effects of the mentioned issues vary across countries and regions worldwide, reflecting their unique characteristics, working practices, ergonomic perspectives, and implementation, thereby affecting applicable laws, regulations, and standards.

Hence, this complexity interferes with and impacts the critical role of the discipline and the HFE specialists. According to the IEA, "Specialists in Human Factors Engineering (HFE) contribute to the planning, evaluation, conception, and evaluation of tasks, jobs, products, organizations, environments, and systems to make them compatible with the needs, capacities, and limitations of people". From this perspective, the HFE requires proper management and development at all levels of the organization to ensure its implementation. On the other hand, proper HFE management involves identifying risks, transforming work, and changing conditions that generate dissonances to adapt the job to the workers. To achieve these purposes, Dul et al. (2012) highlight three fundamental aspects that are crucial for the development of the HFE discipline and profession, as well as its implementation in organizations. These aspects consider that HFE management must adopt a systemic approach. It must be based on a design methodology, and management must prioritize human well-being and performance improvement when defining problems and formulating solutions. However, the lack of applicable laws, regulations, and standards makes HFE interventions and HFE specialists' work difficult. Without these key aspects, the HFE's focus is limited in its ability to accomplish its purposes.

Adding complexity to the difficulties stated earlier is the need to understand the intricate terminology and scope of the legal framework, including the varied laws, regulations, standards, and decrees. These terms differ, and it is important to define them properly to better understand the content and purpose of this chapter. According to the jurist Marcel Planiol, the law is an "obligatory social rule, permanently established by public authority and sanctioned by force" (Alessandri et al., 2007). The law is an obligatory social rule that posits a superior will that commands and a lower one that obeys. It is sanctioned by public force, which implies that, along with setting a rule, a sanction is imposed for noncompliance, which can be preventive or repressive, such as a fine, imprisonment, or penalty.

There is also another set of regulations that govern aspects of social life, whether originating from the executive branch, the legislative branch (parliament), or other government institutions. These can vary in their nature, origin, and scope, within which we can describe norms, decrees, decrees with the force of law, exempt resolutions, regulations, etc. In addition, a different aspect corresponds to "international treaties," which are formal agreements signed by states and are legally binding under international law. It is also important to note that a treaty is also referred to as a pact, convention, or convention (Library of the National Congress of Chile, BCN, n.d.).

Therefore, as has been established, the management of HFE and the labor of HFE specialists depend directly and/or indirectly on the legal framework of the country in which the HFE discipline is practiced. In some countries, regulations govern management and impose sanctions; in others, these are not considered, and consequently, companies are recommended to comply with them voluntarily. Consequently, according to the obligatory aspect, we can distinguish two mandatory ones: laws and decrees, and, finally, ISO standards, which are voluntary.

Accordingly, this chapter addresses the premise that revising regulations is necessary to better understand the conceptual, theoretical, and practical differences in HFE management and application across countries. Regulations reflect societal interests, including the motivations that drive their development. In this way, the regulations are the expression of the issues of interest in the communities. The worldwide interest in sustainable practices and human well-being is increasing noticeably; consequently, we argue that the maturity of HFE management and its application in developed countries, as reflected in their actual legal frameworks for the discipline, differs from that in developing countries. In addition, important differences exist among Latin American countries. These differences should be considered to propose models to compare, evaluate, and measure the progress of HFE in Latin America. A better understanding of the legal framework and the actual applicable laws, decrees, and standards in Latin American countries may contribute to this purpose.

Although advances in HFE have been made, no pre-established, success-guaranteed formulas are available for HFE management and application. Consequently, the legal framework can be considered a support for HFE specialists and companies, depending on the current regulations applicable in each case and country. Thus, every country has a legal system that establishes rules, principles, and values to regulate the organization of power, relations with citizens, and the guarantees of rights and freedoms between them, as well as to direct public policies for the benefit of the general interest, thereby constituting their regulatory framework. Each of these legal documents has a specific regulatory objective. As the articles of a constitution define general aspects, laws and regulations are much more specific. However, in Latin America, there is a heterogeneous reality, characterized by different legal systems, terms, scopes, and rigor, all of which are influenced by their jurisprudence and legal culture, thereby increasing the complexity of analyzing and comparing various regulatory formats within their respective scopes and specific terminology. For example, in some countries, manual material handling procedures (MMHP) are mandated by law, while in others, they are regulated by decree. In some others, it is only mentioned in voluntary regulations based on international standards such as ISO (International Organization for Standardization).

In this context, countries, HFE specialists, and practitioners are commonly concerned about laws and regulations related to occupational health and safety (OSH) issues (Hernández Albrecht, 2016; Niu, 2010; Soares, 2006). Depending on the corresponding regulatory framework, these issues are addressed differently across countries. That is, at the constitutional level, they are described in general terms in the articles. Then, laws are presented to regulate specific aspects, including rules, regulations, or decrees that outline how each topic or issue should be managed and supervised, along with its associated consequences. On the other hand, other voluntary local or organizational regulations collaborate in managing the HFE practice in each country. At the same time, the adoption of international standards and tools, such as ISO standards, is being considered.

The purpose of this chapter is to review and analyze the main laws, regulations, and ergonomic standards applied in Latin American countries, and to compare their scope of application to identify opportunities for the proper management and development of HFE from empirical and theoretical perspectives.

24.1.1 Fundamental Impacts of Ergonomics Laws on Workplace Safety

Several authors support the importance and impact of HFE laws on workplace safety. The following paragraphs summarize these aspects as the theoretical framework.

- *Compliance and standardization*: Ergonomics laws provide a framework for standardizing workplace practices, ensuring that organizations comply with safety regulations and reduce the risk of legal liabilities (Castaño Cardenas, 2023; Niu, 2010; Orr et al., 2024).
- *Reduction in work-related injuries*: Ergonomics laws help minimize the incidence of musculoskeletal disorders and other injuries by promoting proper manual handling techniques and ergonomic workplace designs (Niu, 2010; Putri Cahyani & Humaira Ramadhani, 2024; Stoidner et al., 2023). This is particularly important in industries with high physical demands, such as manufacturing and healthcare (Orr et al., 2024; Salawati & Abbas, 2023).
- *Improved productivity and well-being*: By ensuring workplaces are designed to meet workers' needs, ergonomics laws contribute to increased productivity, reduced fatigue, and enhanced worker satisfaction and well-being (Nedelko, 2022; Putri Cahyani & Humaira Ramadhani, 2024; Salawati & Abbas, 2023).

- *Focus on specific work environments*: Ergonomics laws often target environments such as office settings to address unique challenges, including poor posture and inadequate equipment arrangement, which can lead to health issues and reduced performance (Bailey et al., 2007; Nedelko, 2022; Stoidner et al., 2023).
- *Prevention of occupational diseases*: By addressing ergonomic risk factors, these laws help prevent occupational diseases, thereby reducing long-term health costs and enhancing employee retention (Stoidner et al., 2023).

Accordingly, laws that address HFE-related issues significantly enhance workplace safety by reducing injury risk and improving working conditions, thereby balancing worker performance and well-being. These laws and standards, often influenced by international bodies such as the ILO, aim to protect workers from ergonomic risks and promote safer working environments. The ILO's policies and activities on ergonomics contribute to occupational safety and health by fostering voluntary, participatory, and action-oriented approaches for small- and medium-sized enterprises (Niu, 2010). ILO standards, such as Convention No. 127 and Recommendation No. 128, protect workers against ergonomic risk factors in the workplace and influence the laws and regulations of member states (Niu, 2010; Orr et al., 2024).

24.2 INTERNATIONAL STANDARDS AND REGULATIONS AND THEIR APPLICATION IN LATIN AMERICAN COUNTRIES

The ISO is an organization that develops international standards voluntarily to promote standardized industrial and commercial practices globally. It is the world's largest developer of voluntary international standards, facilitating global trade by providing common standards that enable seamless communication between countries. It brings together global experts to agree on the best approach to tasks, from product creation to process management. The ISO, as one of the oldest nongovernmental international organizations, has enabled trade and cooperation between people and companies worldwide since 1946. The objective of the International Standards published is to make lives easier, safer, and better (ISO). This organization has consistently maintained high standards by ensuring safe, healthy work environments, which are essential for any organization. Occupational safety standards provide comprehensive guidelines for managing workplace hazards, minimizing risks, and ensuring the well-being of the workforce. These standards promote effective safety management practices, enabling organizations to prevent accidents, comply with regulations, and foster a robust safety culture that enhances both safety and productivity (ISO). Additionally, technical reports (TR) are issued by this same organization and serve as a series of application documents that provide additional information on the selection and use of appropriate standards. The ISO has around 175 Ergonomics standards and 8 TRs; however, although these standards are referenced for developing countries in Latin America, they still face challenges in understanding and applying them correctly.

One of these challenges is that, during the homologation process between applicable HFE laws or regulations and ISO standards, a TR can often be mistaken for an ISO standard. This fact can be caused by a misunderstanding of the standard and a lack of appropriate understanding among those who teach, disseminate, and apply it in each country, leading to conceptual errors about how HFE standards should be managed. Therefore, some specialists view HFE solely as the application or evaluation of methodologies, which they consider a limited approach. Thus, those who work with ISO documents, such as TRs, are not applying the entire standard but only the evaluation methodology. This situation has become of great interest to the countries that comprise the Latin American Union of Ergonomics and Human Factors (ULAERGO) as it may cause confusion and lead to a misconception about the specialty.

24.3 CHALLENGES IN ULAERGO COUNTRIES

ULAERGO is an organization that brings together the leading ergonomics associations in Latin America. Its mission is to foster collaboration, knowledge exchange, and the advancement of ergonomics across the region. It was founded in 2002. ULAERGO emerged in response to the need for a body representing the common interests of ergonomics professionals in Latin America. Since its inception, the organization has sought to promote ergonomic principles across various sectors of society, enhance working conditions and quality of life, and foster scientific research in ergonomics.

Over the years, ULAERGO has organized congresses, symposiums, and meetings that serve as platforms for debate and training on topics related to applied ergonomics in industries such as manufacturing, healthcare, and transportation. It also collaborates closely with international organizations, such as the IEA.

The countries that comprise ULAERGO are as follows:

Currently, ULAERGO includes ergonomics associations from several Latin American countries, such as:

1. Argentina – Argentine Association of Ergonomics and Human Factors (ADEA).
2. Bolivia – Bolivian Association of Ergonomics and Human Factors (ASERFHU).
3. Brazil – Brazilian Association of Ergonomics and Human Factors (ABERGO).
4. Chile – Chilean Society of Ergonomics and Human Factors (SOCHERGO).
5. Colombia – Colombian Association of Ergonomics (SCE).
6. Costa Rica – Costa Rican Association of Ergonomics (ACERGO).
7. Cuba – Community of Academic Interest at HFE.
8. Dominican Republic – Dominican Society of Ergonomics (SODOERGO).
9. Ecuador – Ecuadorian Scientific Association of Ergonomics (SOCCERGO).
10. Guatemala – Ergonomics Association of Guatemala (ASERGUA).
11. Honduras – Honduran Association of Ergonomics (ASOHERGO).
12. México – Mexican Society of Ergonomists (SEMAC).
13. Panama – National Association of Ergonomics (ANAERGO).
14. Paraguay – Paraguayan Association of Ergonomics and Human Factors (APE).
15. Peru – Peruvian Society of Ergonomics (SOPERGO).
16. Uruguay – Uruguayan Association of Ergonomics and Human Factors (AUDERGO).
17. Venezuela – Venezuelan Society of Ergonomics (ASOVERG).

These associations' collaborative efforts address the region's challenges, promoting ergonomic practices considering Latin America's cultural, economic, and social particularities. For more information, they can be contacted through the website https://ulaergo.org/inicio/.

Although these 17 associations that comprise the ULAERGO promote ergonomics by disseminating knowledge and supporting the development of standards in their respective countries and have also collaborated in ISO standardization processes, there is still a lack of information about these regulations in ULAERGO member countries worldwide, which hampers the ability to compare and discuss them, thereby weakening the discipline across Latin America. Additionally, the participation of other parties in discussions about regulations without sufficient awareness of and knowledge of the technical details may introduce errors and inconsistencies with existing rules in force. For example, the concept and recommended weight limit are milestones in managing manual material handling (MMH). It presents quite a difference in several countries. In Argentina, the recommended weight limit for women and young people is 15 kg; in contrast, in Chile, the same group of individuals is allowed a limit of 20 kg. Likewise, in Chile, the recommended weight limit for men, as stated in Law 20,949 (Ministerio del Trabajo y Previsión Social, 2016), is 25 kg.

In Bolivia, Ministerial Resolution 1349/2023 (Ministerio del Trabajo, 2023) considers 45 kg the standard weight for men, as established in the General Law on Occupational Health and Safety and Well-being, published on August 2, 1979, and not updated.

Further differences in government regulatory frameworks among Latin American countries pose another challenge to consistently promoting and applying ergonomics and health and safety regulations that benefit the workforce in a highly competitive business environment.

For example, in Chile, the government considers several aspects at all levels of the regulatory framework. It has general recognition of worker protection in the Political Constitution of the Republic of Chile (DTO-100-2005, 2005), which does not explicitly refer to the HFE. However, it does refer to respect for people's physical and intellectual integrity. Then, in Law 16,744, *regarding the Insurance of accidents and occupational diseases*, the importance of safeguarding workers' health is mentioned. In addition, Article 37 of Supreme Decree No. 594 of 2000, issued by the Ministry of Health, a regulation on basic sanitary and environmental conditions in the workplace, states: "Any danger factor that may affect the health or physical integrity of workers must be eliminated in the workplace."

On the other hand, Article 184 of the Labor Code states that "the employer shall be obliged to take all necessary measures to effectively protect the life and health of workers, maintain adequate conditions of hygiene and safety at work, and provide *the necessary implements to prevent accidents and occupational* diseases." Although ergonomics is not explicitly mentioned, it enables the development of a necessary regulatory framework for ergonomics management. Law 20,949 on Maximum Load Handling Regulation sets out general conditions and differentiated weight limits for each sex. As an advance for the ergonomist profession, the technical guide published in 2018 clearly described what an ergonomist is, the requirements they must meet, and the skills they must possess to manage regulations. Finally, the fine for noncompliance with these regulations governing the management of MMH is proportional to the number of workers exposed and can reach USD 30,000. The application of this fine largely depends on the oversight of the government authority.

In Latin America, the development of HFE regulations may be asynchronous because standards development responds to distinct political, social, and economic factors. They are generally a consequence of the following:

1. The requirement is derived from signing agreements with international organizations, such as the ILO. For example, the C127 convention, known as the "Maximum Weight Convention" (1967), has been signed by only 28 of the 187 member countries, including 10 Latin American countries. The details of each agreement are available in the Normex catalog (International Labour Organization, 1967a).
2. The management of stakeholders, from the request of transnational companies to develop a specific business in the region, was added to the bureaucratic structure for the homologation of international standards. So the requested topic concerns economic activity in each country. Thus, the homologation of ISO standards (also called mirror standards) is selected by the applicant from the institutions in charge of this management in each country, such as the Argentine Institute of Standardization and Certification, IRAM; the Bolivian Institute of Standardization and Quality or the National Institute of Standardization in Chile, and the National Institute of Quality, INACAL in Peru, to mention a few.
3. Finally, and less frequently, it is the consequence of the social impact of a situation that the government of the day seeks to regulate through the development of a norm. This is the case of Development Law 21.643, which amends the Labor Code and other legal bodies regarding the Prevention, Investigation, and Punishment of Workplace Harassment, Sexual Harassment, or Violence at Work. This is a result of the case of a public servant and nursing technician who, after suffering a prolonged process of workplace harassment, decided

to take her own life. This fact had significant repercussions in Chile, mobilizing the political structure to undertake normative activities; the press dubbed this regulation the "Karin Law" in memory of the victims (International Labour Organization, 1967a).

In all the scenarios described above, the participation of specialists from various associations in HFE is crucial for providing the necessary technical guidance for effective legal management. Thus, associations from countries such as ADEA (Argentina), Sochergo (Chile), ASERFHU (Bolivia), and ABERGO (Brazil) were included in the technical discussions. In addition, Latin American countries have developed various regulations to address workplace ergonomics, influenced by international standards such as ISO, NIOSH, and ILO guidelines. Accordingly, this chapter examines the regulatory landscape, comparing approaches in Argentina, Bolivia, Brazil, Chile, Colombia, Cuba, the Dominican Republic, Ecuador, Guatemala, Honduras, México, Panama, Paraguay, Peru, Puerto Rico, and Uruguay.

24.4 COMPARATIVE ANALYSIS

24.4.1 Materials and Methods

This analysis is based on information gathered in different instances. Information was first collected through the official website of the digital libraries of standards published by governments. A digital questionnaire has been created to collect and organize the information. The checklist was available from January 9 to March 10, 2025. The application achieved a 50% adhesion rate due to its complexity as the decrees, regulations, and resolutions are not entirely comparable to each other. Therefore, it was necessary to conduct a series of interviews with all the presidents of the ergonomics associations and societies of the ULAERGO countries. Two HFE specialists from countries without a professional association have also participated. Having the expert opinions of representatives from different countries helps observe and understand various scenarios, some of which have regulations. Others give rise to interpretation, so the opinion of the representatives allows us to understand the scope and jurisprudence as the leaders of HFE associations have various essential professions, including doctors, engineers, designers, architects, and physiotherapists. This diversity of perspectives enriches the analysis.

24.4.2 Comparative Analysis Results

It was possible to gather 293 regulations from 18 countries, which have been grouped into three types: the first group, 10.2%, includes laws that generate the conditions to manage HFE; the second group, 27%, includes decrees, resolutions, and regulations, which provide the specific framework for mandatory management; and the third group, 62.8, includes voluntary standards approved by the ISO that determine technical concepts and methods of evaluation, management, and application. The predominant theme of the decrees is the musculoskeletal assessment associated with the authority's inspection, while regulations related to ergonomic design predominate in the approved voluntary ISO standards.

The following figures summarize the primary laws and regulations related to ergonomics in those countries that provided information (Figure 24.1).

Additionally, the volume of regulations by topic was analyzed. Each document was read and analyzed to determine the specific topic to which it refers; this way, it can be determined whether it is a mandatory regulation (as with laws and decrees) or a voluntary standard. Figure 24.2 illustrates the relationship between the volume of regulation for each topic in HFE and the predominance of natural, mandatory, or voluntary regulations. Consequently, the most significant number of rules concerns ergonomics applied to design, where voluntary standards predominate, whereas in MMH, the mandatory standard predominates.

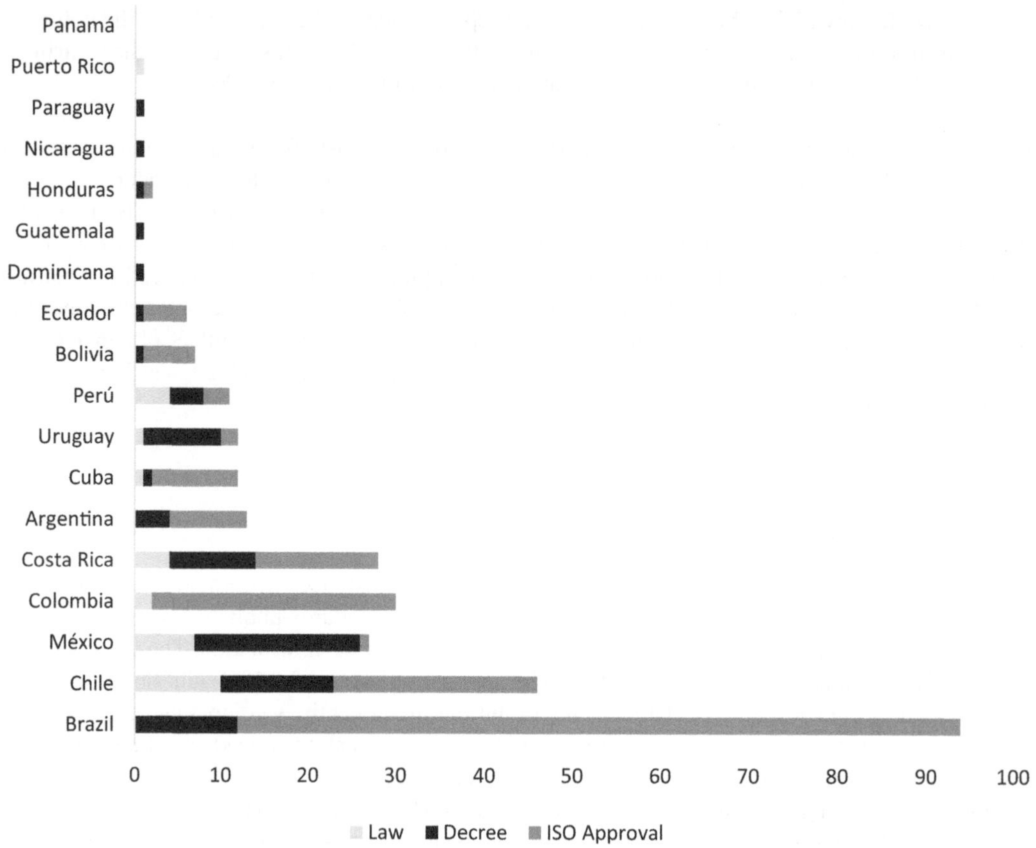

FIGURE 24.1 Volume of HFE regulations by country.

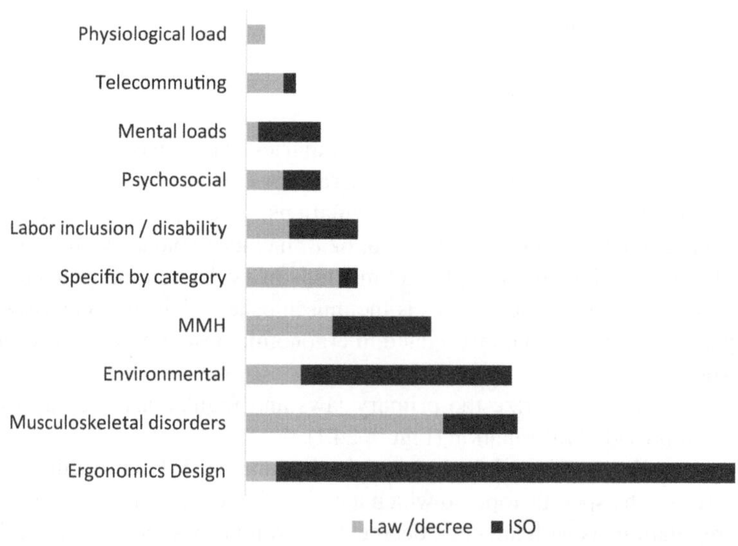

FIGURE 24.2 The volume of regulations as a percentage of the HFE topic, categorized by their voluntary or mandatory nature. (Image by the authors.)

24.5 SCOPE OF COVERAGE

The revised laws are generally transversal and do not delve into specific aspects to be applied across different economic sectors, such as the mining and construction industries.

Approved ISO standards are specific to each topic, such as visual controls, labeling of drug packaging, and anthropometric measurement procedures.

It is interesting to observe the trend in which musculoskeletal topics predominate in laws, regulations, and decrees. On the other hand, in ISO standards, the highest frequency is observed in documents on ergonomic design applications. Those who do not know much about the specialty assume that it only deals with aspects related to occupational health and safety; regulations, decrees, and laws would seem to describe the arguments in that way, but the specialty of engineering has other areas of execution in the development of products, goods, and services, where they are not easily visible because ISO standards are restricted under the purchase of the material. After all, they are strategic as they are a fundamental part of quality, positioning, and market differentiation. The laws, regulations, and decrees are general and are drafted for application in different areas, representing 37.2% of the regulations registered for this review. On the contrary, ISO standards are particular and guide implementation, evaluation, and execution. Consequently, it accounts for the largest share of documents, at 62.8%.

Law 19.404 (1995) in Chile, which permits early retirement based on performance in high-stress work positions, is noteworthy. This Law is unique in the entire register of revised standards and contemplates the certification of a National Ergonomics Commission (CEN), which sanctions the evaluation of jobs by integrating all ergonomic risk factors into four pillars: environmental load, mental load, psychosocial load, and physiological load. Article 1 describes the concept of heavy work: "Heavy work is work whose performance accelerates the physical, intellectual or psychological wear and tear in the majority of those who perform it, causing premature aging, even when they do not generate an occupational disease" (Government of Chile, 2024); consequently, according to the accreditation of the years of exposure of the worker to these positions, this rule allows early retirement. Hence, it requires an overcontribution by the employer and the worker. Although the scope of application of this law is the improvement and management of risk factors, in practice, since its enactment in 1995 to date, a single company has been able to correct and reverse the effects on workers, and, consequently, for all the other companies that have classified their jobs, it operates as an additional social security benefit.

24.6 APPLICATION STANDARDS

The laws reviewed in this chapter do not mention which evaluation standard should be applied because the objective of the law is much more general; on the other hand, the decrees/standards/regulations establish specific criteria and indicate which methods are recognized in the set of ISO standards – for example, the revised equation of NIOSH, OCRA, REBA, etc. However, exceptions and decrees often present checklists of risk factors for HFE, as is the case with the Work-Related Musculoskeletal Disorders (WRMSD) standard in Chile and the Resolution 886/15 Ergonomics Protocol in Argentina.

24.7 REGULATIONS MANUAL LOAD HANDLING (MMH)

These regulations have been the topic of ergonomics and have developed the most among ULAERGO countries. It begins with the ratification of Convention C127 (Ley 19.404 of 1995, 1995), Maximum Weight, 1967 (No. 127) of the ILO. This regulation has been ratified by 28 countries, 10 of which are in Latin America: Brazil, Chile, Costa Rica, Ecuador, Guatemala, Honduras, Nicaragua, Panama, Peru, and Venezuela.

TABLE 24.1
Comparative Summary of Regulations Governing Different Load Limits in Latin America

	Comparison of Load Limits in Regulations of Latin American Countries					
Country	**Men**	**Men, >45 Years**	**Women**	**Women, >45 Years**	**Trained Workers**	**Legislation**
Argentina	25kg	15kg	15kg	15kg	40kg	Resolution 886/15 SRT
Bolivia	45kg	-	22.5kg	-	-	DS2936
Brazil	60kg	-	20kg	-	25kg	CLT DL 5452-1943
Chile	25kg 20kg	20kg	20kg	15kg	-	Technical Guide MMH 2018
Colombia	50kg	-	12.5kg 20kg	-	40kg	RESOLUTION 2400 DE 1979
Costa Rica	-	-	-	-	-	DECREE NO 11074-TSS
Ecuador	35kg	-	23kg	-	-	DE 2393
Guatemala	45kg	-	33.75kg	-	45kg	AG 057-2022
Honduras	40kg 45.4kg	-	15kg 20kg	-	-	AE STSS-053-04
México	25kg	20kg	20kg	15kg	-	NOM-036-1-STPS-2028
Nicaragua	23kg 40kg	-	15kg 23kg	-	55kg 32kg	NIC-2002-R-63575
Peru	25kg	-	15kg	-	40kg	RM 375-2008-TR
Uruguay	25kg	-	25kg	-	-	Ley 19.927
Venezuela	20kg	-	12kg	-	-	RV 9589

Source: Image by the authors.

Consequently, the countries must create regulations in this area. In 1980, Costa Rica published Decree 11074-TSS· "Maximum Weight, Manual Load, and Physical Aptitude Exam." Chile enacted Law 20.001 (International Labour Organization, 1967b) in 2005, "Regulating the Maximum Weight of Human Cargo," which sets a limit of 50kg for loads for men. However, Decree No. 63 (Ley 20.001 of 2005), which regulates the procedure for executing the law, specified a load limit of 23 kilograms, and this discrepancy was corrected and updated in 2016 by Law 20.949 (Decree No. 63, 2005), which corrected the load limit for men to 25kg. This aspect highlights how legislative work influences technical expertise in the development of standards.

We observe differences in how factors related to HF/E are managed, for example, in the manual handling of loads. An updated summary for 14 countries is attached (Table 24.1). When the concept of load limit is applied, it is essential to note that different limits apply to men, women, and individuals over 45 years of age. In some countries, a separate limit is also applied to workers with MMH task training.

The load limits that countries establish vary considerably, and the legal and regulatory management intervenes with other actors that do not necessarily have technical mastery of HFE. It is essential to note that the regulations in Argentina, Chile, Costa Rica, Ecuador, Guatemala, Honduras, México, and Peru all impose charging limits for children under 18 years of age, while only Ecuador establishes a load limit for children under 16 years of age.

24.8 PSYCHOSOCIAL RISK REGULATION

Currently, few countries have regulations related to psychosocial risk. Chile has Resolution No. 1,448, which validates the protocol for monitoring psychosocial risks at work, updating the existing protocol that was first introduced in 2017. These regulations require the management to apply evaluation questionnaires. First, the ISTAS-21 questionnaire was used in its summarized and extended version, and, with the new update, the Questionnaire for the Evaluation of the Work Environment – Mental Health was applied [CEAL-SM (for its Spanish acronym)].

México has the NOM-035-STPS-2018 standard *on psychosocial risk factors in the workplace*. In Peru, Article 56 of Law 29783 stipulates that employers must prevent exposure to psychosocial agents in the workplace to protect workers' health. Article 103 of Supreme Decree No. 005-2012-TR determines a management protocol.

Costa Rica recently enacted Law No. 10412 in 2023, the "National Mental Health Law," which aims to promote the mental health, interests, well-being, and dignity of its people. It even has a national mental health policy from 2024 to 2034 (Ley 20.949, 2016).

Additionally, several countries have ratified the ILO's Convention No. 190 (2019) on *Violence and Harassment in the Workplace*. Today, it has been ratified by 10 Latin American countries: Antigua and Barbuda, Argentina, Bahamas, Barbados, Ecuador, El Salvador, México, Panama, Peru, and Uruguay.

In this vein, Peru has enacted Law No. 27,942 to prevent and punish sexual harassment. Chile Promulgates Law 21.643, "Amending the Labor Code and Other Legal Bodies in Matters of Prevention, Investigation, and Punishment of Workplace Harassment, Sexual Harassment, or Violence at Work." This law is known as Karin's law.

Costa Rica launched Law N° 7,476, "Law Against Sexual Harassment in Employment and Teaching," in 1995. This document aims to prohibit and punish sexual harassment or harassment as a discriminatory practice based on sex, against the dignity of women and men in labor and teaching relationships, and determine sanctions, reparation for moral damage, and even protection for minors, among other things.

24.9 RISK ASSESSMENT TOOLS

The regulations instruct the application of standards validated in the ISO standard. In general, they consider ISO 11.228-1: CMM Lifting/Transport, ISO 11.228-2: CMM Pulling/Pushing, ISO TR 12.296: MMP (Patient/people), ISO 11.228-3: Repetitive movement EESS, and ISO 11.226: Static postures, among others.

In addition, Resolution 886/2015 of the Superintendence of Occupational Risks in Argentina proposes the use of checklists for qualitative evaluation, following the same approach as in Chile, where the WRMSD standard requires the use of a checklist for qualitative evaluation. This procedure, regulated by the corresponding protocol, involves the accreditation of a 40-hour extraordinary course for ergonomics specialists who have already completed a 600-hour diploma, in which the use of the ART Tool, OWAS, REBA, MAC, and RAPP methods is practiced for qualitative evaluation and the OCRA, STRAIN INDEX, NIOSH, LT-ISO, Liberty Mutual, KIM, and MAPO methods for quantitative assessment.

24.10 INSPECTION AND ENFORCEMENT

Accreditation of approved ISO standards corresponds to the audit of certifying houses, a voluntary action on the part of companies and institutions that seek to differentiate their services in the market; therefore, they are willing to pay for the process.

On the other hand, in government regulations, the only motivation for the management process is oversight by the authority. In this sense, the State's intervention is essential to motivate the management of these and other regulations. Because failure to comply with a decree or rule has a pecuniary fine, failure to adhere to a law can empower the inspector to close the company. Unfortunately, audits are not significant due to the reduced number of specialized personnel overseeing vast expanses within their jurisdiction.

24.11 TRAINING REQUIREMENTS

Prestigious authors have identified several challenges for Latin American countries. Among the key factors identified by Hernández Albrecht (2016) are the shortage of professionals and certified ergonomists, the lack of a consistent design for formative programs in HFE, and the preponderance of physical ergonomics and musculoskeletal studies, which neglect an integrated ergonomics perspective in work design. Additionally, significant gaps exist in the laws and regulations of these "developing countries."

In this regard, the training of HFE specialists in Latin America primarily focuses on diploma programs, with an eventual transition to master's programs at universities recognized by governments. This disciplinary training does not necessarily encompass mastering all the tools required for evaluation in all HFE topics, nor does it account for the management necessary to comply with regulations. Therefore, courses are derived from understanding how to apply the regulations and their subsequent updates to extend later training. This creates an opportunity for other institutions to enter and offer courses, even from different countries, that are not recognized and that confuse students about the fundamental concepts of HFE. We have observed this business model in several countries, generating dissonances in all of them. For their part, ergonomics associations and societies in Latin America have promoted the concept of professional certification in HFE as a measure to safeguard the specialty in the described scenario.

In some countries, regulations or decrees require refresher courses to ensure correct application, particularly in application protocols, such as the 40-hour course outlined in the WRMSD standard described above.

24.12 DISCUSSION

24.12.1 Adoption of International Standards

All countries analyzed incorporate international guidelines, such as ISO and NIOSH, into their ergonomic regulations. However, Brazil stands out with its ABNT standards, which adapt ISO principles to local contexts.

24.12.2 Psychosocial Risk Management

México, Peru, Costa Rica, and Chile have made progress in regulations that address psychosocial risks, reflecting global trends in ergonomics. Other countries, such as Argentina, Uruguay, and Colombia, are gradually integrating these aspects, while others, including Paraguay, Bolivia, and Cuba, do not yet have regulations in place on this issue.

24.12.3 Challenges in Implementation

Despite comprehensive regulations, enforcement remains a challenge across the region. Limited resources, inconsistent training, and low employer compliance are common barriers to effective implementation.

The difference in managing legislative and technical visions presents a significant challenge when drafting regulations. Legislative work involves demanding processes and timelines, and legislators may not necessarily master HFE concepts. It is a constant challenge to train legislators through work meetings.

Regulations are a process of constant improvement; once a regulation is promulgated, it is necessary to observe its performance to provide improvements in its next update. According to the ILO, regulations must be updated every 3 years. This poses an interesting challenge to the associations and societies that promote HFE, encouraging them to participate in discussions with the authorities.

24.13 RECOMMENDATIONS

Regional harmonization: Promote the alignment of ergonomics standards across Latin America to facilitate the adoption of best practices. It is necessary to observe how the HFE law operates in each country, for example, through comparative legislation, to systematically implement the best parameters in its implementation and ensure the achievement of the objectives for each regulated theme. The countries of Latin America share a common history; they are related, yet distinct. The idiosyncrasies and internal dynamics pose challenges in management, generating significant opportunities for constant comparative analysis.

Capacity building: Invest in training programs for employers, workers, and inspectors to enhance their skills and capabilities. To help them understand the scope of HFE, it is necessary to establish an HFE-compliant paradigm within companies as many regulations can introduce bias into institutional perceptions and practices.

Research and collaboration: Encourage research partnerships between academia and regulatory bodies to address local ergonomic challenges. The HFE is constantly in motion, and scientific evidence continually challenges the specialist. Therefore, it is necessary to establish technical observatories in HFE in each country to monitor the development, training, and implementation of the specialty's regulations, and to seek opportunities for management and development.

Enhanced enforcement: Strengthen inspection mechanisms and provide incentives for compliance. The authority's inspection is the best motivation for managing regulations in the company. Training inspectors in HFE would enhance the process.

Train the specialist in ergonomics in negotiations with the authority: Prepare the technical representatives in HFE on how to approach legislative processes and support the necessary technical criteria. Combining the legislative and technical visions is key; the partisan political vision can often be a significant obstacle.

24.14 CONCLUSION

The HFE laws standardize management among professionals involved in HFE application and fulfillment. Still, there are essential differences in how similar issues are addressed, for example, in the case of load limits for manual handling of loads between countries. HFE laws are critical for promoting the specialty and initiating collaborative working relationships with companies, as they enable the specialty to be introduced into internal discussions.

The regulatory landscape of ergonomics in Latin America expresses two significant perspectives. First, government HFE laws correlate with occupational health management, and issues related to musculoskeletal risks, psychosocial factors, and other risks arising from manual handling of loads predominate. ISO standards provide guidelines for applying HFE concepts to the design of both industry-specific jobs and interfaces. ISO standards have limited visibility due to their voluntary nature and the associated costs of accessing them, whereas government regulations are mandatory and freely accessible and should be implemented. This contributes to the bias about the specialty paradigm and should be a significant challenge for HFE specialists in harmonizing professional practice across countries.

ACKNOWLEDGMENTS

Paolo Bandeo, President, and Marcela Pellegrino, member of the board of directors of the Association of Ergonomics and Human Factors of Argentina (ADEA); Lucy Mara Baú, President, and Patricia Rossafa Branco, member of the Association of Ergonomics and Human Factors of Brazil, (ABERGO); Nelson Chávez, President of the Association of Ergonomics and Human

Factors of Bolivia (ASERFHU); Jeisy Jofré F., President, and Gabriel Portal director and Daniel Lizama, member of the Chilean Society of Ergonomics and Human Factors (SOCHERGO); Gustavo Barrantes, President, and Marco García, Vice President of the Costa Rican Association of Ergonomics (ACERGTO); Liliana Joaqui, President of the Colombian Society of Ergonomics (SCE); Iván López, President of the Ecuatorian Scientific Society of Ergonomics (SOCEERGO); Ivone Soto, President of the Dominican Society of Ergonomics (SODOERGO); Luis García G, President of the Association of Ergonomics of Guatemala (ASERGUA); Mario Sabillón, President of the Association of Honduran Ergonomics (ASOHERGO); Carlos Navarro, President; Carlos Espejo, former President; Aide Maldonado, and Jean Paul Becker, member of the Society of Ergonomists of México (SEMAC); Javier Córdoba, President of the National Association of Ergonomics of Panama (ANAERGO); Marta Centurión, President of the Paraguayan Association of Ergonomics (APE); Andrés Arce G, President of the Peruvian Society of Ergonomics (SOPERGO); Silvia Díaz, President; Federico Ferreira, Vice President; and Marcelo Scavone, Treasurer of the Uruguayan Association of Ergonomics and Human Factors (AUDERGO), and Erick Vega (Puerto Rico) and Joaquín García and Juan Lázaro Acosta of CUBA. Special thanks to Paulina Hernández Albrecht, former President of SOCHERGO and Former President of ULAERGO, Dra. Cointa Lagunes Cruz of the Ministry of Labor and Social Welfare of México and Fabiola Gómez-Marañon. We acknowledge Dr. Alejandro Barajas for editing, citing, and providing references in this chapter.

REFERENCES

Bailey, Brian. P., Busbey, C. W., & Iqbal, S. T. 2007. TAPRAV: An interactive analysis tool for exploring workload aligned to models of task execution. *Interacting with Computers*, *19*(3), 314–329. https://doi.org/10.1016/j.intcom.2007.01.004

Bolis, Ivan, Sigahi, Tiago Fonseca Albuquerque. Cavalcanti, Thatcher, Andrew, Saltorato, Patricia, & Morioka, S. N. 2023. Contribution of ergonomics and human factors to sustainable development: A systematic literature review. *Ergonomics*, *66*(3), 303–321. https://doi.org/10.1080/00140139.2022.2079729

Castaño Cardenas, Diana. 2023. Health and safety at work: The importance of an ergonomic workplace. *SCT Proceedings in Interdisciplinary Insights and Innovations*, *1*, 83.

Decree No. 63. (2005). *Approve regulations for implementing Law No. 20,001, which governs the maximum weight of human cargo*.

DTO-100-2005. (2005). *Constitución política de República de Chile*.

Dul, Jan, Bruder, Ralph, Buckle, Peter, Carayon, Pascale, Falzon, Pierre, Marras, William S., Wilson, John R., & van der Doelen, B. 2012. A strategy for human factors/ergonomics: Developing the discipline and profession. *Ergonomics*, *55*(4), 377–395. https://doi.org/10.1080/00140139.2012.661087

Government of Chile. (2024). *Today, it comes into force: What does the Karin Law consist of?* https://www.gob.cl/noticias/hoy-entra-en-vigencia-en-que-consiste-ley-karin/

Hernández Albrecht, Paulina. 2016. Principales brechas de la Ergonomía en América Latina: A quince años del siglo XXI. *Revista Ciencias de la Salud*, *14*, 5–10. https://doi.org/10.12804/revistas.urosario.edu.co/revsalud/a.5071

International Ergonomics Association. (2000). *What Is Ergonomics (HFE)?* https://iea.cc/about/what-is-ergonomics/

International Labour Organization. (1967a). *C127—Maximum Weight Convention, 1967 (No. 127)*. https://normlex.ilo.org/

International Labour Organization. (1967b). *Ratification of Convention No. 127: Maximum Weight Convention, 1967*.

ISO. *About ISO*. https://www.iso.org/about

Ley 19.404 of 1995. (1995). *It introduces amendments to Decree Law No.* 3,500 *of 1980 and establishes rules regarding old-age pensions, taking into account the nature of heavy work*.

Ley 20.001 of 2005. (2005). *It regulates the maximum weight of human cargo*.

Ley 20.949. (2016). *Amends the labor code to reduce the weight of manual handling loads*.

Ministerio del Trabajo y Previsión Social. (2016). *Ley 20.949 "Modifica el código del trabajo para reducir el peso de las cargas de manipulación manual" artículo* 211-*j*.

Ministerio del Trabajo, E. y P. S. 2023. *Resolución ministerial N°1349/23. Artículo 7, punto II*.

Nedelko, Zlatko. (2022). *6th FEB International Scientific Conference: Challenges in Economics and Business in the Post-COVID Times*. https://doi.org/10.18690/um.epf.5.2022

Niu, Shengli. 2010. Ergonomics and occupational safety and health: An ILO perspective. *Applied Ergonomics*, *41*(6), 744–753. https://doi.org/10.1016/j.apergo.2010.03.004

Orr, Gary, Tschida, Breca, Carcamo, Ernesto, Goggins, Rick, & Cochran, David. 2024. Update on ergonomics standards in the United States. *Proceedings of the Human Factors and Ergonomics Society Annual Meeting*. https://doi.org/10.1177/10711813241272124

Principles and Guidelines for Human Factors/Ergonomics (HF/E) Design and Management of Work Systems. (s. f.).

Putri Cahyani, Amelia, & Humaira Ramadhani, Salwa. (2024). Optimizing workplace and employee safety through ergonomics: scoping review. *SHS Web of Conferences*, *189*, 01007. https://doi.org/10.1051/shsconf/202418901007

Radjiyev, Ayubkhon, Qiu, Hai, Xiong, Shuping, & Nam, Kyung. 2015. Ergonomics and sustainable development in the past two decades (1992–2011): Research trends and how ergonomics can contribute to sustainable development. *Applied Ergonomics*, *46*, 67–75. https://doi.org/10.1016/j.apergo.2014.07.006

Salawati, Liza, & Abbas, Ibnu. (2023). The application of ergonomics to improve work productivity. *Jurnal Kedokteran Syiah Kuala*. https://doi.org/10.24815/jks.v23i2.33566

Soares, Marcelo M. 2006. Ergonomics in Latin America: Background, trends, and challenges. *Applied Ergonomics*, *37*(4), 555–561. https://doi.org/10.1016/j.apergo.2006.04.014

Stoidner, Micha, Bründl, Patrick, Nguyen, Huong, Baechler, Andrea, & Franke, Jorg. 2023. Workplace analysis and ergonomics in engineer-to-order production sites: A study on the workplace design of control cabinet manufacturing enterprises. *2023 IEEE International Conference on Industrial Engineering and Engineering Management (IEEM)*, 0838–0842. https://doi.org/10.1109/IEEM58616.2023.10406509

Index

For Product Safety Concerns and Information please contact our EU representative GPSR@taylorandfrancis.com Taylor & Francis Verlag GmbH, Kaufingerstraße 24, 80331 München, Germany